P9-CBE-268

Systems Architecture

Systems Architecture

Hardware and Software in Business Information Systems

Stephen D. Burd
University of New Mexico

A DIVISION OF COURSE TECHNOLOGY
ONE MAIN STREET, CAMBRIDGE, MA 02142

an International Thomson Publishing company I(T)P

Albany • Bonn • Boston • Cincinnati • London • Madrid • Melbourne • Mexico City
New York • Paris • San Francisco • Singapore • Tokyo • Toronto • Washington

DEDICATION

To my parents, Marion and Dale. Thanks for your time, patience, and answers to my incessant stream of questions.

Senior Acquisitions Editor	James H. Edwards
Project Manager	Christopher T. Doran
Production Editor	Barbara Worth
Manufacturing Coordinator	Tracy Megison
Marketing Manager	Eileen Pfeffer
Interior Design and Composition	Gex, Inc.
Cover Design	Kevin Meyers
Cover photograph	© Gabriel M. Covian, The Image Bank

For more information contact:

Course Technology
One Main Street
Cambridge, MA 02142

International Thomson Publishing Europe
Berkshire House 168-173
High Holborn
London WCIV 7AA
England

Thomas Nelson Australia
102 Dodds Street
South Melbourne, 3205
Victoria, Australia

Nelson Canada
1120 Birchmount Road
Scarborough, Ontario
Canada M1K 5G4

International Thomson Editores
Campos Eliseos 385, Piso 7
Col. Polanco
11560 Mexico D.F. Mexico

International Thomson Publishing GmbH
Kônigswinterer Strasse 418
53227 Bonn
Germany

International Thomson Publishing Asia
211 Henderson Road
#05-10 Henderson Building
Singapore 0315

International Thomson Publishing Japan
Hirakawacho Kyowa Building, 3F
2-2-1 Hirakawacho
Chiyoda-ku, Tokyo 102
Japan

ISBN 0-87835-876-5

Printed in the United States of America

10 9 8 7 6 5 4 3

Contents in Brief

Contents

Preface

The purpose of this book is to provide a broad technical description of computer hardware and systems software. It is intended as a reference for information system (IS) professionals and as an undergraduate text for students majoring or concentrating in information systems.

INTENDED AUDIENCE

There is a substantial difference between traditional curriculums in computer science and information systems. Students of computer science are generally exposed to a great deal of computer hardware and systems software technology in their normal coursework. The topics covered in this book are typically taught in several different computer science courses. As such, there are many texts designed for computer science courses that cover a subset of these topics. Typically, a computer hardware text will provide only a chapter or two on systems software and a systems software text will provide only a chapter or two on computer hardware.

Exposure to computer hardware and systems software in the typical information systems curriculum is usually limited. A brief overview is provided in an introductory information systems course. Some very specific technical topics may be covered in other courses (e.g., systems design and computer networks). However, the primary emphases of an information systems curriculum are application development and management of computing resources. These emphases are represented in courses covering application programming, systems analysis, systems design, database management systems, and management of computing resources. Due to the number of these courses, students tend to find little available time to take technical computer science courses.

This book addresses the limitations of the traditional IS curriculum by covering a broad range of hardware and software technology in a fully integrated text. Topics that are most useful to IS professionals are stressed at an appropriate level of detail. The book is designed to provide a technical foundation for systems design, hardware and software procurement, and the management of computing resources. In some areas, the reader will gain a sufficient depth of technical knowledge to tackle technical problems alone. In other areas, the reader will gain a sufficient depth of technical knowledge to allow him or her to communicate effectively with technical specialists.

The topical coverage closely follows the outline of course CIS-17 in the 1986 version of the Data Processing Management Association (DPMA) recommended CIS curriculum. However, this is an elective course not offered in many IS programs. In the most recent DPMA recommended curriculum (1991 draft), the nearest course equivalent is IS-3. However, this curriculum has yet to be widely adopted. It is hoped that the mere existence of this book will encourage more IS faculty to offer courses similar to CIS-17 and IS-3.

This book may be used as supplemental reading in courses on systems design and management of computer resources. With respect to systems design, the book covers many technical topics that must be addressed in the selection and configuration of computer hardware and systems software. With respect to computer resource management, the book provides the broad technical foundation necessary for effective management of computer hardware and systems software resources.

ASSUMPTIONS OF READERS' BACKGROUND KNOWLEDGE

The reader is not assumed to have an extensive background in mathematics, physics, or engineering, as is the case in many computer science texts. Where necessary, background information in these areas is presented in simple and easy-to-read terms. Also, these topics are covered only to the extent necessary for understanding subsequent coverage of hardware and software.

It is not assumed that the reader knows any particular programming language. However, classroom or practical experience with at least one programming language is very helpful for fully comprehending the coverage of operating systems and application development software. Programming examples are provided in several programming languages and in pseudocode. All of these examples should be understandable to readers not familiar with the particular programming language used.

It is not assumed that the reader has detailed knowledge of any particular operating system. However, as with programming experience, detailed exposure to at least one operating system is helpful. Lengthy examples from specific operating systems are purposely avoided. The variability between operating system capabilities and user interfaces makes the use of such examples difficult. However, the text includes some examples from MS-DOS and Unix as well as discussions of capabilities typically found in graphically intensive operating systems and in mainframe operating systems. These examples and discussions are designed to be understandable by readers not familiar with the particular operating system discussed.

It is not assumed that the reader possesses any knowledge of low-level machine actions or assembly programming. Low-level machine actions are discussed in great detail in the hardware section of the book. Assembly programming is discussed in Chapter 9, but no specific assembly language is used. In addition, the coverage of assembly language can be skipped entirely without loss of continuity.

ORGANIZATION OF THE BOOK

The book is organized into four groups of chapters. Part One consists of three introductory chapters containing a general overview, a hardware overview, and a software overview. Part Two consists of five chapters covering details of hardware technology. Part Three consists of five chapters covering details of software technology. Part Four consists of three chapters covering networks, architecture other than that of von Neumann, and managerial issues.

Although the chapters were designed to be covered in a linear sequence, a number of alternative orderings are possible. The prerequisite material for each chapter is given in the chapter descriptions below. Alternate orderings may be constructed based upon these prerequisite relationships. In particular, it is possible to intermix the detailed coverage of major hardware components and their related systems software. For any alternative ordering, it is suggested that Chapters 1 through 4 and Chapter 6 be covered first. Also, Chapters 10 and 11 should be covered prior to any other chapters in Part Three.

CHAPTER DESCRIPTIONS

Part One: Introduction

Chapter 1: This chapter briefly describes the uses of technical knowledge about computer hardware and systems software. These uses are described in terms of phases of a system development life cycle and the persons who participate in that life cycle. The chapter also describes various sources of knowledge about computer hardware and systems software and provides a list of recommended periodicals.

Chapter 2: This chapter provides an overview of computer hardware technology and architecture. The primary components and functions of a computer system are described. Various methods of computer system performance measurement and evaluation are discussed. The chapter also describes classes of computer systems and the characteristics that distinguish them.

Chapter 3: This chapter begins by describing the role and function of software. Classes of software, including application, application development, and application support software, are defined. Operating system components and functions are also discussed. A layered model of software is stressed throughout the chapter. Chapter 2 is a necessary prerequisite.

Part Two: Hardware

Chapter 4: This chapter describes data representation in terms of coding formats and digital signals. Binary, octal, and hexadecimal numbering systems are described. Primitive data types, along with common coding conventions for each type, are described. A discussion of data structures is also included. Chapter 2 is a necessary prerequisite to this and all other chapters in Part Two.

Chapter 5: This chapter describes electrical and optical technologies for data transformation (processing) and storage. These include the use of electrical switches for processing and primary storage as well as the use of magnetic and optically based devices for mass storage. Chapter 4 is a necessary prerequisite.

Chapter 6: This chapter describes the architecture of the central processing unit and of primary storage. The description of processor functions includes discussions of primitive data transformations, data movement, and sequence control. Various architectural features of a von Neumann processor, including instruction/execution cycles, registers, bus control, instruction formats, and interrupt processing, are discussed. Memory allocation, addressing, and related topics are also discussed. Chapter 4 is a necessary prerequisite. Chapter 5 is a recommended prerequisite.

Chapter 7: This chapter describes data communication technology. Basic technological issues discussed include signals, signal propagation media, and data coding methods. Various communication and protocol conventions, including serial and parallel transmission, error detection and correction, and other protocol considerations, are discussed. Chapter 4 is a necessary prerequisite. Chapter 5 is a recommended prerequisite.

Chapter 8: This chapter describes the architecture of mass-storage devices and the technology and architecture of input/output (I/O) devices. General considerations of device control, bus access, and communication performance are discussed first. The architecture of mass-storage devices is then discussed, with an emphasis on processor control. Individual classes of I/O devices are then described in terms of their function, basic technology, architecture, and processor control. Chapters 4 and 6 are necessary prerequisites. Chapters 5 and 7 are recommended prerequisites.

Part Three: Software

Chapter 9: This chapter describes the development, structure, and content of assembly language programs. A general discussion of program translation and the related concepts of linking and loading are provided first. These are followed by a detailed discussion of assembly language structure and the assembly process. Chapter 3 is a required prerequisite for this and all other chapters in Part Three. Chapter 6 is a necessary prerequisite and Chapter 8 is a recommended prerequisite.

Chapter 10: This chapter describes the development of application programs using higher level languages and other application development tools. The chapter begins with a discussion of various approaches to application development and the software required to support each approach. A detailed discussion of third- and fourth-generation programming languages is provided and is followed by a detailed discussion of compilation, interpretation, and the use of support libraries. The final sections discuss various forms of advanced application development tools, including CASE tools and code generators. Chapters 4 and 6 are necessary prerequisites.

Chapter 11: This chapter provides an overview of operating systems and detailed descriptions of resource allocation, processor control, and memory management. The chapter begins with a discussion of software layers, with particular attention to operating system layers. The role of the operating system as a resource allocator is then discussed, followed by detailed discussions of the mechanisms by which processor and memory resources are allocated. Chapters 2, 6, 8, and 10 are necessary prerequisites.

Chapter 12: This chapter provides a detailed description of mass-storage device organization and file management. The chapter begins with a description of logical and physical secondary storage accesses. File content, structure, and manipulation are then discussed. Physical considerations for device I/O, storage allocation, and directory structure are described next. Other topics in the chapter include file protection and security, file sharing, file system administration, and database management. Chapters 6, 8, 10, and 11 are necessary prerequisites.

Chapter 13: This chapter describes operating system facilities for process I/O, device I/O, and job control. The chapter begins with a discussion of service functions for memory management and interprocess communication. Support services for interactive I/O are discussed next. This is followed by a detailed discussion of both interactive and batch job control. This coverage includes command languages, forms-based interfaces, and window-based command facilities. Chapters 6, 8, 10, and 11 are necessary prerequisites.

Part Four: Advanced Topics

Chapter 14: This chapter describes the architecture and implementation of computer networks and distributed computing systems. The chapter begins with a discussion of the costs and benefits of distributed computer systems. Next, various architectures for distributed computing are discussed, and a detailed discussion of client-server architecture is provided. In the next section, the OSI network layers model is presented. The final section discusses options for physical network topology and hardware. Chapters 7, 11, 12, and 13 are necessary prerequisites.

Chapter 15: This chapter describes various advanced features of computer hardware architecture and advanced programming paradigms. The discussion of advanced architecture is primarily concerned with various means of parallel processing. SIMD, MISD, and MIMD architectures are discussed in detail, and examples of specific architectures are provided. The implications of parallel processing for application and operating systems software are also discussed. The discussion of advanced programming paradigms concentrates on nonprocedural programming and symbolic reasoning. The implications of these paradigms for hardware architecture are also discussed. Chapters 6, 8, 10, 11, 12, and 13 are necessary prerequisites.

Chapter 16: This chapter discusses various issues of computer system acquisition. It begins with a discussion of the strategic role of hardware and software resources in an organization. The normal acquisition process is then discussed. Next, a detailed discussion of requirements determination and performance modelling is given. This discussion centers on the development and analysis of models of resource demand and availability. The final section discusses various physical aspects of computer operation. Chapters 6, 10, and 11 are necessary prerequisites.

ACKNOWLEDGMENTS

This book originated from a planned revision of another book titled *System Architecture: Software and Hardware Concepts* by William E. Leigh and Dia L. Ali. The project evolved to encompass a broader range of topics and a more specific information system emphasis than the original book. However, many of the figures and some of the textual material here are taken or adapted from that book. The author is indebted to Leigh and Ali and the boyd & fraser publishing company for providing a starting point for this project and for the figures and text included herein.

The author would like to thank everyone who contributed to the development of this book, including the editors and staff of boyd & fraser publishing company and students of the Anderson School of Management at the University of New Mexico. Several drafts of the book were used in the Software and Hardware Concepts course at the Anderson School during 1991 and 1992. Comments from students in these classes were invaluable for the improvement of the material. Thanks also go to the following reviewers for their comments and suggestions during the preparation of this book:

Charles P. Bilbrey
James Madison University

Robert G. Brookshire
James Madison University

William G. Carlson
Palm Beach Atlantic College

Carl Clavadetscher
California State Polytechnic University

Cristal D. Ewald
Iowa State University

G.K. Hutchinson
University of Wisconsin-Milwaukee

Robert C. Loomis
Onondaga Community College

George Sargent
University of Wisconsin-Whitewater

Randy S. Weinberg
St. Cloud State University

PART

1

Introduction

1 General Introduction

Chapter Goals

- *Describe the activities of information systems professionals.*
- *Describe the technical knowledge of computer hardware and systems software needed to develop information systems.*
- *Identify additional sources of information for continuing education in computer hardware and systems software.*

COMPUTER TECHNOLOGY: YOUR NEED TO KNOW

To make any technical device usable by a large number of people, it is necessary to hide its technical complexities. Given the technological progress of the modern world, daily life abounds with examples of this reality. Average people know very little about the inner workings of their cars, their home stereo systems, their microwave ovens, and most of the other technologically complex devices they use. If detailed knowledge of the inner workings were required to use these devices, the average person would either be unable to use them or would require a substantial amount of training and knowledge.

Computers are no exception to this reality of technology. When computers first became available, human operators required a substantial amount of training and knowledge to use them to perform even the simplest of tasks. Since that time computers have become increasingly more complex and powerful machines, yet the amount of knowledge necessary to use them effectively has been reduced. The result has been a wide proliferation of computers beyond their original scientific applications and into business, the classroom, and the home. Why, then, do *you* need to know anything about the innermost workings of a computer?

Acquiring and Configuring Technological Devices

Although the average person needs to know very little about most technical devices in order to use them, the knowledge required to select and configure them is somewhat greater. As an example, consider a home stereo system. The knowledge necessary to operate a stereo is relatively simple. You need to know how to plug it in, where the power switch is, how to select input (tape or radio), how to adjust volume, and so on. The average person can gain this knowledge in a matter of minutes by reading the operating manual.

But what happens when that same person walks into a store to purchase a stereo? The customer is immediately confronted with a wide range of choices ranging from watts of amplifier power to the choice between an integrated system versus a combination of separate components. The knowledge required to make these choices includes a thorough understanding of the purchaser's desires and requirements (type of music, sensitivity to distortion, size of rooms, and the like) and sufficient knowledge of the available alternatives to determine their compatibility with those desires and requirements.

Unfortunately, determining compatibility between alternatives and needs usually requires some knowledge of the technical details of the alternatives. A purchaser with a high degree of sensitivity to distortion may decide that "high-end" equipment is needed. An amplifier with low distortion is desired—but how low? A radio receiver (tuner) with a high signal-to-noise ratio is desired—but how high? A tape player with low distortion and wide dynamic range is desired; does that mean a digital audio tape player is required, or will an analog cassette tape player suffice? An amplifier capable of filling a six-room house with ear-splitting heavy metal music is desired; how many watts of power are required, what speakers are required? A system that is highly reliable and requires very little periodic maintenance is desired; what options fulfill this criteria?

The purchaser may not be familiar with the terminology by which technological capabilities are stated (e.g., what is signal-to-noise ratio and how is it measured and evaluated?). Further complicating the decision is the large number of products of similar capability, the availability of alternative technologies to meet the same need (phonograph records vs. analog cassette tapes vs. compact discs vs. digital audio tape), and the issues of compatibility that arise when configuring a system from multiple components.

Acquiring and Configuring Computers

The purchase of a computer is usually an even more complex endeavor. Issues parallel to those of a stereo purchase include the basic type of computer, storage capacity, processor type and speed, and input/output devices. What about software? What operating system is required? Is a database manager needed? What about a word processor or spreadsheet package? Does the purchaser need tools to develop his own software from scratch? What is the optimum combination of software and hardware components to meet a particular set of needs?

The purchaser of a home computer may deal with these issues infrequently and with a relatively simple set of requirements. The information systems professional may deal with these issues every day and for needs and requirements that vary widely from one situation to the next. The needs and requirements may even change *during* the development or selection of an information system.

INFORMATION SYSTEMS DEVELOPMENT: TECHNOLOGICAL KNOWLEDGE

Within the context of information systems development, what knowledge of computer hardware and software is required and when is it required? As a framework to answer these questions, we will examine steps in the process of information systems development. For each step, we describe the technical knowledge needed to complete the step.

The steps necessary to develop an information system are commonly called a *system development life cycle (SDLC)*. A number of SDLCs have been used over the history of automated information processing. A current popular approach to the development of information systems is the structured SDLC, developed in the 1970s and still in use today. The structured SDLC specifies a set of steps to be followed in system development as well as tools and procedures for each step. The steps (or phases) of the structured SDLC are as follows:

1. Systems survey
2. Systems analysis
3. Systems design
4. Systems implementation
5. Evaluation and maintenance

Each of these is described in detail below.

Systems Survey

The *systems survey* briefly examines user information needs, existing ways in which those needs are currently addressed (if any), and the feasibility of developing or acquiring an information system to address those needs. The survey is normally conducted within a relatively short time frame (days or weeks) and is intended to determine the existence and nature of unmet needs and to provide a basis for deciding whether a solution is worth the investment of further resources.

Technical knowledge of computer hardware and systems software is required to assess the degree to which user needs are currently being met and to estimate the resources required to address unmet needs. For example, an analyst surveying a point-of-sale (POS) system in a retail store might pose questions such as:

- How much time is required to process a sale?
- Is the existing system easy for a salesperson to use?
- Is sufficient information being gathered (for instance, for marketing purposes)?
- Can the existing hardware handle peak sales volumes (for example, holidays such as Christmas)?
- Will the existing hardware be sufficient three years from now?
- Can the existing system be easily expanded?
- What are the current hardware operating costs?
- Are there cheaper hardware alternatives?

Formulation of and response to all of these questions directly or indirectly require some technical knowledge of hardware and systems software.

Processing time and ease of use are dependent upon the capabilities of the hardware and software and upon the structure of the user interface. The system's ability to handle peak demands requires fairly detailed knowledge of the processing and storage capabilities of the hardware, operating system, and the application software. This same knowledge is required to determine sufficiency at any point in the future. Expansion possibilities depend upon unused capacity in the hardware and limitations in both the hardware and software. Determining whether or not cheaper alternatives exist requires all of the aforementioned knowledge, not only with respect to the current system, but also with respect to a wide range of currently available hardware and software options.

The systems survey requires a very broad range of technical knowledge. However, the time limit inherent in the survey usually precludes extremely detailed use of this knowledge. For example, it is rare to see exacting calculations of processor performance or storage capacity made during this phase. Similarly, cost estimates are rarely made "to the dollar."

Systems Analysis

Systems analysis is primarily concerned with the detailed examination of user needs and the extent to which they are being met. This phase of the SDLC may also be called *needs analysis* or *requirements analysis*. In this phase, information requirements are described in minute detail. Processing requirements are precisely defined. Performance requirements for the current and future systems are determined.

Many of the questions posed during the systems survey are addressed in detail during systems analysis. Ease of use and time constraints lead to specifications for interface style and processing speed. The systems analyst uses detailed estimates of current and future transaction volumes to derive both processing and storage requirements. Other needs for the data determine requirements for integrating this system with other information systems and lead to detailed specifications for hardware, software, and database compatibility.

The detailed specification of requirements serves two primary functions. First, it serves as a basis for systems design. Detailed requirements allow the cost and desirability of various system implementations (combinations of hardware and software) to be precisely determined. The requirements specification also serves as a contract with the user. It is a statement of the acceptance criteria for the system that will ultimately be delivered.

Systems Design

During the next phase, *systems design*, the exact configuration of all hardware and software components is specified. Specific design phase tasks include:

- Selection of computer hardware (e.g., processing, storage, input/output, and telecommunication devices)
- Selection of the operating software (e.g., an operating system and/or network communication software)
- Selection of application program development tools (e.g., programming languages and database management systems)

These design decisions have a broad impact on the system and are thus made early in the design phase. Other design decisions are of a more limited scope and are dependent upon earlier decisions. Examples of these include:

- Design of files and/or databases (e.g., the grouping of data elements into records and files, indexing, and sorting)
- Design of individual programs
- Design of user interfaces (e.g., input screen formats, report formats, and command dialogues)
- Design of system backup and recovery mechanisms

The output of these decisions is documented in detail and forms the basis of detailed implementation tasks.

Technical knowledge of computer hardware and systems software is most required during the early stages of systems design. The selection of hardware components requires a detailed knowledge of their capabilities and limitations. The integration of multiple hardware components into a single system requires that all of the devices be compatible and that they work well with one another. Operating system selection is constrained by the hardware system and by the overall performance requirements of the system. Compatibility of both the hardware and the systems software with other systems within the organization must be determined.

The selection of application program development tools is dependent upon both the requirements of the information system and the limitations of the hardware and operating systems. Programming languages, window managers, database managers, and other tools must be selected and integrated. All of these tools (and the software components built with them) vary widely in their efficiency, power, and compatibility. Because these tools are often expensive, their selection will likely affect future systems development projects.

Systems Implementation

During the *systems implementation* phase, the plans formulated during systems design are implemented. Hardware and systems software are installed, configured, and tested. Application programs and files are developed, installed, and tested. The entire system is tested to ensure that all of its components work together and that the system correctly meets the users' needs as defined during the analysis phase. There is generally some overlap of tasks during the design and implementation phases. For example, the installation and testing of hardware and systems software are often performed concurrently with the detailed design of files and application programs.

A great deal of technical knowledge is required for many of the implementation tasks. The installation and configuration of hardware and systems software is a highly specialized task that requires a great deal of understanding of both the components being installed and the purposes for which they will be used. Both hardware and software must usually be configured to a particular operating environment. This requires that adjustments be made to the components and that the system then be tested to determine if the adjustments are sufficient. This process can require many iterations and much time. More mundane tasks such as formatting storage devices, setting up system security, installing applications programs, and establishing accounting and auditing controls also require time and technical expertise.

Maintenance and Evaluation

As the system is used over time, various shortcomings might be noticed. Errors or problems that escaped detection during installation and testing might appear. The system could become overloaded due to inadequate estimates of processing volume. Additional information needs may become apparent, necessitating additional data collection, storage, and/or processing.

Relatively minor system changes (for instance, correcting program errors or minor processing changes) are normally handled as maintenance changes. Maintenance changes might or might not require extensive technical knowledge, but some technical knowledge could be required to properly classify a proposed change as major or minor. Will new (additional) processing requirements be the "straw that breaks the camel's back" in terms of hardware or software capacity? Do the proposed changes require programming tools that are incompatible with the current system design or configuration? The answers to questions such as these determine whether the existing system will be modified or replaced by an entirely new system.

If the existing system is to be modified, the programs and/or files to be changed are identified, modified, and tested. The specific technical knowledge requirements are heavily dependent upon the specific hardware and software components affected by the change. In the event that an entirely new system is required, a new SDLC with all of the technical knowledge requirements described previously will be initiated.

MANAGEMENT OF COMPUTER RESOURCES

Thus far, the need for technological knowledge has been discussed in the context of developing a single information system. But consider the complexities and knowledge required to manage the computer resources within a large organization. Existing information systems might number in the hundreds or thousands. Numerous new development projects or major upgrades of existing systems could be in progress at any one time.

Such an environment requires increased attention to two very important technological issues: compatibility and future trends. Compatibility and integration have become increasingly important issues as the number of computer systems and applications has increased. These are not all stand-alone systems. Most are integrated with other systems through both software and hardware. For example, accounts payable and accounts receivable programs may share a common hardware platform and operating system. Data from both of these systems may be input to a financial reporting system on an entirely different computer. Data from many sources within the organization might be available over a communications network.

The manager of such a disparate collection of information systems must contend with a great deal of technical complexity. The manager must ensure that each new system not only operates correctly by itself, but in conjunction with all of the other systems in the organization. The manager must ensure that hardware and software acquisitions are cost effective. These acquisitions must also provide a good basis for the implementation of future systems.

Given the rapid pace of change in computer technology, the manager must have a broad understanding of both the current technology and its likely trends in the future. Will the computer purchased today be compatible with the hardware likely to be available three years from now? Will the communication network purchased today have sufficient capacity to meet future needs? Should the organization invest only in "tried and true" technologies or should it risk the purchase of "cutting-edge" technologies in hope of greater performance?

Answering these questions requires a great depth of technical knowledge—far more knowledge than any one person may possess. Typically, a manager confronted by such questions will rely on the advice of experts and other sources of information. Even so, the manager responsible for such decisions must have a sufficient base of technical knowledge to understand the information and advice provided by those experts.

ROLES AND JOB TITLES

A large number of people can be loosely classified as "computer professionals," and an even larger number utilize computers in the workplace, school, or home. A bewildering array of job titles, specializations, and professional certifications accompanies this wide range of roles. The following sections attempt to classify these roles into groups, describe some of their common characteristics, and describe the knowledge of computer hardware and systems software required by each group.

Users

By definition, anyone who comes into contact (directly or indirectly) with a computer system can be considered a *user*. The term *user* is generally applied to persons who interact directly with *application software*, which is software designed to accomplish a specific purpose. Examples of application software include payroll programs, statistics programs, and programs to input and edit business transactions such as invoices or customer orders. Examples of the users of such software include data entry clerks, customer service representatives, and managers.

In recent years, a newer class of user has arisen as a result of the proliferation of general-purpose *application tools*. Examples of such tools include spreadsheet programs such as Lotus 1-2-3, database managers such as dBASE IV, and word processors such as WordPerfect. Although this class of software is not geared toward a particular need such as payroll or customer billing, it *is* geared toward specific types of processing needs, such as financial calculation and document preparation.

Application Development Personnel

A large class of computer professionals is employed to create application software for specific processing needs. These professionals go by many job titles, including *programmers*, *programmer/analysts*, *systems analysts*, and *systems designers*, among others. Each of these professionals contributes to a different part of the system development life cycle. Systems analysts are primarily responsible for conducting surveys, determining feasibility, and the definition and documentation of user requirements. Systems designers are primarily responsible for procuring hardware and procuring or designing applications software. *Application programmers* are primarily responsible for implementation and testing of software.

Many professionals engage in activities that don't fit neatly into these job title definitions. For example, it is very common for a person with the systems analyst job title to be responsible for survey, analysis, design, and management of a development project. Programmers are often responsible for analysis and design tasks as well.

Adding further confusion are the types of applications that can be developed and the training required for each type. Applications can be loosely classified into two types: information processing and scientific/technical. Personnel who develop information processing applications typically develop software geared toward the processing of business transactions or the provision of information to managers of organizations. These people usually have college or technical degrees in management or business, with a specialization in information processing.[1]

[1] Other names for the field include management information systems, data processing, and business computer systems.

Personnel who develop scientific/technical applications typically have degrees in *computer science* or some branch of engineering. The applications they develop are oriented toward scientific pursuits such as astronomy, meteorology, and physics. Technical applications are often oriented toward the control of hardware devices, such as robots, flight navigation, and scientific instrumentation.

Systems Software Personnel

Another large class of computer professionals is responsible for the development of systems software. The typical job title for such a person is *systems programmer*. This person's responsibilities include software such as operating systems, compilers, database management systems, and general-purpose application tools. These professionals typically have degrees in computer science or computer engineering and do not develop specific applications. Organizations with large amounts of computer equipment and software will frequently employ systems programmers as hardware/software consultants who perform hardware and software installation and configuration. Larger numbers of systems programmers are employed by organizations that develop and market systems software.

Hardware Personnel

Personnel who interact with computer hardware include those who operate it, install it, and design or build it. A *computer operator* is responsible for executing application software and for other operational tasks such as backing up files to tape and starting and stopping the system. This person usually has a technical degree or has had vendor-specific training. The proliferation of small and medium-sized computers has caused the role of user and computer operator to be combined in many cases. In the past, users relied almost exclusively on operators for tasks such as starting programs and loading files. Today, users of personal and small computers must usually perform such operations themselves.

Personnel responsible for hardware installation and maintenance are typically employed by vendors of computer hardware. Lower level personnel will usually have technical degrees and/or vendor-specific training; higher level personnel will often have degrees in computer science or computer engineering. Personnel who design computer systems typically have degrees in computer science or a related branch of engineering.

SOURCES OF INFORMATION ABOUT COMPUTER TECHNOLOGY

The contents of this book and other material that you will cover during your studies will provide a foundation of technical knowledge for a career in information systems management and development. Unfortunately, that foundation will quickly erode due to the rapid pace of change in computer and information technology. Thus, you will need to constantly update your knowledge simply to stay current.

You can achieve this goal via a number of paths. Newspapers and periodicals can provide a wealth of information regarding current trends and technologies. Training courses through hardware and software vendors or others will teach you the specifics of current products. Additional coursework and self-study can keep you abreast of technologies and trends less specifically geared toward particular products. By far, the most important of these activities is reading the current periodical literature.

Unfortunately, the volume of literature available on computer topics is vast and it is often difficult to determine those sources that are most important and/or appropriate to information systems professionals. Much of the problem arises from differences in training between information systems professionals, computer scientists, and computer engineers. Information systems professionals are extensively trained in application development tasks (for example, systems analysis), managerial tasks (for instance, project management), and the functional areas of business (such as, accounting and marketing). The extensive coverage of these topics leaves less time available for detailed coverage of computer hardware and systems software.

Computer scientists and engineers are more extensively trained in hardware and systems software than are information systems professionals. They are less extensively trained in application development, management, and functional business areas. Thus, many sources of information that are oriented toward one specialty make difficult reading for the others. For example, a detailed discussion of user interaction and validation of requirements models may be beyond the training of a computer scientist or engineer. Similarly, detailed descriptions of the theory of optical telecommunications may be beyond the training of an information systems professional.

These differences pose a problem for the information systems professional who desires current information on hardware and systems software technology. Much of the available literature is oriented toward computer scientists and/or computer engineers. The content and/or detail of this material is often beyond the training of a typical information systems professional. It is necessary to find information sources that have appropriate depth and breadth without overwhelming the reader with technical detail.

The periodicals listed below are recommended as a good source of technical information for the information systems professional. On the whole, they avoid excessive levels of technical depth while still providing sufficient detail to be useful.

- *ACM Computing Surveys.* An excellent source of information on the latest research trends in computer software and hardware. Articles are in-depth summaries of technologies or trends geared toward a readership with a moderate level of familiarity with computer hardware and software.

- *Byte Magazine.* An excellent source of information on computer hardware, software, and trends, primarily oriented toward small computers. Coverage of both specific products and of general trends and technologies. Comparative reviews of specific products, as well as tutorials and theme articles on basic technologies, are regularly provided.
- *ComputerWorld.* A weekly magazine primarily geared toward computer news items. Coverage of product releases, trade shows, and occasional coverage of technologies and trends is provided.
- *Communications of the ACM.* A source of information about research topics in computer science. Many of the articles are too technical and specialized for the general reader, but some are geared toward a less research-oriented audience.
- *Computer.* A publication of the Institute of Electrical and Electronics Engineers (IEEE) devoted to coverage of computer hardware and software. Many of the articles are research-oriented, but occasional coverage of technologies and trends is provided for a less technically oriented audience.

The following periodicals are also good sources of information on computer hardware and systems software. With the exception of *Science* and *Scientific American*, they tend to be more specialized than the periodicals listed above.

- AI Magazine
- Data Base
- Datamation
- Dr. Dobbs Journal
- EDN
- Electronics Week
- IEEE Spectrum
- InfoWorld
- PC Magazine
- PC World
- Personal Computing
- Science
- Scientific American
- Unix Review

SUMMARY

Technical knowledge of computer hardware and systems software is required in the development of information systems. The depth and breadth of required knowledge differ among phases of the structured system development life cycle (SDLC). Breadth of knowledge is required in the survey phase to evaluate feasibility and related issues. Depth of knowledge is most required during systems design, when detailed specifications for hardware and systems software are derived.

Technical knowledge is also required to manage the computer resources of an organization. Particular attention must be given to compatibility and future trends. Compatibility is important because computer hardware and systems software are typically shared among organizational units and subsystems. Future trends must be considered in acquisitions due to the long-term nature of hardware and software investments.

Technical knowledge must be constantly updated due to changes in the hardware and software technology. Information systems professionals must engage in continuing education and study to keep pace with these changes. Training may be obtained through vendors, educational organizations, and self-study. Self-study relies heavily on periodical literature. Care must be exercised in the selection of periodical literature due to differences among them in intended audience and required background and training.

Key Terms

application programmer
application software
application tool
computer operator
computer science
programmer
programmer/analyst
requirements analysis (needs analysis
systems analysis
systems analyst
systems design
systems designer
system development life cycle (SDLC)
systems implementation
systems programmer
systems survey
user

Review Questions

1. In what way(s) is the knowledge needed to operate complex devices different from the knowledge needed to acquire and configure them?

2. What knowledge of computer hardware and systems software is necessary to successfully complete the survey phase of the system development life cycle?

3. What knowledge of computer hardware and systems software is necessary to successfully complete the design phase of the system development life cycle?

4. What additional technical issues must be addressed when managing a computer center as compared to developing a single information systems application?

2 *Introduction to Computer Hardware*

Chapter Goals

- *Describe the functions and capabilities of automated computing devices and the means by which they may be implemented.*
- *Describe the functions, capabilities, and operation of a computer processor.*
- *Describe the components of a computer system and the function of each.*
- *Describe the classes of computer systems and the characteristics by which they are distinguished.*
- *Describe the ways in which computer system power is measured and compared.*

All computation devices share three common functions. These are:

- The ability to perform one or more types of computation
- The ability to store data on a temporary or permanent basis
- The ability to accept input and return output (i.e., the ability to communicate)

Each of these functions may be implemented by many methods and/or devices.

As an example, consider a human operating as a computation device. Computational functions such as addition and subtraction are implemented by neurons within the brain. Neurons operate both chemically and electrically. Storage of data is also implemented via neurons. The exact basis of human memory is not fully understood, but it appears to involve a combination of physical connections between neurons and chemical changes within individual neurons. Input is implemented through the five senses and associated organs. Output is implemented by the physical movement of various parts of the body. The bases of communication include sound waves (hearing and speech), chemical detection (smell and taste), and light.

Automated computation may be implemented by many methods and/or devices. Bases of implementation include mechanics,[1] electronics, and optics. Computational devices may utilize one or more of these bases to implement the functions listed above. For example, a modern computer may implement computation electronically (e.g., using transistors within a microprocessor), implement storage optically (e.g., using a laser and the reflective coating on an optical disk), and implement communication using a combination of electronics and mechanics (e.g., the mechanical and electrical components of a printer).

Mechanical Implementation

Early mechanical computation devices were designed to perform repetitive mathematical calculations. The most famous of these machines, called the *difference engine*, was built by Charles Babbage in 1821. This machine was an entirely mechanical device that computed logarithms through the movement of various gears and other mechanical components. Many other mechanical computation machines were developed well into the twentieth century and were in common use as recently as the 1960s.

The common element in all early computational devices was a mechanical representation of mathematical calculation. As a simple example, consider a mechanical clock driven by a spring and a pendulum. Each swing of the pendulum allows a gear to move one step under pressure from the spring. As time passes, the pendulum swings more and more times and the gears advance the hands of the clock.

[1] The term *mechanics* refers to the use of interconnected (or interacting) moving parts.

A clock can be considered a relatively simple adding machine. Each swing of the pendulum performs addition by allowing the gears to advance their position. The current position of the gears represents their initial position at some time in the past and the accumulated (added) value of the time that has passed since the clock was started. A set of hands attached to one or more of the gears displays this position for the user in relation to the numbers printed on the face of the clock.

A more modern example of simple mechanical calculation is the odometer in an automobile. The wheels of the car are connected to the odometer through a series of gears and mechanical links. As the wheels of the car move forward, the physical movement is transmitted by the gears and links and is accumulated by the odometer in terms of the current position of its display wheels. Each of the display wheels has numbers printed on it, thus providing a display of the current accumulated value (miles the car has driven) to the user.

More complex computational functions can also be represented mechanically. Multiplication of whole numbers, for example, can be represented mechanically as repeated addition. Thus a machine capable of addition can be used to perform multiplication by executing the addition function multiple times. (For example, 6 times 3 can be calculated by adding 6 to 6 and then adding 6 to the result.)

Although a variety of computational functions can be performed by mechanical means, there are inherent limitations and shortcomings of mechanical computation. The first of these is the difficulty in designing and building mechanical devices. Automated computation through the use of gears and wheels requires a complex set of components that must be designed, manufactured, and assembled to exacting specifications. As the complexity of the computational function increases, the complexity of the mechanical device that performs it also increases, thus exacerbating these problems.

Mechanical devices are also subject to wear and breakdown and tend to require frequent maintenance. In a mechanical clock, for example, the internal gears are subject to wear due to friction. Over time, the size of the gear teeth decreases due to this wear, and the gears interact with less and less precision. Thus, the clock starts to run fast or slow and requires adjustment. This problem can be minimized by regular cleaning and lubrication, but that will only slow down the wear, not eliminate it. Eventually the clock will either cease working or become so unreliable as to become unusable.

Another problem with mechanical devices is the speed at which they perform their function. The speed of any device that uses physical movement to implement a function is limited by the maximum possible speed of the moving parts. The speed of a car, for example, is limited not only by the power of the engine but by the physical limits of movement of the wheels, transmission gears, and engine components. Causing any of these components to move too quickly (for instance, revving the engine to 20,000 revolutions per minute) will cause rapid breakdown.

Electrical Implementation

Much as the era of mechanical clocks gave way to the era of electrical clocks, the era of mechanical computation eventually gave way to electrical computers. The change to electricity was motivated by increasing knowledge of how to effectively use it and a desire to overcome the shortcomings of mechanical computation. The big impetus came just before and during World War II. Many types of problems faced by the military (such as navigation and breaking codes) required massive amounts of complex computation to solve. The mechanical devices of the time were simply not up to the task; thus the quest for electrically based computation began.

In an electrical computer the movement of electrons performs essentially the same functions as those performed by the gears and wheels of earlier mechanical computers. Storage of numerical values is accomplished by the storage of magnetic charges rather than by the position of gears and wheels. Where necessary, electricity can be used to cause physical movement, such as in the hands of an electrical clock or the pins of a dot matrix printer.

Electrical computers addressed most of the shortcomings of mechanical computation. They are inherently faster due to the relatively high speed at which electricity flows (or electrons move). With improvements in the design and construction of electrical devices, they became more reliable and easier to build than their mechanical counterparts. Electronic computers made it possible to perform complex calculations at a speed previously thought impossible. This allowed larger and more complex problems to be addressed and made the process of solving simpler problems much faster.

Optical Implementation

There are possibilities for implementing computers besides mechanical and electrical devices. The most promising possibility today is the use of light (optics). The movement of photons can be used to perform computation. Numbers can be represented as pulses of light and stored either directly (an image stored as a hologram) or indirectly by materials that reflect (or don't reflect) a light source (such as the bits of information on an optical disk).

Optical implementations are at the cutting edge of computer hardware technology. Optical data communication is now common in computer networks that cover relatively large distances. Optical devices are currently in use for storing and retrieving large amounts of data. Some input/output devices, such as laser printers and optical scanners, are based on optical technologies and devices. Purely optical and hybrid electro-optical devices have been developed for communication between computer system components and for computer processing.

At this time, these devices are largely experimental. However, it is expected that optical and electro-optical technologies will find much wider application in the computer hardware of the next decade and beyond.

PROCESSORS

A *processor* is a device that performs data manipulation and/or transformation functions. These functions are usually computational in nature but might include other functions, such as comparison and the movement of data among storage locations. A processor is usually capable of many different functions: addition, subtraction, multiplication, division, and comparison. An *instruction* is a signal (or command) to a processor to perform one of its functions. When a processor performs a function in response to an instruction, it is said to be *executing* that instruction.

Each of the functions that a processor can perform is very simple (primitive). An example of a simple processor function is the addition of two numbers. Complex functions are performed as a sequence of simple functions. For example, a typical processor cannot add a set of 10 numbers together in response to a single instruction. The processor must be instructed to add the first number to the second and to temporarily store the result. It must then be instructed to add the stored result to the third number and to temporarily store that result, and so on. Individual instructions must be issued and executed until all 10 numbers have been added.

Most useful computational tasks, such as calculating payroll, require execution of a long sequence of instructions. This sequence of instructions is called a *program.* A program is usually a complex mixture of processing operations. For example, payroll calculation requires a complex series of input, computation, storage, and output functions to be performed on many data items. Some programs, like the addition example above, require the repetitive execution of similar instructions.

A processor can be classified as either general-purpose or special-purpose. A *general-purpose processor* is capable of executing many instructions and can execute instructions in many different sequences or combinations. The task that is performed by such a processor can be altered by changing the program that directs its actions. Thus, a general-purpose processor can be instructed to do many different tasks—payroll calculation, text processing, scientific calculation, and the like—simply by supplying it with an appropriate program.

A *special-purpose processor* is designed to perform only one specific task. Although this processor might be capable of executing many types of instructions, it can execute them in only one sequence. This is analogous to a general-purpose processor that has only one program. Many common devices, such as simple calculators, microwave ovens, video cassette recorders, and computer printers, contain special-purpose processors. Although such processors (or devices that contain them) can be called computers, the term *computer* is more often used to refer to a device containing a general-purpose processor that can execute any program supplied to it.

COMPUTER SYSTEM CAPABILITIES

Although a computer system is an automated computing device, all automated computing devices are not computer systems. However, the distinction between the two terms is not absolute. The primary characteristics that distinguish a computer system from simpler automated computation devices include:

- A general-purpose processor capable of performing computation, data movement, comparison, and branching functions
- Storage capacity sufficient to hold large numbers of programs and data
- Flexible communication capability through the use of multiple communication media and devices

Using these characteristics and criteria, a device such as an IBM Personal Computer system would be classified as a computer. Devices such as adding machines and calculators would fail most or all of these tests.

Processor Capabilities

Some processing tasks require little more than the computation functions of a processor. Examples of these tasks include single-step and multiple-step computations. For example, consider the following calculation:

$$Profit = Sales - Expenses - Taxes$$

The only processor capabilities needed to compute profit are the ability to perform subtraction and to store a single temporary result. Thus, expenses may be subtracted from sales and the result stored in a temporary location. Taxes are then subtracted from the temporarily stored value, and the result is profit. Processing tasks such as this one are called *formulaic problems*. They can be expressed as a formula and solved by a processor using a sequential series of computational operations.

The majority of processing tasks that computers are expected to perform are not formulaic problems. Although many processing tasks contain subtasks that can be solved by formulaic approaches, additional processor capabilities are required. These additional capabilities include the ability to compare data items and the ability to alter the sequence of instruction execution, which is called *branching*. A procedure (or program) for solving such a problem is called an *algorithm*.

For example, consider the computation of taxes within a payroll program. In the United States, most income taxes are computed at progressively higher rates on higher levels of income. Table 2.1 shows a typical computational method for income taxes. Note that different levels of income require the use of different formulas in order to calculate the correct amount of tax. Thus, a program that computes tax based on this table must be capable of using a variety of formulas, depending on the income level supplied to the program. For a program to exercise this capability, the processor that executes the program must be capable of both comparison and branching functions.

TABLE 2.1 A sample table for the computation of federal income tax.

		Tax Rate Schedule	
If Gross_Pay is:		**The tax is:**	
Over:	**But not over:**		**of the amount over:**
$0	$2145	15%	$0
2145	5190	$321.75 + 28%	2145
5190	—	1174.35 + 31%	5190

Figure 2.1 illustrates a program that utilizes comparison and branching functions to calculate taxes. In statements 20 and 50, an explicit comparison of GROSS_PAY to a constant is made to determine the next appropriate instruction to execute. Depending on the results of the comparison, the program might execute the next instruction in the sequence, or it might alter the sequence by issuing a GOTO instruction.

```
10    INPUT GROSS_PAY
20    IF GROSS_PAY > 2145 THEN GOTO 50
30    TAX = GROSS_PAY * 0.15
40    GOTO 90
50    IF GROSS_PAY > 5190 THEN GOTO 80
60    TAX = 321.75 + ( GROSS_PAY - 2145 ) * 0.28
70    GOTO 90
80    TAX = 1174.35 + ( GROSS_PAY - 5190 ) * 0.31
90    OUTPUT TAX
100   END
```

FIGURE 2.1 A program to calculate federal income taxes. Comparison functions (lines 20 and 50) and branching functions (indicated by the term GOTO) are used to select the proper formula.

The general-purpose processor contained within a computer must be capable of many types of comparisons (e.g., equality, less than, and greater than) and must be capable of altering the sequence of instruction execution, depending on the result of such comparisons. These basic capabilities form the basis upon which the processor can solve complex problems. It is these processor capabilities that distinguish a computer processor from the processors of lesser automated computation devices such as adding machines and calculators.

Storage Capabilities

The storage capacity of a computer system must be large for a variety of reasons. These include the need to store:

- Intermediate processing results
- Programs that are currently being executed
- Programs that are not currently being executed
- Data used by currently executing program(s)
- Data that will be needed by programs in the future

The need to store intermediate results was demonstrated in the previous example of a profit formula. Many programs that solve real-world problems require the storage of hundreds, thousands, or millions of intermediate results during program execution. For maximum speed of program execution, storage for this purpose should be accessible by a processor as quickly as possible.

A typical program consists of many instructions. These instructions can number in the thousands for simple programs or in the millions for extremely complex programs. In modern computer systems, the processor may be executing many programs at the same time. For maximum speed, all of these program instructions should be quickly accessible by the processor.

Data must also be stored for present and/or future use. A user could require storage of and access to thousands or millions of data items. A large organization could require storage of and access to billions of data items. This data may be used by currently executing programs, held for future processing needs, or held as an historical record for future reference.

Communication Capabilities

The communication capability of a computer must be flexible and encompass a variety of media and devices. A typical small computer will normally have several communication devices. These may include a video display, a keyboard, a mouse, a printer, and a modem or network interface. A large computer will typically have many such devices and can utilize devices of substantially more power and complexity than a smaller computer.

COMPUTER SYSTEM COMPONENTS

The components of a *computer system* are illustrated in Figure 2.2. Each component addresses one of the computer capabilities described in the previous section. The number, implementation, complexity, and power of these components can vary substantially from one computer system to another. However, the functions performed by these components are generally quite similar.

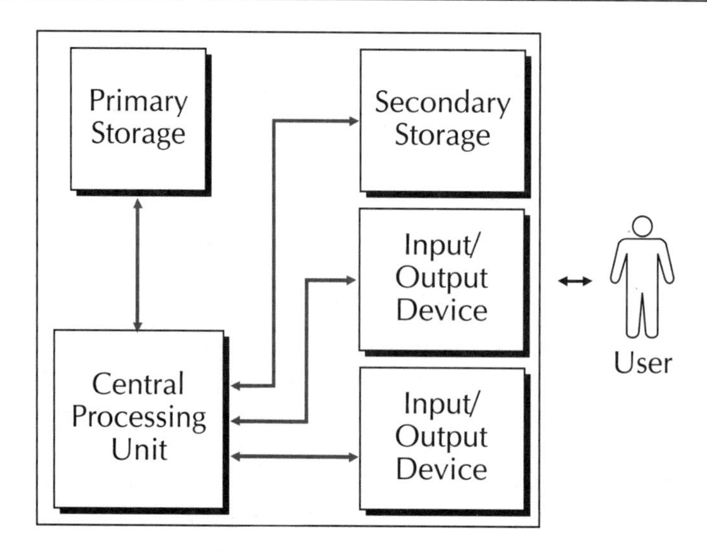

FIGURE 2.2 **The primary components of a computer system.**

The heart of the computer system is the *central processing unit (CPU).* Each of the other components is attached to the CPU via a communication channel. Instructions and data from currently executing programs flow to and from *primary storage.* Programs that are not currently being executed as well as groups of data items that are too large to fit in primary storage are held in *secondary storage.* Secondary storage can be composed of several devices, such as multiple hard disk drives, although only one device is shown in Figure 2.2. The CPU is also connected to a number of *input/output (I/O) devices.* These devices allow the CPU (and the computer system as a whole) to communicate with the outside world—for instance, to a user through a video display and keyboard. Two such devices are shown in Figure 2.2, although the actual number of I/O devices can vary widely from one computer system to another.

Central Processing Unit

A CPU is composed of several components. In most computers these are:

- The arithmetic logic unit
- Registers
- The control unit

The *arithmetic logic unit (ALU)* is composed of electronic circuits that perform the computational and logical functions of the processor. Different portions of this circuitry correspond to different functions, and the execution of a math or logic instruction causes data (electronic signals) to flow through the appropriate portion(s) of this circuitry. The arithmetic instructions typically implemented include addition, subtraction, multiplication, and division. More advanced computational functions such as exponentiation and logarithms can also be implemented. Logic instructions include various forms of comparison operations such as equality, greater than, and less than.

A small set of temporary storage locations (or cells) are generally located within the CPU. These cells, called *registers,* can typically hold a single instruction or data item. Registers are used to store data or instructions that are needed immediately, quickly, and/or frequently. For example, two numbers that are about to be added together are each stored in a register. The ALU reads these numbers from the registers and stores the result of the addition in another register. Because registers are located within the CPU, their contents can be accessed quickly by the other CPU components.

The *control unit* is primarily responsible for the movement of data and instructions to and from primary storage and for controlling the ALU. As program instructions and data are needed by the CPU, they are moved from primary storage to registers by the control unit. The control unit is also responsible for other functions related to input to and output from I/O devices.

Processor operation. A complex chain of events occurs when a computer executes a program. To start, the first instruction of the program is read from primary storage by the control unit. Data inputs for the instruction are also read from primary storage. If the instruction is a computational or comparison instruction, the control unit signals the ALU what function to perform, where the input data is located, and where to store the output data. Other types of instructions, such as input and output to secondary storage or I/O devices, are executed by the control unit itself. When the first instruction has been executed, the next instruction is read and executed and so forth until the final instruction of the program has been executed.

The steps required to process each instruction can be divided into two groups: the instruction cycle and the execution cycle. The operation of the instruction and execution cycles is depicted in Figure 2.3. The sequence of events performed during the *instruction cycle,* also referred to as the *fetch cycle*, or simply *fetch,* is as follows:

1. The control unit retrieves an instruction from primary storage and increments a pointer showing the location of the next instruction.
2. The instruction is separated into its components: the instruction code (or number) and the data inputs to the instruction. Each of these components is stored in a register.
3. The control unit generates an internal signal to the ALU to execute the instruction.

The sequence of events during the *execution cycle* is as follows:

1. The ALU accesses data within registers to determine what function (instruction) to perform and to obtain the input data for the instruction.
2. Data is passed through circuitry to perform the function (instruction).
3. The results of the instruction are placed in a register or returned to the control unit to be written to memory.

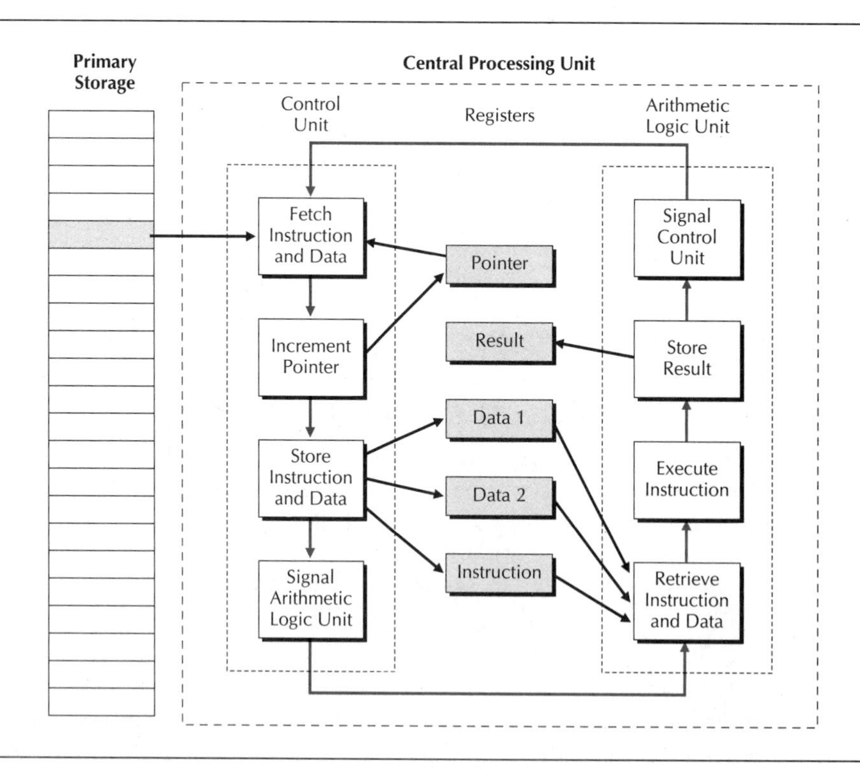

FIGURE 2.3 **Flow of control and data during the instruction and execution cycles. The instruction cycle is performed within the control unit; the execution cycle is performed within the arithmetic logic unit.**

At the conclusion of the execution cycle, a new instruction cycle is started. Thus, a central processing unit executes a program by constantly alternating between the instruction and execution cycles. This switching is precisely regulated by an internal clock that allocates fixed intervals of time to each cycle.

Primary Storage

Primary storage consists of storage cells that hold programs (and their associated data) that are currently being executed. Primary storage is usually referred to as *main memory* or simply *memory*. The execution of a program requires a great deal of movement of both instructions and data between main memory and the CPU. Because the CPU is a relatively fast device, it is desirable to implement main memory using devices that are capable of rapid access.

Primary storage can be thought of as a sequence of contiguous, or adjacent, memory cells, as shown in Figure 2.4. Each of these cells has a unique address corresponding to its physical location in the sequence. The CPU uses these addresses to specify which instructions are to be loaded into the CPU (during the instruction cycle) and what data is to be read or written. Because the CPU may require instructions or data from many different parts of memory, it is necessary to implement primary storage with a device capable of *direct access*, or *random access*, terms that refer to the ability to specify that a specific location be read from or written to, ignoring any other locations that may occur before or after it.

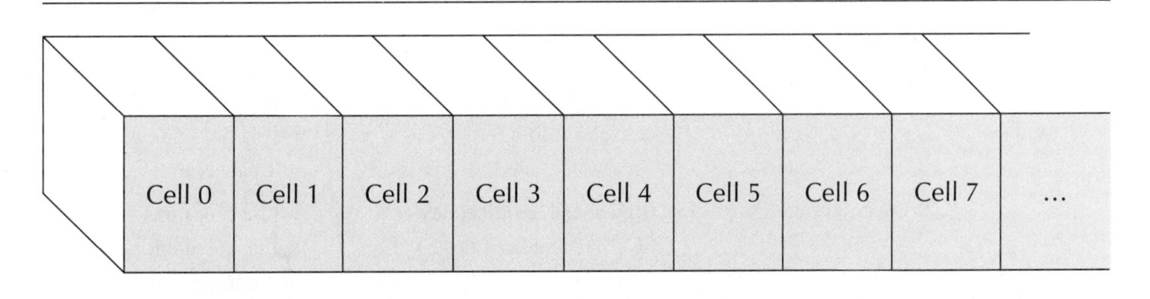

FIGURE 2.4 **The sequential organization of primary storage locations.**

In current computer hardware, main memory is implemented with silicon-based semiconductor devices commonly referred to as *random access memory (RAM)*. These devices provide the access speed the CPU requires and allow the CPU to read or write a specific memory cell by referring to its location. Unfortunately, RAM is costly, and so tends to limit the amount of main memory in a computer system. Another problem with RAM is that it does not provide permanent storage. When power to RAM is turned off, its contents are lost. This characteristic of RAM (or any other storage device) is called *volatility.* Any type of storage device that cannot retain data values indefinitely is said to be *volatile.* In contrast, storage devices that permanently retain data values are said to be *nonvolatile.* Due to both the volatility and limited capacity of primary storage, a computer system requires auxiliary means of storing data and programs over long periods of time.

Secondary Storage

Secondary storage refers to storage devices that are nonvolatile and able to store large volumes of data. Secondary storage is used to store programs that are not currently being executed. It also stores data that is not needed by currently executing programs as well as data that cannot fit into primary storage. In general, the contents of secondary storage must first be moved to primary storage before they can be accessed by the CPU.

Within a typical information system, the number of programs and amount of data are quite large. Thus, a typical computer system must contain a large amount of secondary storage as compared to its primary storage. For example, a typical mid-sized computer might contain 16 million primary storage locations and hundreds of millions (or billions) of secondary storage locations. Differences in utilization and implementation of the various types of storage in a computer system are summarized in Table 2.2.

TABLE 2.2 **Summary of the differences in implementation and usual content among the various types of storage in a computer system.**

Storage Type	Implementation	Contents
Registers	Very high-speed electrical devices within the CPU	Currently executing instruction; a few items of related data
Primary storage	High-speed electrical devices (RAM) outside of the CPU	Entire programs currently being executed; small amounts of data
Secondary storage	Low-speed electromagnetic or optical devices (e.g., magnetic and optical disk)	Programs not currently being executed; large amounts of data

To keep the cost of necessary secondary storage within acceptable limits, it is imperative to use storage devices that provide relatively slow access speeds and/or limited access methods compared to primary storage devices. The most common devices used to implement secondary storage are magnetic disks, optical disks, and magnetic tape. Disk storage provides relatively fast access (compared to magnetic tape) and allows each storage location to be accessed directly via its address (i.e., direct or random access). Magnetic tapes provide a slower but cheaper method of storage.

Input/Output Devices

To be useful to humans, a computer system must have the ability to communicate with users. It may also be desirable to allow the computer system to communicate directly with other computers or processing devices. As you learned earlier, this role is filled by a general class of devices called I/O devices. I/O devices are implemented with a wide range of technologies, depending on the exact nature of the communication they support.

Examples of input devices for human use include keyboards, pointing devices such as a mouse, and voice recognition devices. The purpose of such devices is to accept input from a human on his or her own terms—for example, by voice or keystroke input—and convert that input into something the computer can understand (electrical signals). Output devices for human use include video displays, printers, plotters, and speech output devices. All of these devices perform a conversion from electronic signals to a communication medium that a human can understand, such as pictures, words, or sound.

Other input/output devices include modems, network interface units, and multiplexers. These devices provide communication capabilities between computers or between a computer and a distant I/O device.

von Neumann Computer Architecture

In a broad sense, the basic underlying architecture of computers has not changed since the early 1950s. This architecture was first described by the mathematician John von Neumann and others years before the building of such computers was even attempted. The essential characteristics of this type of architecture are those discussed in the previous sections:

- Use of a single general-purpose processor
- Use of stored programs
- Sequential processing of instructions
- Alternating instruction and execution cycles

Computers that are designed around these basic concepts are called *von Neumann machines*, although the term is not commonly used today.

Until quite recently, virtually all computers used for information processing were von Neumann machines. The increases in speed and power that occurred since the 1950s were primarily a result of improvements in the design and construction of computer circuits and components, not of changes in the basic design. Some changes to this basic design have found their way into current computers, but we defer discussion of these changes until later chapters.

CLASSES OF COMPUTER SYSTEMS

Computer systems can be loosely classified into the following categories, based on performance and capabilities:

- Microcomputers
- Workstations
- Minicomputers
- Mainframes
- Supercomputers

A *microcomputer* is a computer system designed to meet the information processing needs of a single user. Microcomputers are also referred to as *personal computers (PCs)*. The hardware and capabilities of a microcomputer are typically geared toward "ordinary" processing needs such as word processing, computer games, and small to medium-sized application programs, such as programs to compute an individual's income tax and compute payroll for a small business.

A *workstation* is essentially a microcomputer designed to handle very demanding processing tasks. Examples of such tasks include complex mathematical computation, computer-aided design, and the composition and display of high-resolution video (graphic) images. These tasks generally require more processor speed, more storage capabilities (in both quantity and speed), and higher quality video display devices than are provided with an ordinary microcomputer. Particularly with respect to processor speed and storage capabilities, the power of a workstation is often similar to that of a minicomputer. However, the overall design of a workstation is targeted toward a single-user operating environment.

A *minicomputer* is designed to provide information processing for multiple users and to execute many application programs simultaneously. They are typically designed to allow 4 to 64 users to interact with the computer system simultaneously. Supporting multiple users and programs requires fairly extensive capabilities for processor speed, storage, and input/output. It also requires more sophisticated system software than is typically found on microcomputers.

A *mainframe* computer system is designed to handle the information processing needs of a large number of users and applications. These machines are typically capable of interacting with tens or hundreds of users at a time and of executing hundreds of programs at a time. A typical use of a mainframe might include 50 users entering customer orders, several programs generating periodic reports, various users querying the contents of a large corporate database, and an operator making backup copies of disk files—all at the same time! This level of simultaneous processing demands fast processing, large amounts of primary and secondary storage, and extensive I/O capabilities.

A *supercomputer* is designed primarily for one purpose: to perform large amounts of mathematical computation as quickly as possible. These machines are used for the most demanding of computational applications, including simulation, three-dimensional modeling, weather prediction, and computer animation. All these tasks require an extremely large number of complex calculations and thus demand a CPU with the highest possible computational speed. Storage and communication requirements are also extremely high. Because of these requirements, supercomputers are typically implemented using the very latest (and most expensive) devices.

Table 2.3 summarizes the configuration and capabilities of each class of computer system. Each class is represented by a typical model available in 1993. The performance and other specifications of all computer classes are in a constant state of flux. Rapid advancements in computer technology lead to rapid improvements in computer capability as well as a redefinition of both the classes themselves and the expected capabilities of computers within that class.

TABLE 2.3 Comparison of representative models of each class of computer and their typical costs and performance capabilities, circa 1993.

Class	Typical Product	Typical Specifications	Approximate Cost	Approximate Speed
Microcomputer	IBM ValuePoint 486	• 8 million main memory cells • 120 million disk storage cells • Single user	$2,000	5 million instructions per second
Workstation	DECstation 5000/133	• 16 million main memory cells • 400 million disk storage cells • High-performance video • Single user	$12,500	20 million instructions per second
Minicomputer	DECstation 5000/240	• 32 million main memory cells • 1 billion disk storage cells • 1 tape drive • 16 interactive users	$30,000	25 million instructions per second
Mainframe	IBM 9120/320	• 256 million main memory cells • 10 billion disk storage cells • Multiple tape drives • 16 high capacity I/O channels • 512 interactive users • 4 CPUs	$1,000,000	50 million instructions per second
Supercomputer	IBM 9076/SP1	• 1 billion main memory cells • 10 billion disk storage cells • 64 CPUs	$2,500,000	2 billion floating point operations per second

PERFORMANCE MEASUREMENT AND DESIGN FACTORS

The overall performance of a computer system depends on the power and speed of its various components, the speed of communication between the components, and the degree to which the components are "well matched" in speed and capacity. Because so many factors influence the system's performance, there are numerous ways to measure performance and numerous ways in which performance can be tuned (or tailored) to the needs of an individual user or set of applications.

Workload Performance Measurement

When looking at a computer system as a whole, a user is primarily concerned with the amount of work of which it is capable. The primary problem with measuring "work" is that its definition varies from one application or user to another and that larger computer systems must respond to work demands from various user and applications simultaneously. The definition of work to a secretary may involve creating and editing letters and memos; the definition of work to a manager may involve complex retrievals and analysis of data. Each of these kinds of work place different demands on different components of the computer system. To measure the work of which a computer system is capable, it is necessary to define the term *work* fairly precisely and to define a time reference for accomplishing that work.

The definition of work varies not only with the type of processing, but also with the nature of human-computer interaction (or lack thereof) during normal processing. Two broad classes of human-computer interaction are generally recognized. *Batch processing* describes a situation in which users do not normally interact with an application program once execution has begun. *Interactive,* or *on-line, processing* describes a situation in which a user interacts directly with an executing application program. A typical information system contains application programs of both types.

For example, consider an information system that supports mail-order sales. On-line processing in this environment could occur through the use of telephone operators who talk directly with customers by telephone while simultaneously interacting with an order entry/query application program. Orders are entered as they are received, and are stored for later processing. One of these later processing steps might be the generation of purchase orders for the next day or week. An application program might be constructed to examine the goods ordered during the day (by examining all stored orders). Based on this examination, a prediction could be made of goods needed to satisfy future orders. These predictions would then be used to automatically generate purchase orders to suppliers of the goods. Such an application could run once per day and would not require any direct user interaction. All of the application's data inputs would have been previously stored by the order entry/query program.

Several types and units of work can be identified in the above example. One unit of work is the acceptance, editing, and storage of a single customer order. Another unit of work is the query of the status of a single order. The common term for such units of work is *transaction.* By establishing a time reference of one hour, we can state the work capacity of the computer system in terms of transactions processed per hour. Such a measurement is normally called *throughput.* Defining the unit of work for the batch purchase order generation program is more difficult. It could be measured in terms of its inputs (customer orders), outputs (purchase orders), or simply as a single unit of work (execution of the entire program). In the latter case, throughput would be measured in terms of programs executed per unit of time—in this case, one hour.

When we consider computer systems running a mixed variety of application programs, the measurement of total throughput becomes more difficult since we must allow for multiple types (and thus definitions) of work. For instance, we might say that a computer is capable of executing 500 transactions and 10 programs per hour or provide a range of possible combinations (e.g., 1000 transactions and 0 programs to 0 transactions and 45 programs). The tradeoff can also be shown graphically, as in Figure 2.5.

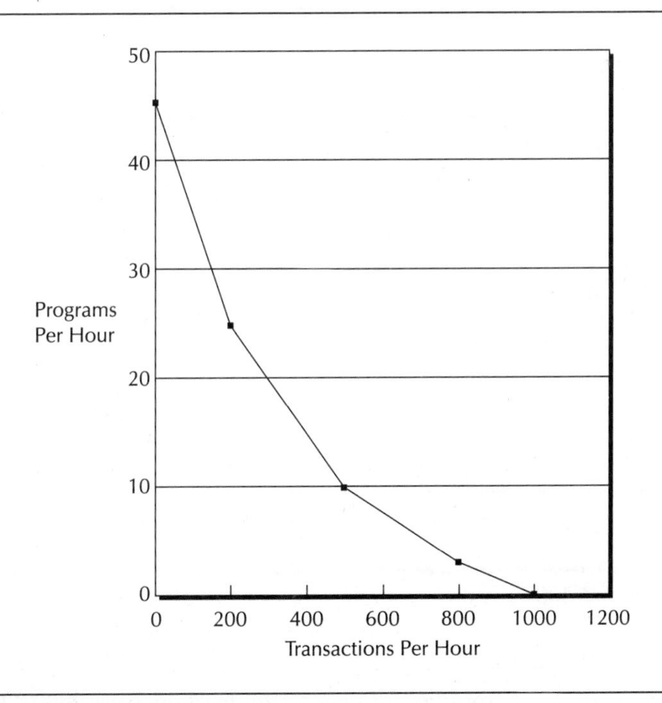

FIGURE 2.5 Throughput measured in programs and transactions per hour.

Another way of measuring the work capacity of a computer system is to examine the interval of time necessary to satisfy a single processing request, such as a query of the status of a single order, by a single user. This measurement is usually described by a number called *response time*. Unfortunately, this measure is also difficult to determine, because it depends on the number of processing requests that the computer is attempting to service at any one time. Response time will typically be fast (a low number) when few processing requests are being serviced and slow (a high number) when many processing requests are being serviced. Therefore, it is not accurate to state response time as a single number.

A complete specification of response time should show a range and an average (for example, best, 0.5 second per transaction; worst, 10 seconds per transaction; average, 2 seconds per transaction). Because this performance measurement varies with total processing demand, it can be determined only by knowing the demand likely to be placed on the computer system. In other words, it isn't possible to state an average response time unless the average demand for processing is known. In the above example, the demand for on-line processing would be expressed in terms of the number of new order and order query transactions per unit of time. The relationship between response time and processing load could be represented graphically, as shown in Figure 2.6.

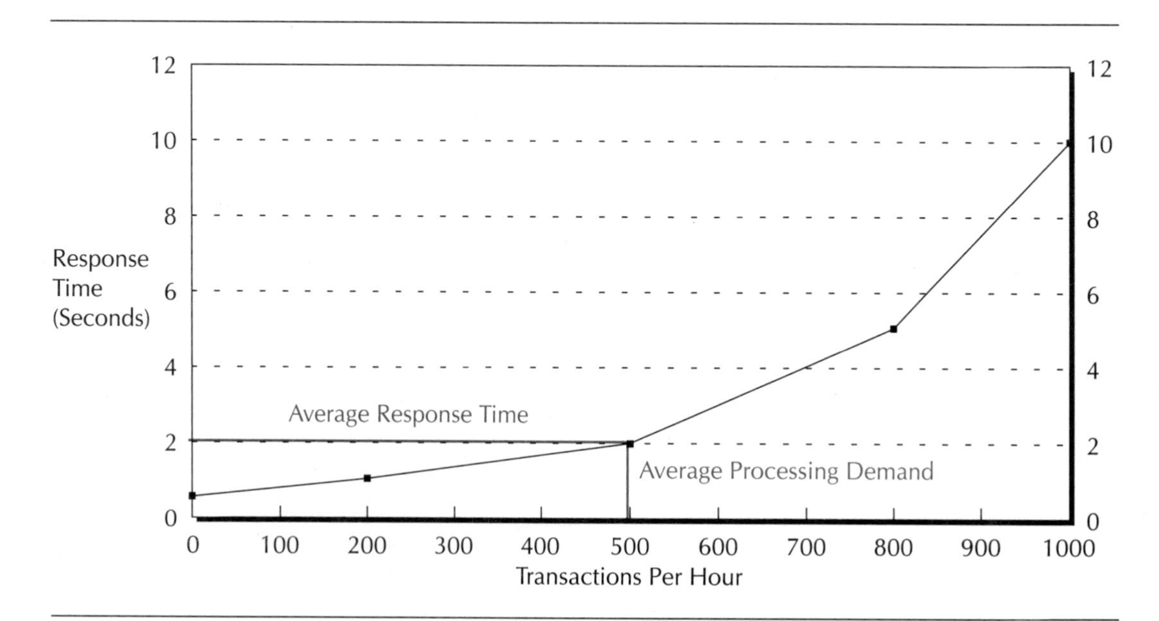

FIGURE 2.6 **Response time as a function of processing load measured in transactions per hour.**

Computer System Performance Measurement

Because computer systems are designed to be general-purpose—in other words, to support a wide variety of application programs—it is not possible for computer vendors to specify the types of workload measurements defined above. Instead, performance measurements will generally be provided for various types of machine actions, such as executing an instruction, accessing main memory, or reading data from a magnetic disk. It is then up to the machine's user to determine how these various measurements can be translated into the measurements of throughput and response time. Performance measurements are generally given for each individual component (or subsystem) of the computer, the communication capacity among the subcomponents, and a general measurement that attempts to summarize all of these performance measurements.

Within the CPU the primary performance consideration is the speed at which an instruction can be executed. This is generally stated in terms of *millions of instructions per second (MIPS).* Unfortunately, not all instructions take the same time to execute. Certain instructions, such as addition of whole numbers, execute quickly. Others—for example, division of fractional numbers—might execute more slowly. A related measurement that is specifically oriented to computation of fractional (floating point) numbers is *millions of floating point operations per second (MFLOPS).* This performance measure is commonly given for workstations and supercomputers because it is expected that applications running on these machines will primarily perform floating-point computations.

The speed at which the CPU can retrieve data from a storage or input/output device is dependent on the *access time*, or speed, of that device and the capacity and speed of the communication channel between the device and the CPU. Access time is normally measured in milliseconds (millionths of a second) or nanoseconds (billionths of a second). Access time is generally the same for both reading (device to CPU) and writing (CPU to device), unless otherwise stated. Access time for some devices, such as disk drives, might vary from one request to another depending on various factors, such as the location of the requested data on the disk. In that case, access time is normally stated as an average value.

When access time is stated for a specific device, there is no guarantee that the communication channel between the device and the CPU is capable of supporting that speed. Other electronic components between the device and the channel can also limit the channel capacity. The ability of the communication channel to support data movement between the CPU and a device (or vice versa) is normally expressed as a *data transfer rate.* This rate states the amount of data that can be moved over the channel in a specified interval of time (e.g., 1 million data items per second).

The overall performance of a computer system is a combination of the CPU speed, the access times of the storage and I/O devices, and the capacity of the communication channel(s) connecting them to the CPU. Because different applications make different demands upon these various subsystems, any overall performance measure for the computer system is valid only for a particular instance or class of application program.

For example, an on-line order entry application will normally utilize the secondary storage and I/O subsystems extensively while utilizing the CPU rather sparingly. A numerical simulation program will heavily utilize primary storage and the CPU while utilizing the other subsystems very little. The overall performance of a computer system with a fast CPU and memory and slow secondary storage will differ greatly depending on which of these applications is run on the system. An overall measure of system performance is often expressed in terms of MIPS. Unfortunately, assumptions about the application are rarely stated with this number.

Cost/Performance Relationships

In 1952, the computer scientist H. A. Grosch asserted that computing power is proportional to the square of the cost of hardware. According to Grosch, large, powerful computers always will be more efficient than smaller ones. The mathematical formula that describes this relationship is called *Grosch's Law*. For many years, computer system managers pointed to this law as justification for investments in ever larger central computers.

In the years since Grosch's law was first asserted, several major changes have occurred in computer technology. These include the emergence of distinct classes of computers, expanded abilities to change the configuration and capabilities of a given computer, an increase in software costs relative to hardware costs, and the emergence of computer networks. These changes have largely invalidated Grosch's law.

In Grosch's time there was basically only one class of computer: the mainframe, although its power was often less than the microcomputers of today. Purchasing larger machines within a single class tends to offer more MIPS for the money; thus Grosch's law appears to hold. This relationship is represented graphically in Figure 2.7. But if all classes of computers are considered as a group, Grosch's law no longer holds (as shown by the dashed curve in Figure 2.7). Instead of following Grosch's prediction, this plot shows that cost per MIPS actually increases with computer power. The implication of this finding is that one should obtain the smallest class of computer that can support an application.

Comparisons are even more difficult when networked computers are considered as an option. In Grosch's time the hardware and software to support computer networking did not exist. But it is now possible to tie together multiple machines and make them operate as if they were a single machine. A dramatic demonstration of this was made by Sandia National Laboratories when 1000 microcomputers were networked and shown to mimic the processing power of a supercomputer on certain applications. Similarly, many organizations have opted

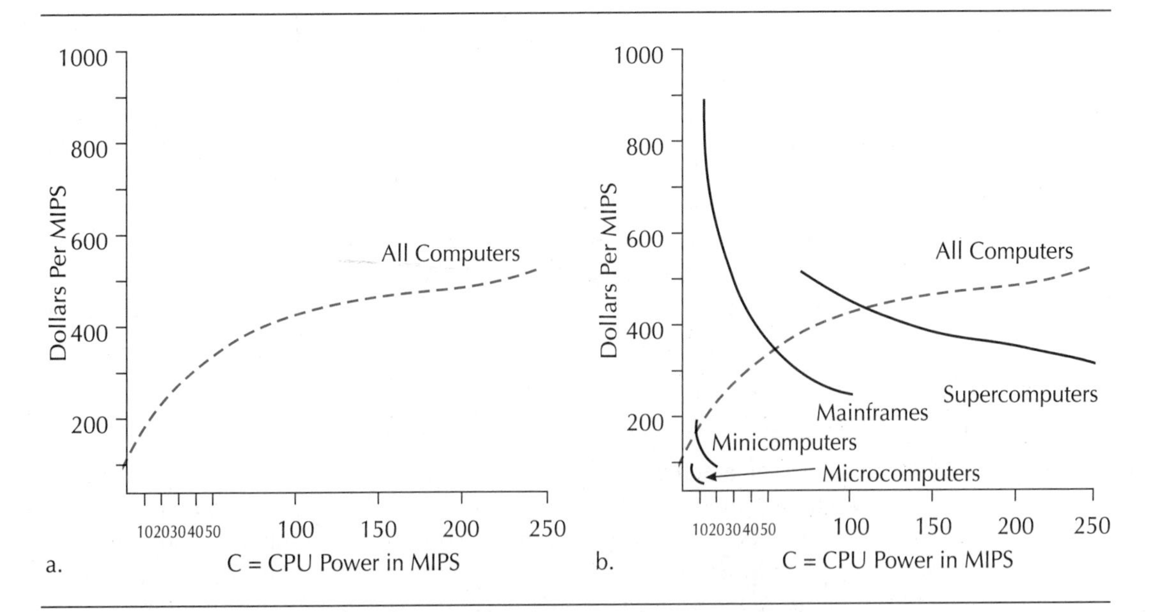

FIGURE 2.7 **These graphs plot ratios of price to performance (in dollars and MIPS) for (a) all computers and (b) by class and size of computer.**

to construct networked systems of microcomputers and minicomputers instead of a single large mainframe.

Within a given class or model of computer system, cost is affected by the power or capacity of the various subsystems. Often a deliberate choice can be made to cost-effectively tailor the performance of the system to a particular application. For example, in a transaction-processing application, the system might be configured with fast secondary storage and I/O subsystems while sacrificing CPU performance (especially as measured in MFLOPS). The same system might be cost-effectively tailored to a simulation application by providing a large amount of high-speed main memory, a very fast CPU, and a high-capacity communication channel between them. Secondary storage and I/O performance would be far less important, and so less costly alternatives could be used.

Such "tuning" of computer systems by substituting subsystems of varying power and cost is very common. Modern computers are designed with this substitution in mind. This fact has led to a significant blurring of the lines between the various classes of computers—particularly between microcomputers, workstations, and low-end minicomputers and between high-end minicomputers and low-end mainframes. Although earlier we defined a microcomputer as a single-user machine, high-powered versions of these machines can often be utilized for multi-user applications simply by upgrading the capacity of a few of the subsystems and utilizing more powerful system software. Similar possibilities exist for upgrading the performance of a minicomputer to that of a mainframe.

Another difficulty in applying Grosch's law today is the reality of costly systems and application software. The power of a computer system as expressed in MIPS and data transfer rates is only a potential as far as the end user is concerned. The realization of this power requires systems and application software, but the development of such software requires a substantial commitment of labor and its associated costs. This fact, coupled with the ever-decreasing cost of computer hardware, has made the cost of computer hardware an ever less significant factor. Thus, although hardware costs are still significant and should be managed effectively, a user or manager is far more likely to maximize the cost effectiveness of an entire information system by concentrating on its software components.

SUMMARY

A computer is an automated device for performing computational tasks. It accepts input data from the external world, performs one or more computations on the data, and returns results to the external world. Early computers had extremely limited capabilities and were implemented with mechanical devices. Modern computers have more extensive computational capabilities and are implemented with electronic devices. The advantages of electrical (as opposed to mechanical) implementation include speed, accuracy, and reliability.

A computer system contains a general-purpose processor with specific capabilities and components. The capabilities include simple arithmetic functions, numeric comparisons, and the movement of data between storage locations. A command to the computer to perform one of these basic functions on one or more data items is called an instruction. Complex manipulations of data can be implemented as a sequence of instructions called a program. The actions of a computer system are determined by the program and can be changed by using a different program.

A computer processor alternates continuously between the instruction cycle and the execution cycle. During the instruction cycle the control unit fetches an instruction and its related data from storage. During the execution cycle the ALU executes the instruction upon the data. The next instruction cycle fetches the next instruction and its data. This process continues until the last instruction in the program has been executed.

A computer system consists of a central processing unit (CPU), primary storage, secondary storage, and input/output (I/O) devices. The CPU contains the ALU, the control unit, and a small set of temporary storage locations called registers. Registers hold both data and instructions for use by the other components. The ALU performs arithmetic and comparison functions. The control unit is responsible for fetching instructions and data from storage and for directing the ALU.

Primary storage consists of a large number of relatively fast storage locations that hold programs and data currently in use by the CPU. Instructions and data are moved continuously between primary storage and the CPU during program execution. Primary storage is generally implemented by electrical devices called random access memory (RAM). These devices provide direct access and high speed and at a high cost.

Secondary storage consists of one or more devices with high capacity and low cost. Access speed is normally much slower than for primary storage. Secondary storage holds programs and data that are not currently in use by the CPU. I/O devices are the means by which the CPU communicates with the external world. They may allow communication with humans or with other computers. I/O devices vary in speed and communication medium.

A computer system may be classified as a microcomputer, a workstation, a minicomputer, a mainframe, or a supercomputer. Microcomputers and workstations are designed for use by a single user. Minicomputers and mainframes are designed to support many programs and users simultaneously. Supercomputers are designed to perform large amounts of numeric computation very quickly.

The performance of a computer system can be measured in terms of its ability to perform basic computing functions or useful work (i.e., workload). Performance measures for basic computing functions include millions of instructions per second (MIPS) and millions of floating-point operations per second (MFLOPS). Performance measures for workload include throughput and response time. Although both types of performance measures are directly related, the relationship is difficult to specify precisely.

Selecting the minimal-cost computer system for a given workload is a complex problem. With earlier generations of computers, economical computing solutions were generally achieved by sharing very large general-purpose computers. With modern computer systems, economical solutions are generally achieved by utilizing the smallest computer system applicable to a particular type of computing task.

Key Terms

access time
algorithm
arithmetic logic unit (ALU)
batch processing
branching
central processing unit (CPU)
computer system
control unit
data transfer rate
direct access (random access)
execution
execution cycle
fetch cycle
formulaic problems
general-purpose processor
Grosch's Law
input/output (I/O) device
instruction
instruction cycle
interactive processing
main memory (memory)
mainframe
microcomputer
millions of floating-point operations per second (MFLOPS)
millions of instructions per second (MIPS)
minicomputer
nonvolatile storage
online processing
personal computer (PC)
primary storage
processor

program
random access
random access memory (RAM)
register
response time
secondary storage
special-purpose processor
supercomputer
throughput
transaction
volatile storage
von Neumann machine
workstation

Vocabulary Exercises

1. A _____ is moved from primary storage to the CPU during the _____ cycle.
2. The speed of a primary or secondary storage device is generally stated in terms of its _____.
3. A device that uses a general-purpose processor and stored programs and operates by alternating instruction and execution cycles is called a _____.
4. The speed and capacity of a communication channel is generally stated in terms of its _____.
5. Generally, a _____supports more simultaneous users than a _____. Both are designed to support more than one user.
6. A _____ is a storage location implemented within the CPU.
7. The processing speed of a supercomputer is generally measured in terms of _____. The processing speed of other classes of computers is generally measured in terms of _____.
8. _____ processing describes a mode of application program execution that does not require input directly from a user.
9. A problem-solving procedure that requires the execution of one or more comparison instructions is called an _____.
10. The term _____ storage describes any storage device that holds its data content indefinitely.
11. The _____ of a computer system may be measured in terms of programs per time interval and/or transactions per time interval.
12. The _____ of an on-line transaction processing application increases as processing load increases.

Review Questions

1. What similarities exist among mechanical, electrical, and optical methods of computation?
2. What shortcomings of mechanical computation were addressed by the introduction of electrical computing devices?
3. What shortcomings of electronic computation will be addressed by the introduction of optical computing devices?
4. What is a CPU? What are its primary components?
5. What are registers? What is/are their function(s)?
6. What is main memory? In what way(s) does it differ from registers?
7. Explain the steps of the instruction cycle. Explain the steps of the execution cycle.
8. What is a von Neumann machine?
9. What are the differences between primary and secondary storage?
10. How does a workstation differ from a microcomputer?
11. How does a supercomputer differ from a mainframe computer?
12. In what ways (by what measures) is the performance of computer system components stated? What problems are encountered when utilizing these measures to acquire computer hardware for specific user needs?
13. What is throughput? What is response time? How are they measured?
14. What is Grosch's law? Does it hold today? Why or why not?
15. How can a computer system be "tuned" to a particular application?

Research Problem

The measurement of computer performance is a difficult task due to the complexity of hardware and assumptions about application program behavior. As discussed in the text, many computer system vendors state system performance in terms of millions of instructions per second (MIPS). However, this measure can be misleading, partly because it does not account for differences between applications. Common differences between applications include the mix of CPU instructions executed, amount of secondary storage access, and amount and type of I/O device access. Three alternative measures of computer system performance are the Norton SI index, the Byte benchmarks, and the LINPACK benchmarks. Investigate each of these performance measures and identify the class(es) of computer and type(s) of applications for which each benchmark is intended.

3 *Introduction to Software*

Chapter Goals

- *Describe the translation between user requests and machine instructions that satisfy those requests.*
- *Define the role and functions of application software and systems software.*
- *Define the role of a programming language and the various types of programming languages.*
- *Describe the mechanisms by which programs are developed, translated, and executed.*
- *Describe the functions and components of an operating system.*
- *Describe the economic advantages of systems software.*

THE ROLE OF SOFTWARE

The process of asking a computer to perform useful work can be thought of as a translation process. This translation is not so much a translation from one language to another as it is a translation from one level of detail to another. The need or idea that motivates a request for computer processing is generally stated at a relatively high level of abstraction (or a low level of detail). The actual computer actions (execution of instructions) necessary to satisfy a request are very specific—in other words, they contain a high level of detail.

The primary role of software is to provide a means for translating needs and requests stated in general terms into a sequence of instructions to a computer that will produce a result satisfying the need or request. Because the instructions the computer is capable of executing are quite primitive, the number of instructions that must be executed to satisfy a need or request is large; the translation process is therefore quite complex. The individual steps in this translation process are illustrated in Figure 3.1, along with the division of responsibility among the user, software, and hardware.

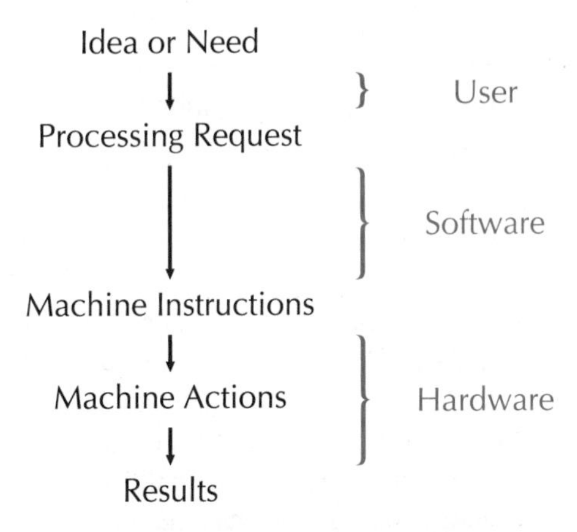

FIGURE 3.1 **The division of responsibility for translation steps among the user, software, and computer hardware.**

From Ideas and Needs to Machine Instructions

Imagine for a moment that you want to ask a person to perform a complex task such as cooking a meal. How would you go about it? If the person were an adult and an experienced cook, the request would be relatively simple. You would tell the person what meal you wanted and he would know what tasks to perform and how to perform them. You would have communicated a request for a complex task ("Would you please make me a plate of spaghetti?") with relatively few words and left it to the person receiving the request to translate it into the series of individual actions necessary to satisfy the request. The translation steps for this example are represented in Figure 3.2.

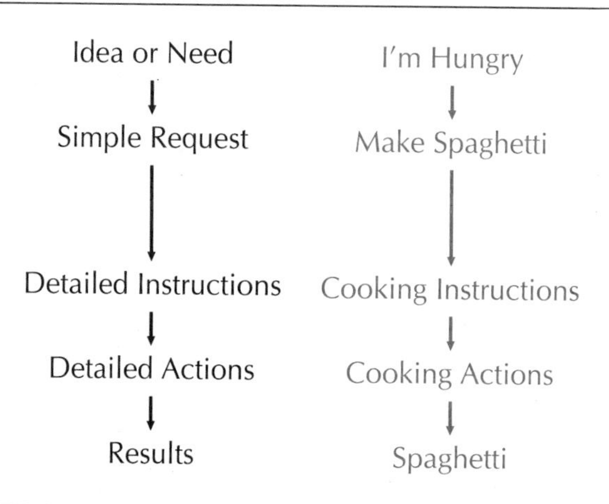

FIGURE 3.2 **The translation from idea or need to results for a cooking example.**

Now imagine that you have made the same request of a six-year-old child. The child might know what a plate of spaghetti is, but have no idea how to prepare it. However, the child does possess a set of primitive skills that could be combined to solve the problem. If you wanted, you could instruct the child in a very detailed manner in the individual steps necessary to prepare the meal. For instance, you could start by telling the child to get the large metal pot from the cabinet on the left, put it in the sink, fill it two thirds full with water, put the pot on the left burner of the stove, turn the left knob on the front of the stove to the position marked MED, and so on. If the child were old enough to read, you could write down the detailed instructions as a recipe.

Although the meal will eventually be prepared in either scenario, the distribution of intelligence and effort required to bring about that result differs substantially. In the first scenario, the requestor has stated the request simply, and the cook is responsible for determining the appropriate detailed instructions and actions. Thus, the intelligence needed to prepare the meal is embedded within the cook. In the second scenario, the requestor cannot assume that the child has a high level of intelligence with respect to the task at hand. Therefore, the child is given detailed instructions to perform specific actions within his or her knowledge and ability. The knowledge of how to perform the task is thus embedded within the requestor.

Programs

The basic abilities of the child in the above example are analogous to the basic abilities of a computer system. The child can perform many simple (primitive) tasks but has little knowledge of how to combine them to respond to a complex demand. Similarly, a computer has a set of primitive capabilities, but it has no direct knowledge of how to combine primitive actions to perform a complex task. It must be instructed in a very detailed and specific manner, either directly by a user or by a previously defined set of instructions—a program.

Much as the child will eventually memorize the procedure for making spaghetti, the computer can "remember" a procedure by storing the corresponding program. This program can then be recalled as needed, much as a human can recall a recipe. The recipe is analogous to an *application program* or *application software*, a stored set of instructions for responding to a very specific request. Examples of application programs in a system devoted to payroll processing include programs to print checks, enter new employee information, and produce annual tax reports.

Other programs can be created and stored for more general purposes. In the example above, the child had to be instructed in the exact mechanics of boiling water in which to cook the spaghetti. But boiling water is not a procedure specific only to the task of cooking spaghetti. It is a general procedure that is frequently a component of many other cooking tasks. It is doubtful that any recipe for spaghetti will provide detailed instructions for boiling water. Rather, the recipe will simply state that a certain quantity of water should be brought to a boil and assume that the user knows how that is done.

The procedure for boiling water is analogous to a *utility program*, a set of instructions for performing a relatively basic task that is a necessary component of many application-specific tasks. Examples of utility programs include programs to print text files, log onto a multi-user computer, and copy a file from one disk to another. Each of these tasks is relatively basic (although it may still require thousands of CPU instructions to complete) and is performed in conjunction with many different applications.

ECONOMICS OF SOFTWARE DEVELOPMENT

The process of developing application software is lengthy, complex, and costly. When a user wants a specific task to be performed by a computer, a program, or set of machine instructions, must be developed for that purpose. This development process follows a system development life cycle (SDLC), as defined in Chapter 1. The development process consumes a substantial amount of human effort for survey, analysis, design, programming, and the like. Thus, an investment of resources is required before the user can receive any benefit from the computer.

The benefits of this investment are realized by repetitive use of the programs developed. If the costs of operating the computer program are less than the costs of performing that same function manually, the initial investment of resources will eventually be repaid through savings. If a sufficient difference in operating costs does not exist, software development is not worth undertaking.

To make the use of computers economically feasible, it is necessary to minimize the costs of developing software. There are numerous ways to approach this basic problem, including:

- Reusing software as much as possible
- Minimizing labor inputs to the software development process

Most of the software that is commonly called *systems software* is designed to further one or both of these approaches.

The issue of software reuse is addressed primarily through the development of utility programs. These programs are designed to perform functions that are frequently used and/or needed by many different application programs. For example, most application programs need to access permanent data on secondary storage devices. Therefore, it is economically advantageous to provide a set of utility programs for that purpose. These programs perform functions such as creating and deleting files, reading and writing data to/from files, and initializing storage devices.

The availability of utility programs allows application software developers to avoid "reinventing the wheel" each time they develop a new system. Thus, a programmer can concentrate on developing software only for the functions that are unique to a particular application; they can use the utility programs as necessary. Application programmers no longer must write programs for these tasks, so the amount of time and labor needed to develop application software is decreased.

Figure 3.3 illustrates the interaction among the user, application software, utility programs within systems software, and computer hardware. Note the similarity between this diagram and the translation model depicted in Figure 3.1. User input and output are limited to direct communication with the application program. Application programs in turn communicate with systems software to request basic services such as opening a file or reading data from a file. Systems software translates a service request into a sequence of machine instructions, passes those instructions to the hardware for execution, and receives the results of that execution.

Figure 3.3 also illustrates what is commonly called a *layered approach* to software. Application software is layered above systems software, which is in turn layered above the hardware. A key advantage of the layered approach to software is that users and application programmers do not need to know the technical details of physical processing. Instead, they interact with hardware through a set of standardized service requests or commands. Those requests are then translated into the actual hardware instructions needed to satisfy the requests. Knowledge of the machine's physical details and many basic processing tasks is thus embedded within systems software and hidden from the user and application programmers. This advantage is commonly referred to as *machine independence* or *hardware independence*.

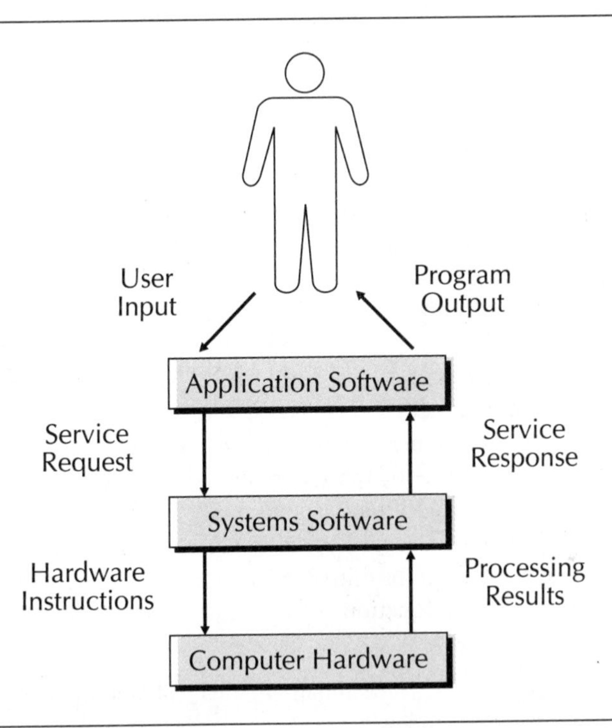

FIGURE 3.3 **The interactions among the user, application software, systems software, and computer hardware.**

PROGRAMMING LANGUAGES

Another method of reducing the labor input for application software development is to minimize the number of instructions a programmer must develop to perform a given task. Even simple programs (application tasks) often require hundreds of thousands of machine instructions. To a large degree, the time and effort required to develop software is proportional to the number of instructions a programmer must write. Thus, minimizing the number of these instructions reduces the labor input to the software development process.

A *programming language* is a language for instructing a computer. A complete program in such a language is often referred to as *code,* and the programmer is often referred to as a *coder*. The most primitive type of programming language is the language that the computer understands directly—that is, the instructions that can be executed by the CPU. This is commonly referred to as *machine language*. Machine language was the earliest form of programming and is also the most tedious and time consuming. Programmers are required to determine the exact set of primitive machine actions (instructions) necessary to perform a complex task and to write down each instruction in a form the computer can understand.

Machine Language Programming

Most of the work performed by a computer consists of the execution of relatively simple computational functions as well as the movement of data between various computer system components. The capabilities of the computer system are defined by (and limited to) the set of data manipulation and movement functions of which the CPU is capable. This set of capabilities is referred to as the computer's *instruction set*.

Computer instruction sets vary widely in capability. In general, the size of the instruction set—in other words, the number of functions that can be performed—increases as the class of the computer system increases. Thus, mainframes and supercomputers usually have significantly larger instruction sets than do microcomputers. However, regardless of the class of the computer, the instructions tend to be quite similar and primitive.

The processing functions of a typical CPU can be classified into the following three categories:

1. *Computation*: Perform a mathematical calculation using two numbers as input and store the result
2. *Comparison*: Compare the values of two stored numbers and determine if they are equivalent
3. *Data movement*: Move an individual item of data from one storage location to another storage location

A *simple machine instruction* performs a single function that can be classified as one of these types. A CPU can also provide *complex machine instructions* that combine two or more processing functions. For example, an instruction can be provided to perform data movement only if two stored values are not equal to one another. Such an instruction would be a combination of a simple comparison and a simple data movement. Large computers such as mainframes are more likely to provide complex machine instructions than are smaller computers. However, even if complex instructions are provided, many instructions are still necessary to perform any significantly complex task.

Each instruction is known to the CPU by a unique number called the *instruction code*, the *operation code*, or the *op code*. The execution of an instruction requires that the instruction code and related data be moved into the CPU from primary storage. Both the instructions and data are stored in main memory and are accessed via their addresses (locations). A program in machine language is, therefore, little more than an extremely long sequence of numbers, some of which represent instructions and some of which represent data items or the main memory addresses of data items or other instructions. This sequence of numbers is generally called *executable code* (or simply an *executable*) because it is code that is "ready to be executed."

To write a program in machine language, a programmer must know all of the available instructions by their instruction codes. In addition, the programmer must know where each of these instructions and all of the data items are located in memory. All program statements consist of numbers representing instruction codes, data items, and memory locations. A typical program will consist of many thousands of these statements. Developing this mass of numbers is a tedious and complex task. This complexity and its resultant costs (human programming time) make machine language programming a very inefficient method of implementing software. To address this inefficiency, a new class of programming languages called *assembly language* was developed.

Assembly Language Programming

The primary difference between assembly language and machine language is the way in which instructions and memory locations are written. Machine language requires that these be specified by number or address. Assembly language allows the programmer to use names composed of alphabetic characters. For example, an instruction to add two numbers might be known to the CPU as instruction code 4. In assembly language, this instruction is given a character name such as ADD.

An add instruction might require the addresses in memory of the two numbers to be added. In machine language, the programmer has to know where these numbers are stored in memory and must supply the memory addresses to the CPU with the instruction. In assembly language, the programmer can specify names like N1 and N2 for these storage locations and use these names instead of the addresses. Thus, the assembly language equivalent of an instruction such as 04 0020 0024 might be ADD N1 N2.

The use of assembly language requires a means of translating it into machine language. This translation is performed by a program called an *assembler*. The assembler reads each assembly language instruction and creates an equivalent machine language instruction. After all of the assembly language instructions have been translated, the entire set of corresponding machine instructions can then be loaded into memory and executed by the CPU.

Although the use of assembly language still requires the programmer to specify each individual CPU instruction, it provides a more convenient way in which to write those instructions. Humans are better at remembering and manipulating character names than long strings of numbers. Thus, the effort required to develop an assembly language program is substantially less than the effort required to develop an equivalent program in machine language. Programmer productivity improved dramatically as a result of the development of assembly languages.

Higher Level Programming Languages

Further increases in programmer productivity required a fundamentally different approach to programming—one that eliminated the need to specify each individual CPU instruction. This was accomplished by the development of *high-level programming languages.* A high-level language allows the programmer to specify a processing action requiring many CPU instructions in a relatively small number of statements. Examples of such languages include FORTRAN, COBOL, Pascal, C, and many others.

Figure 3.4 shows portions of a COBOL program and a C program. Although the program fragments differ substantially, they accomplish the same purpose. Depending on the CPU's capabilities, a machine language program to perform this same function would require hundreds or thousands of instructions. An assembly language program would require a similar number of instructions.

Much as assembly language programs are translated into machine instructions by an assembler, a high-level language program is translated into machine instructions by a *compiler.* In essence, the compiler translates each statement of the high-level language into an equivalent set of machine instructions. The compiler is also responsible for keeping track of instruction and data locations in memory and various other "housekeeping" chores. Because translation from high-level language to machine code is highly complex, the compiler itself is a highly complex program. The statements of the assembly or high-level language program are called *source code.*

(a) COBOL example:

```
DETAIL-PARAGRAPH.
    READ CARD-FILE AT END GO TO END-PARAGRAPH.
    MOVE CORRESPONDING CARD-IN TO LINE-OUT.
    ADD CURRENT-MONTH-SALES IN CARD-IN, YEAR-TO-DATE-SALES IN
        CARD-IN GIVING TOTAL-SALES-OUT IN LINE-OUT.
        WRITE LINE-OUT BEFORE ADVANCING 2 LINES AT EOP
        PERFORM HEADER-PARAGRAPH
```

(b) C example:

```
detail_procedure()
{
    int file_status, current_line, id_num;
    char *name;
    float current_sales, ytd_sales, total_sales;

    current_line=64;
    file_status=fscanf(card_file,"%d%s%f%f",&id_num,&name,&current_sales,
        &ytd_sales);
    while (file_status != EOF){
    if (current_line > 63){
        header_procedure();
        current_line=3;
    }
    file_status=fscanf(card_file,"%d%s%f%f",&id_num,&name,&current_sales,
        &ytd_sales);
    total_sales = current_sales + ytd _ sales;
    printf("%4d%30s%6.2f%6.2f%7.2f\n\n", id_num, name,current_sales,
        ytd_sales, total_sales);
    current_line ++; current_line ++;
    }
end_procedure();
```

FIGURE 3.4 **Sample program fragments in (a) the COBOL and (b) the C high-level programming languages.**

TYPES OF SYSTEMS SOFTWARE

The previous sections discussed system software in the contexts of translation and software development. This section will provide a more pragmatic classification of systems software into the following categories:

- Application development software
- Application tools
- Operating software

The functions of systems software illustrated in Figure 3.3 fall primarily within the category of operating software. The first two categories cannot realistically be called either application or systems software using the definitions given thus far. They are not directed toward a specific need, nor do they provide an interface to hardware services during the execution of an application program. They do, however, provide "general-purpose" capabilities. These distinctions will become clear in the following sections.

Application Development Software

Application development software includes any program used to develop application programs. In years past this was a relatively small class of software that included programs such as assemblers, compilers, and text editors. Due to advances in computer hardware and software, as well as the practice of application development, a much wider range of programs now fills this class. Application development software can be further classified into the following categories:

- Program translators
- Program development tools
- System development tools
- Data manipulation tools
- Input/output tools

Program translators include two types of software already discussed: assemblers and compilers. These both translate programming language statements into instructions that can be executed by the operating system and/or the hardware. Another type of software that fills this role is an *interpreter*. An interpreter differs from a compiler or an assembler primarily in when and how program statements are translated. Assemblers and compilers both translate an entire program into executable code that can be loaded into memory and run as a unit. An interpreter translates one program statement at a time and executes it immediately before translating the next statement. It is, in essence, a program designed to execute other programs one statement at a time.

Program development tools include a wide range of programs to aid programmers in developing application programs. Examples of these include text editors, program verifiers, and debuggers. *Text editors* are basic tools that allow a programmer to create files containing assembly or high-level language programs. *Program verifiers* (also called *code checkers*) are programmers' aids that help evaluate program correctness and quality. These programs read an application program (source code) as input and check for such things as syntax errors, logic errors, and other actual or potential errors. Similar functions are often embedded within compilers and interpreters, especially those for the more recent high-level languages.

Debuggers are also tools for verifying program correctness. They differ from program verifiers in that they are used after a program has been compiled rather than before. That is, they verify executable code instead of source code. In general, they allow a programmer to simulate the execution of a compiled program and to check for errors or problems during the simulated execution. They allow programmers to do things such as examine the contents of variables or restart the program in various places.

System development tools are a relatively new class of application development software. They are designed to support the earlier phases of the systems development life cycle (survey, analysis, and design) and to support the development of groups of application programs that comprise an entire information system. They address the complexity of developing an entire system of application programs, which is substantially greater than the complexity of developing a single program. System development tools are often called *computer-assisted software engineering (CASE) tools.*

Data manipulation tools are designed to extend the basic capabilities of programming languages and operating systems to manipulate data stored on secondary storage devices. These are commonly called *database management systems (DBMSs).* The need for these tools has developed as a result of ever larger and more complex information systems. Such systems require complex methods of storing, retrieving, defining, and redefining data. They also require methods of sharing data among competing programs and users. Although operating systems and high-level programming languages provide some support for these functions, DBMSs are designed to provide a much higher level of support. These tools can be considered a form of systems software due to their role in supporting application software and due to the interface service they provide to secondary storage.

Input/output tools are similar to data manipulation tools in that they extend the basic capabilities of operating systems and programming languages. Complex input/output has become commonplace in application programs today. Examples include the use of full-screen menus, windows, and high-resolution graphics. These tools can be considered systems software because they provide input/output support for application programs and because they provide an interface to input/output hardware.

Application Tools

Application tools include two broad types of software: tools designed to automate repetitive clerical or technical tasks and tools designed to take the place of application programs. Examples of the first type include word processors such as

WordPerfect and automated drawing packages such as Harvard Graphics and AutoCAD. Rather than address a specific output or processing result, they address a specific type of processing.

Tools designed to take the place of (or augment) application programs allow users to perform processing tasks without the need for an application program developed specifically for their processing requirements. Examples of these tools include spreadsheet processors such as Lotus 1-2-3 and some database management systems—for example, dBASE IV. These tools either allow a user to perform information processing without an application program or provide the user the ability to generate simple application programs.

Some application tools can be classified as *application generators.* These tools provide an integrated set of systems software, including a high-level programming language (compiler and/or interpreter), a DBMS, and advanced I/O tools. An integrated set of application development tools is often called a *fourth-generation language* (*4GL*) or *fourth-generation development environment.*

OPERATING SOFTWARE

The final class of systems software is *operating software*. This class encompasses a very wide range of functions, including the following:

- Process control
- File control
- Secondary storage control
- I/O device control
- User control

The common term for a set of software that performs all of these functions is *operating system*. Examples of commonly available operating systems include MS-DOS, OS/2, Unix, and VMS.

The Operating System Model and Its Functions

All of the functions listed above provide an interface between computer hardware and the user of application programs. The primary components of an operating system are illustrated in Figure 3.5. Note that this diagram (and the operating system) is organized according to the principle of software layers. Input from a user or an application program must travel through the various layers to reach the hardware and results from the hardware must travel back to the user through those same layers.

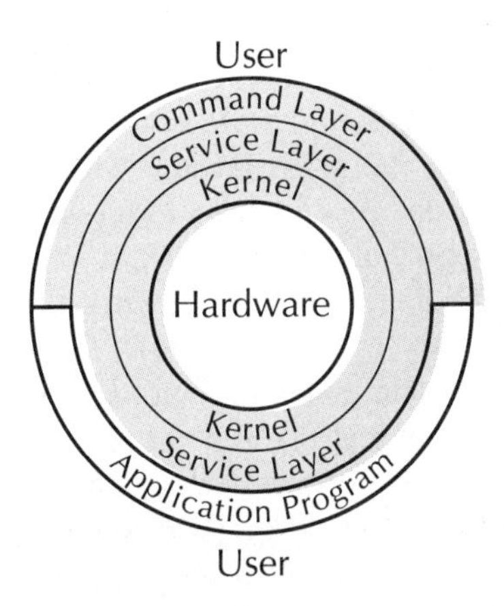

FIGURE 3.5 **Operating system layers and their relationship to the user, application software, and computer hardware.**

The *kernel* is the only layer that interacts directly with the hardware. This provides a measure of machine independence within the operating system. At least in theory, an operating system can be altered to interact with a different set of computer hardware by making changes to the kernel only. This layering also insulates users of the application-level and command-level interfaces from direct knowledge of hardware specifics, thus simplifying these interfaces considerably. The name *kernel* is derived from the fact that this is the innermost layer of the operating system.

The *service layer,* also called the *systems service interface*, is the layer that most closely resembles the systems software description given earlier in this chapter and illustrated in Figure 3.3. This layer accepts service requests from application programs and/or the command layer and translates them into detailed instructions to the kernel. Processing results, if any, are passed back to the program that requested the service.

The *command layer* is the only portion of the operating system with which the user directly interacts. Because it is the outermost layer it is sometimes referred to as the *shell.* The command layer responds to commands given directly by the user or by a previously stored list (or file) of commands. The set of commands and their syntax requirements are referred to as a *command language* or sometimes a *job control language (JCL).*

The Command Layer

The command layer is the user's interface to the operating system. The user can directly request system services via a command language or a command interpreter. The command interpreter decides which service calls must be executed to process the request and passes these to the service layer. Processing results are displayed to the user in some human-readable form.

Although command languages are still the most common interface to the command layer, they are rapidly being replaced by alternative methods. Command languages tend to be difficult to learn and use. They are similar to programming languages in that the user (or programmer) must know the syntax and semantics of the language in order to communicate. The MS-DOS command interpreter (COMMAND.COM), IBM MVS JCL, and the Unix Bourne Shell command set are all examples of command languages in common use today.

Alternatives to command processors include forms-based (full-screen) and graphically based interfaces. Forms-based interfaces provide the user full-screen forms or prompts for operating system commands. These interfaces alleviate some of the difficulties in remembering command syntax because command parameters and options are explicitly requested (prompted) from the user. Examples of operating systems that use forms-based command interfaces include portions of VMS and certain IBM interactive operating systems such as TSO and SPF.

Graphically based interfaces are the newest form of interface. These interfaces allow users to execute many operating system commands by manipulating graphical images on a video display using a pointing device such as a mouse. For example, files may be represented by graphical images of file folders. Moving a file from one directory to another might be accomplished by moving the associated folder image. The visual metaphor for commands and functions provides an easy-to-learn interface. Examples of these types of interfaces include the Macintosh operating system and Microsoft Windows.[1]

Regardless of the type of interface, the functions of the command layer are similar. They provide an interface that allows a user to perform common operating functions without writing application programs specifically for those purposes. Examples of these common operating functions include loading and executing application programs, manipulating files, and initializing (formatting) secondary storage devices.

[1] Strictly speaking, Microsoft Windows is not an operating system. It is actually an extension to the MS-DOS operating system. Its command interface provides an alternative interaction method to the MS-DOS command layer, along with additional commands not available in MS-DOS.

The Service Layer

Service calls implement commonly used functions in process control, file control, and I/O control. Examples of process control service calls include loading a *process* (a program or program segment) into memory, starting its execution, and terminating it. Examples of file control service calls include opening, closing, creating, deleting, and renaming files as well as reading data from and writing data to files. Examples of I/O control service calls include initializing printers and display devices, sending characters to a printer, sending characters or graphic images to a video display, and reading characters from a keyboard.

The number of system service requests, called *service calls*, provided by the service layer is generally quite large. It is also dependent upon the capabilities of the operating system and of the hardware that it controls. In general, more operating system capabilities and/or more powerful and complex hardware require more service calls. This relationship also applies to the amount of code needed to implement the kernel.

As an example of this relationship, consider the difference between MS-DOS, OS/2, and Unix. MS-DOS was originally designed to operate the IBM Personal Computer. This machine was a relatively simple microcomputer. In addition, MS-DOS was designed to provide a minimal level of capability—for instance, it included no multi-user or graphical input/output capabilities. As such, the number of service calls available in MS-DOS is relatively small (fewer than 100) and the size (the amount of memory consumed by the kernel) is also small.

OS/2 was designed to operate higher powered microcomputers. Partly due to the more powerful hardware, OS/2 was also designed to provide more capabilities than MS-DOS: window interfaces, multiple program execution, and the like. The additional power and complexity of the hardware and the more advanced capabilities of the operating system lead to a larger number of system calls (approximately 250). The kernel is also substantially larger and, as a result, a computer needs more memory to use OS/2 than it does to use MS-DOS.

The Unix operating system was designed to operate minicomputers and later augmented to operate computers ranging from workstations to supercomputers. Primarily because of the wide range and power of hardware that it is designed to operate, the Unix operating system has a large number of system calls (approximately 300). This number would be even larger if the user-oriented input/output capabilities were extended to be similar to those of OS/2. The kernel is substantially larger than OS/2, and a computer using Unix requires more memory.

The Kernel

The kernel is primarily responsible for direct control of hardware devices. It provides a set of interface programs for each hardware device in the computer system. These interface programs are commonly called *device drivers*. Examples include device drivers for input keyboards, video display devices, disk drives, tape drives, and printers.

The service layer uses the device drivers in much the same way that application programs use the utility programs within the service layer. That is, the service layer makes specific requests to kernel device drivers for access to hardware devices. In general, there are many different service layer functions that use each device driver. For example, there may be several dozen service functions for input and output to files. All of these will use the same device driver to perform required disk operations.

The advantages of this organization of service layer and kernel functions are similar to the advantages of the organization of application and service layer functions. The kernel provides a set of capabilities that are reused by different elements of the service layer. If these were not available, they would be redundantly implemented within the service layer—for example, each file manipulation service program would contain redundant code for control of disk hardware.

The use of kernel device drivers also provides a degree of modularity and hardware independence. As new hardware devices are added to the computer system, new or revised device drivers can be added. This modularity allows the operating system to incorporate capabilities to control new hardware with a minimum amount of disruption to existing code and capabilities. In addition, existing capabilities can be modified without affecting the service layer. For example, a change in type of hard disk drive can be incorporated into the operating system by modifying the hard disk device driver. The service layer (and all layers above it) should be completely unaffected by this change.

Resource Allocation

One of the most important and least visible of the operating system's functions is that of resource allocation. This function is normally implemented within the kernel. Resources to be allocated include access to the CPU, access to secondary storage devices, access to input/output devices, and access to memory. Resource allocation is a direct consequence of program execution and an indirect consequence of a program's requests to the service layer. The execution of a program requires that CPU time be allocated to it. Service calls generally require access to hardware resources for their satisfaction (for example, access to secondary storage devices for file manipulation).

Resource allocation is relatively straightforward if the operating system supports the execution only one program at a time, as is true for MS-DOS. The currently executing program is simply allocated whatever resources it requests. Any service call issued by an executing program is executed immediately. The only complexities are determining if the resources requested are available and checking for error conditions. For example, when a program requests a file to be opened, the operating system must determine if the file exists and monitor the actual operation for errors, such as the inability to read the file from disk.

The complexities of resource allocation rise substantially as the number of users, programs, and hardware resources increase. Under such conditions the kernel cannot immediately provide any resources requested by a program or user. It must first determine whether or not those resources are currently being used by another user or process, and must decide if and when to satisfy the service request. The kernel must, therefore, balance the competing demands of users and processes for access to the CPU and other hardware resources. This complexity reaches its zenith in modern mainframe computers, in which hundreds of users and thousands of programs can simultaneously contend for access to the CPU, memory, communication channels and devices, and multiple secondary storage devices.

PARALLEL HARDWARE/SOFTWARE DEVELOPMENT

The system software in use today bears little resemblance to that used in the past. In fact, system software as we know it today did not exist until many years after computers came into use. Programming languages other than machine and assembly language did not appear until the late 1950s. Operating systems were also not developed until that time, and their capabilities were extremely limited compared to modern operating systems.

Why is this so? Are we simply better at developing system software today than we were 40 years ago? The answers to these questions follow from two basic facts of computer hardware and software:

1. System software requires hardware resources.
2. The cost per unit of computing power has decreased at nearly an exponential rate.

These facts underlie a shifting tradeoff between the costs of developing and supporting application programs and the cost of computer hardware resources.

When computers were introduced, they were extremely expensive to purchase and operate. Because of this, computer resources were only expended on "high-value" applications and only where the alternatives to their use were more expensive.

For example, consider the operations of a typical accounts payable department of a large firm in the 1950s. Such an operation would typically employ hundreds of clerks to pay the firm's bills and maintain manual accounting records. Although the labor costs associated with such an operation were high, they were usually lower than the cost of computer resources and software development necessary to automate accounts payable processing tasks. As the cost of computer power decreased, the tradeoff in cost between automated and manual processing shifted. Repetitive and high-volume processing was now cheaper if automated. Thus, computer resources were substituted for manual labor in common transaction and information processing tasks. This trend has continued to the present, with an ever greater substitution of computer resources for labor.

Computer resources can also be substituted for labor in the development of application programs. In the early years of computers, application development was almost entirely a manual process. Computer resources were too expensive to allow them to be used to directly support application development. Thus, tools to support the development or execution of application programs did not exist.

As the cost of computer resources decreased, the economic tradeoff shifted. Tedious programming in machine language and extensive manual checks of code were more expensive than application development using advanced tools. Thus, a trend toward increased automated support of application development began. This trend took three basic paths through the development of:

- Application development tools
- Operating system service functions
- Operating software

Application Development Tools

The use of application development tools began with the introduction of programming languages and their associated tools. Assembly language and assemblers appeared first. These were soon followed by high-level languages and their associated interpreters and compilers. Text editors, debuggers, code verifiers, CASE tools, and a host of other software tools are also examples of this trend. All of these tools reduce the labor input to programming by substituting computer resources.

The use of application tools is a continuation of this trend. Tools such as word processors and spreadsheet programs minimize application development labor to nearly zero. The number and power of such tools continue to grow and will likely do so for the foreseeable future.

Service-Layer Functions

A parallel trend in supporting application programs occurred in the development of service layer functions. Early programmers had to specify every primitive machine action. This included very basic functions that occurred in virtually every application program. Examples of the functions include access to I/O devices, input and output to secondary storage, file management, and memory management. Programmers often rewrote these same functions over and over again in different application programs.

The trend toward utilizing service-layer functions started with these basic functions. A set of utility programs was provided within the operating system, and application programmers would use these utilities rather than writing equivalent functions for each application program. This resulted in reduced labor input for application development. However, there are some hidden costs to this approach.

Consider, for example, the machine instructions that comprise a program to read data from and write data to a file. These instructions are contained within one or more programs within the service layer. Because they can be used for many types of files and/or storage devices, they must be general in nature. That is, they must allow for variations in file content, file size, and other factors.

If the same function were implemented within an application program, the instructions could be more specifically oriented toward the file content and size for that application. This specificity would likely lead to a more efficient program (i.e., reduced use of computer hardware resources). This situation is representative of a basic tradeoff in software. General-purpose utility programs tend to be less efficient than programs written for a specialized function. However, by their very nature, utility programs and service layer functions must be general.

When computer resources were relatively expensive, this difference in efficiency was very important. At that time, it was more cost effective to implement specialized functions within application programs as compared to utilizing general-purpose system software. The extra cost in application development (e.g., programmer labor) was more than offset by the reduced operating costs of the application program due to increased efficiency. As computer hardware became cheaper (and programming labor more expensive), this situation reversed, and general-purpose system software became widespread.

This trend continues to the present day. It is driven both by cheaper hardware resources and by increased expectations for the functionality of application software. Examples of systems software that addresses both of these issues include database management systems and service-layer support for graphical interfaces. Both of these types of software provide a set of utility programs that can be used by application programs. Both also address advanced requirements—for instance, the management of large databases and "user-friendly" program interfaces—for application program functionality. As computer technology advances and cost decreases, the development of such tools will continue.

Operating Software

It is difficult for the users of today's computers to imagine operating a computer without the aid of software. But such software did not exist when computers were introduced. As an example, consider the loading of a program into memory and its subsequent execution. A user of a personal computer running MS-DOS simply types the name of the program at the command-layer prompt, and systems software handles program loading and execution. When the program terminates, the prompt is redisplayed and the command layer waits for the next user command.

In the early 1950s this conceptually simple task was a much more complex matter. A deck of punched cards containing the program instructions was first loaded into a card reader. The operator then manually positioned a set of switches on the computer. These switches instructed the machine to execute a small hardware-based program that consisted of a small loop of instructions that sequentially read each card and loaded the corresponding instruction into sequential memory locations. After all the cards were read, the machine stopped and displayed a status code through a set of lights.

Next, the computer operator loaded a deck of punched cards containing the input data into the card reader. Switches on the machine were then set to indicate the location in memory of the first program instruction. Another switch would then be set to instruct the computer to load and execute that first instruction. After that, the application program would "take control" of the machine. When the last program instruction was executed, the machine would stop and display another status code. If an error occurred during program loading or execution, the machine stopped and a status code was displayed. The operator had to consult a reference manual to interpret the meaning of the status code and correct it.

These procedures were used partly because a typical computer did not contain enough memory to hold both operating and application software. They were also used because manual computer operations were cheaper than diverting hardware resources from application to operating software. Thus, a human operator, in concert with a few hardware-based operating programs, served as the operating system. As hardware became cheaper and more powerful, operating software was developed to automate such mundane tasks. This trend continues to the present day with the development of sophisticated operating software that is powerful and easy to use. As a result, the job of computer operator has virtually disappeared.

As computers became more powerful, it became possible for multiple application programs to execute at the same time on a single computer system. However, multiple program execution required a new type of operating software that had extensive capabilities for resource allocation and management. The complexity and function of this software was further increased with the introduction of online interactive applications and computer networks.

The resource management functions of a complex operating system consume a substantial amount of computer hardware resources. These resources do not directly support application programs. They do, however, make resource sharing possible. A major goal in designing such operating systems is to minimize the amount of resources consumed by management functions.

Future Trends

The use of system software will continue to increase as computer hardware becomes cheaper and more powerful. There is constant push and pull between hardware advances and software advances. Cheaper hardware allows a greater substitution of hardware for labor in the development and execution of application programs.

Advanced hardware functions alter user expectations of application programs. These eventually result in increased functionality of systems software for application development, support, and computer operations.

Consider graphical interfaces as a modern example. Windowing, icons, and high-resolution graphics did not suddenly appear in the last few years. They were under development at Xerox Corporation in the early 1970s. Why then did they take so long to come into use? Because they consume huge amounts of computer resources. Displaying a window or graphic image requires the movement of a large amount of data from main memory to the graphic display device. Moving or resizing a window requires a great deal of I/O and a large number of CPU instructions.

Until recently, the cost of these resources was prohibitive. In the past it was far more cost efficient to train users in the obscure syntax and semantics of a command language than it was to pay for the extra computing resources needed to implement a graphical interface. Current hardware capabilities and costs make graphical interfaces cost effective. This creates a demand for them in application and operating software.

Because the demand for such interfaces occurs across many application programs, a demand for systems software to support such interfaces is also created. This demand has been answered by the development of systems services for graphical and windowing functions—for example, X Windows and Microsoft Windows. It also requires an advance in operating software to control the new hardware resources and to make them easy to use and transparent to the user.

Mathematicians, physicists, computer scientists, and other researchers ponder new possibilities of automated computation—even automated thinking. Engineers give physical reality to these ideas by building devices based on them. Systems and application programmers then ponder what to do with those devices. As time passes, manufacturing experience accumulates and the devices become cheaper and more powerful. Users demand the benefits of this hardware. This in turn causes programmers to look for new ways to exploit the additional power. Such has been the push and pull of hardware and software advances to date, and such is it likely to be for the foreseeable future.

SUMMARY

The role of software is to translate user processing requests into machine instructions that will satisfy the request when executed. This translation is highly complex due to the wide disparity in detail and content between a user processing request and a machine language program. The two primary types of software are application software and systems software. Application software consists of programs that satisfy specific user processing needs. Examples of these programs include payroll calculation, accounts payable, and report generation programs.

Systems software consists of utility programs that satisfy a class of processing needs. These programs are designed to be general in nature and may be used many times by many different application programs and/or users.

A program consists of instructions that can be directly executed by the CPU. Such programs are called *machine language programs*. Writing machine language programs is a tedious and labor-intensive activity. Alternative languages for writing programs include assembly language and high-level programming languages. Assembly language programs are similar to machine language programs except that their instructions and data can be written using names composed of characters instead of numbers. This difference makes assembly language programs easier to write. High-level language programs are designed to allow a programmer to write a small number of statements that correspond to many machine instructions. These languages require an interpreter or compiler to translate the high-level language statements into machine instructions. They improve programmer productivity by allowing complex processing to be stated in a compact form.

Systems software can be classified into three categories: application development tools, application tools, and operating software. Application development tools are programs used to create (or help create) other programs. These include compilers, interpreters, CASE tools, and other programs commonly used by programmers. Application tools are programs that allow a user to perform certain types of processing without an application program. They are designed to address a specific type of processing such as document preparation, drawing, and data management. Operating software (or an operating system) exists to manage hardware resources and to perform functions that are common to many application programs. These functions include process, file, I/O, and user control.

An operating system is composed of the command layer, the service layer, and the kernel. The command layer is the user interface to the operating system. It is the means by which the user controls computer hardware and executes other programs. Functions such as file manipulation, program execution, and secondary storage management are performed using command layer facilities. The user can interact with the command layer via a command (or job control) language. Alternate command mechanisms include forms-, menu-, and graphically-based interfaces.

The service layer consists of a set of utility programs that are used by application programs and the command layer. These programs implement commonly used functions such as input and output to files and I/O devices. Access to hardware by users and/or application programs is directed through the service layer. This provides a measure of independence between application software and physical hardware. This feature is commonly called hardware or machine independence.

The kernel is responsible for the control of hardware devices and the management of resources. Direct manipulation of the CPU, primary and secondary storage, and I/O devices is implemented within the kernel. The kernel contains a set of device drivers for each hardware device that serve as interface and control programs.

The kernel also allocates hardware resources to application programs and users. This is a complex and relatively invisible process from the user's point of view.

The development of systems software has been motivated by advances in hardware technology. These advances have resulted in computers with lower cost and greater power. Systems software serves several economic purposes. Application tools and application development tools reduce the cost of meeting users' processing needs. Service-layer functions also reduce this cost by reducing the amount of code that application programmers must write. Command-layer functions automate the operation of a computer and allow it to be shared by many users and programs. Each of these uses represents a substitution of relatively cheap computer hardware for relatively expensive human labor.

Key Terms

application development software
application generator
application program
application software
application tool
assembler
assembly language
code
code checker
coder
command (job control) language (or JCL)
command layer
compiler
complex machine instruction
computer-assisted software engineering (CASE) tool
database management system (DBMS)
data manipulation tool
debugger
device driver
executable code
fourth-generation language (4GL)
hardware independence
high-level programming language
instruction code (operation code or op code)
instruction set
interpreter
job control language (JCL)
kernel
machine independence
machine language
operating software
operating system
process
program development tool
programming language
program translator
program verifier
service call
service layer (system service interface)
shell
simple machine instruction
source code
system development tool
systems software
text editor
utility program

Vocabulary Exercises

1. The translation of a high-level language into machine instructions is performed by a _____ or a _____.
2. A _____ extends the ability of an operating system or programming language to manipulate data stored on secondary storage devices.

3. _____ must be translated into _____ before it is ready for execution by the CPU.
4. A _____ is a processing request made by an application program to the operating system service layer.
5. The term _____ describes a programming language in which one statement can be translated into many CPU instructions.
6. _____ differs from machine language primarily in the use of short alphabetic names for instructions and data.
7. _____ is an alternative name for the operating system command layer.
8. A component of the kernel that controls access and interface to a single hardware device is called a _____.
9. Resource allocation and direct hardware control is the responsibility of the _____.
10. The language that is used to direct the actions of an operating system is called a _____ or a _____.
11. The _____ is the user interface to an operating system.
12. A _____ is a numeric code that directs the CPU to perform one primitive processing function.
13. A programming language that incorporates database and advanced user interface capabilities is called a _____.

Review Questions

1. What is the instruction set of a computer system?
2. What are the three basic types of instructions?
3. How are user requests for processing translated into machine instructions?
4. Define the term *application software.* Define the term *systems software.*
5. In what ways does systems software make the development of application software easier?
6. In what ways does systems software make application software more portable?
7. How does assembly language programming differ from machine language programming? Which is easier and why?

8. What is a program translator? What is its purpose?
9. What are application tools? In what way(s) do they differ from application development tools?
10. What are the primary components of an operating system?
11. What is the service layer? With what does it interact?
12. What is the kernel? What functions does it perform?
13. Why has the development of systems software paralleled the development of computer hardware?

Research Problem

The Windows NT operating system is unusual in that it incorporates support for several sets of service calls. These include service calls supported by Windows version 3.1, the POSIX standard, and MS-DOS. In addition, these various service-layer definitions can interact with several CPU instruction sets, including Intel 80X86 series and MIPS microprocessors. Such capabilities require a more complex scheme of operating system layers than is described in this chapter. What operating system layers and interactions between layers are used by Windows NT to support multiple service-layer definitions and CPU instruction sets?

PART

2

Hardware

4 *Data Representation*

Chapter Goals

- *Describe the methods by which data can be represented and transmitted within a computer.*
- *Describe the advantages and disadvantages of various methods of data representation.*
- *Describe alternative numbering systems and their use in data representation.*
- *Describe the representation and manipulation of non-numeric data.*
- *Describe the representation and manipulation of data structures.*

DATA CODING

To be used and manipulated by people, data is represented as symbols or groups of symbols. These symbolic representations can be of many types. For example, data can be composed of letters of the alphabet, numbers, punctuation marks, special symbols, and even blank spaces. Types of numbers include numerals (as in addresses and telephone numbers) and numeric digits that are used in calculations. Numeric digits can be grouped to represent either whole numbers or numbers with fractional parts. Another type of data is indicated by values that are either true or false. For computer programmers and users, such categories of data correspond with various *data types.*

The electronic circuitry of computers is not capable of processing data in any of these symbolic forms. For sensing and manipulation by a computer, data must be represented, or coded, as electronic signals. Binary coding formats, composed entirely of ones and zeros, have been adopted because such codes can be manipulated readily by the bistable (or two-state) devices of electronic computer circuits. Binary coding can be used to directly represent whole numbers. Other types of data, such as fractional numbers and characters, must first be converted to a numeric representation. Computers that represent and manipulate data in this manner are called *digital computers*.

Digital computers use a variety of binary coding formats to represent data. Within a single computer, data values can be represented in many ways, depending on the type of data and the transformation that is being applied. For example, consider the decimal number 9. In the binary (or base 2) number system, this value is represented as

$$\text{Value}_{\text{base}} = 9_{10} = 1001_2$$

Each numeral in this binary number is referred to as a *bit* (a shortened form of the term ***binary digit***). A bit is the fundamental unit of data storage and transmission within a digital computer.

The number 9 might not be represented as 1001 within a digital computer, even though that is a valid binary equivalent. If the value is being used in a subtraction or division operation involving whole numbers, it might be represented as 0111. If it is being used in a calculation with fractional numbers (as 9.0), it might be represented as 0000010010000001. If, instead, the numeral, or character, 9 is being used, it might be encoded as 0111001.

The reasons for such a variety of binary formats stem from the variety of data representation objectives, which include:

- Ease of manipulation
- Conciseness and machine efficiency

- Accuracy
- Standardized data communication

Each of these objectives is discussed in detail below.

Ease of Manipulation

Some binary coding formats are tailored closely to specific types of data transformation. For example, in performing arithmetic operations, it is helpful to indicate whether a numeric value is negative; in other words, to designate the sign of the value. If numeric data is coded to include a sign, all arithmetic operations can be performed as addition.

Subtraction can be handled as the addition of a negative number. An example of subtraction in the decimal system of notation is:

$$4 - 2 = 2$$

The same result can be obtained by adding the value –2 to 4:

$$4 + (-2) = 2$$

Multiplication can be performed as a series of addition operations. For example,

$$4 \times 4 =$$
$$4 + 4 + 4 + 4 = 16$$

Division can be performed as a series of subtractions—that is, the quotient, or result of division, is the number of times that the divisor can be subtracted from the dividend without producing a value that is less than 0. For example:

$$16 \div 4 = 4$$
$$16 - 4 - 4 - 4 - 4 = 0$$

$$18 \div 4 = 4, \text{ remainder } 2$$
$$18 - 4 - 4 - 4 - 4 = 2$$

Because subtraction can be performed as the addition of a negative value, division also can be done as a series of additions:

$$16 + (-4) + (-4) + (-4) + (-4) = 0$$

Thus, manipulating signed (negative or non-negative) numbers makes it possible to perform all basic arithmetic operations as addition. This greatly simplifies the logic circuitry within the CPU.

Conciseness and Machine Efficiency

From an architectural viewpoint, data representation (coding) and circuit design are closely related. That is, the method of data representation in electronic signals affects the design of the circuits that transmit, process, and store those signals. For efficient machine design and optimum performance, it is desirable to code data in as few binary digits as possible. Concise data representation minimizes the cost of electronic circuitry, including the width of communication channels, the number of registers in the CPU, the number of memory cells in RAM, and so on. Concise coding also maximizes processing and data transfer speed.

Thus, binary codes should be as concise as possible for efficient data transfer, processing, and storage. The processing operations on those codes also should be as straightforward as possible.

Accuracy

Although concise coding minimizes the complexity and cost of hardware elements, it might do so at the expense of accurate data representation. The accuracy or precision of representation increases with the number of data bits that are used. The number of data bits also limits the largest (and smallest) number that can be directly represented within the machine.

It is possible for routine calculations to generate quantities that are either too large or too small to be contained within the finite circuitry of a machine (i.e., within a fixed number of bits). For example, the fraction $^1/_3$ cannot be represented within a fixed number of bits, because it is a nonterminating fractional quantity (0.333333333, with an infinite number of 3s). In such cases, the quantities must be manipulated and stored as approximations. Each approximation introduces a degree of error. If approximate results, in turn, are used as inputs for other operations, errors can be compounded. In some cases, the extent of error can be significant. Thus, it is possible for a computer program to lack apparent logical flaws and yet produce incorrect results.

If all data types were represented in the most concise form possible, the approximations required in some instances would introduce unacceptable margins of error. If, instead, a large number of bits were allocated to each data value, machine efficiency and performance would be sacrificed and hardware cost would be substantially increased. The most favorable tradeoffs can be achieved by using an optimal coding method for each type of data and/or for each type of transformation that must be performed. This is the main reason for the variety of binary coding formats around which processing machinery must be designed.

Standardized Data Communication

Data must be communicated between devices in a single computer system and between computer systems. To ensure correct and efficient data transmission, it is desirable to code data in a format suitable to a wide variety of devices and computer systems. This is especially true with the alphanumeric data that forms the bulk of human-readable data communication. For this reason, various standards organizations have proposed data-encoding methods for communication between computer system components. Adherence to such standards provides computer users with the flexibility to configure systems of "mixed" equipment, with minimal problems in data communication.

SIGNALS

A *signal* is a means of communication between sender and receiver using a *communication channel*. Computers use electrical signals to communicate data and instructions between computer system components—for example, between memory and CPU—and between computers and other computers. The communication channel is implemented as one or more wires over which electricity travels from the sender to the receiver. For communication to take place, the message must somehow be encoded in the signal, and both the sender and receiver must agree as to the method of encoding. There are two basic methods of encoding messages (numeric values) with electricity. These methods are called *analog signals* and *digital signals*.

Analog Signals

Analog signals are capable of representing a continuous range of values. Many signals that we encounter in everyday life are analog in nature. Sound is one such signal. Sound can vary in intensity (loudness) and in frequency (pitch). Either of these can assume a virtually infinite range of values. Within a limited range of values, the number of possible values is still infinite if infinite precision in measurement is assumed.

For example, if we limit the range of frequency to between 20 Hz[1] and 20,000 Hz (the range of normal human hearing), there are still an infinite number of possible pitches. The frequency of a given sound could be 440 Hz, 441 Hz, or an infinite number of possibilities in between, such as 440.1, 440.0001, or

[1] The frequency of sound is measured in cycles per second or Hertz (abbreviated Hz). For example, the note A below middle C on a piano corresponds to a pitch of 440 cycles per second (440 Hz).

440.000100004. The number of possible values of frequency for a sound in this range is limited only by the receiver's ability to distinguish among them.

Digital Signals

Digital signals are capable of representing only a finite number of possible values. A more precise term for digital signal is *discrete signal,* which is a signal with a limited number of values. An inherently analog signal such as sound can be used to represent a digital signal by dividing its range of possible values into a discrete number of intervals.

For example, the range of frequency between 20 Hz and 20,000 Hz could be divided into two intervals: less than 10,000 Hz and greater than or equal to 10,000 Hz. The dividing line between the two ranges is called a *threshold.* A digital interpretation of an analog signal is defined by assigning a meaning to each range of values—for instance, "zero" for the lower range and "one" for the upper range. In this example there are only two possible values (meanings), so the signal is called a *binary signal.* The assignment of signal values or ranges of signal values to specific meanings is one form of a signal *coding scheme.*

Electricity can be used to carry either analog or digital signals. Variations in voltage or amperage can be used as the message medium. For example, we could send the number 32.39 as an analog signal by sending 32.39 volts of electricity through a wire. As in the previous example, we could also use electricity to transmit binary signals. We could, for example, define voltages of less than 10 to mean zero and voltages of greater than or equal to 10 to mean one.

To send multiple messages across an electrical medium, a timing convention must be used. That is, an interval of time must be defined during which a single value of the signal will be present on the wire. For example, we could adopt one second as the time interval and send first "one" and then "zero" across the wire (as a digital signal) by sending 20 volts across the wire for one second and then immediately lowering the voltage to 0 for the next second. As with coding schemes, both sender and receiver must agree on the timing convention. The set of all signalling conventions—correspondence between signal values and message meanings, timing of signals, and so on—to which both sender and receiver must agree is called a *communication protocol.*

Signal Capacity and Errors

Analog and digital signals each have strengths and weaknesses relative to one another. The most important of these are channel capacity and susceptibility to error. In general, analog signals can carry a greater amount of information than digital signals within a fixed time interval. The reason for this lies in the number and variety of messages that can be communicated during one time period.

For example, assume that electrical voltage is used as the signal medium and that the duration of one signal is one second. The number of signal values that can be transmitted using analog encoding is limited only by the range of voltages that can be carried on the wire. The number of signal values that can be transmitted using binary encoding is only two.

To transmit more than two possible values with binary coding, we must combine adjacent signals. For example, we could send four different binary signals in succession and combine them to form one message. It is thus possible to transmit 16 different messages ($2^4 = 16$ possible combinations of four binary signals) during four consecutive time periods. But an analog signal could have communicated that many possible values (and considerably more) in a single time period. Thus, analog signals have an inherently higher message-carrying capacity than do digital signals.

Although analog signals have an advantage in message-carrying capacity, they are at a distinct disadvantage in their susceptibility to transmission error. If the mechanisms for encoding, transmitting, and decoding electrical analog signals were perfect, this would not be a problem, but errors are always possible. Electrical signals are subject to noise and disruption because of electrical and magnetic disturbances. The noise one hears on a telephone during a thunderstorm is a direct result of this phenomenon. The voltage or amperage of electrical signals can be altered during transmission by such interference.

Consider the following extreme scenario: Your bank's computer communicates with its automated teller machines via analog electrical signals. You are in the process of making a withdrawal of $100. The value "$100" is sent from the automated teller to the bank's computer by a signal of 100 millivolts. The computer checks your balance and decides that you have enough money, so it sends an analog signal of 100 millivolts back to the automated teller as a confirmation. However, during the transmission of this signal, a bolt of lightning strikes near the wire that carries the signal, inducing current in the wire of 5 volts (5,000 millivolts). When the signal reaches the automated teller, it dutifully dispenses $5,000 to you in clean crisp $10 and $20 bills.

Susceptibility of electrical signals to noise and interference can never be completely eliminated. Computer equipment is normally heavily shielded to prevent noise from interfering with internal signals, but external communications might not be so well protected. In addition, errors can be introduced by the resistance present in internal wiring or by the magnetic fields generated within the device.

A digital electrical signal is not nearly as susceptible to noise and interference. Consider the previous example with digital encoding used instead of analog encoding. If we use the previous protocol of 0 volts for 0, 20 volts for 1, and a threshold of 10 volts, we can tolerate almost 10 volts of noise without misinterpretation of a signal. If a zero was being sent at the instant the lightning struck, the automated teller would still interpret the signal as a zero because the 5 volts induced by the lightning is still below the threshold value of 10. If a one was being sent, the lightning would raise the voltage from 20 to 25 volts, which is above the threshold. Resistance in the wire or other factors degrading voltage would not be a problem as long as the strength of the signal remained greater than 10 volts.

The accuracy desired in computer computation and communication virtually requires the use of digital signals. However, analog signals are sometimes used when message-carrying capacity is of high importance, such as with high-speed communication between computers. Analog devices also remain important and useful in handling inputs and outputs within digital systems. To connect these two technologies, a conversion step is required. The transformation of continuous, analog signals into discrete, digital data is called *analog-to-digital (A/D) conversion.* An input device that transforms the movements of an electronic drawing stylus into a stream of numeric values performs A/D conversion. Similarly, digital signals must be converted to analog motor control signals *(D/A conversion)* to direct the physical motions of a plotter in producing a drawing as computer output.

BINARY DATA REPRESENTATION

Although digital computers are binary devices, there is nothing special or mandatory about the choice of two-state design. Binary signals are convenient because they correspond with clearly identifiable states of electron flow. However, it would not have been impossible to create four-state, six-state, or even eight- or ten-state computing devices. In fact, many nonbinary digital devices have been developed during the evolution of the computer industry. However, binary devices proved to be highly efficient, practical, and economical.

The representation of data in binary digital code is highly compatible with the basic types of operations computers perform. In a sense, the only type of calculation a computer can perform in a single step is addition. Subtraction is handled as the addition of a negative number. Multiplication is performed as a sequence of additions. Division can be performed through successive subtractions and multiplications. Thus, for performing all types of numeric calculation, a computer can be regarded as a very fast adder.

Binary codes are also well suited to computer processing because they correspond directly with values in Boolean logic. This form of logic is named for nineteenth-century mathematician George Boole. He developed methods of reasoning and logical proof using sequences of statements that could be evaluated only as true or false. Similarly, a computer can perform logical comparisons of two binary data values to determine whether it is true or false that a presented value is greater than, equal to, less than, less than or equal to, not equal to, or greater than or equal to another value.

Both decimal and binary notation are forms of *positional number systems.* In a positional number system, numeric values are represented as patterns of symbols—that is, strings of numeric digits. The symbol used for each digit and its position within the string determine its value. The sum of the values of all positions in the string equals the value of the string.

For example, in the decimal number system, the number 5,689 is interpreted as follows:

$$(5 \times 1000) + (6 \times 100) + (8 \times 10) + 9 =$$
$$5{,}000 + 600 + 80 + 9 = 5{,}689$$

The same series of operations can be represented in columnar form, or with positions of the same value aligned in columns:

$$\begin{array}{r} 5{,}000 \\ 600 \\ 80 \\ +\quad 9 \\ \hline 5{,}689 \end{array}$$

For whole numbers, values are accumulated from right to left. In the answer to our example, the digit 9 is in the first position, 8 is in the second position, 6 in the third, and 5 in the fourth.

The maximum value, or weight, of each position is a multiple of the weight of the position to its right. In the decimal number system, the second position is 10 times that of the "ones," or first, position. The third is 10 times the second, or 100. The fourth is 10 times the third, or 1,000, and so on.

This multiplier corresponds with the base of the number system. Another term for base is *radix.* The base, or radix, of the decimal number system is 10. For fractional parts of real numbers, values are accumulated from left to right, beginning with the digit immediately to the right of the (.) notation, or the *radix point.* In the decimal number system, the radix point usually is referred to as the decimal point. An example is:

$$5{,}689.368$$

The fractional portion of this real number is .368. The digits are interpreted as follows:

$$(3 \times .1) + (6 \times .01) + (8 \times .001) =$$
$$0.3 + 0.06 + 0.008 = 0.368$$

Note that, proceeding rightward from the radix point, the weight of each position is a fraction of the position to its left. In the decimal number system, the first position to the right of the decimal point represents tenths; the second position represents hundredths, the third thousandths, and so on. As with whole numbers, each fractional position has a weight that is 10 times that of the position to its right. Table 4.1 shows a comparison between decimal and binary representations of the numbers 0 through 10.

TABLE 4.1 Equivalent notations for the values 0 through 10 under the binary and digital numbering systems.

	Binary System (Base 2)					Decimal System (Base 10)			
Place Values	**2^3 8**	**2^2 4**	**2^1 2**	**2^0 1**		**10^3 1000**	**10^2 100**	**10^1 10**	**10^0 1**
	0	0	0	0	=	0	0	0	0
	0	0	0	1	=	0	0	0	1
	0	0	1	0	=	0	0	0	2
	0	0	1	1	=	0	0	0	3
	0	1	0	0	=	0	0	0	4
	0	1	0	1	=	0	0	0	5
	0	1	1	0	=	0	0	0	6
	0	1	1	1	=	0	0	0	7
	1	0	0	0	=	0	0	0	8
	1	0	0	1	=	0	0	0	9
	1	0	1	0	=	0	0	1	0

The electronic signals that are processed by computers are encoded under the binary (or base 2) number system. In the binary number system, as in the decimal number system, real numbers can have whole and fractional parts. In binary notation, whole numbers are indicated by the digits to the left of the radix point, and the fractional portion is shown to the right. An example is:

101101.101

The position immediately to the left of the radix point represents "ones." Each position has a weight that is two times greater than that of the position to the right. Thus, starting at the rightmost digit of the whole portion, the maximum values of the positions are 1, 2, 4, 8, 16, 32, and so on. The values of the fractional positions, starting immediately to the right of the radix point, are $^1/_2$, $^1/_4$, $^1/_8$, $^1/_{16}$, $^1/_{32}$, and so on.

Converting a number in binary notation to its decimal equivalent can be done by multiplying the value of each position by the decimal weight of that position, then summing the results. For the example given above, the decimal equivalent is shown in the diagram in Figure 4.1.

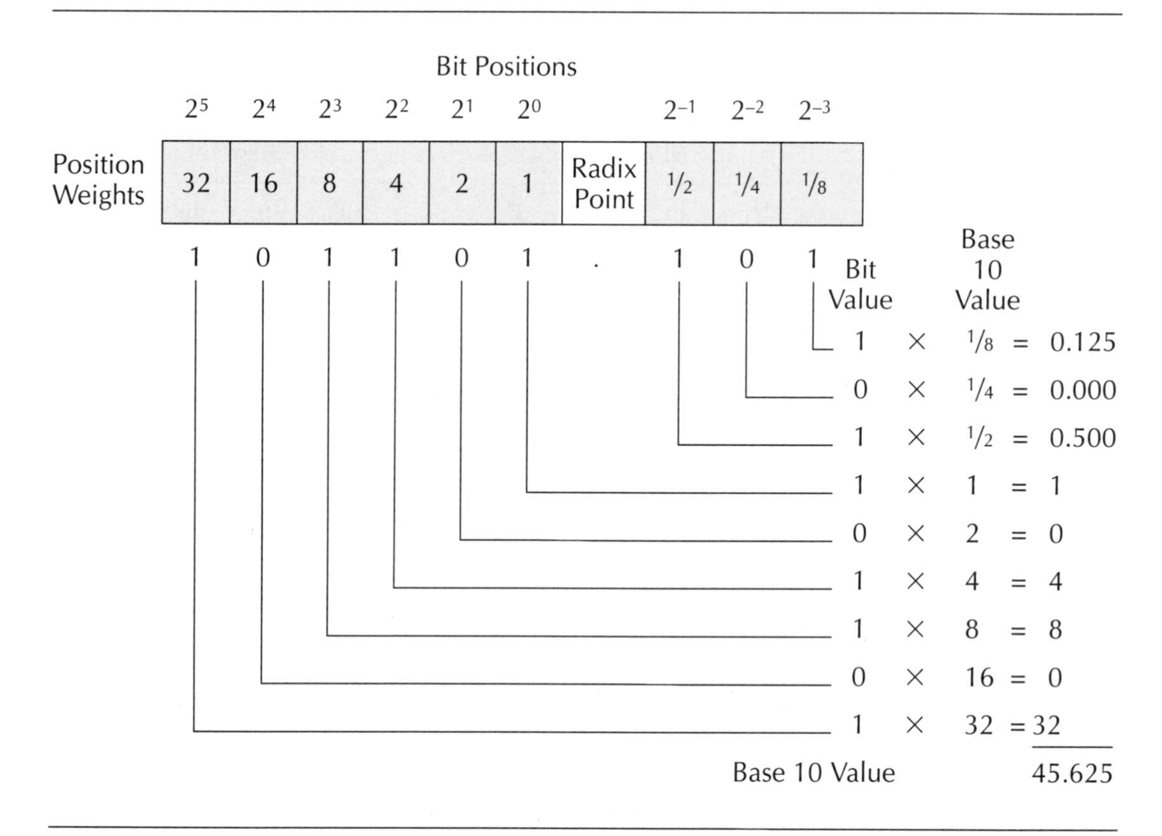

FIGURE 4.1 The procedure for computing the decimal equivalent of a binary number.

In computer terminology, each bit occupies a unique *bit position.* A meaningful sequence of binary digits, or a group of bits, is called a *bit string*. The leftmost digit, which has the greatest weight, is called the *most significant digit,* or *high-order bit.* Conversely, the rightmost digit is the *least significant digit*, or *low-order bit.* A string of eight bits is called a *byte.*[2] A byte is generally the smallest unit of data that can be read or written.

Besides determining the numeric value of bit strings, these positional relationships also affect the ways in which data is transmitted and stored within a computer. The number of binary digits required to represent decimal values is shown in Table 4.2.

[2] Although relatively few in number, some computers utilize bytes of fewer than eight bits. For example, certain computers manufactured by Control Data Corporation use bytes consisting of only six bits.

TABLE 4.2 Binary data representations for bit strings ranging from one through 16-bit positions.

Number of Bits (n)	Number of Values (2^n)	Numeric Range
1	2	0..1
2	4	0..3
3	8	0..7
4	16	0..15
5	32	0..31
6	64	0..63
7	128	0..127
8	256	0..255
9	512	0..511
10	1024	0..1023
11	2048	0..2047
12	4096	0..4095
13	8192	0..8191
14	16384	0..16383
15	32768	0..32767
16	65536	0..65535

Addition operations involving positive, binary numbers can be performed by applying the following four mathematical rules, or binary addition facts:

$$
\begin{aligned}
0 + 0 &= 0 \\
1 + 0 &= 1 \\
0 + 1 &= 1 \\
1 + 1 &= 10
\end{aligned}
$$

Addition of two binary, real numbers is done by first aligning the respective radix points:

$$
\begin{array}{r}
101101.101 \\
+\ 10100.0010 \\
\hline
\end{array}
$$

At each position, an addition operation is performed, the result of which is determined by applying the appropriate addition rule. Values at each position or in each column are added, starting with the rightmost, or least significant, digit. Columnar addition proceeds just as it does for decimal numbers: If the result exceeds the weight assigned to that position, the excess value must be "carried" to the next column or position. The excess value then is added to the values at the next position. The result of adding the two numbers above is:

```
 111      1
 101101.101
+ 10100.0010
------------
1000001.1100
```

Note that the result is the same as that obtained by adding the values in base 10 notation:

Binary		Real Fractions		Real Decimal
101101.101	=	$45\frac{5}{8}$	=	45.625
+ 10100.0010	=	$20\frac{1}{8}$	=	20.125
1000001.1100	=	$65\frac{3}{4}$	=	65.750

OTHER NUMBER SYSTEMS

The representation of data as binary numbers poses problems for human programmers. Binary numbers require many digits and their use by humans leads to a large number of errors. To minimize these errors, alternate number systems are sometimes used in computer hardware design and machine or assembly language programming. These number systems include:

- Hexadecimal notation
- Octal notation

Hexadecimal Notation

In base 16, digit values range from 0 to 9 and from A to F, which correspond with decimal values of 0 to 15. Hexadecimal notation also is highly compatible with units of storage within addressable memory. Each memory cell typically contains one byte, or character, that is stored as two groups of four bits each. Thus, each four-bit group can store values in the range of 0000 to 1111 (binary), or 0 to 15 (decimal). These same values can be represented by a single digit in hexadecimal notation, or 0 to F "hex." Byte values that can be stored within a single memory cell, then, will be within the range of 00 to FF hex.

Hexadecimal numbers also are used to designate memory addresses. An area that contains 64Kbytes memory cells can hold 65,536 bytes (64 × 1,024 bytes/K). Thus, the range of possible memory addresses must be 0 to 65,535 (decimal). In binary notation, 16 bits will be required to cover this range:

00000000 00000000 to 11111111 11111111

Such lengthy bit strings, however, are cumbersome for programmers to manipulate. A much more concise and understandable representation is to code the addresses in four-bit groups, using one hex digit for each group. In this scheme of notation, the full range of addresses within 64K of memory is 0000 to FFFF hex. This is the accepted numbering scheme for memory addresses.

To differentiate binary and hexadecimal numbers in printed text, it is necessary to indicate the type of number being represented. In mathematics it is common to denote the base of a number with a subscript. Thus,

$$1001_2$$

would indicate to the reader that 1001 should be interpreted as a binary number. Similarly,

$$6044_{16}$$

would indicate that 6044 should be interpreted as a hexadecimal number.

When writing about computers it is more common to designate the base of a number by affixing a letter to the end of the number. Thus,

1001B

would indicate a binary number;

6044H

would indicate a hexadecimal number. No letter is normally used to indicate a decimal (base 10) number.

Unfortunately, these conventions are not always followed. Often it is left to the reader to guess the base by the content of the number or the context in which it appears. Bit strings, for example, are generally assumed to be expressed in binary; memory addresses are assumed to be expressed in hexadecimal (base 16). The content can also guide the reader. Any numeral other than a 0 or a 1 indicates that the number cannot be binary. Similarly, the letters A through F indicate that the contents are expressed in hexadecimal.

Octal

In some machine language applications, it is convenient to work with groups of three bits, or 000 to 111 (binary). An alternate method of coding these groups is in base 8, or octal, notation. Octal digits range from 0 to 7, which correspond exactly with the minimum and maximum values that can be represented in three bits. For binary values that extend beyond three bits, decimal 8 is octal 10, decimal 9 is octal 11, and so on.

NUMERIC DATA REPRESENTATION SCHEMES

There are several different representation schemes for numeric data. These include

- Sign-magnitude notation
- Excess notation
- Two's-complement notation

Each of these methods represents a different tradeoff among accuracy, conciseness, and machine efficiency. A hardware designer will typically choose the representation most appropriate to the type of data being manipulated, subject to cost constraints. Thus, representation schemes vary among data types within the same machine. They can also vary among computers due to cost tradeoffs made by the computer designer or manufacturer.

Sign-Magnitude Notation

The objective of *sign-magnitude notation* is to indicate the sign of a binary value. It is preferable to include the sign within the bit string that represents the value so that subtraction can be performed as a special case of addition. In sign-magnitude notation, the high order (or leftmost) bit within a bit string is designated the *sign bit.* Non-negative values, or values that are either 0 or positive, are preceded by a sign bit of 0. Negative values, or values less than 0, have a sign bit of 1. Sign-magnitude notation is independent of the method used to code the rest of the bit string. That is, the sign bit is merely affixed to the bit string; it does not affect its content.

Excess Notation

Another coding method that can represent signed values is *excess notation.* Under this method, a fixed number of bit positions is used. For example, bit strings might be restricted to four positions. Thus, values could range from 0000 to 1111. Within this range, the lowest value with a 1 as its leftmost digit—1000—is designated to represent zero (0). In ascending order, bit strings that are greater than 1000 are used to represent positive values. As shown in Table 4.3, in excess notation all non-negative values have a 1 as the high-order bit; negative values have a 0 in this position. Thus, in descending order, bit strings that are less than 1000 represent negative values. Recall that the reverse is true of sign-magnitude notation.

TABLE 4.3 Formats for excess notation in four-bit patterns known as "excess eight."

Bit String	Decimal Value	
1111	7	Non-negative Values
1110	6	
1101	5	
1100	4	
1011	3	
1010	2	
1001	1	
1000	0	
0111	–1	Negative Values
0110	–2	
0101	–3	
0100	–4	
0011	–5	
0010	–6	
0001	–7	
0000	–8	

A system of excess notation that uses four-bit positions is called *excess eight* because the value for zero is binary 1000, which otherwise would be the equivalent of decimal 8. Because this value is used as a starting point for counting, the binary representations of all values exceed their actual values by decimal 8, or binary 1000. Similarly, a five-bit excess notation scheme would use 10000 as a starting point, or zero value, which would be the equivalent of decimal 16. Thus, excess notation in five bits is called *excess 16.* The *absolute value*, or unsigned value, of a number expressed in excess notation can be derived by dropping the leftmost digit and interpreting the remaining string in base 2.

Two's-Complement Notation

In the binary number system, the complement of 0 is 1, and the complement of 1 is 0. The complement of a bit string is formed by substituting 0 for all values of 1, and a 1 for each 0. This transformation is the basis of *two's-complement notation.* As with excess number schemes, two's-complement notation uses a fixed number of bit positions. Non-negative values are rendered in base 2 notation. For example, in four-bit, two's-complement notation, decimal 7 would be coded as

$$0111$$

In two's-complement notation, a negative value is represented as follows:

Complement of positive value + 1 = Negative representation

Thus, in four-bit, two's-complement representation, decimal –7 would be:

$$\begin{aligned} (0111) + 0001 &= \\ 1000 + 0001 &= \\ 1001 &= -7_{10} \end{aligned}$$

In this example, the positive binary value for decimal 7 (0111) is converted to its complement, 1000. Then a binary value of 1 (0001) is added to the least significant (or rightmost) digit. The result, 1001, is the two's-complement equivalent of –7 (decimal).

Two's-complement notation is highly compatible with digital electronic circuitry for reasons that include these:

- The leftmost bit, though part of the data value, can be interpreted as a sign bit.
- A fixed number of bit positions is used, requiring a minimum number of electronic circuits.
- Subtraction can be performed as addition of a negative value, which simplifies processing circuitry.
- The number of logic circuits required to perform addition is reduced to two.

For these reasons, which are explained more fully below, arithmetic operations on integers within computer processors typically are performed on two's-complement values.

BASIC DATA TYPES

At the level of computer software, various data types represent the programmer's specification for the interpretation and processing of categories of inputs. Differences in implementation exist among the compilers and interpreters of various computer programming languages.

At the machine level, for reasons discussed above, each data type represents a different binary coding format. Coding formats correspond with different sets of transformations and the degree of accuracy required for each. Although implementations differ, the data types handled by most computers include

- Integers
- Real numbers
- Character data types
- Boolean data types

Integers

An *integer* is a whole number, or a value that does not have a fractional part. Thus, the values 2, 3, 9, and 12,964 are integers. Arithmetic operations on integers, or *integer arithmetic*, have some specific characteristics. If addition, subtraction, or multiplication are performed entirely with integers, the result will also be an integer. In the case of dividing one integer by another, it is possible for fractional values to be generated, as in the example already presented:

$$\begin{aligned} 18 \div 4 &= 4\text{, remainder } 2 \\ &= 4^{2}/_{4} \\ &= 4^{1}/_{2} \end{aligned}$$

If integer arithmetic is used, the fractional portion, the remainder of 2 in this case, is discarded. The result, 4, is an approximation.

Within the CPU of most computers, arithmetic operations on whole numbers, or integers, are done in *complementary arithmetic*, or sequences of addition operations on two's-complement values. Only two basic logic circuits are required to perform arithmetic transformations: an adder and an inverter. The adder merely combines two binary inputs by applying the fundamental binary addition facts. The inverter transforms base 2 values into two's-complement values of the opposite sign.

In integer addition and multiplication operations, only the adder is used:

$$0101 + 0001 = 0110$$

In subtraction, the value to be subtracted is passed through the inverter, where it is transformed into a negative value. This value is then combined with the first value in the adder. For example:

$$\begin{aligned} 0101 - 0001 &= \\ 0101 + (-0001) &= \\ 0101 + 1111 &= 0100 \text{ (truncated to 4 bits)} \end{aligned}$$

In this example, four-bit notation is used. Note that the addition operation normally would yield a value of 10100. However, the result is truncated to four bits by dropping the leftmost digit.

In a division operation, the divisor is passed through the inverter and transformed into a negative value. This value is then added to the dividend repeatedly until the result is negative. The number of times this operation is performed before the result becomes negative is output as the quotient. (Again, in integer arithmetic, any remainder from a division operation is discarded.)

The operations of raising a value to a power or extracting a root also can be performed with sequences of addition operations. An exponential value can be computed as a series of multiplication operations:

$$\begin{aligned} 2^4_{10} &= 2_{10} \times 2_{10} \times 2_{10} \times 2_{10} \\ &= 0010_2 \times 0010_2 \times 0010_2 \times 0010_2 \end{aligned}$$

These multiplications, in turn, can be done as a series of addition operations:

$$\begin{aligned} 2^4_{10} &= (0010 + 0010) + (0010 + 0010) + (0010 + 0010) + (0010 + 0010) \\ &= 1000 \end{aligned}$$

The extraction of roots follows a similar process of successive division operations. Thus, all arithmetic operations on two's-complement values can be reduced to simple addition.

Overflow. In complementary arithmetic, the number of bit positions in any value is fixed. That is, the result is limited to the same number of bit positions as the inputs. If the result exceeds this number of digits, a condition of *overflow* exists. In case of overflow, the result is truncated, or the leftmost digit is discarded. For example, in four-bit notation, a result of 10011 would be truncated to 0011. This truncation can be a source of error.

In a two's-complement system, the CPU can detect overflow conditions by testing the sign bit of the result of each arithmetic operation. Overflow can occur only if the operation involves the addition of two negative or two positive values. Thus, overflow has occurred if the result of adding two positive values has a sign bit of 1 (negative) or if the result of adding two negative values has a sign bit of 0 (non-negative).

In integer arithmetic, the primary method of preventing overflow and truncation is to use a relatively large number of bit positions to store two's-complement values. Many computers manipulate integers in fixed-length strings of 32 bits. This number of bit positions permits the representation of integers within a numeric range of –2,147,483,648 to 2,147,483,647 (decimal) without overflow.

To avoid overflow and to increase accuracy, some computers and programming language implementations permit the use of *double-precision* numeric data types. In double-precision representation, two fixed-length units of storage are combined to hold each value. In the case of whole numbers, double-precision representations are sometimes called *long integers*.

Overflow also can be prevented through care in programming. If the programmer anticipates overflow is possible, the units of measure for program variables, or data items, can be made larger. For example, if calculations were being performed in centimeters, the values might be represented instead in meters.

Converting to other units of measure, as well as performing some types of calculations with precision, might not be practical if data values are restricted to whole numbers. Numbers with fractional parts, or real numbers, might need to be introduced. Manipulating such quantities within computers requires a different scheme of data representation.

Real Numbers

In mathematics, real numbers can contain both whole and fractional components. In the decimal number system, the fractional portion is indicated by digits appearing to the right of the decimal point. Thus, the equation above could be expressed in real decimal numbers, such as

$$18.0 \div 4.0 = 4.5$$

The fractional portion (.5) in this result is interpreted as $^{5}/_{10}$. In program source code, real numbers usually are written in decimal notation. This system of notation is discussed in greater depth below.

In arithmetic operations on real numbers, the results of addition, subtraction, multiplication, and division also will be real. Exceptions include operations that are undefined with the real number system, which also cannot be performed in integer arithmetic:

- Division by 0
- Roots of negative numbers

In mathematics, the result of dividing a number by 0 is undefined; in other words, it is a function that cannot be implemented. In computer processing, division by 0 will cause an error. Within some computers, controls can be built into system software that can detect such an error before processing is disrupted completely. In such cases, an error message will be reported to the user or system operator. An important quality control step in programming is to check source code for the possibility that any divisors might be assigned the value of 0.

In arithmetic with integers or real numbers, even numbered roots of negative numbers are undefined. For example, attempting to find the square root of –1 will cause an error. Odd-numbered roots of negative numbers (for example, cube roots) are valid. For example, the cube root of –1 is –1.

Within computers, real numbers are generally represented in *floating-point notation.* Under this scheme, a radix point can be located anywhere within a bit string. Thus, the position of the radix point is not fixed, but "floats" among the data bit positions. The ability to move the radix point within the string permits flexibility of representation and manipulation.

Coding schemes for floating-point values vary considerably among computers. Values can be coded in excess notation, two's-complement notation, or other methods. There is also variation in the interpretation and position of sign bits.

The discussion here ignores these differences and presents floating-point concepts at a general, overview level.

The diagram in Figure 4.2 shows a typical approach. This example uses an eight-bit string, although actual representations within computers also vary in length. Many computers use 24- or 32-bit representations. Supercomputers can handle strings that contain more than 100 bits. In any case, the number of bit positions is fixed.

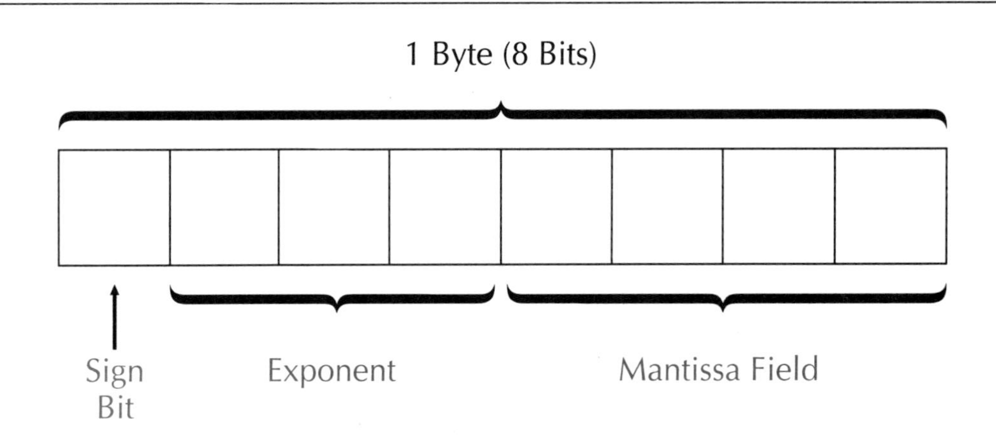

FIGURE 4.2 **A possible representation of a signed floating-point number.**

In many implementations, the leftmost bit is a sign bit for the real number. This bit is interpreted according to sign-magnitude notation: 0 if the number is non-negative (zero or positive), and 1 for negative. In other implementations, the sign bit is embedded within the string. The rest of the string is divided into two fields: the *exponent field* and the *mantissa field.* In Figure 4.2 the exponent field contains three bits, and the mantissa contains four. Again, these lengths will vary among implementations.

A real value is derived from a floating-point bit string according to the following formula:

$$\text{Value} = \text{Mantissa} \times \text{Base}^{\text{exp}}$$

The mantissa holds the bits that are interpreted to derive the digits of the real number. By convention, the mantissa is assumed to be preceded by a radix point. The exponent field indicates the position of the radix point.

In binary notation, the base in the formula is 2. The exponent is an integer value, or power of 2. Multiplying the mantissa by 2^{exp} yields the same result as shifting the radix point by the number of bit positions indicated in the exponent. Thus, the exponent can be regarded as a count of bit positions by which the radix point must be shifted within the mantissa to yield the intended real value.

The leftmost bit of the exponent field is interpreted as a sign bit. The sign of the exponent field indicates the direction in which the radix point must be moved. If the sign bit of the exponent is non-negative, the radix point is moved to the right. If the sign bit is negative, the radix point is moved to the left.

Assuming the eight-bit format discussed above, consider the bit string

00101011

In this example, the sign bit has a value of 0, the exponent is 010, and the mantissa is 1011. Recall that the mantissa, which holds the significant digits of the real value, is assumed to be preceded by the radix point:

.1011

The exponent indicates how the radix point will be shifted to yield the real value. The value of the exponent in this case is 010, or decimal 2. The value is positive, so the radix point in the mantissa must be moved two bit positions to the right. In binary notation, the resulting real value is

10.11

The decimal equivalent of this value is 2.75.

The floating-point range and its limitations. The number of bits in a floating-point string and the formats of the mantissa and exponent fields impose limitations on the range of values that can be represented. The number of significant digits in the mantissa corresponds to the number of digits in the largest integer that can be represented in the mantissa. In turn, the number of digits in the exponent determines the number of possible bit positions to the right or left of the radix point.

As with integers, the results of some arithmetic operations might be too large to be stored within the number of binary digits and radix-point places that are available in a given floating-point representation scheme. In such cases, a condition of overflow will occur. Similarly, results might be too small to be represented accurately. In such cases, the radix point cannot be moved far enough to the left, and a condition of *underflow* is the result.

As with long integers, many computer and programming language implementations permit the use of *double-precision real representations*, or twice the usual number of bit positions for accuracy in floating-point calculations. Double-precision representations are costly in storage space and in processing overhead, however.

Real numbers that often create problems of representation within computers are nonterminating numbers. An example of such a real number is the decimal equivalent of the fraction $^1/_3$:

$$^1/_3 = 0.33333333...$$

In mathematics, the number of digits to the right of the decimal point in this example is assumed to be infinite. Within a computer, the representation of this data value would have to be terminated (or truncated) to a specific number of digits.

An added difficulty is that more values have nonterminating representations in the binary system than in the decimal system. For example, the fraction $^1/_{10}$ is nonterminating in binary notation. The representation of this value in floating-point notation would be a rounded, or approximated, value. Such problems can usually be avoided, however, through careful programming. As a rule of thumb, programmers reserve floating-point calculations for quantities that might vary continuously over wide ranges. Measurements made by scientific instruments, for example, might be handled in floating-point notation.

If floating-point calculations are used, some general guidelines can help reduce the potential for error. For example, to minimize rounding errors in a series of addition operations, the smaller values should be added first. Also, as in integer arithmetic, the programmer must avoid all situations in which a value of 0 might be used as a divisor.

Because dollars and cents are usually shown as decimals in source code and in printouts, novice programmers sometimes assume that floating-point representations are appropriate for financial calculations in binary. However, even if overflow and underflow are avoided, rounding errors due to nonterminating representations of tenths will persist. Cumulative errors will mount as approximations are input to subsequent calculations. For this reason, careful programmers use integer arithmetic for accounting and financial applications. In such applications, using integer representations is equivalent to expressing all values in pennies. Then a decimal point is inserted when the result is converted to decimal notation for output.

Character Data

As a data type, characters include alphabetic letters, numerals, punctuation marks, and special symbols. A sequence of characters that can be manipulated as a single unit or data item is called a *string* or *character array*. For example, the address "349 MAPLE STREET" is a string of alphanumeric characters. By convention in computer programming, quotation marks are used to enclose character string data.

In this example, the digits of the street number, 349, are numerals, or indicators, rather than numeric values. That is, the numeric component of the address serves as a label or name rather than as a data value to be used in arithmetic operations.

Alphanumeric data, such as names and addresses, is usually represented within computers as strings of characters. As discussed previously, the character set used for such purposes includes the letters of the alphabet, numerals, punctuation marks, and special symbols. Because these characters represent conventional symbols that can be printed in human-readable text, they are called printable characters. The character set also may include *control codes*, or nonprintable characters, that are used for machine signalling.

Character-coding methods differ in their character sets, coding patterns for characters, length of bit strings, and processing applications. Common coding methods include

- BCD
- ASCII
- EBCDIC

BCD. *Binary-coded decimal (BCD)* is an early coding method that uses strings of six bits. The character set includes alphabetic letters in uppercase, or capitals, only. Also included are the numerals 0 through 9 and some punctuation marks.

BCD can be considered obsolete as a general-purpose code for data processing and transmission. However, some computers store numeric data, and particularly integers, in BCD notation. As shown in Table 4.4, the BCD notation for numerals is the same as a four-bit, base 2 representation preceded by 00. If BCD notation is used for numeric data, the most significant bit can be interpreted as a sign bit.

TABLE 4.4 A partial list of the ASCII, EBCDIC, and BCD binary codes for numerals, uppercase letters, and lowercase letters.

Symbol	ASCII	EBCDIC	BCD
0	0110000	11110000	000000
1	0110001	11110001	000001
2	0110010	11110010	000010
3	0110011	11110011	000011
4	0110100	11110100	000100
5	0110101	11110101	000101
6	0110110	11110110	000110
7	0110111	11110111	000111
8	0111000	11111000	001000
9	0111001	11111001	001001
A	1000001	11000001	010001
B	1000010	11000010	010010
C	1000011	11000011	010011
a	1100001	10000001	
b	1100010	10000010	
c	1100011	10000011	

ASCII. The *American Standard Code for Information Interchange (ASCII)* is a coding method that has been adopted in the United States and that is used widely in data communication and within many computers. A subset of ASCII codes are included in Table 4.4. The full set of codes is listed in Appendix A. The international equivalent of this character set, which is used widely to transmit text in languages other than English, is *International Alphabet Number 5 (IA5)*. This standard is promulgated by the International Standards Organization (ISO).

As used in data communication, ASCII is a seven-bit code. Seven-bit character representations in ASCII are said to be in *ASCII-7 notation*. Within computer hardware, however, data is typically handled in eight-bit bytes. Thus, if ASCII is used as a character code within a computer, a 0 is added in the leftmost (or high-order) bit position. This scheme is called *ASCII-8 notation*.

Besides its binary representation, each ASCII character has a numeric equivalent in decimal notation. The ASCII-7 character set corresponds with decimal values 0 to 127. Characters with decimal values 0 through 32 and 127 are control codes that have been standardized for signalling between devices in data communication. These codes are shown in Table 4.5. In ASCII-8, addition of the eighth bit permits inclusion of decimal values 128 through 255. Characters in this range are undefined within the standard ASCII set. However, they are often used either as additional control characters or for alternate character sets such as graphics, line drawing, and multinational characters.

TABLE 4.5 **Control codes from the ASCII character set.**

Decimal Code	Control Character	Description
000	NUL	Null
001	SOH	Start of heading
002	STX	Start of text
003	ETX	End of text
004	EOT	End of transmission
005	ENQ	Enquiry
006	ACK	Acknowledge
007	BEL	Bell
008	BS	Backspace
009	HT	Horizontal tabulation
010	LF	Line feed
011	VT	Vertical tabulation

(continued)

TABLE 4.5 **Control codes from the ASCII character set. (continued)**

Decimal Code	Control Character	Description
012	FF	Form feed
013	CR	Carriage return
014	SO	Shift out
015	SI	Shift in
016	DLE	Data link escape
017	DC1	Device control 1
018	DC2	Device control 2
019	DC3	Device control 3
020	DC4	Device control 4
021	NAK	Negative acknowledge
022	SYN	Synchronous idle
023	ETB	End of transmission block
024	CAN	Cancel
025	EM	End of medium
026	SUB	Substitute
027	ESC	Escape
028	FS	File separator
029	GS	Group separator
030	RS	Record separator
031	US	Unit separator
127	DEL	Delete

The values and numeric order of ASCII characters are designed to support certain processing functions. These functions include

- Data typing
- Validity checking
- Collating

A programmer can use numeric values of ASCII characters to suspend data typing. For example, ordinarily it would not be desirable or permissible to perform numeric calculations on fields containing character data. One common exception to this rule occurs in the manipulation of dates stored as sequences of characters (numerals). For example, an accounts payable program might need to determine whether the current date is less than 30 days from an invoice date. If the digits of the invoice date are stored as characters, their arithmetic difference may be computed using their ASCII values. For example, the time interval between January 31 and January 20 can be calculated as:

$$\begin{aligned} \text{"31"} - \text{"20"} &= ((\text{'3'} - \text{'2'}) \times 10) + (\text{'1'} - \text{'0'}) \\ &= ((051 - 050) \times 10) + (049 - 048) \\ &= 10 + 1 \\ &= 11 \end{aligned}$$

The result of the subtraction is a numeric data item that can be used in subsequent arithmetic operations or numeric comparisons.

In dealing with character inputs, the programmer often wants to test a presented value to assure that it is an appropriate data type for a given field. Such *validity checking*, for example, might attempt to detect the entry of numeric data into an alphabetic field. If extracted from the ASCII character code, the decimal equivalent can be used to determine whether the character is alphabetic or numeric, with a minimum of calculation. For example, a character variable called FIELD can be tested for numeric content by the following statement:

```
if (FIELD > 047) and (FIELD < 058) then
```

where the numeric constants are compared to the ASCII code of the character stored in FIELD.

The numeric ordering of the character set also determines the order in which character data will be sorted. This is commonly referred to as the *collating sequence* that is defined for the processor. In ASCII character coding, numerals precede all alphabetic letters, and uppercase letters precede lowercase letters.

EBCDIC. An eight-bit method of character coding is *Extended Binary-Coded Decimal Interchange Code (EBCDIC)*. IBM developed this code on the basis of its BCD format. It is used primarily within medium and large-scale computers. Conversion between ASCII and EBCDIC is often required for data communication between microcomputers and mainframes. EBCDIC codes are included in Table 4.4.

A major difference between ASCII and EBCDIC coding schemes lies in their respective collating sequences. In EBCDIC, numerals correspond with four-bit, base 2 representations preceded by 1111. Because the most significant digits contain high values (ones), numerals are found at the end of the collating sequence, or after alphabetic characters. Unlike most ASCII implementations, EBCDIC numerals also can be used in arithmetic operations.

Boolean Data Types

The *Boolean data type* contains only two data values: true and false. The Boolean data type is the most concise in coding format because only a single bit is required (e.g., binary 1 can represent true and 0 can represent false). Boolean values do not exist as a separate data type for all computers or implementations of programming languages. Boolean values are often coded in the same manner as integers for which zero indicates false and any nonzero value indicates true. The reverse interpretation of true and false can also be used.

DATA STRUCTURES

The previous sections outlined the various basic data types that a computer can manipulate directly. However, it would be difficult to develop useful programs of any kind if these were the only possibilities for data representation. Even simple applications require that individual data elements be combined to form useful aggregations of data. For example, a set of data about a customer or employee could be considered a disconnected set of basic data elements. However, many application programs that manipulate this set of data might want to manipulate it as an integrated unit. It may be desirable to read, write, or display all of these elements simultaneously.

A *data structure* is a set of basic data elements organized for some type of common processing. Data structures are defined and manipulated within software. Computer hardware cannot manipulate them directly, but must deal with them in terms of their basic components—integers, floating-point numbers, single characters, or the like. Thus, software must provide the means for translating operations on data structures into equivalent sets of machine instructions that operate on individual basic data elements.

The complexity of data structures is limited only by the imagination and skill of systems and application programmers. As a practical matter, however, certain data structures are useful in a wide variety of situations and are thus commonly implemented. Examples of such data structures include character strings (or arrays), records, and files. Because of their frequent use and programmers' desire to implement them efficiently, support for the representation and manipulation of such data structures is commonly provided in systems software. For example, the operating system will normally provide service routines for inputting and outputting records to a file and for I/O operations with character strings.

Other types of data structures might or might not be supported within systems software. Examples of these include numeric arrays, indexed files, and complex database structures. Indexed files are supported in some, but not all, operating systems. Numeric arrays are normally supported within a programming language, but not within the operating system. Database structures are normally supported by a database management system (separate from the operating system), although there is an increasing tendency to support some database processing within the operating system.

Pointers and Addresses

Whether implemented within systems or application software, virtually all data structures make extensive use of pointers and addresses. A *pointer* is a data element that contains the address of another data element. An *address* is the location of some data element within a storage device. The storage device can be memory, a disk drive, or some other device.

Addresses vary in content and representation, depending on the storage device addressed. Secondary storage devices are normally organized as a sequence of data blocks. A *block* is a group of bytes that is read or written as a unit by the device. For example, diskette drives on personal computers often read and write data in 512-byte blocks. For storage organized as a sequence of blocks, an address can normally be represented as an integer containing the sequential position of the block. Integers can also represent the address of an individual byte within a block.

Memory addresses can be substantially more complex. If memory is organized as a sequence of contiguous bytes, an integer can be used to store the address. However, many computer systems use much more complex methods of organizing and addressing memory. For these computers, a separate basic data type is provided for memory addresses. For the purpose of discussing data structures, we will assume a sequential byte-oriented organization for memory. More complex memory-addressing schemes are covered in a later chapter.

Arrays and Lists

Many sets of data can be considered lists. A *list* is a set of related data values. In mathematics, a list is considered unordered. That is, there is no concept of the first, second, or last element of a list. When writing software a programmer usually prefers to impose some ordering on the list. For example, a list of the days of the week might be ordered sequentially starting with Monday.

An *array* is an ordered list in which each element can be referenced by an index to its position. An example of an array for the first five letters of the alphabet appears in Figure 4.3. Note that the index values are numbered starting at zero, as is common (although not universal) practice in computer programming. Although the index values appear in the diagram, they are not stored. Instead, they are inferred from the location of the data value within the storage allocated to the array. Within a high-level programming language, individual array elements are normally referenced by the name of the array and the index value. For example, the third letter of the alphabet stored in an array might be referenced as

alphabet[2]

where *alphabet* is the name of the array and *2* is the index value (numbered from zero).

Index	Value
0	A
1	B
2	C
3	D
4	E

FIGURE 4.3 Array elements in contiguous storage locations.

Figure 4.4 shows a character array stored in contiguous, or sequential, memory locations. In this example, each character of the name "John Doe" is stored within a single byte of memory, and the characters are ordered in sequential byte locations starting at byte 1000. An equivalent organization might be used to store the name on a secondary storage device. Access to an individual element of the array can be achieved using the starting address of the array and the index of the element. For example, if we want to retrieve the third character in the array, we can compute its address as the sum of the address of the first element plus the index (assuming index values start at zero).

Data Values	J	o	h	n		D	o	e
Memory Address	1000	1001	1002	1003	1004	1005	1006	1007

FIGURE 4.4 **A character array stored in contiguous (sequential) memory locations.**

The use of contiguous storage locations (especially in secondary storage devices) complicates the allocation of those locations somewhat. If an array needs to be enlarged in order to add more data, the locations at the end of the existing array might already be allocated to other purposes. Because of this possibility, contiguous allocation is generally used only for arrays of fixed length.

For a variety of reasons it may be desirable to store individual elements of the array in widely dispersed storage locations. One common reason is to make it easier to expand or shrink an array. In such a situation, each element must be stored along with a pointer to the next element of the array. An example is depicted in Figure 4.5. The letters of the name "John Doe" are stored in noncontiguous locations, and each letter is followed by the address of the next letter in the array. (The empty box represents the space between "John" and "Doe.") In the figure, the addresses are depicted as arrows instead of numeric addresses.

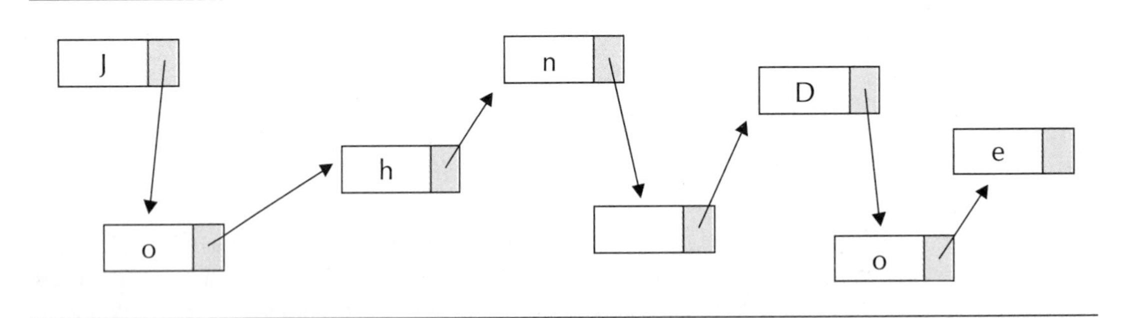

FIGURE 4.5 **A character array stored in noncontiguous memory locations.**

Note that the use of pointers increases the amount of storage needed for the array. In addition, it complicates the task of locating individual array elements. Instead of a simple address computation, as above, references to array elements must be resolved by following the chain of pointers, starting with the first array element. This process can be highly inefficient if the number of array elements is large.

The type of data structure used in Figure 4.5 is called a *linked list.* Linked lists can be used for many things besides the storage of arrays. Figure 4.6 shows a generic example of *a singly linked list.* The term is derived from the use of one pointer (or link) per list element.

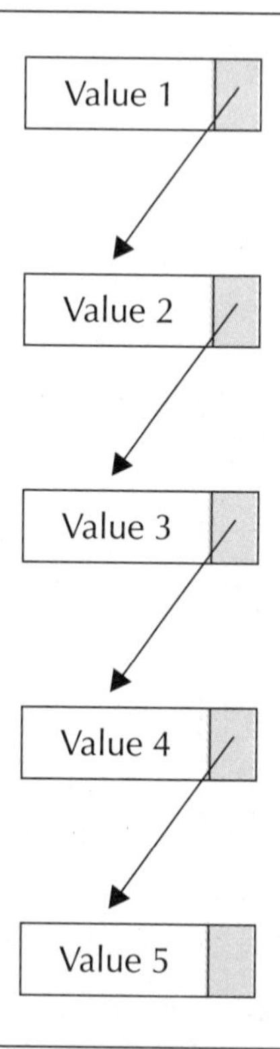

FIGURE 4.6 The value and pointer fields of a singly linked list.

Although access to individual elements in a linked list may be slow, the addition or deletion of elements is relatively fast. The procedure to add a new element is as follows:

1. Allocate storage for the new element.
2. Copy the pointer from the element preceding the new element into the pointer field of the new element.
3. Write the address of the new element into the pointer field of the preceding element.

An example of the procedure is depicted in Figure 4.7.

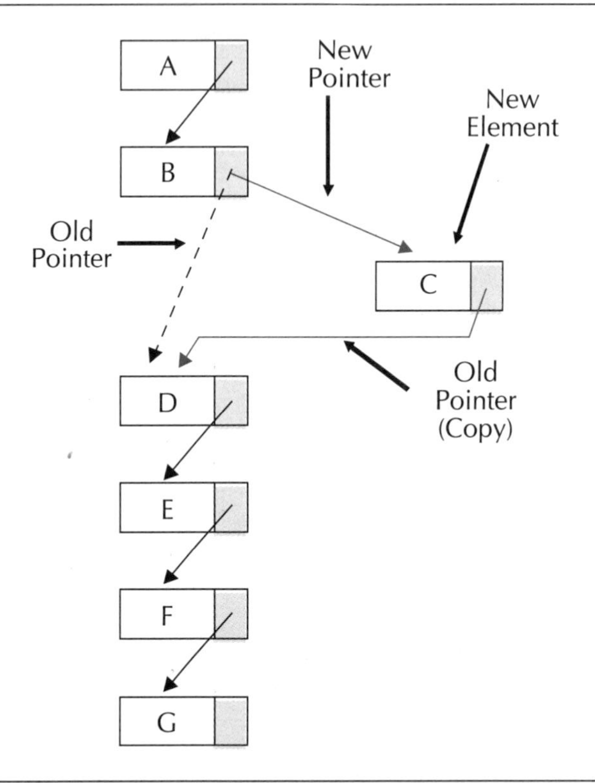

FIGURE 4.7 **The insertion of a new element into a singly linked list.**

In contrast, insertion of an element into a list stored in contiguous memory locations can be very time consuming. The procedure is as follows:

1. Allocate a new storage location to the end of the list.
2. For each element past the insertion point, copy the element value to the next storage location.
3. Write the new element value in the storage location at the insertion point.

The procedure is depicted in Figure 4.8. For insertion near the beginning of the list, efficiency is degraded by the large number of copy operations required.

Figure 4.9 depicts a more complicated form of linked list called a *doubly linked list*. As the term implies, elements of a doubly linked list have two pointers each. One pointer points to the next element in the list, and the other pointer points to the previous element in the list. The primary advantage of this storage method is that lists can be traversed in either direction with equal efficiency. The primary disadvantage is that more pointers must be updated each time an element is inserted into or deleted from the list. In addition, more storage locations will be required by the extra set of pointers.

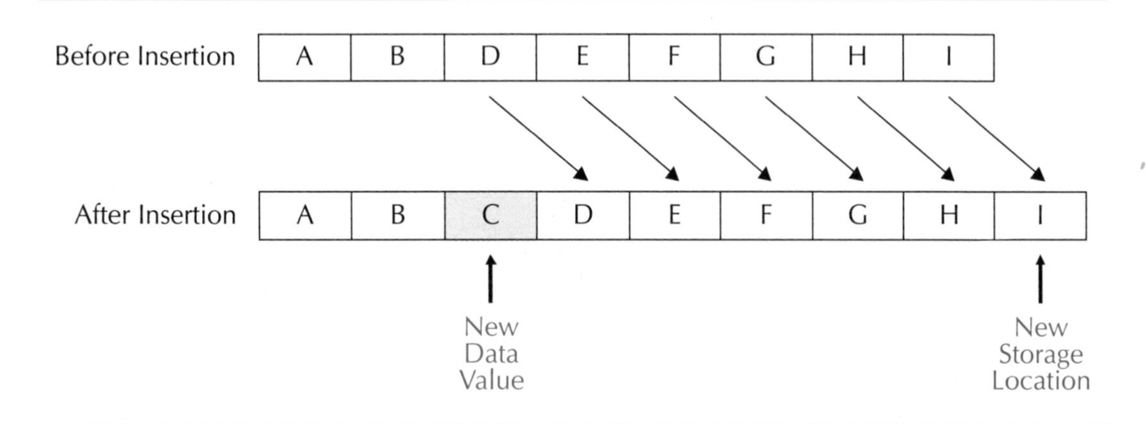

FIGURE 4.8 **The procedure for inserting an element into an array stored in contiguous memory locations.**

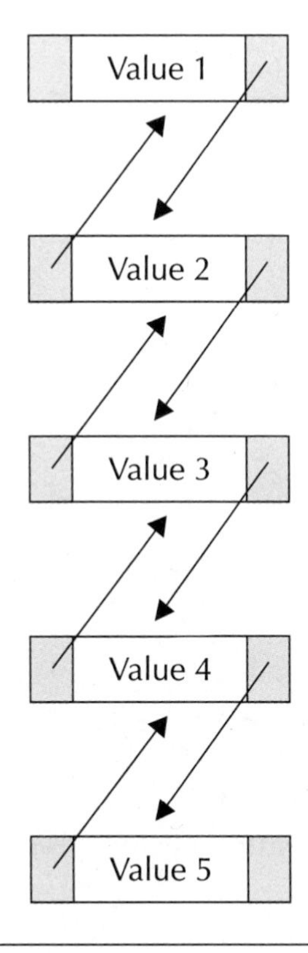

FIGURE 4.9 **A doubly linked list.**

Records and Files

A *record* is a data structure composed of other data structures or basic data types. Records are commonly used as a unit of input and output to files as well as for grouping related data components together into a single named unit. As an example, consider the following items of data:

Account-Number

First-Name

Last-Name

Middle-Initial

Street-Address

City

State

Zip-Code

This set of data might be the contents of the data structure for a customer record, as shown in Figure 4.10. Each component of the record is either a basic data element (such as Middle-Initial) or another data structure (such as a character array for Street-Address). To speed input and output, records are normally stored in contiguous storage locations. This restricts array components of the record to a fixed length.

Account-Number	Last-Name	First-Name	Middle-Initial	Street-Address	City	State	Zip-Code

FIGURE 4.10 A data structure for a record.

A sequence of records on secondary storage is called a *file.* A sequence of records stored within main memory is normally called a *table,* although the structure of the storage can be similar. Files can be organized in many ways, the most common being sequential and indexed.

In a sequential file, the records are stored in contiguous storage locations. As with arrays stored in contiguous storage, accessing a specific record is relatively simple. The address of the nth record in a file can be computed as:

$$\text{address-of-first-record} + ((n - 1) \times \text{record-size})$$

Thus, if the first byte of the first record is at address 1 and the size of a record is 200 bytes, the address of the fourth record is 601.

Sequential files suffer the same problems as contiguous arrays when inserting and deleting records. A copy procedure similar to that depicted in Figure 4.8 must be executed to add a record to the file. The efficiency of this procedure is even less for files than for arrays due to the relatively large size of the data (records) that must be copied.

One method of solving this problem is to use linked lists. With files, the data elements of a linked list are entire records instead of basic data elements. The methods for searching, record insertion, and record deletion are essentially the same as previously described.

Another method of organizing files involves the use of an *index,* which is an array of pointers to records. The pointers can be ordered in any sequence the user prefers. For example, a file of customer records might be ordered by ascending account number, as shown in Figure 4.11.

The advantage of using an index lies in the efficiency of record insertion, deletion, and retrieval. When a record is added to a file, storage is allocated for the record and the data placed therein. The index is then updated by inserting the address of the new record. The index update is accomplished by the same procedure as an array update. Because the array contains only pointers, it is relatively small in size and thus relatively fast to update.

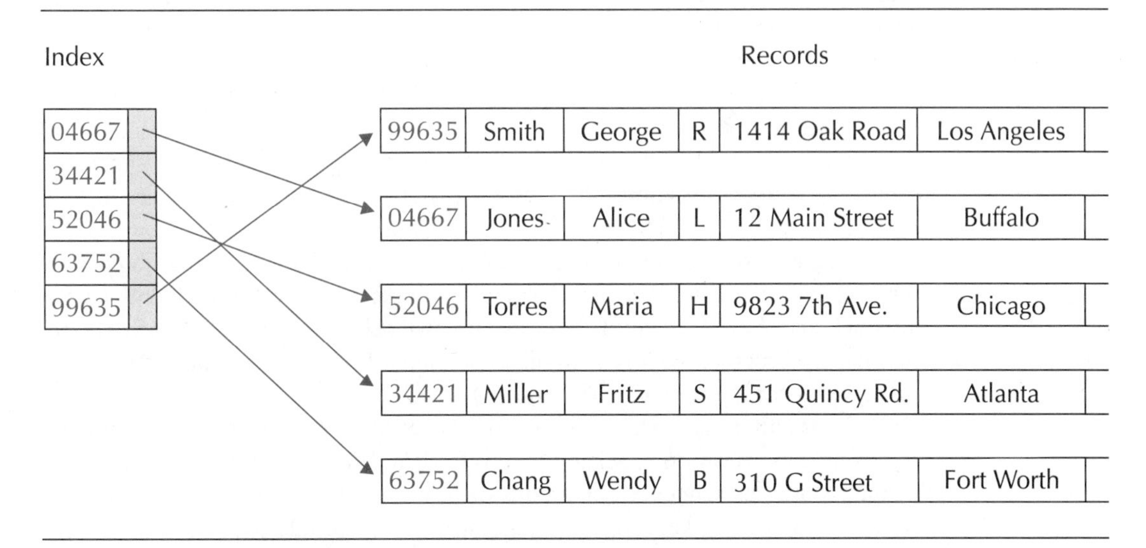

FIGURE 4.11 An indexed file.

SUMMARY

A computer is capable of manipulating data only if it is encoded in pulses of electricity that represent numbers. Thus, all data must be encoded numerically and converted into pulses of electricity for storage, transmission, or processing.

Data can be encoded in electricity using analog or digital signalling methods. Analog signals offer superior message-carrying capacity at the expense of high susceptibility to noise and interference. Digital signals offer superior reliability and are thus used exclusively within modern digital computers.

The electrical representation of numbers within digital computers is based on the binary number system. This method is chosen because it corresponds with clearly identifiable states of electron flow. Devices that transmit, store, and process binary signals are reliable and inexpensive.

The various data types that can be represented within a computer include integer, real, character, and Boolean data. Integers are whole numbers. Real data consists of numbers that can contain fractional components. Characters consist of printable symbols such as the letters of the alphabet. Characters can also include special control codes for controlling the behavior of I/O devices. Boolean data elements can have only the values "true" or "false."

A variety of methods, or coding formats, exist by which data can be encoded as binary numbers. These methods vary in ease of manipulation, conciseness, accuracy, and standardization among devices and computers. Various coding methods represent a range of tradeoffs between these characteristics and the cost of computer hardware.

Numeric data can be encoded in binary form using sign-magnitude, excess, or two's-complement notation. Two's-complement notation is the preferred method of encoding integer values because it simplifies processing circuitry within the CPU. Fractional numbers require a specialized representation method in which the whole and fractional portions of the number are stored separately.

Characters are encoded as binary numbers by assigning a unique number to represent each character. Various character-coding schemes exist, including BCD, EBCDIC, and ASCII. Each of these has its own relative advantages. ASCII, the most widely accepted coding method, is a standard method of data communication between computers and peripheral I/O devices.

Data structures are structured sets of related data items. They are useful in processing and transmitting related groups of data. Within a computer, data structures are stored as sequences or groups of primitive data items such as integers, characters, and the like. Common data structures include arrays, linked lists, records, and files.

Key Terms

absolute value
address
American Standard Code for Information Interchange (ASCII)
analog signal
analog-to-digital (A/D) conversion
array
binary-coded decimal (BCD)
binary signal
bit
bit position
bit string
block
Boolean data type
byte
coding scheme
collating sequence
communication channel
communication protocol
complementary arithmetic
control code
data structure
data type
digital computer
digital signal
digital-to-analog (D/A) conversion
discrete signal
double-precision representation
double-precision real representation
doubly linked list
excess notation
exponent field
Extended Binary-Coded Decimal Interchange Code (EBCDIC)
file
floating-point notation
index
integer
integer arithmetic
International Alphabet Number 5 (IA5)
least significant digit (low-order bit)
linked list
list
literal
long integer
mantissa field
most-significant digit (high-order bit)
overflow
pointer
positional number system
radix
radix point
record
sign bit
sign-magnitude notation
signal
signal protocol
singly linked list
string (character array)
table
threshold
two's-complement notation
underflow
validity checking

Vocabulary Exercises

1. _____ notation encodes a real number as a mantissa multiplied by a power (exponent) of two.
2. Integers are represented in _____ notation, and arithmetic operations on them are performed using _____ arithmetic.
3. A _____ is a data item composed of multiple primitive data items.
4. In IBM mainframe computers, characters are coded as numbers according to the _____ coding scheme. In most other computers, the _____ character coding scheme is used.

5. A _____ is the address of another data item or structure.
6. The _____ separates bit positions that represent whole number quantities from bit positions that represent fractional quantities.
7. A _____ is an array of characters.
8. A set of data items that are related in a specified order using a set of pointers is called a _____.
9. A _____ is a sequence of primitive data elements stored in sequential storage locations.
10. A ____ data item can contain only the values true or false.
11. A _____ signal can carry one of an infinite number of messages during one unit of time.
12. Many computers implement _____ numeric data types to provide greater accuracy and prevent overflow and underflow.
13. The _____ is the bit of lowest magnitude within a byte or bit string.
14. The value of a binary signal parameter is compared to the _____ value to determine whether it represents a zero or a one.
15. _____ occurs when the result of an arithmetic operation exceeds the number of bits available to store it.
16. Within a CPU, _____ arithmetic is generally simpler to implement than _____ arithmetic due to a simpler data coding scheme and data manipulation circuitry.
17. A _____ is a means of carrying a message between two points.
18. A _____ signal can carry only one of two possible messages during one unit of time.

Review Questions

1. How does a digital signal differ from an analog signal? What are their comparative advantages and disadvantages for data communication?
2. What is the binary representation of the decimal number 10? What is the octal representation? What is the hexadecimal representation?
3. Why is binary data representation and signalling the preferred method of computer hardware implementation?

4. What is sign-magnitude notation? What is excess notation? What is two's-complement notation? What criteria determine which notation is used by computer hardware?
5. How can subtraction, multiplication, and division be implemented using only addition?
6. What is overflow? What is underflow? How can the probability of their occurrence be minimized?
7. How and why are real numbers more difficult to represent and process than integers?
8. What are the differences between ASCII and EBCDIC?
9. What basic data types can normally be represented and processed by a CPU?
10. What is a data structure? List several types of common data structures. What software layer(s) support their definition and manipulation?
11. What is an address? What is a pointer? For what are they used?
12. How is an array stored in main memory? How is a linked list stored in main memory? What are their comparative advantages and disadvantages?

Research Problem

Choose a common, simple microprocessor such as the Intel 8086 or the Motorola 68000. What data types are supported? How many bits are used to store each data type? How is each data type internally represented? How are operations on real (floating-point) numbers implemented?

5 *Data Manipulation and Storage Technologies*

Chapter Goals

- *Describe the electrical components used to implement processors and primary storage devices.*
- *Describe the historical, current, and future implementation of processing and primary storage devices.*
- *Describe the use of magnetism and optics for secondary storage.*
- *Describe the components and capabilities of modern secondary storage devices.*

PROCESSING DEVICES

Chapter 4 discussed the representation of various types of data in terms of binary signals. Chapter 5 concentrates on the devices and technologies by which those signals are processed and stored. Recall that the CPU implements three basic types of instructions: computation, comparison, and movement of data. Processing devices for computation and comparison instructions are contained within the arithmetic logic unit (ALU) of the CPU. These devices perform their functions by performing one or more primitive transformations on individual binary signals (digits).

The primitive transformations that can be performed on binary signals were originally stated by George Boole. These include the operations NOT, AND, OR, and XOR. They form the basis of what is commonly called *Boolean logic*. The result of these operations for all possible binary inputs is summarized in Table 5.1. Note that each operation is applied to either a single binary digit or a pair of binary digits.

TABLE 5.1 **The results of logical operations on binary digits.**

Logical Transformation	Result
NOT 1	0
NOT 0	1
1 AND 1	1
1 AND 0	0
0 AND 1	0
0 AND 0	0
1 OR 1	1
1 OR 0	1
0 OR 1	1
0 OR 0	0
1 XOR 1	0
1 XOR 0	1
0 XOR 1	1
0 XOR 0	1

Typically, a CPU's instruction set is defined on larger groups of bits that comprise a character or number. The size of these bit groups ranges from 8 to 64 bits and beyond. Primitive operations on groups of bits can be performed by applying single-bit (or bit-pair) transformations on each bit position in parallel. Thus, to perform a NOT operation on a data value stored in eight bits, each bit is simultaneously input to eight separate devices that perform a NOT operation on a single bit. The outputs of these eight devices are then recombined in their original sequence to form an eight-bit result.

A single or parallel set of primitive transformation devices is all that is necessary to implement some CPU instructions. For example, the negation of an integer value stored in sign-magnitude notation can be implemented by applying the NOT operation to the sign bit. The parallel NOT operation described in the previous paragraph can be used to generate the complement of an integer. However, most CPU instructions must be implemented as combinations of primitive Boolean operations. For example, a greater-than (>) comparison on two-bit values (A and B) can be performed by combining the operations NOT and AND (NAND):

A AND (NOT B)

The result of this combined operation will be 1 (interpreted as "true") only if A = 1 and B = 0. All other combinations will produce a result of 0 (interpreted as "false"). More complex instructions and instructions applied on larger bit groups require substantially more complex combinations of primitive Boolean transformations.

Switches and Gates

The basic building blocks of computer processing circuits are electronic *switches,* or *gates.* In computer processing circuitry, sequences of gates are used to transform binary digital signals according to the patterns required to perform computation, comparison, or data movement operations. As shown in Figure 5.1, a computer system's CPU can be thought of as a set of logic gates that can be reconfigured on command to perform a variety of processing operations. Basic processing functions on binary digits are performed with the logical functions AND, OR, XOR, and NOT. Each of these functions can be implemented as a switch or gate, as discussed below. Processing circuitry is thus composed largely of groups of gates that represent these basic transformations.

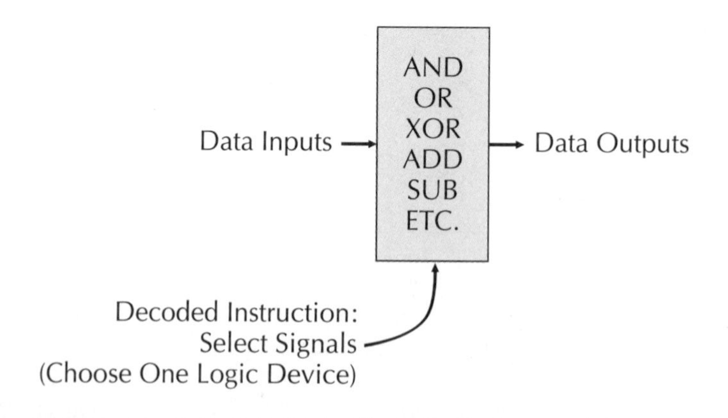

FIGURE 5.1 **The operation of a CPU. The instruction input selects the set of circuits through which data inputs will be passed to generate data outputs.**

A basic type of logic gate merely inverts the value of the signal that is presented to it, as shown in Figure 5.2a. This gate, or signal inverter, would transform a value of 0 into a 1, and vice versa. Logically, the gate can be used to represent a transformation of A to $\overline{A}$, or NOT A. Thus, a signal inverter performs the electronic equivalent of a logical NOT statement.

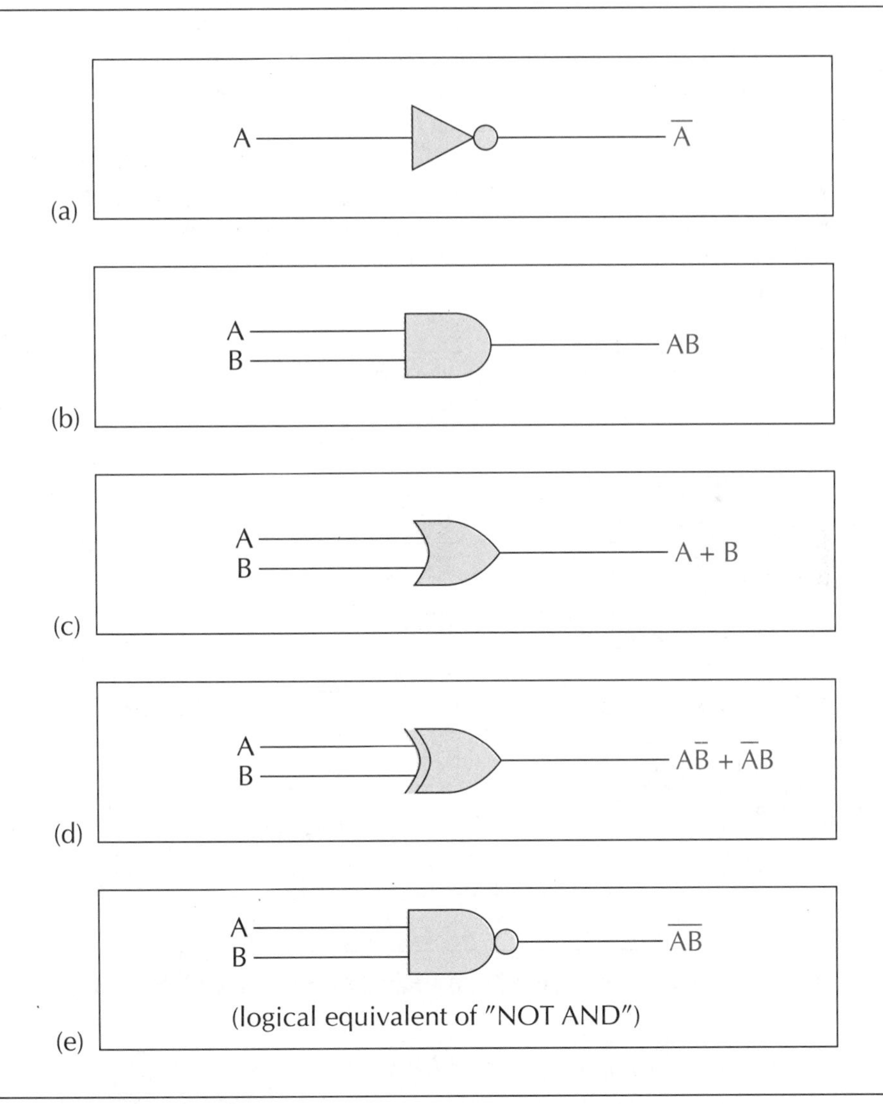

FIGURE 5.2 **Components and current flow for logic gates to implement (a) a signal inverter, (b) an AND function, (c) an OR function, (d) an XOR function, and (e) a NAND function.**

As shown in Figure 5.2b, another type of gate combines two incoming data values. At this gate, separate inputs of A and B values are transformed to become AB, or A AND B. This type of gate, therefore, is known as an AND gate. Another type of gate can be used to select one of two values presented to it, as shown in Figure 5.2c. At this OR gate, inputs of A and B will generate an output of A OR B. A different result is produced by an exclusive-OR, or XOR, gate. For inputs of A and B, this gate produces A AND $\overline{B}$ (NOT B), or $\overline{A}$(NOT A) AND B, as shown in Figure 5.2d.

A type of gate that combines the logical transformations of NOT and AND is shown in Figure 5.2e. Inputs of A and B are transformed into A$\overline{B}$ (A AND NOT B), or $\overline{A}$B (NOT A and B). The actual implementation of the gate in electronic circuitry combines both functions in a NAND gate. Thus, electronic circuits that process digital signals can be built by arranging logic gates in the proper sequence for producing a desired type of operation. For example, the circuits shown in Figure 5.3 are adders, or combinations of logic half-adder and full-adder gates that are used to perform integer addition in the ALU.

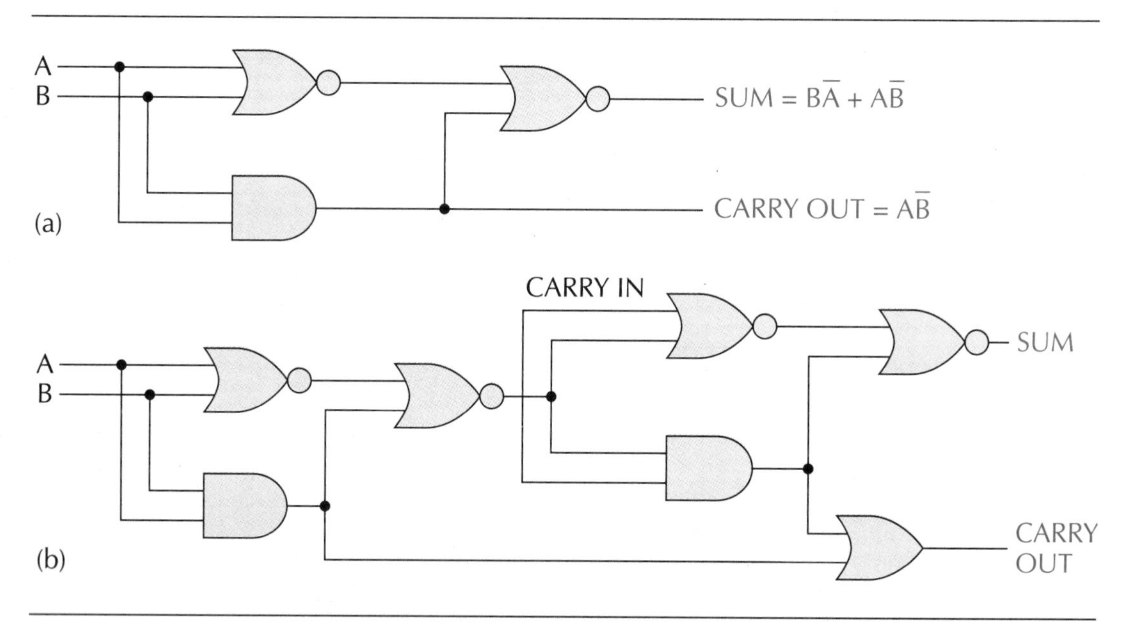

FIGURE 5.3 **Configurations of logic gates in (a) a half-adder circuit and (b) a full-adder circuit.**

Electrical Properties and Devices

The logic gates and memory cells in computer hardware are implemented with bistable electronic devices. A bistable device has only two stable states, or conditions, and may be switched on command from one state to the other. Thus, bistable electronic devices are ideal for performing the binary switching required to process digital signals.

Electrical devices all share common problems and capabilities associated with basic characteristics of electricity. These include

- Speed
- Resistance
- Heat generation

When electricity is used for processing functions, the flow of electrons is harnessed to perform useful work. The flow of electricity and the routes it takes through electrical circuits perform the electrical equivalent of moving the gears and wheels of mechanical computation devices. As with mechanical devices, there is a limit to the speed of movement (electron flow) and, thus, an upper bound on the speed of any device that functions as a result of electrical flow.

An electrical circuit is a set of pathways over which electricity can flow. Circuits are composed of devices that use electrical energy to perform work (e.g., switches) and connections between those devices (e.g., wires). The speed of electricity through a circuit is constant and depends on the length of the circuit and the materials of which it is composed. The amount of electricity that flows through a circuit can be increased (within limits), but the time it takes electrons to move through the circuit cannot. The length of the circuit is fixed and, thus, the time it takes for the circuit to perform its function is constant. The only way this time can be reduced is to reduce the size of the circuit.

Much as mechanical devices experience friction between moving parts, electrical devices experience friction between electrons and the pathways that conduct them. The common term for this phenomenon is *resistance.* The amount of resistance in a circuit is dependent on temperature, circuit length, and the material(s) of which the circuit is constructed. Resistance in an electrical circuit causes the amount of electricity that flows to be reduced. The lost power is converted to heat.

Because all materials have some level of resistance, all electrical devices produce some level of heat. If this heat is not dissipated, the increased temperature within the circuit could create additional resistance. This in turn generates additional heat, and a vicious cycle is thus created. In sufficient quantities, heat can reduce the flow of electrons to nearly zero or can cause the physical breakdown of electrical conductors and devices (e.g., a "blown" fuse). These problems are especially prevalent when high-resistance conductors are used or when electrical circuits are placed in close proximity to one another.

The problems of heat and resistance can be addressed in three ways:

- Low-resistance materials
- Auxiliary cooling
- Reduced circuit size or packing density

Because conductive materials have various levels of resistance, one way to minimize heat is to use low-resistance materials, which reduces the power consumption of an electrical circuit and increases its reliability. However, all materials have at least some degree of electrical resistance, so other methods of minimizing resistance must also be considered.

Auxiliary cooling can be used in two ways to reduce heat-related problems. The first is simply to cool the circuits in order to dissipate heat. This is, in essence, air conditioning for electrical circuits. Various methods of cooling include the use of heat sinks, vents, fans, and refrigeration. Most computer equipment employs one or more of these methods.

The other use for auxiliary cooling is to reduce the resistance within electrical circuits. Most materials have less resistance at lower temperatures. This phenomenon is especially noticed at extremely low temperatures. At such temperatures, many materials become *superconductors,* in which resistance is virtually unmeasurable.

Reducing circuit size is another method of minimizing heat-related problems. Because the total resistance of a circuit increases with its length, total resistance (and its related heat generation) can be reduced by reducing circuit size. However, the reduction in heat is offset if the smaller circuits are packed more tightly together.

The evolution of electronics and computer hardware technologies has been marked by a series of transitions in the types of components and materials that were used as logic gates and memory cells. Early components were relatively large and constructed of relatively high-resistance materials. They were slow, consumed large quantities of electricity, generated a large amount of heat, and were prone to frequent failure.

Performance and cost breakthroughs have been achieved as electronic components and circuits have become progressively smaller and as the materials of which they are constructed have been improved. Major milestones in the evolution of computer electronics include the following:

- Vacuum tubes
- Transistors
- Integrated circuits (ICs)
- Microchips
- Microprocessors

Vacuum tubes. *Vacuum tubes* and relays, or electro-mechanical switches, were used as logic gates in early computers. As shown in Figure 5.4, vacuum tubes are similar in construction to conventional light bulbs. They contain a filament, a cathode, and an anode. The filament is composed of a high-resistance material that heats up as electricity flows through it. The heat generated by the filament is transmitted to the cathode, enabling it to emit electrons into the vacuum.

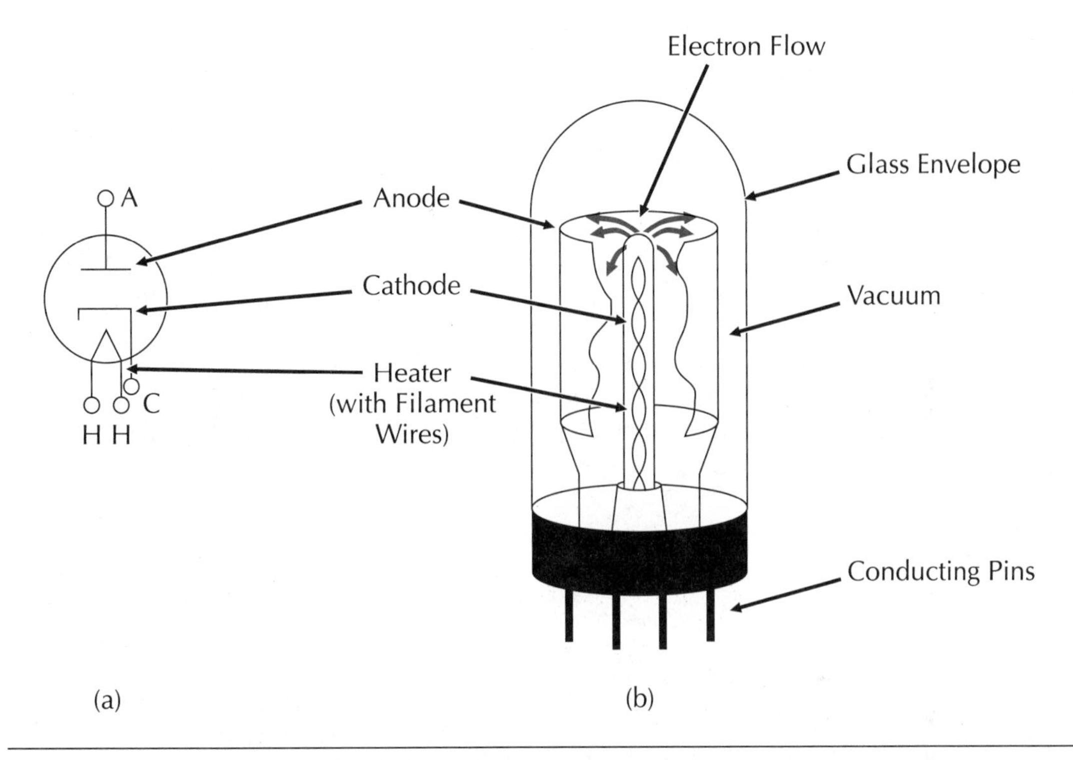

FIGURE 5.4 Symbolic representation (a) and a functional diagram (b) of a vacuum tube and its components.

The anode is a metal plate. If a positive voltage is applied to this plate, electrons are drawn to it from the cathode and can flow out of the vacuum tube through connecting wires. If a negative voltage is applied to the anode, electrons are repelled and no electricity flows out. Thus, the basic action of this type of vacuum tube is that of a switch. The switch is on when positive voltage is applied to the anode and off when negative voltage is applied.

A major problem within early computers was the heat generated by their vacuum tubes. The ENIAC, perhaps the first large-scale electronic computer, was built in 1946 and contained about 19,000 vacuum tubes. The machine filled a room and had to be cooled continually by heavy-duty refrigeration equipment. Due to the destructive effects of heat, computers with vacuum tubes were prone to frequent failure. They were also quite slow due to the relatively long paths over which electricity had to travel.

Transistors. In 1947, research engineers at Bell Laboratories discovered a class of materials called *semiconductors.* The electrical conducting properties of these materials vary in response to the electrical inputs applied. *Transistors* are made of silicon or germanium that has been treated (or doped) with chemical impurities to enhance these semiconducting effects. Both silicon and germanium are basic elements with resistance characteristics that can be tailored to an application through use of chemicals called dopants.

As shown in Figure 5.5, transistors can be made to function as switches. The emitter and the collector of a transistor are similar in function to the cathode and the anode of a vacuum tube. Transistors and other semiconductors are called solid-state electronic devices because a solid material, called the *base*, is used as the medium through which electrons are transferred from emitter to collector. Solid-state devices consume less electricity and generate far less heat than do vacuum tubes. Partly for these reasons, the introduction of transistors as logic gates in computers in the 1950s greatly increased their reliability. The smaller size of the transistor also helped to diminish bulk and increase computer operating speeds.

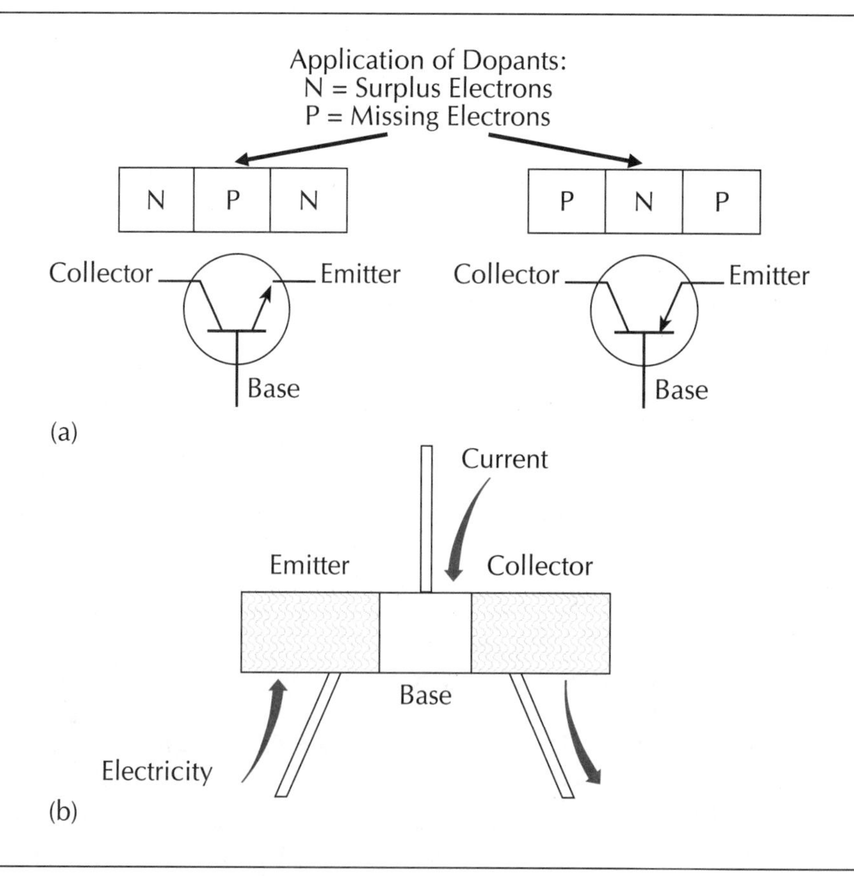

FIGURE 5.5 **The flow of current through the components of a transistor.**

Integrated circuits. In the early 1960s, techniques were developed for fabricating miniature electronic circuits from multiple layers of materials that have differing electrical properties. These layers are built up on small, thin chips. Some of these devices are named *metal oxide semiconductors (MOS)*, a term derived from the materials used in each of three layers on the chip.

MOS technology made it possible to implement several logic gates and their interconnections on a single chip to form an *integrated circuit (IC)*. One of the first ICs, shown in Figure 5.6, was a "quadruple two-input positive NAND buffer," which contained four NAND gates.

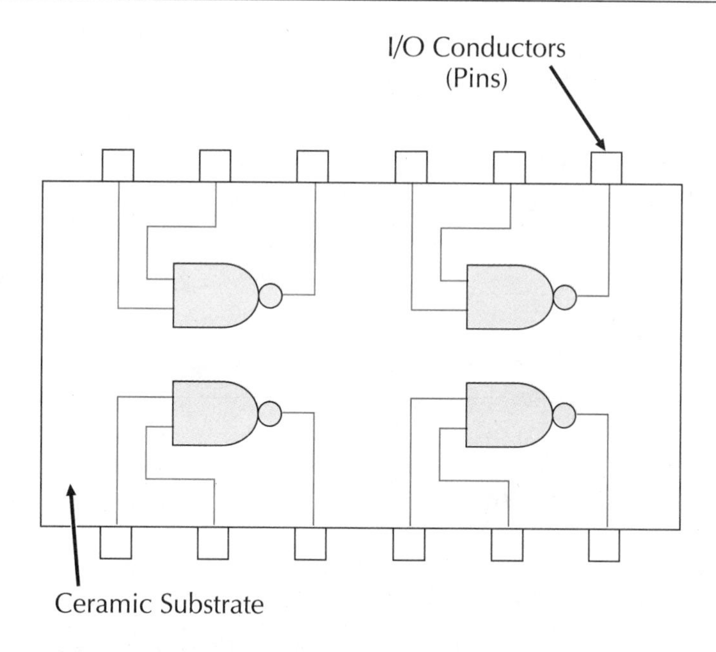

FIGURE 5.6 **An early integrated circuit (IC): a quadruple two-input positive NAND buffer.**

Integrated circuits represent advances in reduction of manufacturing cost, space utilization, processing performance, and reliability. Costs are reduced because many chips can be manufactured in a single sheet, or wafer. The wafer is then cut apart and chips are mounted on dual in-line packages (DIPs) or chip carriers with dual rows of pins for interconnection with larger circuits on printed circuit boards (PCBs). Combining multiple gates on a single chip also reduces the manufacturing cost per gate and creates a compact, modular package that simplifies circuit design and manufacturing. Processing speed is improved as the physical space between gates decreases. A chip is more reliable than a circuit that is built of separate, or discrete, components, because its circuits are sealed within a single package.

Integrated circuits also marked a breakthrough in the cost of computer memory. With transistor technology, providing storage locations within processing hardware was extremely expensive. A typical memory cell for storing a single bit required two transistors in a *flip-flop circuit,* as shown in Figure 5.7. However, with IC technology, it became possible to combine many memory cells on a single chip, greatly reducing the storage cost per bit.

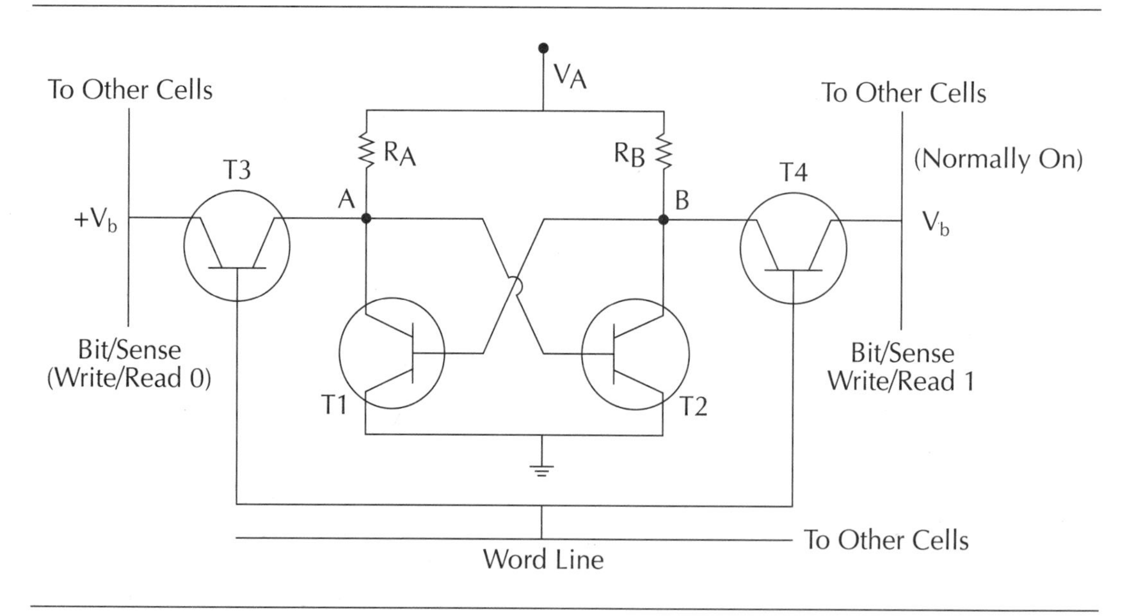

FIGURE 5.7 **A flip-flop circuit with four transistors of the type used as a memory cell in computers of the 1950s.**

Microchips. Further economies were realized as it became possible to put increasing numbers of gates and/or memory cells on IC chips. Early integrated circuits housed 4, 10, then 100 gates on a single chip. Later generations of devices that contained 100 to 1000 gates came to represent medium-scale integration (MSI). At the levels above 1000 gates, a chip is said to implement large-scale integration (LSI). More recently, chips with more than 10,000 gates, representing very large-scale integration (VLSI), have become possible. As the size of these integrated circuits continues to shrink, processing speed and storage capacity increase. At these higher levels of integration, the term *microchip* was coined to refer to this new class of electronic devices.

One impact of VLSI has been to greatly reduce the cost per bit of electronic random access memory (RAM). Groups of memory cells in RAM are organized to hold data in eight-bit bytes. By the mid- to late 1980s, it became possible to put enough memory cells on a chip to achieve 1,024,000 bytes of RAM. These one-megabyte (Mbyte) microchips cost about the same to manufacture as 64K chips that had appeared a decade earlier.

As VLSI technology increased the packing density of electronic components, the problem of heat dissipation intensified again. Even though individual devices on a microchip generate relatively little heat, their number and close proximity led to problems with heat dissipation. As a result, some current models of large mainframes incorporate their own cooling systems.

Another application of the microchip was to implement all the gates, memory cells, and interconnections for the CPU of a small computer on a single chip about the size of a thumbnail. The invention of the *microprocessor* in 1969 is attributed to Dr. Edward Hoff at Intel Corporation.

Microprocessors ushered in a completely new era in computer design and manufacture. The most important part of a computer system could now be produced or purchased as a single package. This negated the need for computer designers to construct processors from smaller components and, thus, simplified the task of computer system design. It also led to a degree of standardization as a few specific microprocessors became widely used. The revolution of compatible microcomputers would not have been possible without standardized microprocessors.

Future Trends

A potential replacement for silicon in microchips is *gallium arsenide*. Because of its molecular structure, it is possible to fabricate much smaller electrical circuits with this substance than with silicon. Thus, chips with larger numbers of circuits can be manufactured within the same size constraints. The speed of these devices is also faster due to the shorter distances over which the electricity must travel.

Although gallium arsenide is a promising material, it has yet to see wide application. It is substantially more difficult to fabricate electrical components with gallium arsenide than with silicon, partly because it is more brittle. In addition, materials scientists and engineers have gained a substantial amount of experience in the production of silicon-based devices over the past 30 years. It will take some time before enough experience is gained with gallium arsenide to reach the production yields and costs of silicon today.

Much as electrons flow across a conductive pathway, so too can light flow across an optical pathway. The science of fiber optics and optical switching has already been used with much success in communication devices. Its next logical application is to more complex switching devices and, thus, to computer processors. Because light flows at a much higher rate than electricity, the theoretical upper bound for optical processing speed is substantially greater than for electrical processing. Light can travel in a straight, focused path without the benefit of conductive pathways such as wires. Because of this, it is possible to construct processors in which individual components do not have to be directly "wired together."

Gallium arsenide has both electrical and optical properties. As stated earlier, gallium arsenide is a potential replacement for silicon in conventional semiconductors (microchips). However, it may also be used to implement devices that combine aspects of electrical and optical processing, such as opto-electrical devices. For example, gallium arsenide is capable of switching the flow of electricity based on optical input. In essence, it can be used to implement a transistor that allows electrical current to flow when an optical input is applied. Thus, gallium arsenide devices can form a bridge between optical data communication and electrical processing.

As described in later portions of the text, optical technologies are already in use for data communication and storage. In comparison, the field of optical processing is in its infancy. However, a considerable amount of resources is being invested in research and development of optical and opto-electrical processing devices. Thus, it is likely that optical processing will become a commercial reality at some point in the future.

STORAGE ALTERNATIVES AND TRADEOFFS

Because of the relatively high cost of the semiconductor devices that are used as memory cells, memory capacity, though growing, has finite limits. By contrast, mass-storage devices, including magnetic tape and disk, provide far greater storage capacities and relatively permanent retention of data or programs. Although memory and mass storage are usually considered distinct classes of devices, there are overlaps in their functionality.

Some types of semiconductor memory provide ongoing, or relatively permanent, storage of programs or data. Furthermore, the cost of memory devices is dropping to a point at which even the memories of microcomputers are measured in megabytes. Distinctions are blurred further by the use of mass-storage devices as extensions of memory within virtual memory systems.

Despite these changes, semiconductor memory remains comparatively more costly per unit of data than other storage methods. If the focus of attention is the main memory of a computer, this cost premium can be justified because of the high data access speed of semiconductors. The fast input/output offered by semiconductor memory cells is needed to make the most productive use of processor time. However, at some point, expanding main memory space will become impractical. Paging or swapping programs or data between semiconductor memory and a mass-storage device can provide equivalent performance at lower cost.

The data storage capacity of a computer system (including memory) is a major architectural consideration. Alternatives represent various levels of cost, capacity, and data access speed. As a reference for design decisions, the alternatives can be arranged in a *memory/storage hierarchy*, as shown in Figure 5.8.

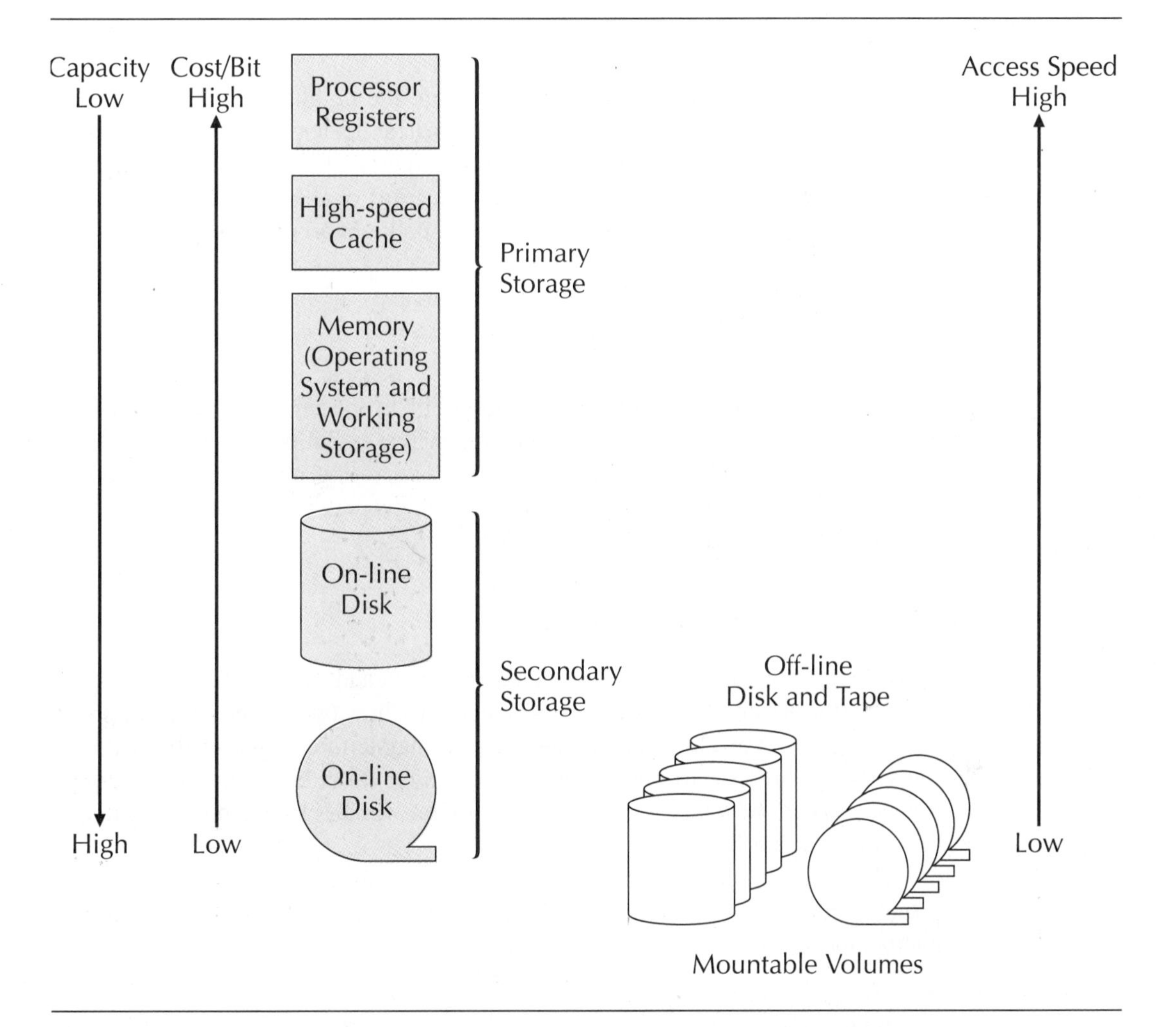

FIGURE 5.8 **Various storage options within a computer system compared in terms of cost, capacity, and access speed.**

From the viewpoint of the computer hardware designer or integrator, memory and storage represent a common pool of data retention capacity for the processor. Implementation tradeoffs vary with intended use requirements at multiple levels. For processor support, the main requirement is data access speed. At the opposite extreme, long-term storage of large amounts of data demands low cost.

The options shown in Figure 5.8 represent various tradeoffs of cost and performance. Memory/storage devices require capacity and data access speed. The total storage area for a given system will represent a mix of devices that attempts to balance cost and performance. This mix will be the memory/storage hierarchy for the system.

In designing a memory/storage hierarchy, the computer system designer tries to provide the data access speed of the fastest device (cache) at the cost of the least expensive device (tape). This objective can be achieved by optimizing the capacity and type of storage at each level of the hierarchy. Tradeoffs include the cost of each option, performance, and overall efficiency.

PRIMARY STORAGE DEVICES

The critical characteristic of primary storage devices is their access speed. Because they must supply data and instructions to the processor, they must be capable of matching the speed of the processor. The evolution of memory devices has, thus, closely mirrored the evolution of processing devices. Technologies applied to the construction of processors have, in general, been applied simultaneously to the construction of primary storage devices.

Electronic Memory Devices

Early computers used rings of ferrous material (iron and iron compounds) to store bits of information. These rings were embedded in a two-dimensional mesh of wires. Signals sent along the wires induced magnetic charges in the metallic rings. The polarity of the charge was dependent on the direction in which current flowed through the wires. Thus the value of a bit was set depending on the direction of the current flow.

Modern computers use memory composed of flip-flop circuits that are implemented with semiconductors. Types of memory that are built from semiconductor microchips include

- RAM
- Bubble memory
- ROM

RAM. *Random access memory (RAM)* usually constitutes the bulk of a computer's main memory. Devices commonly used for this purpose include *complementary metal oxide semiconductors (CMOS)*, which contain arrays of flip-flop circuits that function as memory cells. Groups of memory cells are assigned specific numeric addresses that correspond with storage locations within the memory array.

This type of memory is said to be *volatile,* because data content can be changed dynamically and rapidly as processing is under way. This volatility is both an advantage and a disadvantage. The advantage is that volatile memory can serve as a high-speed "scratchpad" for the processor's calculations. Each address in RAM can be accessed with equal speed. The relatively fast I/O of CMOS devices is needed to keep pace with processor instruction/execution cycles. Because it supports the processor directly, RAM is often called *working storage.* The main disadvantage of volatile memory is that data content is retained only as long as electrical power is applied to semiconductor devices. When power is turned off, the data content of RAM is lost.

The most widely used materials for CMOS RAM chips are silicon and germanium. At this writing, four- and 16-megabyte RAM chips are commercially available, and 64-megabit chips are in the final stage of research and development. As with most other types of electronic technology, rapid advances in size and reductions in cost are expected in the future.

The access speed of current CMOS RAM is generally in the range of 20 to 200 nanoseconds (ns). Cost rises nonlinearly as a function of speed, such that 20 ns devices may cost 50 times as much as 200 ns devices. Current processor speeds range as high as 100 megahertz (MHz). The speed of the memory devices should match the speed of the processor, yet current processors demand memory speeds that require the fastest and most expensive memory devices available. Providing the amount of memory needed by current hardware and software can be prohibitively expensive.

When memory access speeds are slower than processor speed, the processor must remain idle until memory completes the access. A processor cycle spent waiting for memory (or other device) access is called a *wait state.* Wait states represent wasted processor resources and, thus, are to be avoided whenever possible.

Bubble memory. Microchips that use *bubble memory* technology are similar in function to RAM chips. However, data is held within bubble memory indefinitely, even in the absence of electrical power.

The physical structure of a bubble memory device is shown in Figure 5.9. The device has a substrate of semiconducting crystal made of synthetic garnet. This crystal has been impregnated with tiny pockets, or bubbles, of magnetic material. As electrical current is applied to a region of the device, the bubbles in that region migrate to a different position. The positions of magnetic bubbles correspond with data values of 0 or 1. Bubble positions remain unchanged until subsequent electrical signals are applied. Thus, data is retained permanently, even in the absence of electrical power.

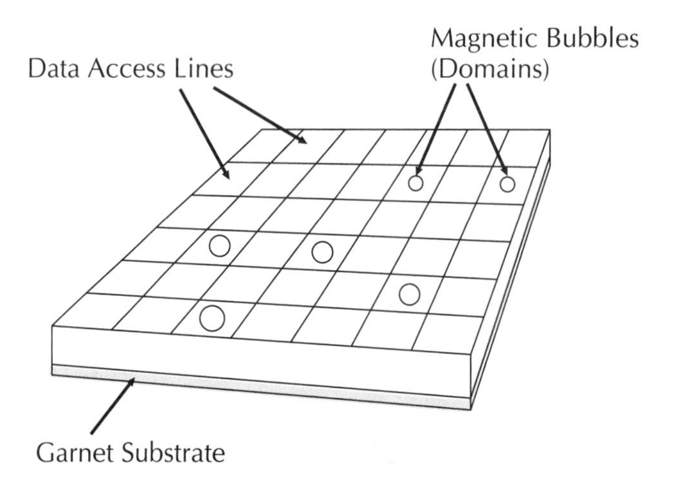

FIGURE 5.9 **The physical structure and operation of a bubble memory device.**

Despite optimistic predictions by some industry observers, use of bubble memory devices for auxiliary storage remains limited. At one time, it was predicted that the cost of these devices would fall rapidly and that bubble memories would replace CMOS devices in RAM implementations. However, the cost of silicon germanium CMOS chips also has fallen, and much more rapidly than bubble memory devices. So, bubble memory devices remain more costly than silicon germanium RAM chips.

In some applications, the relative permanence of data retention within bubble memory is a benefit that justifies its higher cost. For example, many briefcase-sized, portable microcomputers have bubble memory RAM so that data can be retained there, with power off, until the user has the opportunity to upload its content to another computer or storage device. This is also a convenience feature, because programs can remain in main memory and need not be reloaded at the beginning of each work session.

ROM. Microchips within which data is embedded permanently are called *read-only memory (ROM)*. Thus, ROM is a type of nonvolatile memory. A common application of ROM chips is to store sequences of instructions. These instructions may be loaded at high speed into main memory from ROM for execution by the processor. Instructions that reside in ROM are called firmware. In general, ROM content is fixed during manufacturing and cannot be altered.

Another type of microchip, a *programmable read-only memory (PROM)*, is delivered to the system integrator or application programmer as a functional "blank slate." Instructions or data can be transferred to the PROM electronically. This process is called "burning in" the PROM and usually takes many minutes of repeated data transfers. Once data has been burned into the PROM, it cannot be altered.

An *erasable programmable read only (EPROM)* memory chip is identical in function to a PROM, except that its content can be modified. On an EPROM microchip, a tiny glass window is mounted above the chip on the dual-inline package (DIP) carrier, as shown in Figure 5.10. Through this window, the chip can be exposed to a strong ultraviolet light, which, after a period of minutes, erases data content. Once erased, the EPROM can be reprogrammed in another burning-in operation. After this step, the glass window is usually covered with a paper label or sticker to prevent stray light from striking the chip and altering the data.

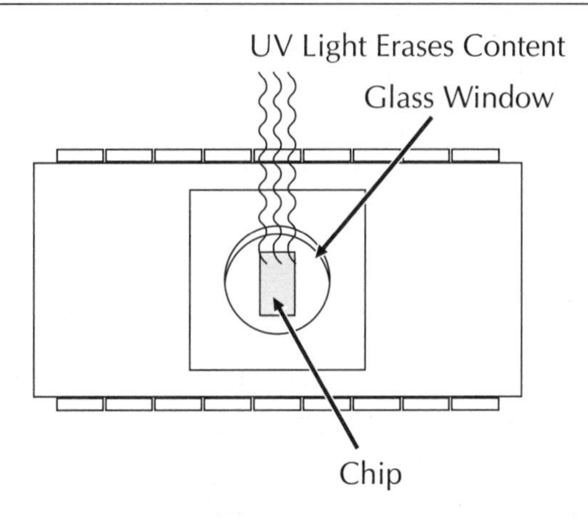

FIGURE 5.10 **An erasable programmable read-only memory (EPROM) chip includes a transparent window that permits its contents to be erased with ultraviolet light.**

A more recent development is the *electronically erasable programmable read-only memory (EEPROM)* chip. EEPROMs can be programmed, erased, and reprogrammed under processor control. Once the EEPROM is programmed, instructions are retained indefinitely or until an erasure triggering signal is sensed.

A common application of EEPROMs is to implement "soft" keyboards. Under this approach, the function of each key on a keyboard is defined within a lookup table. The table resides on the keyboard printed circuit board in EEPROM. As new programs are loaded into the host system, keyboard functions are downloaded to the EEPROM. Thus, keyboard functions can be customized for each application and program.

MASS-STORAGE DEVICES

The earliest computer had no mass-storage devices (in the modern sense of the term). Data to be stored was output onto punched cards or tape. These were stored until needed again, at which time they were read back into the computer.

As computers became faster and more powerful, this method of data storage became too cumbersome.

Modern computers use either magnetic or optical mass-storage devices. Magnetic devices are the more common of the two, although optical devices are rapidly gaining popularity. Both of these types of devices are described in the following sections.

Magnetic Mass Storage

As shown in Figure 5.11, magnetic data recording is based on the principle that individual particles of chargeable (usually ferrous) material will become aligned, or polarized, along the lines of force of a nearby electromagnetic field. A surface that is coated with chargeable particles is called a recording medium. Polarization can be negative or positive, conditions that can correspond with bit values of 0 or 1.

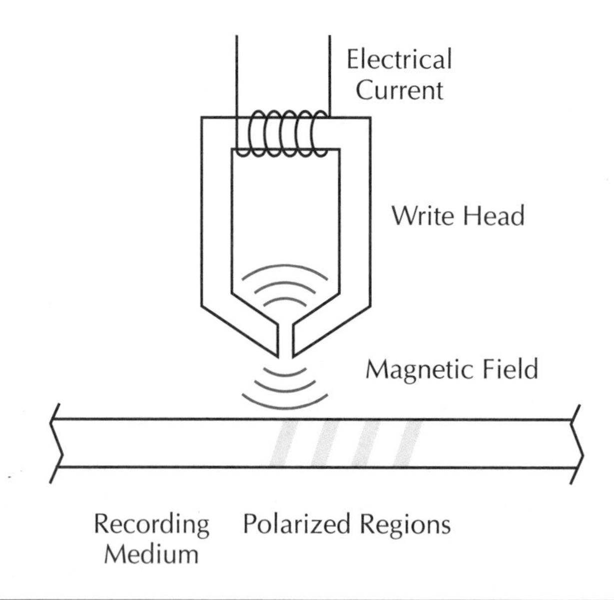

FIGURE 5.11 **The principle behind magnetic recording involves applying current to an electromagnet. The resulting magnetic field polarizes particles embedded in a coating applied to a plastic or metal surface.**

The process of polarizing a region to correspond with a given bit value is called a *write operation*, and the field-generating device is called a write head. Once polarized, a region will retain its magnetism indefinitely or until it is altered by being passed through a magnetic field of opposite polarity. Thus, magnetic data recording is relatively permanent.

In a *read operation*, the reverse of the recording process is used to sense data on previously recorded magnetic media. Regions that have been polarized magnetically will induce electrical changes in sensing circuits through which they are passed. In magnetic data storage devices, the recording medium is passed or rotated beneath a read head. Electrical currents induced in the head are sensed as bit values. Typically, the functions of writing and reading are combined in a single device called a *read/write head.*

Magnetic storage devices must all compensate for some basic (undesirable) characteristics of magnetism. These include the tendency of magnetism to decay and leak, the amount of charge required for reading and writing, and the environmental conditions required to maintain charge and media integrity.

Magnetically charged particles tend to lose their charge over time. This decay is relatively constant and proportional to the power of the charge. The read head requires a minimum, or threshold, level of magnetic charge for successful operation. Thus, over a sufficient length of time, magnetic decay will cause the amount of charge to fall below the threshold and, thus, data will become unreadable.

To counter this problem, data bits must be written with sufficient charge to compensate for decay. Thus, write operations typically charge bit areas of the medium at a substantially higher level than is required for proper reading. As long as the charge does not decay below the threshold for reliable reading, the data will still be recoverable. Magnetic charges also tend to leak into nearby chargeable areas. This phenomenon can be clearly heard in audio recordings made on analog magnetic tape.[1] Two physically adjacent areas of opposite charge will tend to cancel one another out over time due to this effect.

Various materials have different capacities to hold magnetic charge. In general, metals and metallic compounds offer the highest "chargeability" at reasonable cost. The total amount of charge that can be held is directly proportional to the mass of chargeable material. In magnetic recording media, this mass is a function of the surface area in which a bit is stored. Larger areas (higher mass) are capable of holding higher amounts of charge.

Most users want to store as much data as possible on the medium. The simplest way to do this is to reduce the amount of surface area used to store a single bit value, thus increasing the total number of bits that can be stored. The amount of surface area allocated to a bit is referred to as the *recording density.* Recording density is typically expressed (measured) in bits, bytes, or tracks per inch (generally abbreviated as bpi, Bpi, and tpi, respectively).

[1] Try the following experiment. Find a phonograph record or analog audio tape containing music that was originally recorded on analog master tape (almost any music recorded before 1980). Start the playback in the blank section preceding a musical selection with the volume turned up fairly high. You will hear the song "start" at a very low volume a few moments before it actually "starts" at full volume. This effect is due to the leakage of magnetic charge between portions of the master tape that were wound upon one another on the tape reel.

In some magnetic devices recording density is fixed, but in others, such as tape, it is variable. However, reductions in area per bit reduce chargeable material per bit, so the problems of magnetic decay and leakage are increased as recording density is increased. Designers of magnetic media and devices must find a balance between high recording density and the reliability of the media (especially over relatively long periods of time).

Magnetic tape. Magnetic tapes are ribbons of acetate with surface coatings in which particles of metal oxide are suspended. Until recently, tapes used on minicomputers and mainframes were 0.5 inch in width and were mounted on open reels. Tape length on reels varies between 300 and 2400 feet. A typical size is 1200 or 2400 feet of tape mounted on a 10-inch reel.

Open reel tapes are being rapidly replaced by tapes that are packaged in plastic cartridges or cassettes. Tape widths of cartridges vary between 0.125 and 0.5 inch. In general, cartridge tapes are easier to store and to mount or dismount on a tape drive unit. Tape reels and cartridges are mounted manually on tape drives, which turn the reels to unwind the tape and pass it beneath a read/write head (Figure 5.12).

The speed of tape travel, or transport speed, of tape drives ranges from 75 to 200 inches per second. A tape drive is called a *sequential access device* because data records must be accessed in the physical sequence in which they are recorded. For example, to find the data record that is in the nth position on the tape, it is necessary to search sequentially through the preceding $n - 1$ records.

The advantages of magnetic tape are high storage capacity and low cost. The storage capacity of a given tape device depends on the length of the tape and the recording density of the tape and device. Typical recording densities for 0.5-inch magnetic tape include 200, 556, 800, 1600, 3200, and 6250 bytes per inch (Bpi). Current maximum recording densities range as high as 36,000 Bpi.

Tape drives have the slowest data access speeds of all mass-storage devices. Accordingly, tapes are used for low-demand applications and are appropriate media for keeping backup and archival copies of programs and data files. Tape files also are used for sequential processing applications in which I/O speed is not critical to system performance.

The data-recording pattern for 0.5-inch magnetic tape is shown in Figure 5.13. Bits are recorded in nine parallel channels. Each column can hold one data bit in each of eight channels, representing one byte or character. The ninth position is assigned to a parity bit.[2] Each channel in this parallel recording pattern is called a *track.*

[2] The concept of parity checking and parity bits is fully discussed in Chapter 7.

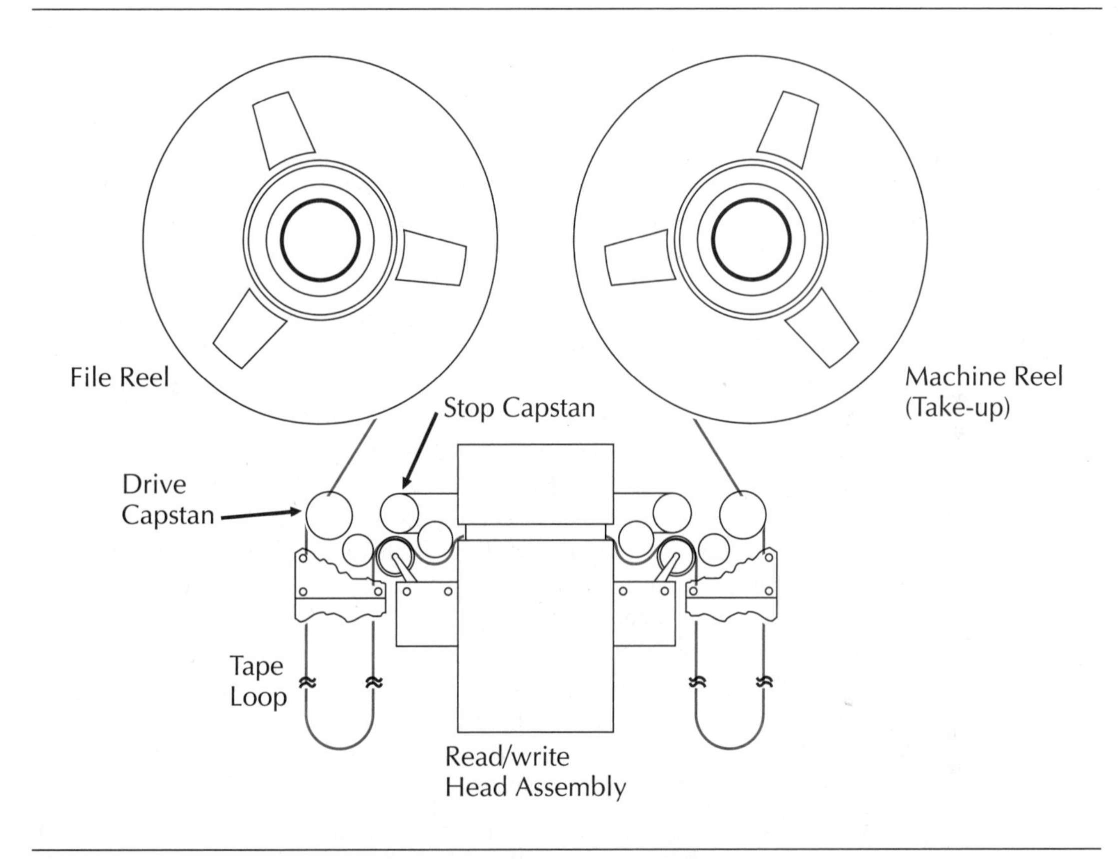

FIGURE 5.12 **The components and configuration of a typical tape drive mechanism.**

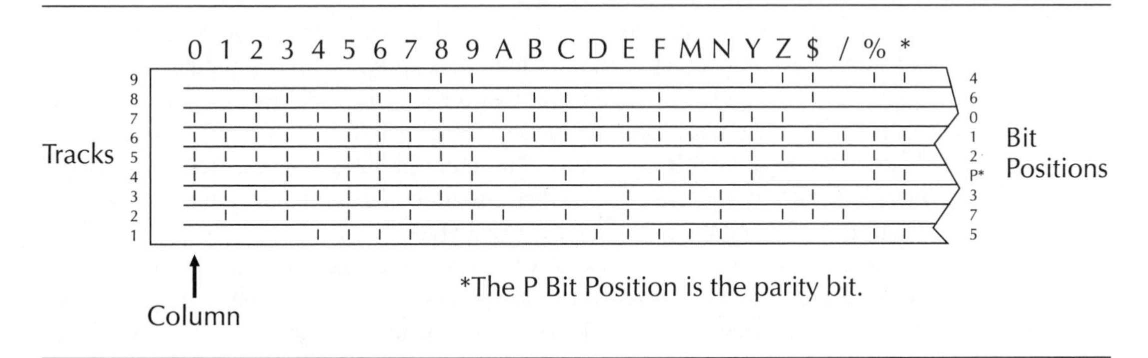

FIGURE 5.13 **A data recording pattern for a nine-track tape.**

Tape is subject to all of the problems of magnetic decay and leakage discussed earlier. It compounds the problem of leakage by winding upon itself. Thus, leakage can occur from adjacent bit positions on the same area of the tape as well as from the layer of tape wound above or below on the reel. Tapes are also susceptible to problems arising from stretching, friction, and temperature variations. As a tape is wound and unwound, the base layer of acetate tends to stretch in length. If enough stretching occurs, the distance between bit positions can be altered, making individual bits difficult or impossible to read.

In many tape drives, the read/write head is in direct contact with the tape as it passes. This contact causes friction as the tape moves. Chargeable material is literally scraped off the tape each time it is read or written, resulting in a lower capacity to hold charge and, therefore, information. This problem is accelerated by dirt on the tape or on the read/write head. For this reason, old tapes are routinely discarded after a certain number of read/write operations.

The loss of chargeable material and, to a lesser extent, of charge itself can be increased by adverse environmental conditions. Excess heat tends to reduce the strength of the bond between the coating of chargeable material and the base layer of the tape. The result is that chargeable material is more easily scraped away during read/write operations.

Excessive heat also reduces the strength of the base layer, thus making it more susceptible to stretching. Cold can also adversely affect tape integrity by making it more brittle and, thus, easier to break. Wide variations in humidity can cause similar problems. For these reasons, tapes should be stored and used in environmentally stable areas, at or near room temperature.

Magnetic disks. Magnetic disk media are flat, circular platters with ferrous coatings that are rotated beneath read/write heads. Data is recorded in concentric tracks. Magnetic disks are classified into two groups, depending on whether their magnetic recording surfaces are rigid or not.

The term *hard disk* refers to magnetic disk media with rigid metal bases (or substrates). A typical platter size for hard disks used in large computer systems is 14 inches in diameter. Storage capacities for *hard disk drives* that contain multiple platters can be measured in thousands of megabytes or tens of gigabytes. Hard disks also are available for small computers in sizes ranging from 3.25 to 10 inches. Typical capacities of these disks might be 20 to 500 megabytes, with maximum capacity ranging as high as 1.5 gigabytes.

Read/write heads do not make contact with the recording surface of hard disk media. Rather, the head floats above the platter on a thin cushion of air as the disk spins at high speed, 3500 revolutions per minute (rpm) or more. This rotational speed is a major factor in the high data access speed of hard disk devices.

Disks can be mounted on disk drive devices, either singly or as a stack of platters. The platters can be either fixed to the spindle (or rotor) of the drive or be removable. Removable disk media include disk cartridges containing single platters and disk packs containing a stack. Devices with removable platters usually have a single read/write head per disk surface; this head moves in and out to seek desired tracks. Fixed-disk devices may have multiple read/write heads in fixed, or permanently mounted, positions above each recording position, or track.

A particularly important performance consideration is that magnetic disks are direct access devices. Data records can be directly accessed, or in any desired order, because of physical relationships between read/write heads and disk recording patterns (Figure 5.14). On stacks of disk platters, tracks are organized into concentric cylinders that correspond with head positions on the stack. Any data record can be sought directly by specifying its platter (or read/write head) number, track (or cylinder) number, and sequence within the track.

The average access time of a hard disk is dependent on several factors. These include

- Time required to position read/write heads
- Rotational delay
- Controller and communication channel throughput

On devices with movable heads, delays will be incurred in positioning the head and its *access arm* over the desired track. The speed at which an access arm can be positioned is subject to physical (mechanical) limitations. No mechanical delays are incurred in fixed-head hard disk drives. Accordingly, access time on fixed-head devices are shorter.

Another performance factor of hard disk devices is *rotational delay*, which is the time between the moment a read/write head is positioned over a track and the time that desired portion of that track rotates beneath the head. On average, rotational delay will be one half of the rotational speed of the disk platter(s). For example, average rotational delay for a platter spinning at 10,000 revolutions per minute will be 0.0005 second (1 ÷ 10000 ÷ 2). Rotational delay is measured in milliseconds (ms) and is encountered on all types of disk devices.

Drive performance also will be affected by the *data transfer rate* of the I/O circuits to the processor. This includes the data transfer capacity of disk drive controller circuitry as well as the communication channel between the drive controller and the CPU. These data transfer rates are usually measured in megabytes per second and should be matched with the physical performance characteristics of the drive.

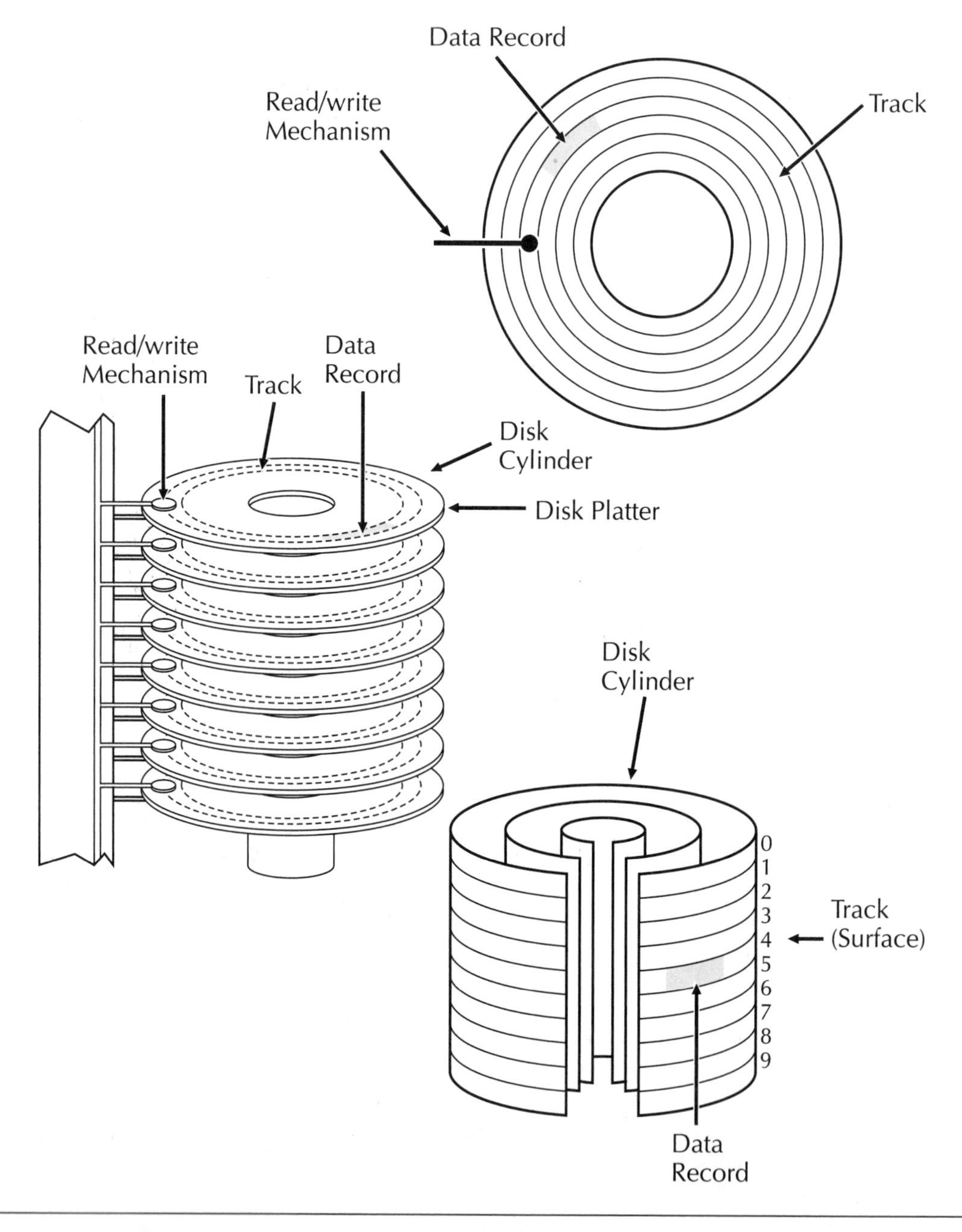

FIGURE 5.14 **Recording patterns used for hard disk packs. Shown are patterns for individual surfaces, positions of read/write mechanisms, and the configuration of recording cylinders.**

Data storage media, called *floppy disks* or *diskettes* use bases of flexible plastic material and are otherwise similar in function to hard disks. Technical differences between hard and floppy disks exist in the areas of storage capacity, recording pattern, and data access speed. Floppy disks are typically used with smaller computers to hold a few programs or data files. All floppy disks are removable and are encased in flexible plastic envelopes or in rigid plastic sleeves to prevent users from touching recording surfaces. Sizes and storage capacities vary. Conventional sizes are 3.5, 5.25, and 8 inches in diameter. Storage capacities range from 360 kilobytes (Kbytes) to about 5 Mbytes.

A typical data recording pattern for a floppy disk is shown in Figure 5.15. As with hard disks, data is recorded in concentric tracks. Tracks are subdivided into sectors, or wedge-shaped areas. For a floppy disk, the sector is a fixed-length unit of storage. Read/write heads within floppy disk drives are movable and, unlike hard disk heads, remain in physical contact with the recording medium during access operations. The rotational speed of the drive is 350 rpm, or one tenth that of hard disk drives. Accordingly, the data access speed of hard disk drives is about 10 times that of floppy disk drives. Floppy disk drives also are subject to rotational delay, delay due to head movement, and limitations in the data transfer rate of I/O channels.

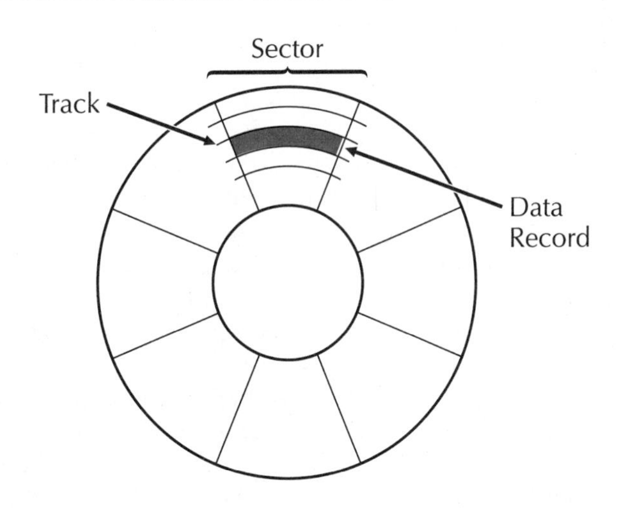

FIGURE 5.15 **The data-recording pattern for an individual disk surface, including relationships of sectors, tracks, and records.**

A practical consideration for computer hardware designers and system integrators is that there can be considerable variation among floppy disk recording formats. As a result, diskettes for different makes and models of computers or drives are likely to be incompatible.

For example, floppy disk media and devices can be either single or double sided. For double-sided media, separate read/write heads are positioned above and below the diskette. Floppy disks also can be either hard or soft sectored. The physical locations of sectors on hard-sectored diskettes are fixed during manufacture. Sector locations on soft-sectored diskettes are determined when the diskette is formatted, or initialized, by the operating system.

Recording densities of floppy disks also vary. Single-density diskettes are recorded at 50 tracks per inch (tpi). Double-density diskettes have 100 tpi, and quad-density diskettes have 400 tpi. Current maximum storage capacity for floppy disks ranges as high as 3 Mbytes.

Optical Mass-Storage Devices

Whereas magnetic mass-storage devices store data as variations in magnetic polarity, optical mass-storage devices store bits as variations in light reflection. The storage medium is a disk of highly reflective material organized in concentric tracks, much like a magnetic disk. The reflectivity of individual bit positions can be reduced by burning a pit at the location. The structure of optical media is illustrated in Figure 5.16.

The read mechanism consists of a *laser* light source and a *photo-electric cell.* The laser beam is focused to an area equal to the size of a single bit position. As the light is guided along the track, variations in the reflective property of individual bit positions are sensed as variations in the reflected light detected by the photo-electric cell. These variations are interpreted as either zero or one.

Optical storage has two primary advantages over magnetic storage: storage density and permanence. A laser can be focused to a very tight beam. Thus, the area on the media surface needed to store a bit is very small (approximately 1 micron in diameter). This is at least 10 times as small as the chargeable area needed to store a bit magnetically. Thus, optical media can store at least 10 times as much data as magnetic media per unit of media area. A standard 120 millimeter *compact disk (CD)* can store 550 Mbytes of data.

Optical media are far more permanent than their magnetic counterparts. The plastic material of which they are constructed is stable across a wide range of temperatures. They are not subject to the problems of magnetic decay and leakage, nor are they affected by electromagnetic interference. They are susceptible to scratches on the media surface and to warping of the media itself, but neither of these problems are likely in normal use.

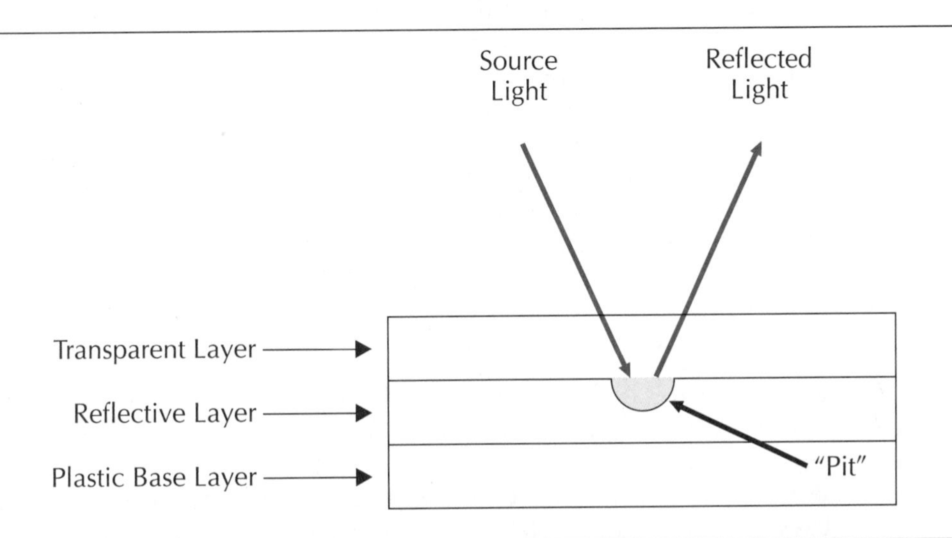

FIGURE 5.16 **A cross section of the layers of an optical disk. A laser is focused onto the reflective surface and the reflected light is sensed by a photo-electric cell as a bit value.**

Most optical media are removable. That is, an individual disk can be easily removed from the drive and replaced by another. Such is not the case with most hard drives. Thus, it is possible to leverage the investment in a single drive mechanism by using it with many disks, effectively increasing its storage capacity.

Optical storage does have some shortcomings as compared to magnetic storage. The most important of these are access speed and the relative difficulty in writing to the media. Current optical drives have average access times in the range of 50 to 100 milliseconds. This is substantially higher than the 10- to 30-millisecond access times typical of magnetic hard drives. However, access speed is improving as the technology matures. Although the more complex read head of an optical drive is a limiting factor in access speed, it is reasonable to expect access speeds approaching those of magnetic drives in the future.

The difficulty in writing to optical media arises from the use of heat to burn pits at bit locations. Laser light must be focused tightly, at sufficient power, and for a sufficient length of time to complete the operation. This results in write times that are substantially longer than read times. The process is also destructive. That is, once written (burned), a bit position cannot be changed. One answer to this problem has been to combine optical and magnetic technology in a *magneto-optical disk.*

CD-ROM. *Compact disk read-only memory (CD-ROM)* is the optical medium with which most people are familiar. Audio compact discs are an example of this type of media. The contents of the media are written during manufacture and cannot be altered.

These devices are useful when large amounts of relatively permanent data need to be stored on-line. Examples of such data include books and indices. It is not surprising, therefore, that one of the largest applications of CD-ROM technology has occurred in libraries. Other applications include the storage of systems software. Because such software does not change as frequently as application software, the inability to write to the media is not much of a problem. New versions of software can simply be distributed as replacement disks that are ready to use.

WORM disks. A *write-once, read-many (WORM)* disk is the optical storage equivalent of a programmable read-only memory. Applications for this technology are similar to those for CD-ROM. They are primarily used by organizations that need to generate large quantities of data for distribution on an infrequent basis. Automobile parts catalogs, for instance, can be updated annually or quarterly. If the distribution volume is large (several thousand copies or more), it is generally cheaper to manufacture CD-ROMs. For lower volumes, WORM drives and media are more cost effective.

Read/write optical disks. Optical storage devices will never completely replace magnetic storage devices until they support both reading and writing at speeds equivalent to those of magnetic hard disks. This is unlikely with purely optical drives given the problems inherent in writing bit values. The current solution to this is a combination of the two technologies: magneto-optical disks.

A *magneto-optical disk* uses a laser and reflected light to sense bit values. However, the variations in reflectivity of the disk surface are achieved by applying a magnetic charge to the bit area. There are no pits burned into the surface as with purely optical disks.

The disk surface is a material that is resistant to change at normal temperatures. At higher temperatures (approximately 150 degrees Celsius), the material can be charged by a magnetic field in much the same way as purely magnetic media. To bring the material to a chargeable temperature, a laser is focused on a bit position. A magnetic field then induces the appropriate charge. Because the laser beam focuses on a single bit position, surrounding bits are unaffected by the magnetic field.

The write operation is still relatively slow compared to the read operation, but it is considerably faster than with purely optical drives. However, rewriting bit values is slow. In general, rewriting must be performed in two passes: the first to reorient a bit to its original (neutral) state, and the second to write the new bit value.

Magneto-optical technology is coming into wide application. However, the cost of the drives is considerably higher than those of conventional magnetic media due to the immature state of the technology and the relatively high complexity of the devices. These costs can be partially recovered by using a single drive with many removable media. They are also likely to decrease as the technology matures and economies of scale in production are realized.

SUMMARY

The processing and memory circuitry within a modern computer is composed of switches and gates. Logic gates implement primitive processing functions on electrically encoded data bits. The primitive processing functions include negation (NOT), combination (AND), and difference (OR and XOR). The processing circuitry of a CPU contains many of these gates. Complex operations are performed by causing data bits to pass through one or more gates.

Modern computer processors and memory are implemented with electrical components. Early computers used vacuum tubes to implement electrical switches and gates. Vacuum tubes suffered problems of heat generation and reliability. They were also slow due to the long conductive pathways over which electricity travelled within them.

Vacuum tubes were eventually replaced with transistors. These devices were smaller and more reliable than vacuum tubes due to smaller size, solid-state implementation, and the use of low-resistance materials. Transistors and other electronic components are currently implemented using semiconductors. These materials allow each component to be extremely small. Many components can be fabricated as a single unit called a microchip. A microchip that contains all of the components of a CPU is called a microprocessor.

Potential replacements for the current generation of microchips include microchips fabricated from gallium arsenide and optically based switching devices. Gallium arsenide promises smaller circuits (and higher speed) than are possible with silicon germanium microchips. Optical switches promise increased speed and the elimination of problems associated with electromagnetic interference, component connections (wiring), and heat generation.

Primary storage is currently implemented with semiconductor devices. These include random access memory (RAM), bubble memory, and read-only memory (ROM). RAM is composed of flip-flop circuits packaged within microchips. ROM is implemented in a manner similar to RAM. However, once data is stored in ROM it is difficult or impossible to erase. Access speeds for RAM and ROM are as low as 20 nanoseconds. This speed is barely able to keep pace with the demands of current microprocessors.

Bubble memory stores data bits in magnetic "pockets." Unlike RAM, storage is permanent because power is not required to maintain data content. Bubble memory is comparable to RAM in access speed but is more expensive. It is used only where the need for fast and permanent storage outweighs its extra cost.

Secondary storage is implemented using magnetic or optical storage devices. Magnetic storage devices store data bits as polarized metal. Data bits are read from and written to magnetic recording media by a read/write head. Types of magnetic media include magnetic tape and magnetic disk. Magnetic tapes are ribbons of acetate coated with a magnetically chargeable coating. Data is written to (or read from) tapes by passing it over the read/write head of a tape drive. Tape is stored on open reels or in enclosed cartridges. Tapes are considered sequential access devices because data can be read or written in physical order on the tape surface.

Magnetic disks are platters coated with magnetically chargeable material. The platters are rotated within a disk drive and a read/write head accesses data at various locations on the platter(s). Magnetic disk drives are random access devices because the read/write head can be freely moved directly to any location on a disk platter. Hard disks are rigid platters coated with highly chargeable material. Floppy disks are flexible platters and may generally be removed from the disk drive. Hard disks offer faster access speeds and higher capacity than floppy disks.

Optical disks store data bits as variations in the ability to reflect light. An optical disk drive reads data bits by shining a laser beam onto a small disk location. High and low reflections of the laser are interpreted as ones and zeros. Optical disks are generally read-only storage media. Their storage cost per bit is generally less than magnetic disks at the expense of slower access speed. Types of optical disks (and disk drives) include compact disk read-only memory (CD-ROM); write-once, read-many (WORM) disks; and magneto-optical drives.

CD-ROMs are written with data during manufacture. Once written, the data cannot be altered. WORM drives are manufactured without data content. They can be written to once by a special disk drive. Magneto-optical drives combine optical and magnetic storage technology. Reading is accomplished by optical means. Writing is accomplished by a combination of optical and magnetic means. Magneto-optical disks may be read and written multiple times.

Key Terms

access arm
Boolean logic
bubble memory
compact disk (CD)
compact disk read-only memory (CD-ROM)
complementary metal oxide semiconductor (CMOS)
data transfer rate
electronically erasable programmable read-only memory (EEPROM)
erasable programmable read-only memory (EPROM)
flip-flop circuit
floppy disk (diskette)
gallium arsenide
hard disk
hard disk drive
integrated circuit (IC)
laser
magneto-optical disk
memory/storage hierarchy
metal oxide semiconductor (MOS)
microchip
microprocessor
photo-electric cell
programmable read-only memory (PROM)
random access memory (RAM)
read-only memory (ROM)
read/write head
recording density
resistance
rotational delay
semiconductor
superconductor
switch (gate)
track
transistor
vacuum tube
volatility
wait state
working storage
write-once, read-many (WORM) disk

Vocabulary Exercises

1. Modern primary storage is implemented electrically with _____ circuits.
2. Within an electrical circuit, _____ causes a portion of the electrical current to be converted to heat.
3. The speed at which data can be moved to or from a storage device over a communication channel is described in terms of its _____ rate.
4. A _____ is a material that has no measurable resistance to the flow of electricity.
5. _____ is a material that has both electrical and optical properties.
6. An electrical processing device that implements a single Boolean transformation function is called a _____ or a _____.
7. _____ stores bit values within magnetic pits embedded in a garnet substrate.
8. An optical storage medium that is written to only during manufacture is called a _____.
9. A _____ stores data in magnetically charged areas on a rigid platter.

10. A _____ is a semiconductor storage device that can be erased using only ultraviolet light.
11. A _____ consists of multiple electrical processing devices implemented within a single semiconductor package.
12. A _____ optical storage device/medium is manufactured blank and can be written to once.
13. A _____ uses a combination of optical and magnetic storage technologies.
14. The _____ and head movement speed of a hard disk drive limit the average access time to data stored on disk platter.
15. The _____ of a magnetic or optical storage medium is the ratio of bits stored to units of medium surface area.

Review Questions

1. What is the function of an ADD gate, an OR gate, an XOR gate, and a NOT gate?
2. What factor(s) limit the speed of an electrically based processing device?
3. What is resistance? How is it related to heat and distance?
4. What shortcomings of vacuum tubes were overcome by transistors?
5. What is a microchip? Of what material(s) is it normally constructed?
6. What advantages does gallium arsenide offer for the implementation of microchips?
7. Why is the recording density of optical disks higher than the recording density of magnetic disks? What factor(s) limit this recording density?
8. What is RAM? How is it currently manufactured? What characteristic(s) of RAM should match the CPU?
9. What is bubble memory? How does it differ from conventional RAM?
10. How is data stored and retrieved on a magnetic mass-storage device?
11. What problems contribute to read/write errors on magnetic tapes?

Research Problems

1. Investigate a current magneto-optical disk drive product offering. How does the cost per unit of storage compare to that of traditional magnetic hard disk drives? How does the access time compare (for both reading and writing)? How does the recording density compare? How is the Kerr effect used to read individual bit values?
2. Current microprocessors are often described in terms of the value of the "micron technology" used to manufacture them. This term describes the width of electrical conductors that are etched onto the semiconductor surface during manufacture. For example, early versions of the Intel 80386 microprocessor were implemented with one micron technology. Later versions are implemented using 0.8 micron technology. Choose a recent microprocessor such as the Intel 80486 or the Motorola 68030. What is the theoretical lower bound on the conductor size of a silicon-based microprocessor? What is the theoretical lower bound on a gallium arsenide-based microprocessor? What micron technology is used to produce it? What upper bound on processor speed is implied by that technology?
3. In current computers, primary storage is always implemented with semiconductor devices. There are, however, optical devices that could conceivably be used for primary storage. Experimental devices have been developed that store data in an optically reactive surface or solid. In a write operation, one or more lasers are focused on a single point to change the reflectivity of that point. In a read operation, one or more lasers are focused on a point, and the reflection from or transparency through that point is used to indicate the data value stored there. What types of materials are used to construct such devices? How is random access (addressing) achieved? What is the access time of such devices? Is storage by this method nonvolatile?

6 *Processor and Memory Architecture*

Chapter Goals

- *Describe the instruction set of a typical CPU.*
- *Describe the components of a CPU and the operation of each.*
- *Describe the key design features of a CPU.*
- *Describe memory organization, allocation, and access.*
- *Describe reduced instruction set computer processors.*

PROCESSORS AND INSTRUCTION SETS

The basic architecture of von Neumann processors was covered in Chapter 2. Chapter 6 moves in for a closer look at the architectural elements of processors and the methods by which processor hardware can be controlled to produce useful results.

The design of a computer processor centers around its instruction set. Recall that an instruction set encompasses the basic set of operations that a computer can perform. Any operation that is not defined within a processor's instruction set must be implemented in software as a sequence of these predefined instructions.

A fundamental design tradeoff involves the size and complexity of the instruction set. The fewer and simpler the available instructions, the faster the computer's execution cycles. Tradeoffs for this efficiency will be the size of machine code programs and the flexibility of programming options. For a machine with a relatively limited instruction set, program code must be more extensive to describe complex operations. Programming flexibility is sacrificed because a complex routine must be defined as a sequence of simple instructions rather than being referenced by a single more powerful instruction.

In many respects, the design of processor hardware merely implements a predefined instruction set. In technical terms, the architecture of a specific processor is defined by

- An instruction set
- The implementation of that instruction set in hardware

This chapter describes the architectural elements that can be used to build an instruction set and its implementation in hardware.

INSTRUCTIONS

The instruction set of a relatively simple processor might include a dozen different instructions. As stored in memory, the instruction is a sequence of bits. The first group of bits represents the unique number of the instruction, commonly called its *op code*. Subsequent groups of bits, called *operands,* hold the input values for the instruction. Depending on the type of instruction, the content of an operand can represent a data item (e.g., an integer value), a memory address, a register address, or the address of a secondary storage or I/O device. The control unit of the CPU tests the bits of the op code to determine the operation or data transformation to be performed. Based on this test, the control unit causes signals that represent operand value(s) to be routed through CPU circuitry that performs the desired operation. These operations represent the smallest increments of machine action that can be specified to the processor. Thus, each instruction is called a *machine primitive*.

Although an instruction set can include many op codes, a von Neumann processor actually performs only three basic types of actions:

- Data copying operations
- Data transformations
- Sequence control

Within these broad categories, a variety of instructions can be implemented, as described below.

Data-Copying Operations

Data-copying operations cause the transfer of data bits from one location within the computer to another. The following operations are normally supported:

- Load
- Store
- Move

Before applying a data transformation operation, the processor must cause data to be fetched from a specific location in memory to a temporary storage area (register) within the CPU. Data transfers from main memory into the processor's internal registers are called *load operations*. In performing output functions, the processor must cause data to be sent from an internal register to a specified location in main memory. The writing of data from a processor register back to a location in main memory is called a *store operation*.

To support processing functions, intermediate results often must be transferred among registers within the CPU. Transfers among registers are referred to as *move operations*. The term can also be used to describe transfers between memory locations. However, this type of move operation can also be implemented as a combination of a load operation and a store operation. This implementation method is required in a CPU that does not implement a direct memory-to-memory move instruction.

In all these instances, the essential operation to be performed is copying a data value, bit by bit, from one storage location to another. In such copying operations, data bits are not moved physically. The bit value in a storage location at the point of origin is tested, and an electronic signal of the same value is sent to the same position in a storage location at the destination. This signal causes the storage location at the destination to be set to the same value. At the completion of the copying operation, both sending and receiving locations hold a copy of the same data byte(s).

Data Transformations

In data transformation operations, data content is modified according to a specific arithmetic or logical rule. The rule that is applied depends on the op code. The op code, in turn, specifies the set of logic gates through which data must be passed to achieve the desired transformation. Some transformations are applied to two separate inputs, or operands, to produce a single result. Instructions that cause transformations on two operands are called *dyadic instructions*. Such operations include arithmetic calculations and logical comparisons. Other operations are used to transform single-bit strings. Instructions that cause transformations on single operands are called *monadic instructions*.

At the machine level, passing a binary signal through a logic gate merely determines whether a data bit will be changed to its complement or will retain its value. That is, a value of 0 can be changed to a 1, a 1 can be changed to a 0, or the value can go unchanged. Each of the dyadic operations below applies a different transformation rule to two separate operands:

- AND
- OR
- XOR
- ADD
- Shift
- Rotate

These primitive data transformation operations may be used for a number of purposes. They may be combined to implement both comparison and arithmetic operations. Several examples are provided in a later section.

AND. One type of transformation is represented by the *AND operation*. The rule applied is that both operands must be 1 to yield a result of 1. Any other combination of values produces a result of 0. That is, 1 AND 1 yields 1. Value combinations of 0 AND 0, 1 AND 0, and 0 AND 1 all yield results of 0. The effect of applying this rule to two separate bit strings, or byte-length operands, is shown in the following example:

```
    10001011
AND 11101100
------------
    10001000
```

The AND operation can be used to fill a designated portion of a bit string with values of 0. For example, assume you wanted to affix a bit of value 0 to an

ASCII-7 character. Using the AND operation, the value 01111111 can be used as the second operand. In an AND operation, this operand is called a *mask*. In transforming the character and the mask with AND, the leftmost bit of the result will always be 0, and the following seven bits will always match the first operand, or character. Thus, a bit of 0 can be affixed to the ASCII-7 binary representation for the character *C* as follows:

```
    1000011
AND 01111111
------------
    01000011
```

The term *mask* is used for the second operand (01111111) because it controls which bits of the first operand will appear (be seen) in the result. Any pattern of bit values can be used as a mask. In an AND operation, values in the first operand that correspond with 1 values in the mask (the second operand) will be copied in the result. Other bit positions will be filled with values of 0.

OR. The inverse of AND is the *OR operation*. If the OR rule is applied, both operands must be 0 to yield a result of 0. Any other combination of values produces a result of 1. That is, 0 OR 0 is 0. Value combinations of 1 OR 1, 1 OR 0, and 0 OR 1 all yield results of 1.

In applying an OR transformation to two-bit strings, the second operand can be used as a mask, just as in AND operations. In an OR operation, values in the first operand that correspond with 0 values in the mask (the second operand) will be copied in the result. Other bit positions will be filled with values of 1. Thus, an OR operation could be used to generate a parity bit of 1 in the example above by using the value 10000000 as a mask.

Exclusive OR (XOR). Under the transformation rule *exclusive OR (XOR)*, a result of 1 is produced only if one of the operand bits has a value of 1. That is, 1 XOR 0 and 0 XOR 1 yield values of 1. Other combinations, 1 XOR 1 and 0 XOR 0, yield values of 0. A common use of the XOR operation is to generate the complement of a bit string. To produce this result, each bit position in the second operand must contain a value of 1. For example,

```
    01010010
XOR 11111111
------------
    10101101
```

Recall that generating the complement of a binary value is the function of a logic gate called an *inverter*. This transformation is necessary, for example, as a first step in performing subtraction or division by complementary arithmetic.

ADD. In an *ADD operation*, the rules applied represent binary addition facts. For values of 0 and 1 (or 1 and 0), the result will be 1. For two values of 0, the result will be 0. For two values of 1, the result will be 10. As covered in Chapter 4, if the data representation format permits the use of signed values, all arithmetic operations can be reduced to sequences of addition operations.

The operations of AND, OR, XOR, and ADD are dyadic; they involve two operands. Note that, in all these operations except ADD, the number of bits in the result will never exceed the number of bits in the longest operand. However, in an ADD operation, two values of 1 will yield a result of 10, perhaps causing an overflow at the leftmost bit position. To handle such situations, and to manipulate bit strings within processor registers, other types of operations must be available.

Shift. The effect of a *shift operation* is shown in Figure 6.1. In Figure 6.1a, the value 01101011 occupies an eight-bit storage location. A shift can be implemented to the right or left and the number of positions shifted can be greater than one. Typically, a second operand is used to hold an integer value that indicates the number of bit positions by which the value will be shifted. Positive or negative values of this operand can be used to indicate right or left shifting.

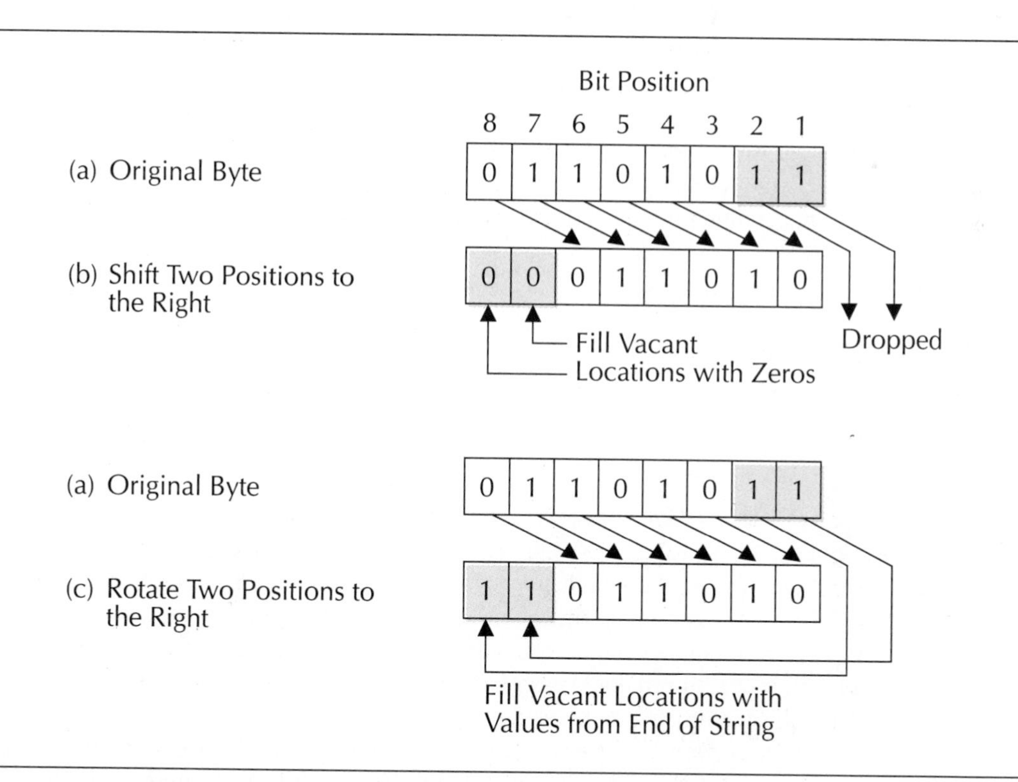

FIGURE 6.1 **The results of shift and rotate operations: (a) the original (unaltered) data input; (b) the result of a shift operation; and (c) the result of a rotate operation.**

Figure 6.1b shows the result of shifting the value two positions to the right. The resulting value is 00011010. In this case, the values in the two least significant positions of the original string have been dropped. Shift operations also can be used to perform arithmetic operations. In a positional number system of base *N*, each position represents *N* times the value of the position to the right. Thus, in a binary number system, shifting a digit to the left has the effect of multiplying its value by two. Similarly, shifting a digit to the right is the same as dividing by two. In such *arithmetic shifts*, it is important to preserve the value of the sign bit. Thus, after an arithmetic shift, the sign bit is reset to its original value.

Rotate. Figure 6.1c shows the result if a *rotate operation* had been performed on the string in the example above. The bits that were dropped in the shift operation have been affixed to the leftmost bit positions of the result, 11011010, as if the ends of the string were joined to form a circle. This is the essential distinction between shift and rotate operations. In a shift operation, strings simply are truncated if the shift causes digits to extend beyond the available positions. In a rotate operation, values are permitted to "wrap around" the available bit positions. Rather than being dropped, digits that are moved beyond the end of the string are placed in the vacant positions at the opposite end.

As with shifts, an operand of a rotate operation can specify the number of bit positions by which the value is to be displaced. As the string is rotated, the values reappear at, or wrap around, the end of the string, opposite the direction of rotation. Thus, rotating a value to the right by seven positions will yield the same result as rotating it to the left by one position (on an eight-bit data item).

Sequence Control

The third category of machine instructions permits the current sequence of instruction execution to be altered. These instructions include

- Branch
- Halt

Branch. A *branch instruction* causes the processor to depart from the normal sequence of a program. Recall that the control unit fetches the next instruction from memory at the conclusion of each execution cycle. A branch instruction requires an operand referencing an address in memory where the next instruction is located. In some machines, a branch is implemented as a move instruction, in which an operand (the address of the next instruction) is loaded into the register that holds the instruction pointer.

In an *unconditional branch*, the processor always departs from the normal sequence and fetches the next instruction from a different address. This instruction is sometimes referred to as a *jump*. On a *conditional branch*, the jump will occur only if a specified condition is met. Typically, the condition (e.g., the equivalence of two numeric variables) is evaluated and the result is stored as a Boolean value in a register. The CPU checks the value of this register to determine if the branch should be executed. In some architectures, the register containing the comparison result is predetermined; in others it can be specified as an operand. A conditional branch instruction can be used to implement an unconditional branch instruction by loading the register with the value "TRUE" prior to executing the conditional branch instruction.

Halt. If the control unit senses a *halt instruction* in the instruction op code, the current sequence of execution will terminate. At this point, control of the processor may be returned to the operating system by executing a branch to a predefined address. If no operating system is present, a status lamp can be lit on the front panel of the unit or at the operator console.

A Short Programming Example

The instructions listed above are sufficient to implement a general-purpose processor. However, most processors will provide a substantially larger set of instructions. Additional instructions that might be provided include subtraction, multiplication, and division. Such instructions are provided as a programming convenience but are not absolutely necessary, because each of these arithmetic operations can be performed using a sequence of the simpler instructions. Similarly, additional instructions for comparison operations can be provided to test for equality, greater than, and less than. These too may be implemented using sequences of simpler operations. The next few paragraphs illustrate some of these implementations.[1]

Subtraction can be implemented as complementary addition. That is, the operation A – B can be implemented as A + (–B). As described above, the XOR operation can be used to derive the complement of a value as the result of an XOR operation applied to the value and a string of binary 1 digits. If the result is to be stored in a two's-complement notation (as is generally the case for integer data), a one is added to the result of the XOR operation. Thus, for example, the complement of 0111, represented as a two's-complement value can be derived as:

$$\text{XOR}(0111,1111) + 0001 = 1000 + 0001 = 1001$$

This result can then be added to implement a subtraction operation. Thus, for example, the result of subtracting 0111 from 1000 can be calculated as:

[1] The reader may want to review the discussion of two's complement notation in Chapter 4 before studying these examples.

ADD(ADD(XOR(0111,1111),0001),1000)
ADD(ADD(1000,0001),1000)
ADD(1001,1000)
10001

Because four-bit values are used, the result of 10001 is truncated from the left, resulting in a value of 0001.

Comparison operations can be implemented in much the same way as subtraction. In general, the purpose of a comparison operation is to generate an output that is interpreted as a Boolean (true or false) value. Typically, an integer value of zero is interpreted as true and any nonzero value is interpreted as false. The comparison $A = B$ can be implemented by generating the complement of B and adding it to A. If the two numbers are equal, the result of the addition will be zero (interpreted as "true").

Greater-than and less-than comparisons can also be implemented as subtraction (complementary addition) followed by extraction of the sign bit. For the condition $A < B$, the subtraction of B from A will generate a negative result if the condition is true. In two's-complement notation, a negative value will always have a one in the leftmost position (i.e., the sign bit). The shift operation can be used to extract the sign bit from the remainder of the value. For example, the two's-complement value 10000111 is a negative number. The sign bit can be extracted by shifting the value seven bits to the right, resulting in the string 00000001. Note that the left seven bits are filled with zeros as a result of the shift operation. The result of the shift can be interpreted as a Boolean value (nonzero, or false in this case). Thus, for example, the comparison

$$00010000 < 00011111$$

can be evaluated as:

SHIFT(ADD(00010000,ADD(XOR(00011111),1)),7)
SHIFT(ADD(00010000,ADD(11100000,1)),7)
SHIFT(ADD(00010000,11100001),7)
SHIFT(11110001,7)
00000001

Consider the following high-level programming language program:

```
IF (BALANCE < 1000) THEN
        INTEREST = BALANCE × RATE1
ELSE
        INTEREST = BALANCE × RATE2
ENDIF
```

Such a computation might be used in a program that computes interest on a savings account in which variable interest rates are applied, depending on the account balance. A program that implements this computation using only the previously defined low-level CPU instructions is shown in Figure 6.2.

```
1    LOAD     M1,R1                  'load BALANCE
2    LOAD     0000001111101000,R2    'load 1000_10
3    XOR      R2,1111111111111111,R2 'start < comparison
4    ADD      R2,0000000000000001,R2 '
5    ADD      R1,R2,R0               '
6    SHIFT    R0,15                  'end < comparison
7    BRANCH   11                     'branch if < is true
8    LOAD     M2,R2                  'load RATE1
9    LOAD     0000000000000000,R0    '
10   BRANCH   12                     'branch past next load
11   LOAD     M3,R2                  'load RATE2
12   STORE    0000000000000000,R3    'initialize accumulator
13   XOR      R2,1111111111111111,R2 'start counter=0 compare
14   ADD      R2,0000000000000001,R2 '
15   ADD      R2,0000000000000000,R2 'end counter=0 compare
16   BRANCH   22                     'branch if counter=0
17   ADD      R1,R3,R3               'add to accumulator
18   XOR      R2,1111111111111111,R2 'start counter decrement
19   ADD      R2,0000000000000001,R2 '
20   ADD      R1,0000000000000001,R2 'end counter decrement
21   BRANCH   13                     'branch to start of loop
22   STORE    R3,M4                  'store INTEREST
23   HALT                            'terminate execution
```

FIGURE 6.2 A simple program that uses primitive CPU instructions.

Instructions 1 and 2 load the balance and the constant 1000_{10} from memory locations M1 and M2 into registers (note that 16-bit, two's-complement notation is assumed). Instructions 3 through 6 implement a less-than comparison with the Boolean result stored in register R0. The result of the comparison determines which load instruction (instruction 8 or instruction 11) will be executed. All branch statements operate conditionally on the content of register R0. A nonzero value causes the branch to be ignored. Thus, the instructions between 7 and 11 implement the selection (loading) of either RATE1 or RATE2.

Instructions 12 through 21 implement the multiplication of BALANCE by whichever RATE was loaded. The multiplication is implemented by a loop. One operand (the RATE, in this example) is used as the counter and the other is used as an addition value. The loop is initialized by storing a zero in the accumulator (instruction 12). On each pass through the loop, the addition value is added to the accumulator and the counter is decremented by one (instructions 17 through 20).[2] The counter is tested for equality with zero at the beginning of the loop (instructions 13 through 15) and a branch out of the loop is taken when the equality test succeeds (instruction 16). At this point, the result is stored to a memory location (instruction 22) and execution is halted (instruction 23).

ARCHITECTURAL ELEMENTS

The relationship of the CPU to other elements of the computer system is shown in Figure 6.3. For the purposes of this chapter, the primary elements of concern are the CPU, primary storage (main memory), and the system bus. An overview of the functions and implementation of the CPU and of memory was given in Chapter 2 and will be expanded on in the following sections. The system bus was only indirectly referred to in Chapter 2. It is the communication channel that connects all of the computer system components. It, too, will be discussed in detail in a later section.

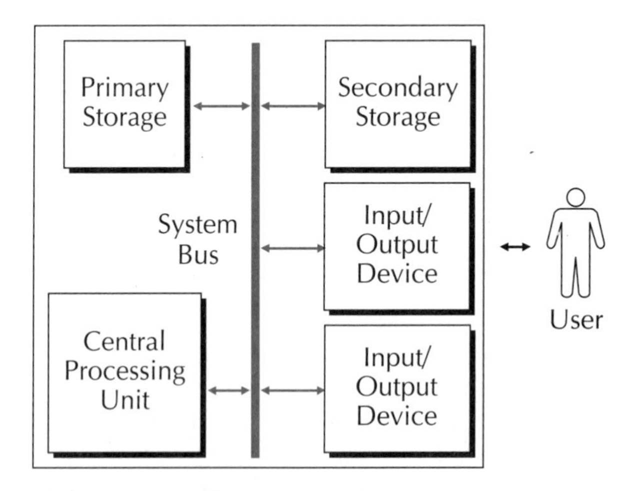

FIGURE 6.3 **The primary components of a computer systems and their relationship to one another.**

[2] This method is applicable only if the counter is an integer. Interest rates are normally expressed as fractional (real) quantities. However, the method can be used in this example if both the rate and balance are scaled upward (e.g., multiplied by 10,000) to eliminate fractional quantities.

A more detailed view of the CPU is provided in Figure 6.4. As discussed in Chapter 3, the three primary components of a processor are

- The control unit
- The arithmetic logic unit
- Processor registers

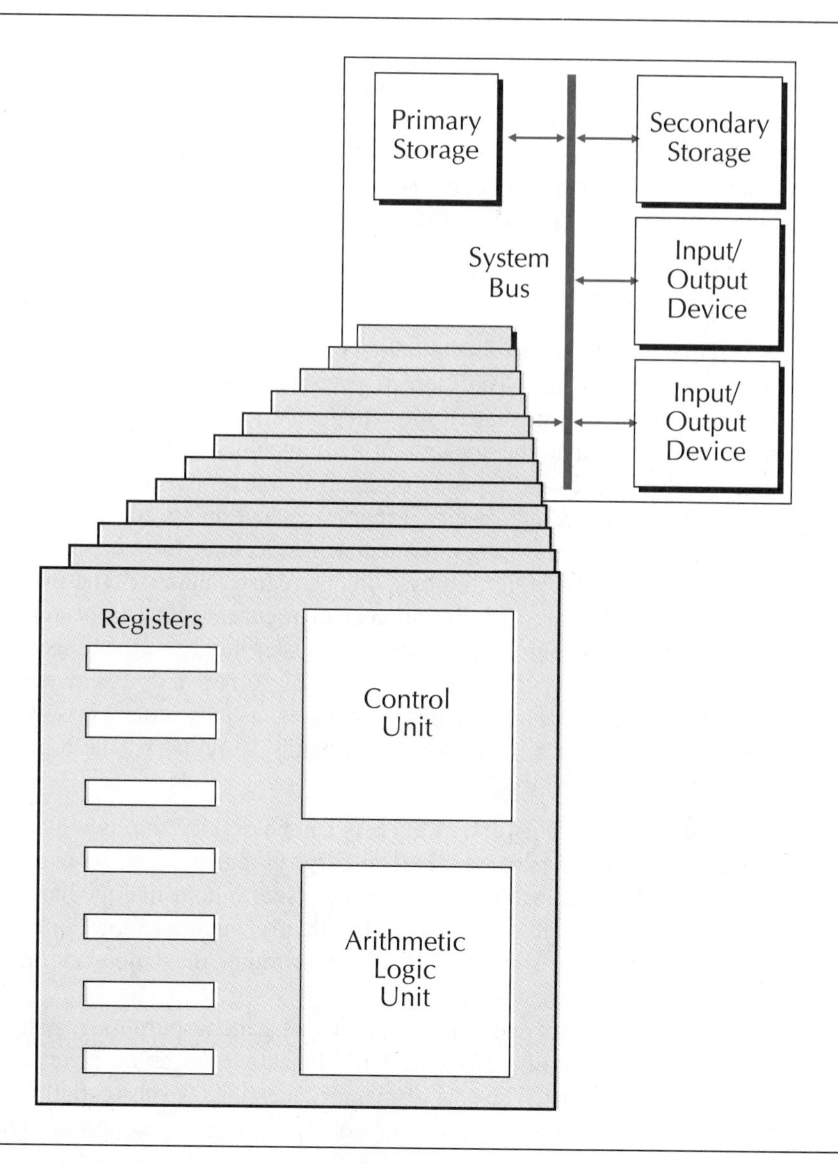

FIGURE 6.4 The primary components of the CPU.

The control unit is responsible for fetching an instruction from memory and for directing other processing elements in the execution of that instruction. This direction might require it to send commands to the arithmetic logic unit, load or store the contents of processor registers, or move data to or from other computer system components.

The ALU is responsible for executing the primitive processing steps necessary to perform mathematical or logical functions. Signals received from the control unit direct the ALU to pass signals that represent data through the appropriate circuitry. Circuitry exists to implement each of the data transformation functions described earlier. If higher level instructions such as multiplication and equality comparison are provided, circuitry also exists to implement those functions. The path of data through the circuitry depends on the format in which the data is represented (two's-complement and so on) and the instruction being executed (addition, subtraction, comparison, and the like).

Processor Registers

A variable in CPU design is the number of registers, or temporary holding areas, within the control unit. The more registers within the processor, the greater its capacity to hold intermediate results. In general, register contents can be accessed far more quickly than the contents of main memory. If the need to store intermediate results exceeds the capacity of the available registers, memory must be used to hold the intermediate results. Program execution speed will decrease due to delays in loading and storing intermediate results to or from memory.

Each register is referenced by a unique register number. The maximum value is a constraint imposed by the number of registers that can be referenced. For example, if four bits are used to specify register numbers, the range of identifiers would be from 0000_2 to 1111_2, for a total of 16 registers. For purposes of notation, register numbers, like memory addresses, usually are expressed in hexadecimal form. Thus, for a processor that contains 16 registers, the register numbers would range from 0H to FH.

General-purpose registers. Registers can be divided into two classes: general-purpose and special-purpose. General-purpose registers can be used in a variety of ways by a program. Typically they are used to hold intermediate results or to hold data values that will be used frequently, such as loop counters or array indices. By holding such data in registers rather than memory, the execution speed of a program can be increased.

In general, an increase in the number of general-purpose registers increases the execution speed of a program. Unfortunately, registers are very expensive in addition to being fast. The cost of memory locations is substantially cheaper and, thus, a tradeoff between processor cost and program execution speed exists.

There is a diminishing rate of return for additional registers beyond a certain point. Although a CPU with eight general-purpose registers can provide substantial improvements in program execution speed as compared to a CPU with four registers, the same amount of speed difference is unlikely when increasing the number from eight to 12.

In order for the potential speed increase of additional registers to be realized, they must be used effectively. Yet there are only a limited number of intermediate results or frequently used data items in any given process or program. Thus, a primary CPU design decision is the optimal tradeoff between the number of general-purpose registers, the extent to which those registers will be used by a typical process, and the cost of implementing those registers.

Special-purpose registers. Every processor uses a number of special-purpose registers. The content and use of each register are specified as part of the CPU design. The implementation of processor circuitry and the instruction set are integrally connected with these registers. Several of the more important of these registers are

- The instruction register
- The instruction pointer
- The program status word

The *instruction register* is used by the control unit to hold an instruction just loaded from memory. Once loaded, the instruction is decoded and then executed. The instruction register thus serves as the data input to the decoding process, and the circuitry of the control unit is designed accordingly. Decoding refers to the separation of the instruction into its op code and operands, the movement of data (such as loading data into a register from a memory address in one of the operands), and the generation of control signals to the ALU for instruction execution.

The *instruction pointer* is also called the *program counter*. Recall that the CPU constantly alternates between the instruction (fetch and decode) and execution (data movement and/or transformation) cycles. At the end of each execution cycle the control unit starts the next instruction cycle by retrieving the next instruction from memory. The instruction retrieved is the instruction at the memory address in the instruction pointer. Because a von Neumann processor assumes sequential execution of program instructions, the instruction pointer is incremented by the control unit, either during or immediately after the instruction cycle.

The only time the CPU deviates from sequential execution is if a branch instruction is executed. A branch instruction is implemented by overwriting the value of the instruction pointer with the address of the instruction to which the branch is directed. An unconditional branch is thus implemented as a copy from one of the operands in the instruction register (containing the branch address) to the instruction pointer.

The *program status word (PSW)* contains numerous individual bit fields that indicate the current status of program execution. The content, format, and use of these bit fields varies considerably from one processor to another. In general, they have two primary uses: to direct the execution of a conditional branch instruction and to indicate actual or potential error conditions.

A single bit within the PSW is generally used to indicate the results of comparison instructions. Recall that a conditional branch is a branch instruction that is executed only if a specified condition (numerical comparison) is true. The truth or falsity of the condition is determined by the ALU and the result is stored in a bit within the PSW.[3] The control unit then checks this bit to determine whether or not to overwrite the contents of the instruction pointer to implement the branch.

Other bits can be used to represent status conditions resulting from instruction execution by the ALU. Conditions such as overflow, underflow, or an attempt to perform an undefined operation, such as division by zero, are represented in status bits within the PSW. These bits are tested by the control unit at the conclusion of the execution cycle to determine if an error has occurred.

Instruction Format

Computers differ in the format used to specify machine instructions. Figure 6.5 shows an instruction format that consists of three bytes. The first byte holds the op code; the second and third bytes contain the operands. The instruction format of a particular CPU will depend on a number of factors, including the number of instructions, the data types used as operands, and the length and format of each type of operand. The total number of instructions (i.e., the largest op code value) will determine the number of bits that must be used to represent the op code. The format and length of primitive data types will determine the format and number of bits required to represent many operands.

OP CODE	OPERAND 1	OPERAND 2
Byte 1	Byte 2	Byte 3

FIGURE 6.5 The format of a three-byte instruction.

[3] Recall that the example program in Figure 6.2 implemented comparison using the XOR, ADD, and SHIFT instructions. Results were explicitly stored to a general-purpose register. This was necessary because no compare instructions were directly implemented within the CPU. If compare instructions are directly implemented, the Boolean result is usually stored in the PSW.

Instruction formats can be either fixed or variable in length and might or might not use fixed-length components for operands. *Fixed-length instructions* and instruction components simplify the circuitry of the control unit. Recall that the control unit is responsible for incrementing the instruction pointer. If the instruction format is of fixed length, the amount by which the instruction pointer must be incremented is a constant (the length of an instruction).

If instruction length is variable (i.e., a *variable-length instruction*), the amount by which the instruction pointer is incremented is the length of the current instruction. Thus, the CPU must check the op code in the instruction register to determine the proper incrementation value. Variable-length instructions also complicate the operation of fetching an instruction, because the number of bytes to be retrieved is dependent on the op code stored in the first field. The use of fixed-length fields for op codes and operands simplifies the circuitry of the control unit by fixing the position of each field within the instruction register. Thus, the location of operands (the starting bit position) is constant and the control unit need not refer to the op code to determine the correct position.

Although fixed-length instructions and fields simplify control unit functions, they do so at the expense of efficient use of memory. Recall that some instructions have no operands (e.g., a HALT), and those with operands can have either one or two (i.e., monadic or dyadic instructions). Some mainframe computers implement instructions with three operands. If fixed-length instructions are used, the instruction length must be the length of the longest instruction—for example, an instruction with two or three large operands. Smaller instructions stored in memory must be "padded" with empty bit positions to extend their length to the maximum. This is an inefficient use of memory due to the amount of space wasted by the "padding." Programs are larger than necessary and will require more memory and time to load into memory. Execution can also be slowed by the movement of empty instruction fields to and from the CPU. The tradeoff between these inefficiencies and simplification of control unit circuitry is a fundamental CPU design decision.

The Bus

Within a computer, the *bus* is a set of parallel communication lines that connect the CPU with main memory and other system components, as shown in Figure 6.6. A constraint on the I/O capability of a processor or computer system is the width of the data bus, or the number of lines over which data transfers can be performed. For example, eight data lines would permit the transfer of one byte in a single memory access cycle. The time interval required to move one set of bits across a bus is referred to as a *bus cycle*.

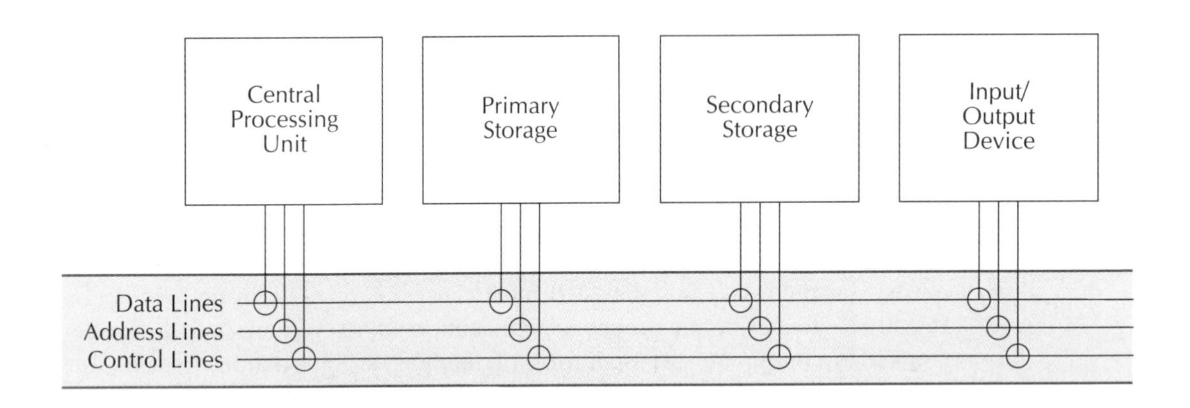

FIGURE 6.6 **The system bus, its sets of communication lines, and the devices attached to it.**

Data lines can be used to load instructions or data from main memory to processor registers or to store results from registers back to memory. Other lines within the bus are used to transmit control and status signals, register numbers, and memory addresses. In general, these control lines are dedicated to that function and are not available to carry data. In many computer architectures, separate sets of lines are used for control and status signals, addresses, and data. These sets of lines are called the *control bus*, the *address bus*, and the *data bus*, respectively.

To prevent confusion and chaos that could arise from, say, two devices trying to communicate across the bus at once, explicit rules of communication and bus access are required. These rules, referred to as the *bus protocol*, are implemented largely using the control/status lines (signals) and exact timing. Every device attached to the bus must understand and obey the rules of protocol. Thus, a bus is a combination of data transfer circuitry, a communication protocol, and the circuitry within the bus and the attached devices that implement that protocol.

The number of bus lines is a cost factor because of the circuitry involved. Specifying a relatively small number of lines might be economical but would have an adverse effect on performance. For example, several memory access cycles might be required to load a single instruction. Ideally, the bus width should be matched to the number of bit positions within the machine's instruction format so that an entire instruction, including op code and operands, can be loaded in a single memory access cycle.

Another factor in bus design is the method of bus access provided to the attached devices. In von Neumann architecture, the CPU is assumed to be the focus of all computer activity. As part of this role, the CPU is also the *bus master*. All other attached devices are *bus slaves*. This master/slave relationship has two important consequences. The first is that no device other than the CPU can access the bus except in response to an explicit instruction from the CPU. The second consequence is that all data communication on the bus must be routed through the CPU.

The first consequence allows the bus protocol (and bus interface circuitry) to be simplified. However, the second consequence can be a substantial limitation on overall system performance. The requirement that all data pass through the CPU can cause a communication bottleneck when large amounts of data need to be transferred from one device to another—for instance, from secondary storage to main memory. It is also a poor use of an expensive device (the CPU) to utilize it largely as a data conduit.

For this reason, most modern computers modify the von Neumann assumptions somewhat to allow direct data transfers between devices without routing those transfers through the CPU. This aspect of bus design is sometimes called *direct memory access (DMA)* or a *multiple-master bus*. However, the terms are not equivalent. Direct memory access is sometimes used generically to refer to the capability of I/O and mass-storage devices to move data directly to or from memory. It is also, however, a specific technology that can be used in conjunction with a bus that does not have true multimaster capability. Typically, a device called a *DMA controller* is attached to the bus and to main memory. The device provides a direct interface between the bus data lines and main memory, bypassing the CPU.

Multimaster capability implies that devices other than the CPU can assume control of the bus—in other words, act as a bus master. Such a capability requires substantially more complex circuitry in the bus and in attached devices in order to mediate and control demands from competing "masters." The complexity of the bus protocol and the number of control/status signals are also increased.

Regardless of which approach is used, the primary benefit is the same. Large data transfers to and from memory can be made without the assistance of the CPU. This allows the CPU to perform other functions at the same time the transfer is taking place. It is, in effect, a form of parallel processing, because the computer system is doing two things at once.

Word Size

A *word* is a unit of data that consists of a fixed number of bytes or bits. A word can be loosely defined as the amount of data that a CPU "processes" at one time. Depending on the CPU, this processing can include arithmetic, logic, fetch, store, and copy operations. Word size will normally match the size of CPU registers, because these are generally the sources and targets of the aforementioned operations. Word size is a fundamental CPU design decision, with implications for most of the other components of a computer system as well.

In general, a CPU with a large word size will be capable of performing a given amount of work faster than a CPU with a small word size. As an example, consider the addition or comparison of two 32-bit integers. A processor with a word size of 32 bits will be able to perform addition or comparison by executing a single instruction. This is because the registers that hold the operands as well as the circuitry within the ALU are designed to process 32-bit values simultaneously.

Now consider the same operation on the same data performed by a processor with a word size of 16 bits. Because the size of the operands exceeds the word size of the processor, they must be partitioned and the operation(s) carried out on the pieces. In a comparison operation, the processor must first compare the first 16 bits of the operands and then, as a separate execution cycle, compare the latter 16 bits. Inefficiency arises as a result of loading and storing multiple operands to and from memory. Inefficiency also arises from executing multiple instructions to accomplish what is logically a single operation. Because of these inefficiencies, a 32-bit processor is usually more than twice as fast as a 16-bit processor when processing 32-bit data values. These inefficiencies are compounded as the complexity of the operation increases. Operations such as division and exponentiation are extremely complex to perform in a piece-wise fashion, and program execution slows considerably as a result.

Processor word size also has implications for the design of the system bus. In general, maximal processor performance is achieved when the width of the data bus is at least as large as the processor word size. If the bus width is any smaller, every load or store operation requires multiple bus cycles. For example, the movement of four bytes of character data to contiguous memory locations will require two separate data movement operations on a 16-bit bus, even if the word size of the processor is 32 bits. Similarly, fetching a 32-bit instruction will require two bus cycles.

Processor word size also has implications for the physical implementation of memory. Although the storage capacity of memory is always measured in bytes, data movement between memory and the CPU is measured in words. For a 32-bit CPU, it is desirable to organize memory such that four contiguous bytes can be accessed in a single memory cycle. Lesser capabilities will cause the CPU to incur wait states.

Clock Rate

The *system clock* is a digital circuit that generates timing pulses and transmits them to other devices within the computer over a separate line on the control bus. This signal is the "heartbeat" of the computer. All actions, especially the instruction and execution cycles of the CPU, are timed according to this clock. The frequency of this clock, typically expressed in millions of cycles per second (megahertz, or MHz), is the *clock rate* of the system.

Instruction and execution cycles usually represent some fraction of the clock rate. Thus, the clock rate is a major factor in measuring the performance of the processor, as measured in millions of instructions per second (MIPS). For maximum efficiency and performance, the clock rate must be balanced with the data transfer rate over the bus. The data transfer rate, in turn, depends on the length of instruction formats and the width of the bus.

MEMORY ACCESS

Although main memory is not a part of the CPU, the need for the CPU to load instructions and data from memory and to store processing results requires a close coordination between the two components. Specifically, the physical organization of memory, the organization of programs and data within memory, and the method(s) of referencing specific memory locations are critical CPU design issues.

Physical Memory Organization

The main memory of any computer can be regarded as a sequence of contiguous, or adjacent, memory cells, as shown in Figure 6.7. Addresses of these memory locations are assigned sequentially so that available addresses proceed from zero (low memory) to the maximum available address (high memory). Thus, if eight bits were used to define each address, the memory sequence would proceed from 00H (low memory) to FFH (high memory). Typically, one byte of data or program instructions is held at each address. By convention, memory capacity is expressed in units of 1024 memory addresses, or 1K units. Thus, a microcomputer with 640K of main memory has 655,360 (640 × 1,024) storage locations in memory.

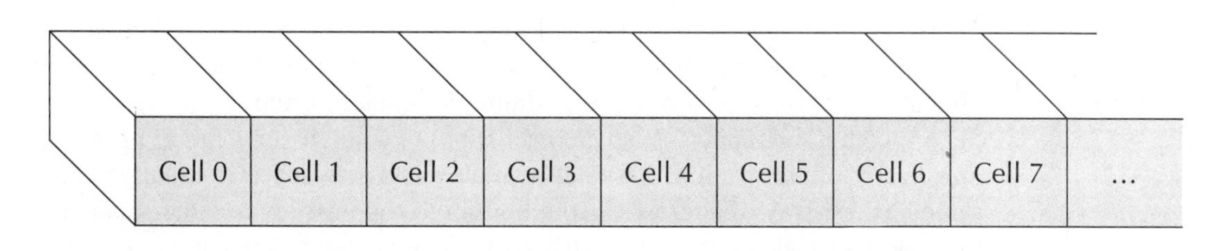

FIGURE 6.7 The sequential physical organization of memory cells.

In most computers, data values require multiple bytes of storage. An integer, for example, can be 16 bits in length, thus requiring two consecutive bytes of storage. When written in a program, the bits (and bytes) of a numeric value are typically ordered from highest position weight on the left to lowest position weight on the right. Thus, if a 16-bit integer is written as 11110000111000, it is assumed that the left position weight is the largest (2^{15}) and the right position is the smallest (2^0). When considered as a sequence of bytes, the leftmost byte is called the *most significant byte*; the rightmost byte is called the *least significant byte*.

In most CPU and memory architectures, the least significant byte of a data item is placed at the lower memory address. For example, the 16-bit word FF00 would be stored in memory with 00 at the lower memory address and FF at the next higher address. Because this storage sequence is the opposite of the order in which the word would be written in program text, this allocation scheme sometimes is called "back-words" storage.

The number of bits used to define memory addresses determines the maximum number of locations that can be referenced. This potential capacity is called the *addressable memory* of the processor. For example, in a computer that uses 24-bit addresses, there are 16 megabytes of addressable memory. By contrast, the computer's *physical memory* corresponds to the number of memory cells that actually exist in a given hardware configuration. Thus, a computer's physical memory capacity can be less than or equal to the size of its addressable memory. Addressable memory limits should be the same for both the CPU and the bus. That is, the number of bits used to represent an address within the CPU should correspond to the number of lines used to implement the address bus.

Memory Allocation and Addressing

Memory allocation refers to the assignment of specific memory addresses to elements of systems software, application programs, and data. The memory allocation scheme for a relatively simple computer is shown in Figure 6.8. Such a computer supports a single user running a single application program. For clarity of illustration, the total available memory space is usually represented as an array of addresses from bottom to top, as shown, rather than as a long (left to right) continuous string of memory cells. The lowest memory addresses are generally allocated to systems software. Working RAM, the space that is available for application programs and data, occupies the space above this area.

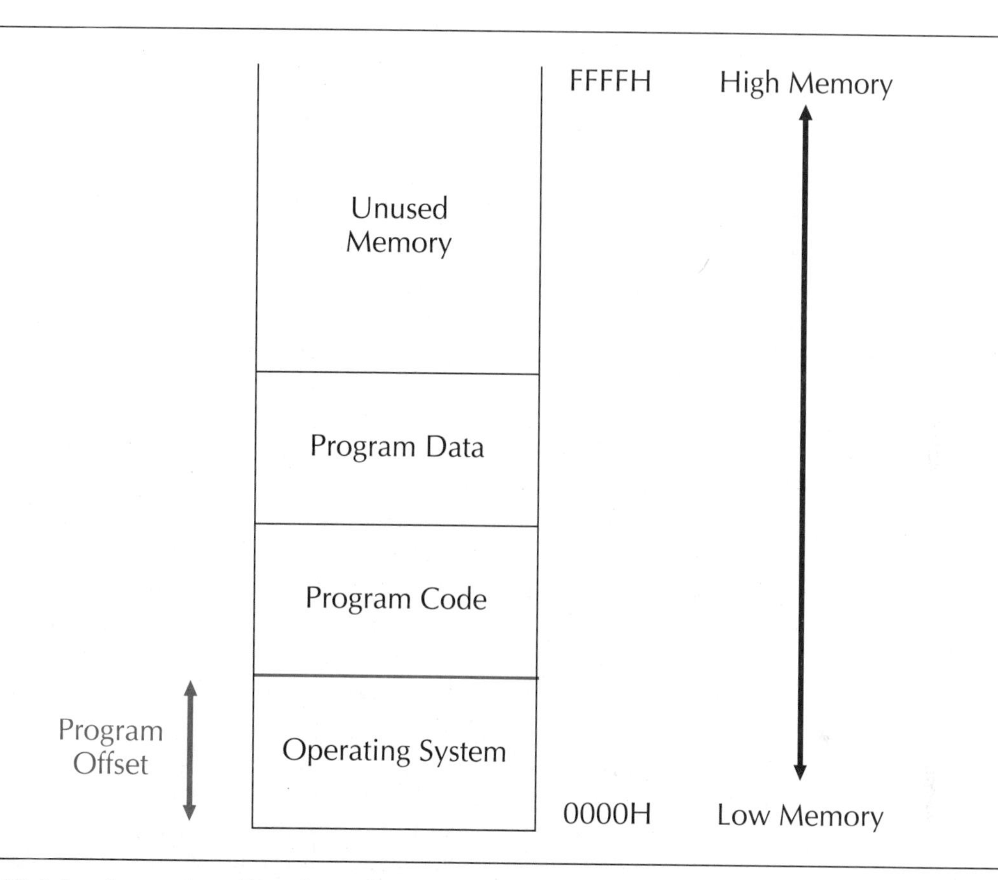

FIGURE 6.8 A memory allocation scheme for a single program.

Application programs are generally written as though their instructions started at address zero. This approach assumes that the program's instructions will be loaded into memory in sequence, starting at the lowest memory location and proceeding up through higher addresses, until the entire program has been loaded. Loading is usually performed from a secondary storage device, such as tape or disk. The utility program that performs this function is called a *loader*.

However, under the memory allocation shown in Figure 6.8, the application program cannot be loaded starting at address zero. Rather, the first available memory cell follows the highest address occupied by systems software in that area. The number of memory cells that must be skipped before the program can be loaded is the *offset* by which the program's starting address must be adjusted. Under the memory allocation scheme shown in the diagram, calculating the correct addresses for program instructions is relatively straightforward. The offset must be added to any memory addresses explicitly stated within the program (e.g., addresses used in load, store, and branch instructions).

One way of implementing an offset is to account for it at the time a program is created. Although program compilation proceeds under the assumption that the program will start at address zero, the instruction addresses could all be changed after compilation and before the program is loaded. This process, called *relocating* a program, is performed by a systems software program called a *relocator*. Essentially, the relocator searches through the program looking for any explicitly stated addresses and adds the offset value to each such address. During program execution, the CPU is said to be using *absolute addressing*, because the memory references within the program refer to actual physical memory locations.

The advantage to this approach is that the address computations (addition of offset) are performed once. Computing addresses at the time the program is executed can add complexity and inefficiency to program execution. The disadvantage is that the location of the program in memory is still fixed. Recall that the offset is essentially the size of the operating system. Should that size change—for instance, if we updated to a new, larger version of the operating system—all programs would have to be relocated.

The more common method of implementing an offset is to perform address calculations while the program is running. These calculations are typically performed within the control unit because it is responsible for memory accesses and for updating the instruction pointer. To speed execution, the offset can be stored in a special-purpose register and added to each explicit memory reference. Thus, values that are placed in the instruction pointer already will be incremented by the offset. Such a method is referred to as *indirect addressing*; the register that contains the offset value is called an *offset register*. This method is also called *relative addressing*, because each program address is determined relative to the content of the offset register.

Memory addressing and allocation schemes become much more complex in architectures that permit multiple programs to be held in memory. Such computer architectures can support multiple users, multitasking, virtual memory access, and so on. Figure 6.9 shows systems software and two programs stored in memory simultaneously. Note that the starting address of each program is different. Thus, the offset value for each program is also different (3000H for program #1, 9000H for program #2).

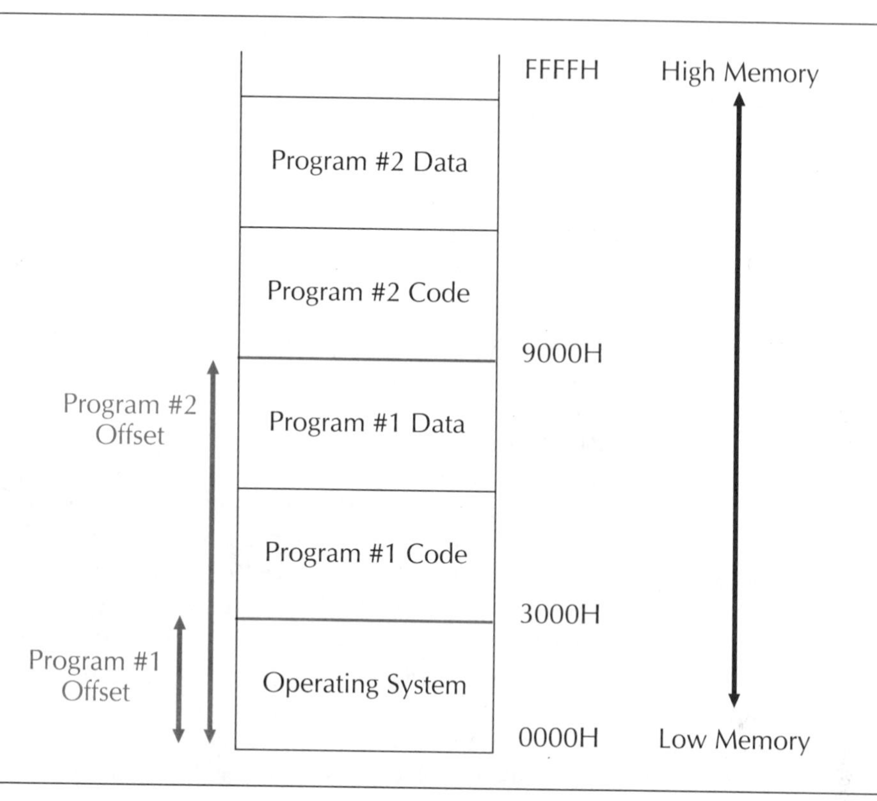

FIGURE 6.9 **Memory allocation for multiple programs.**

In this case, relative address calculations must use the offset applicable to the process currently being executed. This rule implies either the existence of multiple-offset registers or the use of a table of offset values in memory. The latter approach is more common because it is less likely to restrict the maximum number of processes in memory and is less costly. As CPU attention shifts from one process to another, the offset value for that process is loaded into the offset register from the table in memory. Processes can also be moved in memory, if necessary, and the appropriate entry in the offset table updated to reflect the move.

Cached Memory

Recall from Chapter 5 that current processor speeds generally meet or exceed RAM's ability to keep them supplied with data and instructions. This results in a delay whenever the CPU loads data from memory. A processor cycle spent waiting for data input is called a *wait state*. The number of wait states incurred for a memory access by the CPU depends on the relative speed of the CPU and memory. If memory access speed matches processor cycle time—for example, 20-nanosecond RAM and a 50-MHz processor—no wait states are incurred. A mismatch in speed leads to one or more wait states, depending on the severity of the mismatch. For example, a 50-MHz processor will incur two wait states each time it reads from 50-nanosecond RAM.

One way to address the problem of wait states is to provide a limited amount of very fast RAM and use slower (less expensive) RAM for the remainder of primary storage. The small area of fast RAM is called a *cache*. The cache holds a copy of the primary storage area that is currently being used by the processor. Accesses to the cache do not incur wait states. However, at some point the processor will require access to storage locations not held within the cache. When this happens, the contents of the cache must be copied to primary storage and a new area of primary storage (containing the data needed by the processor) must be copied into the cache.

When the data accessed by the processor is contained within the cache, the access is called a *cache hit*. When the data needed is not in the cache, the access is called a *cache miss*. Cache misses require that a *cache swap* to primary storage be performed. This swap usually consumes many processor cycles (and, thus, wait states), but the wait states incurred in a cache swap should be more than offset by the elimination of wait states for cache hits. This tradeoff is dependent on the size of the cache and the number of wait states incurred by a cache swap. A surprisingly small amount of cache memory is required to gain substantial performance gains. Typical memory cache sizes range from 8 to 256 kilobytes.

Early implementations of cached memory generally implemented cache control, such as content checks and cache swaps, through systems software. However, this resulted in a large amount of system overhead for every memory access. Modern implementations of cached memory rely exclusively on hardware-based cache control. Thus, the CPU and systems software are unaware of the use of a cache or of the cache management activities performed by the *cache controller*. Current microprocessors such as the Intel 80486 sometimes include both a cache controller and a small area of cache memory on the same chip as the CPU. Cache memory can also be used in conjunction with accesses to secondary storage, as described in Chapter 8.

I/O PROCESSING

Peripheral devices are connected to the CPU through *input/output (I/O) ports*. Such devices can include keyboards, scanners, display units, disk and tape drives, printers, modems, and so on. To each I/O port, the computer designer must assign a port number. The number of bits available for referencing ports determines the maximum number of ports that can be connected to the CPU. For example, a computer that uses four-bit port numbers could support up to 16 peripheral devices.

CPU access to I/O ports is obtained through the computer's bus. If the CPU has data or control information to send to a port, it transmits the port number over the control lines of the bus. This signal has the effect of selecting (or activating) a specific peripheral device. If a port senses its number on the bus, it responds by sending a *status code* back to the CPU to indicate whether it is ready or busy. If the CPU senses that the device is ready, it sends a command code. On receiving the command, the device performs the required I/O operation and sends the data or status result back over the bus.

In modern computers, each I/O port usually has its own processing circuitry. This processing capability is used for decoding commands from the CPU and for handling the sequence and timing of data transfer and control operations. Performing such processing at the port helps reduce the overhead of input/output operations for the CPU. The technical term that encompasses the access lines over the bus, the port, and its processing circuitry is *I/O channel.*

Interrupt Processing

In most I/O operations, data transfer rates are typically much slower than the speed of the processor. The time interval between a CPU request for input and the moment that input is received can span many CPU cycles. This is primarily due to physical (often mechanical) limitations of secondary storage and I/O devices. If the CPU waits for the completion of the request, the CPU cycles that could have been devoted to instruction execution are wasted. These wasted CPU cycles are referred to as *I/O wait states.*

To prevent such inefficient use of the processor, access to the CPU is usually controlled by *interrupt signals.* In its simplest sense, an interrupt is a signal to the CPU that some event has occurred that requires processor action. It is a coded signal that is generated and sent over the bus by a hardware device (such as an I/O device) to request access to the CPU or to trigger some processing action.

The implementation of interrupts requires a means for both recognizing and responding to the signal. Recognition is implemented by a storage location for the interrupt signal and by checking this storage location at the conclusion of each execution cycle. Each type of interrupt is assigned a unique code; an interrupt is made by placing that code on the bus. The bus interface logic of the CPU continuously monitors the bus for such signals. When an interrupt is detected on the bus, a corresponding *interrupt code*, or number, is placed either in a separate *interrupt register* or within a field of the PSW.

At the conclusion of each execution cycle the control unit checks the interrupt register for a nonzero value. If one is present, the CPU suspends execution of the current process (i.e., it does not fetch its next instruction) and proceeds to process the interrupt. When the interrupt has been processed, the interrupt register is reset to zero and execution of the suspended process resumes.

Communication with I/O devices via interrupts is advantageous only if the CPU has something else to do while it is waiting for an I/O-related interrupt to occur. If the CPU is executing only a single process or program, I/O wait states are not avoided. However, if the CPU is sharing its processing cycles among many processes, performance improvement can be substantial. While one process waits for data to be returned by an I/O device, the CPU can process instructions of another process. When an interrupt is received indicating that the I/O operation is complete, the CPU can process the interrupt (e.g., retrieve the data) and then return to the original (suspended) process.

Interrupt handlers. The interrupt-handling mechanism is less a hardware feature than a means of interfacing to systems software. Recall that the operating system service layer and kernel provide a set of routines, or service calls, for performing low-level processing operations. I/O operations comprise a large portion of these routines. Viewed in this light, an interrupt is actually a request by an I/O or storage device for service from the operating system. For example, an interrupt signal that indicates requested input is ready is actually a request to the operating system to retrieve the data and place it where the program that requested it can access it—for instance, in a register or in memory.

A service routine exists to process each possible type of interrupt. These service routines are called *interrupt handlers*. Interrupt handlers that process requests from storage and I/O devices are implemented within device drivers in the operating system kernel. In order to process an interrupt, the CPU must load the interrupt handler's first instruction for execution. But each interrupt uses a different interrupt handler and, thus, a different address for the first instruction.

The usual method for determining the proper set of instructions to process a given interrupt is to use a master interrupt handler and an *interrupt table*. The master interrupt handler is a service routine that examines the value in the interrupt register and determines the proper service routine to process the interrupt. An interrupt table is maintained; it contains each interrupt code and the interrupt handler's starting address in memory. The value in the interrupt register is used as an index into this table, and a branch instruction to the corresponding address is executed.

Multiple interrupts. The interrupt-handling mechanism described above seems adequate until one considers the possibility of multiple competing interrupts. What happens when an interrupt is received while the CPU is busy processing a previous interrupt? Which interrupt has priority? What is done with the process that doesn't have priority? The answers to these questions require the use of interrupt priorities and a stack.

Interrupts can be roughly classified into the following categories:

- I/O event
- Error condition
- Service request

The interrupts discussed thus far fall primarily in the category of I/O events. These are interrupts that notify the operating system that an I/O request has been processed and that data is ready for transfer. Error condition interrupts indicate error conditions resulting from processing. They can be generated by software (for example, an attempt to open a nonexistent file) or by hardware (for instance, division by zero).

In many operating systems, the interrupt-processing mechanism is the means by which application programs make requests to the operating system service layer. In such systems, interrupt codes are assigned to each service call. An application program requests a service by placing the corresponding interrupt number in the interrupt register. The interrupt code is detected at the conclusion of the execution cycle, the requesting process is suspended, and the service program is executed.

An operating system must establish a hierarchy of interrupts in terms of their importance (priority). For example, error conditions will normally be given higher precedence than other interrupts, and critical hardware errors (e.g., power failure) will be given the highest priority of all. The priorities are used to determine whether an interrupt that arrives while another interrupt is being processed will be handled immediately or suspended until current processing is finished. For example, if a hardware error interrupt code was detected while an I/O interrupt was being processed, the I/O processing would be suspended and the hardware error would be processed immediately.

Stack processing. Consider the following problem. While your computer is executing instructions in an application program, an interrupt is received from a pending I/O request. The interrupt is detected and the appropriate interrupt handler is called. During the execution of the interrupt handler, several values in general-purpose registers are overwritten by intermediate results generated by the interrupt handler. When the interrupt handler terminates, processing of the application program resumes, but an error occurs because a value in a register was altered during the execution of the interrupt handler.

How could this error have been prevented? How did the CPU know which instruction from the application program to load after the interrupt handler terminated? These problems arise from the need to "pick up where you left off" after a process or program is interrupted and suspended. It would be impractical (due to wasted resources) to restart the suspended process from the beginning. Therefore, a mechanism must be provided whereby processing can begin from the point of interruption. This mechanism is called a *stack*.

A stack is an area of storage that is accessed in a last-in, first-out (LIFO) basis. It is analogous to a stack of plates or the bullets in the ammunition clip of an automatic pistol or rifle. As items are added to the stack, they are placed on the top. As items are removed, they are removed from the top and only from the top.

For the purposes of interrupt processing, the stack is a stack of register values. Whenever a process is interrupted, values in the CPU registers are added to the stack. This operation is called a *push*; the set of register values saved is referred to as the *machine state*. When an interrupt handler finishes, the set of register values on the top of the stack is removed and loaded back into the appropriate registers. This operation is called a *pop*.

It is not always necessary to push the values of all registers onto the stack. At a minimum, the current value of the instruction pointer must be pushed, because the instruction at that address will be needed to restart the interrupted process where it left off. If indirect addressing is used, the offset register values must also be pushed. The general-purpose register values are also pushed, because they can contain intermediate results that are needed for further processing. They might also contain values that need to be output to a storage or I/O device.

Multiple machine states can be pushed onto the stack as interrupts of high precedence occur while processing interrupts of lower precedence. It is possible for the stack to fill to capacity, in which case further attempts to push values onto the stack result in a *stack overflow* error. The size of the stack thus represents a limitation on the number of suspended processes.

The stack is physically implemented as a set of extra registers or as a special area of main memory. If the stack is implemented in memory, a separate register within the CPU called the *stack pointer* must be provided. This register always contains the memory address of the value on the top of the stack. Its value is incremented or decremented each time the stack is pushed or popped.

REDUCED AND COMPLEX INSTRUCTION SET COMPUTING

Reduced instruction set computing (RISC) is a relatively new philosophy of processor and computer systems design. To explain it fully, we must contrast it with the opposite design philosophy: *complex instruction set computing (CISC).* As discussed in Chapter 5, early computers were limited in both memory and processing power. Memory, in particular, was extremely expensive, and many computers had barely enough of it to hold an entire application program in memory.

One response to lack of memory in early computers was to implement *complex instructions*. A complex instruction is one that implements the combined function of several simpler instructions. Examples of highly useful complex instructions include those that implement complex integer arithmetic operations (e.g., division) and those that implement floating-point computations. As shown in Figure 6.2, complex arithmetic operations such as multiplication can be implemented by a sequence of simpler instructions. Floating-point computations can be performed as a series of integer math computations on the whole and fractional parts of a real number. However, implementation in this manner tends to be slow.

Recall that every instruction must be loaded and decoded during the instruction cycle, then executed during the execution cycle. Assume for a moment that a floating-point computation, implemented as a single complex instruction, can be executed in one processor cycle. Assume further that an equivalent operation can be performed with five integer math instructions, each of which requires one processor cycle. In this example, a direct implementation of the (complex) floating-point instruction saves four processor cycles each time it is executed.

The execution of other types of complex instructions results in similar processor cycle savings. Thus, programs that make use of many complex instructions will generally run faster than programs that implement these same functions by executing only simple instructions. Such programs will also require less memory as the complex functions are expressed more compactly (i.e., in a single instruction) than an equivalent set of simple instructions.

The primary architectural feature of a RISC processor is the absence of complex instructions in the instruction set. In particular, instructions that combine data transformation and data movement operations are usually avoided. For example, CISC processors often provide data transformation instructions of the form:

Transform(ADDRESS1,ADDRESS2,ADDRESS3)

where ADDRESS1 and ADDRESS2 are the memory addresses of data inputs and ADDRESS3 is the memory address to which the result will be stored. Depending on the processor implementation, this instruction might require as little as one or as many as four execution cycles to complete. In a RISC processor, transformation operations always use register inputs and outputs. Thus, the single complex instruction shown above would require four separate RISC instructions:

```
Load(ADDRESS1,R1)
Load(ADDRESS2,R2)
Transform(R1,R2,R3)
Store(R3,ADDRESS3)
```

Each of these instructions would be independently loaded and executed, thus consuming at least four execution cycles.

Although the lack of complex instructions is the primary distinguishing feature of a RISC processor, other differences are also typical. These include the use of fixed-length instructions and shorter instruction lengths than CISC processors. As an example, consider the differences between the complex instruction shown above and its comparable sequence of four simple instructions. Assume that 6 bits are needed to represent an op code, 4 bits for a register number, and 16 bits for an address. The complex instruction thus requires 54 bits of storage ($3 \times 16 + 6$).

In contrast, the simple load and store instructions require 26 bits and the transform instruction requires 18 bits. Because data transformations in a RISC processor are defined on registers, no more than one address will ever be used as an operand (e.g., for load or store operations). In general, simple instructions require relatively simple (and short) instruction formats; complex instructions require complex (and long) instruction formats. This difference has a significant impact on the design of the control unit and the interaction between the control unit and primary storage.

So why would a CPU designer choose to eliminate complex instructions from a processor's instruction set? The answer lies in the realities of modern processor implementation and the availability of memory. With respect to processors, recall that modern processors are implemented in ever smaller semiconductor implementations. Also, recall that the speed of a processor is inversely proportional to its size.

Complex instruction sets require a complicated processor. To implement complex instructions, one must add processing circuitry to the processor to handle them. Additional circuitry is required for both decoding and execution. Another complicating factor is that complex instructions require longer and/or variable instruction formats. These increase the time necessary to decode an instruction and can increase the complexity of instruction fetches.

Thus, the implementation of complex instructions is not free. The price is substantially more complex processor circuitry. This was not much of a problem when processors and memory were slow. However, as engineers attempt to produce ever smaller (and thus faster) processors, the complexity associated with complex instructions becomes a barrier. It is far easier to reduce the size of the simpler circuitry of a RISC processor than to reduce the size of the more complex circuitry of a CISC processor.

Because of its reduced circuitry, a RISC processor can generally operate at higher speeds (lower cycle times) than can a CISC processor. Thus, *every* instruction executed by a CISC processor runs more slowly than the same instruction executed on a RISC processor. Thus, a tradeoff exists. Is program execution faster when complex instructions are provided, but the processor is relatively slow? Or does a program run faster when no complex instructions are provided, but the instructions that are provided execute relatively quickly? The answer, based on experience with RISC processors over the last decade, seems to be that RISC is preferable in most situations.

As mentioned earlier, another tradeoff that must be reconciled is the extra memory required by programs that cannot use complex instructions. Once again, the realities of modern technology have shifted the balance. Although memory was once quite expensive, it is now relatively cheap. The cost of memory as a percentage of the cost of the entire computer system and of the processor has declined substantially. The result is that architectural features that save memory—for example, complex processor instructions—are no longer as valuable as they once were. Thus, the cost of extra memory associated with RISC processing is relatively insignificant compared to the value gained by improving overall execution speed.

SAMPLE ARCHITECTURE: IBM 360/370 SERIES COMPUTERS

Mainframe computers that use the instruction set of the IBM 360/370 CPU comprise the majority of mainframe computers currently in use in business organizations. The IBM 360 was introduced in 1964 as a general-purpose computer. In the early 1970s, the 370 Series appeared. This machine was based on much the same design as its predecessor, but the implementation used integrated circuitry more extensively. Other enhanced features included expanded systems control facilities and memory management systems.

In the late 1970s, the 3000 Series was introduced, followed by the 4300 Series. These machines provided greater flexibility of configuration and expanded memory-addressing capabilities. Also during this time, other companies such as Wang, Amdahl, Fujitsu, and Nixdorf marketed machines that, although different in hardware configuration, were largely software compatible with the IBM models. All of these machines have been designed around the same instruction set.

The result has been that, for more than 30 years, the instruction set of the original IBM 360 has served as a de facto standard for mainframe computers. In many ways, this architecture represents the pinnacle of CISC processor and memory architecture. Its complexity represents the tradeoffs in memory cost, execution speed, and ease of programming that were typical of computer hardware in the 1960s.

Although the architecture has many shortcomings with respect to current hardware and systems software technology, it is still widely used. In large part, this is due to the huge base of existing systems and application programs that were designed for the architecture. Adherence to the architectural features of the original IBM 360 guarantees compatibility between this base of installed software and later generations of mainframe hardware.

Registers

The IBM 360 architecture specifies 16 registers, each of which can hold up to 32 bits. This capacity is needed to accommodate the relatively large addresses and instruction formats that are used by the system. Some of the larger machines in this family have additional registers that are used to hold data in floating-point notation.

Instruction Formats

The IBM 360 architecture uses five instruction formats, as shown in Figure 6.10. Among these formats, lengths of instructions are 16, 32, or 48 bits. Within the CPU, up to three registers might be required to hold a single instruction. Although lengths vary among the formats, the length of each type of instruction is fixed. In all instructions, the first eight bits contain the op code. The remaining bits specify registers or are used to derive addresses. Each instruction format corresponds to a category of operations:

- RR Instructions in the RR format specify transfers between registers.
- RX The RX format specifies an action between a register and a memory location.
- RS The instruction format RS specifies actions among two registers and a memory location. For example, the contents of two registers might be compared as a condition for branching to the memory location.

SI Instructions in the SI format specify an operation on the content of a single memory location.

SS The only 48-bit format, SS specifies an operation that involves two memory locations, as in copying a string of data from one address in memory to another.

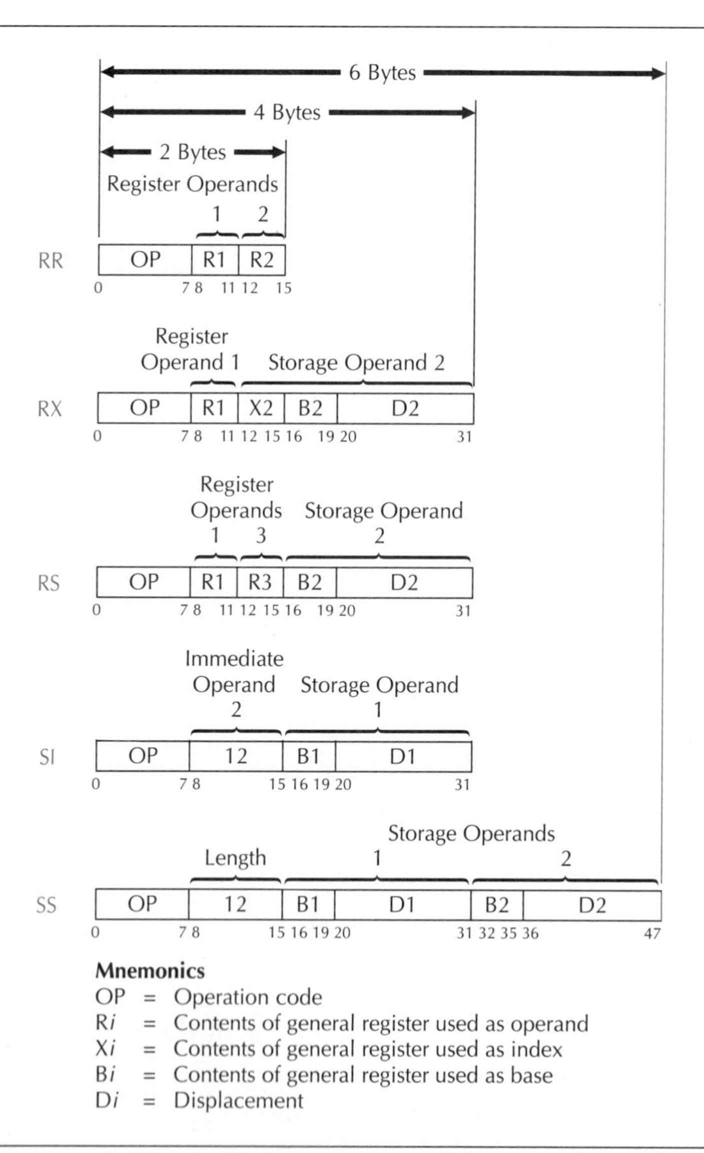

FIGURE 6.10 **The five instruction formats for the IBM 360/370 processor.**

The IBM 360 Instruction Set

The instruction set of the IBM 360 includes more than 140 op codes, as shown in Table 6.1. This relatively large number of instruction types contrasts sharply with the minimal instruction set described earlier in the chapter. Variations in instruction type and format provide flexibility of coding and addressing. This flexibility is needed for efficient, concurrent allocation of systems resources among multiple users and application programs.

TABLE 6.1 The IBM 360 instruction set.

Name	Mnemonic	Op Code	Format
Add	AR	1A	RR
Add	A	5A	RX
Add Decimal	AP	FA	SS
Add Halfword	AH	4A	RX
Add Logical	ALR	1E	RR
Add Logical	AL	5E	RX
Add Normalized, Extended (c, x)	AXR	36	RR
Add Normalized, Long (c)	ADR	2A	RR
Add Normalized, Long (c)	AD	6A	RX
Add Normalized, Short (c)	AER	3A	RR
Add Normalized, Short (c)	AE	7A	RX
Add Unnormalized, Long (c)	AWR	2E	RR
Add Unnormalized, Long (c)	AW	6E	RX
Add Unnormalized, Short (c)	AUR	3E	RR
Add Unnormalized, Short (c)	AU	7E	RX
AND	NR	14	RR
AND	N	54	RX
AND	N1	94	SI
AND	NC	D4	SS
Branch and Link	BALR	05	RR
Branch and Link	BAL	45	RX
Branch on Condition	BCR	07	RR
Branch on Condition	BC	47	RX

(continued)

TABLE 6.1 The IBM 360 instruction set. (continued)

Name	Mnemonic	Op Code	Format
Branch on Count	BCTR	06	RR
Branch on Count	BCT	46	RX
Branch on Index High	BXH	86	RS
Branch on Index Low or Equal	BXLE	87	RS
Clear I/O	CLRIO	9D01	S
Compare	CR	19	RR
Compare	C	59	RX
Compare and Swap	CS	BA	RS
Compare Decimal	CP	F9	SS
Compare Double and Swap	CDS	BB	RS
Compare Halfword	CH	49	RX
Compare Logical	CLR	15	RR
Compare Logical	CL	55	RX
Compare Logical	CLC	D5	SS
Compare Logical	CLI	95	SI
Compare Logical Characters under Mask	CLM	6D	RS
Compare Logical Long	CLCL	0F	RR
Convert to Binary	CVB	4F	RX
Convert to Decimal	CVD	4E	RX
Diagnose		83	
Divide	DR	1D	RR
Divide	D	5D	RX
Divide Decimal	DP	FD	SS
Edit	ED	DE	SS
Edit and Mark	EDMK	DF	SS
Exclusive OR	XR	17	RR
Exclusive OR	X	57	RX
Exclusive OR	XI	97	SI
Exclusive OR	XC	D7	SS

(continued)

TABLE 6.1 The IBM 360 instruction set. (continued)

Name	Mnemonic	Op Code	Format
Execute	EX	44	RX
Halt I/O	HIO	9E00	S
Halt Device	HDV	9E01	S
Insert Character	IC	43	RX
Insert Characters under Mask	ICM	BF	RS
Insert PSW Key	IPK	B208	S
Insert Storage Key	ISK	09	RR
Load Register	LR	18	RR
Load	L	58	RX
Load Address	LA	41	RX
Load and Test	LTR	12	RR
Load Complement	LCR	13	RR
Load Control	LCTL	B7	RS
Load Halfword	LH	48	RX
Load Multiple	LM	98	RS
Load Negative	LNR	11	RR
Load Positive	LPR	10	RR
Load PSW	LPSW	82	S
Load Real Address	LRA	81	RX
Monitor Call	MC	AF	SI
Move	MV1	92	SI
Move Characters	MVC	D2	SS
Move Long	MVCL	0E	RR
Move Numerics	MVN	01	SS
Move with Offset	MVO	F1	SS
Move Zones	MVZ	D3	SS
Multiply	MR	1C	RR
Multiply	M	5C	RX
Multiply Decimal	MP	FC	SS

(continued)

TABLE 6.1 **The IBM 360 instruction set. (continued)**

Name	Mnemonic	Op Code	Format
Multiply Halfword	MH	4C	RX
OR	OR	16	RR
OR	O	56	RX
OR	O1	96	SI
OR	OC	D6	SS
Pack	PACK	F2	SS
Purge TLB	PTLB	B20D	S
Read Direct	ROD	85	SI
Reset Reference Bit	RRB	8213	S
Set Clock	SCK	8204	S
Set Clock Comparator	SOKC	8206	S
Set CPU Timer	SPT	8208	S
Set Prefix	SPX	8210	S
Set Program Mask	SPM	04	RR
Set PSW Key from Address	SPKA	B20A	S
Set Storage Key	SSK	08	RR
Set System Mask	SSM	80	S
Shift and Round Decimal	SRP	F0	SS
Shift Left Double	SLDA	8F	RS
Shift Left Double Logical	SLDL	8D	RS
Shift Left Single	SLA	8B	RS
Shift Left Single Logical	SLL	89	RS
Shift Right Double	SRDA	8E	RS
Shift Right Double Logical	SRDL	8C	RS
Shift Right Single	SRA	8A	RS
Shift Right Single Logical	SRL	88	RS
Signal Processor	SIGP	AE	RS
Start I/O	SIO	9C00	S
Start I/O Fast Release	SIOF	9C01	S
Store	ST	50	RX

(continued)

TABLE 6.1 The IBM 360 instruction set. (continued)

Name	Mnemonic	Op Code	Format
Store Channel ID	STIDC	8203	S
Store Character	STC	42	RX
Store Characters under Mask	STCM	BE	RS
Store Clock	STICK	B205	S
Store Clock Comparator	STICKC	B207	S
Store Control	STCTL	B6	RS
Store CPU Address	STAP	B212	S
Store CPU ID	STIDP	B202	S
Store CPU Timer	STPT	B209	S
Store Halfword	STH	40	RX
Store Multiple	STM	90	RS
Store Prefix	STPX	B211	S
Store Then AND System Mask	STNSM	AC	SI
Store Then OR System Mask	STOSM	AD	SI
Subtract	SR	18	RR
Subtract	S	58	RX
Subtract Decimal	SP	FB	SS
Subtract Halfword	SH	4B	RX
Subtract Logical	SLR	1F	RR
Subtract Logical	SL	5F	RX
Supervisor Call	SVC	0A	RR
Test and Set	TS	93	S
Test Channel	TCH	9F00	S
Test I/O	TIO	9D00	S
Test under Mask	TM	91	SI
Translate	TR	DC	SS
Translate and Test	TRT	DD	SS
Unpack	UNPK	F3	SS
Write Direct	WRD	84	SI
Zero and Add Decimal	ZAP	F8	SS

Memory Addressing and Allocation

Memory configurations and capacities vary widely among computer models in this family. However, the addressing scheme remains the same. Computers in the 360 family use indirect memory addressing. Memory is organized into *blocks* of 4,096 bytes. Subdivision into blocks facilitates partitioning of memory among users or among programs. To specify the address of any byte in memory, two values must be given in the instruction. The first value is the *base address*, or the starting address of the block in which the byte is located. The second value is the *displacement* (offset) of the byte relative to the base address. When an instruction is loaded into the CPU, the register at which the starting address of the block is found is called the *base register*.

Data Formats and Arithmetic Functions

The IBM 360 CPU can handle data in a variety of formats. Characters are represented in EBCDIC. Numeric values can be manipulated in EBCDIC, binary-coded decimal (BCD), or two's-complement binary. Extensive support is provided for floating-point calculations, including representation of values in double-precision (i.e., 64-bit) notation.

SUMMARY

The CPU of a computer system is composed of a control unit, an arithmetic logic unit, and processor registers. The basic capabilities of a CPU are defined by its instruction set. Other CPU design considerations include instruction format, word size, register use, bus capabilities, clock rate, memory organization, and memory access methods. The overall capability and power of a computer system is dependent on the interaction of all these architectural considerations.

The instruction set of a CPU contains data transformation, data movement, and sequence control instructions. The minimal set of data transformation instructions include, AND, OR, exclusive OR, ADD, shift, and rotate instructions. The first four operate on two inputs (operands) to produce a separate result. The latter two instructions modify the content of a single operand.

Data movement instructions include load, store, and move. A load instruction copies the contents of a memory location to a register. A store instruction copies the contents of a register to a memory location. A move instruction copies the data from one register to another or from one memory location to another.

Sequence control instructions include unconditional branch (jump), conditional branch, and halt. A jump instruction causes the instruction sequence of the current program to be altered. A conditional branch alters instruction sequence only if a stated condition is satisfied. A halt instruction terminates a program and generally returns control to the operating system.

CPU registers may be either general or special purpose. General-purpose registers are used to store intermediate processing results. Their use improves program efficiency by avoiding CPU access to memory. Within practical limits, a higher number of general-purpose registers implies a more powerful CPU.

There are a number of special-purpose registers within a CPU. These include the instruction register, the instruction pointer, the program status word, the offset register, the interrupt register, and the stack pointer. The instruction register holds the current instruction (op code and operands) for decoding by the control unit. The instruction pointer holds the memory address of the next instruction to be fetched. Altering the content of the instruction pointer is the mechanism by which branch and halt instructions are implemented. The program status word holds status and error codes.

The CPU's instruction format refers to the size of an instruction and the number and location of parameters within the instruction. Instructions generally consist of an op code (instruction number) and zero, one, or two operands (data inputs). An instruction can be of fixed or variable length. Fixed-length instructions contain fixed-length components in predetermined positions. Variable-length instructions allow instruction components to vary in size and/or position. Variable-length instructions economize on memory used to store instructions at the expense of more complex decoding by the control unit.

The bus is the communication pathway that connects the CPU with memory and other devices. The bus is implemented as a set of data lines, control lines, and status lines. The number and use of these lines and the procedures for controlling access to the bus are called the bus protocol. The CPU is generally assumed to control access to the bus. However, the implementation of a multimaster bus or direct memory access allows devices other than the CPU to temporarily control access to the bus. This generally results in a performance improvement.

The word size of a CPU is the number of bits processed by the CPU during a single execution cycle. It is also the size of processor registers and is usually equivalent to the number of bus data lines. Matching word size to the number of bus data lines ensures that the CPU can be efficiently supplied with instructions and data. Within limits, larger word size implies greater processing power and efficiency.

Memory addressing is implemented through the control unit. The number of bits allocated to hold memory addresses determines the maximum amount of addressable memory for the CPU. This could be less than the amount of memory physically present in the CPU.

Programs are generally created as though they occupied contiguous memory locations, starting at the first location (i.e., low memory). Programs are rarely physically located in low memory due to systems software or other programs that can occupy those locations. Offset addressing is a mechanism to reconcile the difference between where a program "thinks" it is located and where it is actually located. The actual address of a program's first instruction is stored in an offset register; this address is added to each address reference made by the program. This mechanism allows multiple programs to reside in memory and allows programs to be moved during execution.

An interrupt is a signal to the CPU that some condition requires its attention. Interrupts are identified by unique codes (numbers) and can be classified as I/O events, error conditions, or program service requests. Interrupts can be placed on the bus by an I/O device, or they can be placed directly in the interrupt register by a program or CPU component. The control unit checks for interrupts at the conclusion of each execution cycle. If an interrupt is detected, a systems software program called an interrupt handler is executed to service the interrupt. The starting address of the interrupt handler for each interrupt is maintained in an interrupt table.

The detection of an interrupt causes the currently executing process to be suspended. The stack is a mechanism that allows a suspended process to be restarted at the point where it was interrupted. When an interrupt is detected, the current values of processor registers are placed on the stack. These values describe the current state of the process. When an interrupt handler finishes execution, the values on the stack are copied back into the processor registers, and the suspended process resumes execution.

A reduced instruction set computer (RISC) processor is a CPU with several specific architectural features. These include fixed-length instructions and the absence of complex instructions. These design features allow CPU circuitry to be substantially simplified, compared to a complex instruction set computer (CISC) processor. As a result of this simplicity, it is possible to make these circuits smaller and faster. Thus, the clock rate of a RISC processor is generally higher than that of a CISC processor. Although the absence of complex instructions makes some operations slower, the faster clock rate for simple instructions usually offsets this loss, resulting in a net performance gain for RISC processors.

Key Terms

absolute addressing
ADD operation
address bus
addressable memory
AND operation
arithmetic shift
base address
base register
block
branch instruction
bus
bus cycle
bus master
bus protocol
bus slave
cache
cache controller
cache hit
cache miss
cache swap
clock rate
complex instruction
complex instruction set computing (CISC)
conditional branch
control bus
data bus
direct memory access (DMA)
displacement
DMA controller
dyadic instruction
exclusive OR (XOR) operation
fixed-length instruction
HALT instruction
indirect addressing
input/output (I/O) port
I/O channel
I/O wait state
instruction pointer (program counter)
instruction register
interrupt code
interrupt handler
interrupt register
interrupt signal
interrupt table
jump
least significant byte
LOAD operation
loader
machine primitive
machine state
memory allocation
monadic instruction
most significant byte
move operation
multiple-master bus
offset
offset register
op code
operand
OR operation
physical memory
pop operation
program status word (PSW)
push operation
reduced instruction set computing (RISC)
relative addressing
relocation
relocator
rotate operation
shift operation
stack
stack overflow
stack pointer
status code
store operation
system clock
unconditional branch
variable-length instruction
wait state
word

Vocabulary Exercises

1. The _____ is the communication channel that connects all computer system components.
2. In a _____, bit pairs of 1/1, 1/0, and 0/1 produce a result bit of 1.
3. The address of the next instruction to be fetched by the CPU is held in the _____.
4. The CPU is always capable of being a _____, thus controlling access to the bus by all other devices in the computer system.
5. Under _____, a program's memory references correspond to physical memory locations. Under _____, the CPU must calculate the physical memory locations that correspond to a program's memory reference.
6. A _____ is a small area of fast memory that can be used to minimize wait states.
7. A cache controller is a hardware device that initiates a _____ when it detects a _____.
8. The _____ is used to transmit command, timing, and status signals between devices in a computer system.
9. A _____ has only one operand. A _____ has two operands.
10. During a _____, bits of data values are moved right or left, and excess bits are "wrapped around" to the other side of the value.
11. The contents of a memory location are copied to a register while executing a _____.
12. The set of register values placed on the stack while processing an interrupt is also called the _____.
13. A _____ alters the sequence of instruction execution. A _____ alters the sequence of instruction execution only if a specified condition is true.
14. A _____ is a software program that is executed in response to a specific interrupt.
15. A _____ processor does not directly implement complex instructions.

16. During interrupt processing, register values of a suspended process are held on the _____.
17. A _____ is a signal to the CPU or operating system that some device or program requires processing services.
18. The _____ holds the bits of lowest magnitude in a data item.
19. The time necessary to complete an instruction cycle or execution cycle is a fraction of a processor's _____.
20. The _____ is used to store the amount by which a program's starting address in memory differs from zero.
21. The _____ of a computer system is determined by the number of bits used by the CPU and bus to represent a primary storage location.
22. The CPU incurs one or more _____ when it is idle, pending the completion of an operation by another device within the computer system.
23. A _____ is the number of bits processed by a CPU in a single instruction. It also describes the content of a single register.
24. During a _____, one or more register values are copied to the top of the stack. During a _____, one or more values are copied from the top of the stack to registers.
25. In many CPUs, a register called the _____ stores condition codes, including those representing processing errors and the results of comparison operations.
26. The components of an instruction are its _____ and one or more _____.

Review Questions

1. Explain the execution of the load, move, and store instructions. Why is the name "move" a misnomer?
2. Explain the operation of the shift and rotate instructions.
3. What is the difference between a conditional branch instruction and an unconditional branch instruction?

4. Why does the execution speed of an application program generally increase as the number of general-purpose registers increases?
5. What are special-purpose registers? Give three examples of special-purpose registers and explain how each is used.
6. What are the advantages and disadvantages of fixed-length instructions, compared to variable-length instructions? Which type are generally used in a RISC processor?
7. What is the system bus? What are its primary components?
8. What is a bus master? What is the advantage of allowing devices other than the CPU to be a bus master?
9. Define the term *word size*. What are the advantages and disadvantages of increasing word size?
10. What characteristics of the CPU and of main memory should be "balanced" to obtain maximum system performance? What characteristics of the CPU and of the system bus should be "balanced" to obtain maximum system performance?
11. What is direct addressing? What is relative addressing?
12. What are the costs and benefits of indirect addressing?
13. What is an interrupt? How is an interrupt generated? How is it processed?
14. What is a stack? Why is it needed?
15. Describe the execution of the push and pop operations?
16. How does a RISC processor differ from a CISC processor?
17. Under what assumptions will an application execute faster on a RISC processor than on a CISC processor?

Research Problems

1. Research the instruction set and architectural features of a modern RISC processor such as the MIPS 4000 or the Digital Equipment Corporation Alpha. In what ways do they differ from the architecture of the IBM 360/370 processor?

2. The original IBM Personal Computer used a bus architecture that has become known as the Industry Standard Architecture (ISA) bus. Two alternatives to this bus architecture are the Extended Industry Standard Architecture (EISA) and Microchannel Architecture (MCA) buses. Describe the architectural features and differences of these three buses. What shortcomings of the ISA bus are addressed by the other two? Which promises better performance for personal computers using modern microprocessors such as the Intel 80486 or Intel Pentium?

7 Data Communication Technology

Chapter Goals

- *Describe the signals and media used to transmit data.*
- *Describe the methods of encoding and transmitting data using analog or digital methods.*
- *Describe methods for detecting and correcting data transmission errors.*
- *Describe methods for efficiently utilizing communication channels.*

In digital communication, a unit of information to be transferred is called a *message*. The communication path used is the *transmission medium*. Copper wire and optical fiber are two types of transmission medium. The atmosphere or space can be used as a medium for radio or microwave transmissions. For data transfer to be achieved, messages must be encoded in signals that can be conducted or propagated through the transmission medium. In copper wires, signals are carried as streams of electrons. In optical fiber, signals are pulses of light. Microwave and satellite transmissions are broadcast as *radio frequency (RF)* radiation through space. Optical transmissions through space can be performed with beams of laser-generated light.

Channel Characteristics

A *communication channel* is composed of a message, a sending device, a receiving device, and a transmission medium, as depicted in Figure 7.1. In a strict, technical sense, a communication channel provides for the transfer of messages in one direction only. For two-way, or bidirectional, communication, two channels must be used, unless some mechanism is provided for sharing a single channel.

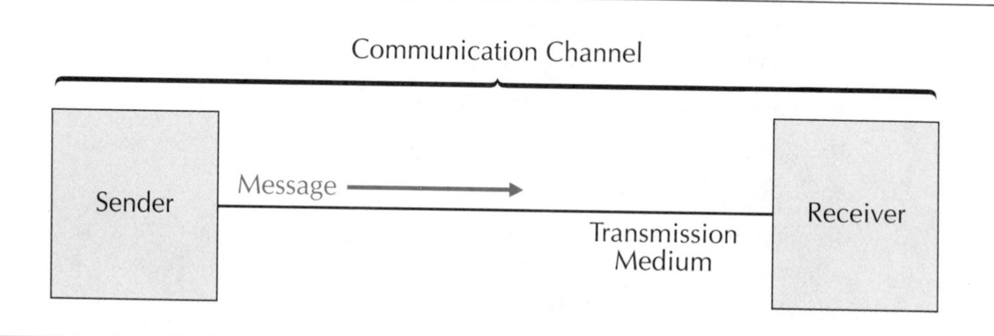

FIGURE 7.1 **The basic elements of a communication channel: a transmitter, a transmission medium, a receiver, and a message.**

Fundamental communication channel tradeoffs involve speed and cost. For a given type of communication channel, a gain in transmission speed usually brings a corresponding increase in cost. However, due to the characteristics of communication channels, absolute speed is an unreliable measurement. A more useful measure of communication capacity is the overall *data transfer rate* that is possible within a given channel. The data transfer rate of a communication channel corresponds to the number of data bits per second (bps) that are received correctly.

Effectively, the data transfer rate of a channel is the transmission speed minus the rate at which errors are encountered. Errors in data transmission can be caused by channel characteristics, including

- Noise
- Distortion

Within a communication channel, *noise*[1] refers to any extraneous signals that are interpreted incorrectly as data. Transmission media such as copper wires are affected by *electromagnetic interference (EMI)*. EMI can be produced by a variety of sources, including electric motors, radio equipment, and other communication lines. EMI can induce noise in nearby electrical circuits, including communication wires and equipment. In an area that is dense with wires and cables, this problem can be compounded because each transmission path can both radiate and respond to EMI. Another source of errors in communication is *distortion*, or any characteristic of a communication channel that causes a data signal to be altered. Altered signals might be misinterpreted or not detected at all.

MEASURING INFORMATION-HANDLING CAPACITY

For a given channel or network, the basic measure of information-handling capacity is the data transfer rate, although possible efficiencies of resource sharing and loading also must be considered. For a given communication channel, factors that affect the overall data transfer rate include

- Signal-to-noise ratio
- Bandwidth
- Coding methods

In combination with these factors, the overall data transfer rate also will be affected by channel organization and transmission mode. These topics are discussed later in this chapter.

Signal-to-Noise Ratio

For a receiving device to interpret data signals correctly, it must be able to distinguish those signals from noise that could be present in the channel. Distinguishing valid signals from extraneous noise becomes increasingly difficult as the speed of

[1] Noise can be heard in many common household devices. Turn on a radio and tune it to a frequency (channel) on which no local broadcasting station is transmitting. The hissing sound is produced by the radio receiving and amplifying background radio frequency noise. Noise can also be heard on any home stereo system. Set the amplifier or receiver input to a device that isn't turned on (e.g., a tape or CD player) and turn the volume up relatively high. (Be sure to turn it back down when you're finished!) The hissing sound is the result of amplifying noise in the signal transmission and amplification circuitry.

transmission increases. The actual "speed limit" of any given channel is determined by the power of the message-carrying signal in relation to the power of the noise in the channel. This relationship is called the *signal-to-noise ratio (S/N)* of the channel.

As an example, consider a human listener's receipt of spoken words during a speech. The difficulty (or ease) of understanding the speech is directly related to the speed at which it is delivered (spoken) and the volume (i.e., strength or power) of the speech in relation to background noise. Accurate reception of speech is impaired by sources of noise, such as other people talking, an air conditioning fan, or a construction project across the street. The speaker can compensate for noise by increasing the volume of speech—in other words, increasing the signal-to-noise ratio. Accurate reception is also impaired if the speaker speaks too quickly. This is due to insufficient time for the listener to interpret one signal (a word or speech fragment) before the next is received.

In digital signal transmission, a message is composed of individual bits. Each bit is a period of time during which a message representing a zero or one is present on the channel. As transmission speed is increased, the duration of each bit in the signal (called the *bit time*) decreases. Eventually, a limit is reached at which the bit time is so short that it cannot be distinguished from noise. The optimum transmission speed within a channel will generally be considerably lower than the maximum implied by the signal-to-noise ratio. The optimum speed is usually greater than, but nearly equal to, the data transfer rate. That is, transmission speed should not exceed the maximum rate at which data can be sensed reliably.

If a higher speed is attempted, the rate of errors will also increase. An error in transmission represents a wasted opportunity to transmit a message, thus reducing the effective (or net) data transfer rate. A further difficulty is that noise usually is not constant. In electronic signals, noise usually occurs in short intermittent bursts due to, for example, a nearby lightning strike or the startup of an electric motor. As discussed below, the receiver can request retransmission of a message if errors are detected. If noise bursts are infrequent, retransmissions might not diminish the overall data transfer rate significantly, and a relatively high transmission speed could be used.

An ideal communication channel would exhibit little or no noise. In copper wire or radio frequency channels, some noise usually is present, and the noise level increases with distance. To compensate for these effects, line conditioning equipment such as amplifiers and repeaters can be used to boost the power of the signal and/or to suppress noise. Of course, this equipment adds to the cost of installing and maintaining the channel. Not surprisingly, telecommunication vendors charge premium rates for high-quality service. One reason telecommunication utilities use fiber optic cables for long-distance links is that optical channels neither respond to nor radiate EMI and exhibit extremely low noise levels.

Bandwidth

As stated above, increasing the speed of transmission causes a corresponding decrease in the bit time of the data signal. A measure of the number of times that a signal can change over a unit of time is its *frequency.* Frequency is usually expressed in cycles per second, or *Hertz (Hz).* Thus, the *frequency range* of a channel imposes lower and upper limits on signal frequencies and on bit times and speeds. The difference between the lower and upper frequencies to which a channel will respond is called its *bandwidth.*

For example, the frequency range of an analog, voice-grade telephone line is from about 300 Hz to 3400 Hz. Signals below 300 Hz and above 3400 Hz cannot be propagated across these lines. Thus, the bandwidth of the channel is the difference between these frequencies, or 3100 Hz. In contrast, the frequency range of human speech is approximately 200 Hz to 5000 Hz (4800 Hz bandwidth), and the frequency range of the human ear is approximately 20 Hz to 20,000 Hz (19,980 Hz bandwidth). Thus, the use of analog telephone lines provides only an approximation of human speech—some high and low frequencies are lost—and is incapable of carrying many signals (for instance, the signals of high-fidelity music) that can be received by humans.

In general, bandwidth, like noise reduction, is a cost factor. That is, increasing the bandwidth of a communication channel increases its cost, often at a nonlinear rate. However, partitioning a channel into multiple subchannels can serve to allocate the available bandwidth efficiently. Channels with wide bandwidth may be subdivided into multiple subchannels, as discussed later in this chapter.

Coding Method

Regardless of channel characteristics, data transfer rate also is affected by the number of bits used to encode the message. For example, coding can be either

- Character-oriented
- Binary

Transmissions composed of character data typically are coded in ASCII-7 with an additional bit for error checking. Other coding schemes include ASCII-8 and EBCDIC. Such methods are collectively referred to as *character-oriented transmission methods*, because the characters are a basic unit of message transmission and error detection. As an alternative, transmissions may be sent in *transparent mode*, also called *binary mode*. Binary transmissions are handled as bit streams, usually combined into relatively large groups called *blocks*. Transparent mode can be used to send virtually any type of data—characters, floating-point numbers, and executable programs.

Transmission speed. Whether transmissions are character-oriented or binary, computer-to-computer communication generally consists of two-state (or binary) signals to represent data bits. In analog channels, a reference signal, or *subcarrier,* is used to carry the two-state data signal. One subcarrier frequency represents the on (or high) state, and another represents the off (or low) state. However, some specialized communication systems use multilevel coding schemes, which might use four-state signals. This distinction becomes important in connection with the *baud rate* of a communication channel.

The transmission speed of computer devices sometimes is given in baud. For example, a low-speed device might operate at 1200 baud. The mathematical definition of baud rate is:

$$\text{Baud rate} = 1 \div \text{Binary signal event duration}$$

The *binary signal event duration* is the time required to send the smallest unit of information. In the case of binary coding, the binary signal event duration is the same as one bit time. So, for a 1200 baud device, the bit time is about 0.0008 second.

For communication among most types of computer devices, the transmission speed expressed in baud is the same as the speed expressed in bits per second. This is always the case if a two-state signal is used, because each state corresponds to a single on or off bit. However, if multilevel coding is used, the baud rate would be slower than the bps rate. For example, multilevel coding with four-state modulation results in a bps rate double the baud rate, because each state can represent two bits ($2^2 = 4$) instead of only one. For this reason, it is preferred practice to express data transmission speeds **and** data transfer rates in bps, rather than in baud.

Data compression. Techniques for *data compression* can reduce the number of bits within either character-oriented or binary messages. These same techniques are often used to reduce data storage requirements, as discussed in Chapter 8. Typically, data compression schemes attempt to reduce the transmission or storage capacity used by repetitive messages. Such messages are coded in a more compact form by the sending device. The receiving device performs an inverse coding operation (decompression).

A simple method of data compression is the compression of repeated characters. Under this scheme, a series of identical characters is replaced with a two-byte message containing the number of times the character occurs and the character code. Such a message must be preceded by another byte (e.g., Null–ASCII 0) to indicate to the receiver that a coded (compressed) message follows. Thus, a sequence of 10 spaces would be encoded as:

00000000 00001010 00100000

where the first byte is ASCII NULL, the second byte is the number 10 (binary representation), and the third is ASCII SPACE. In this example, compression has saved seven characters worth of transmission capacity.

Data compression represents a deliberate tradeoff of additional processing overhead at each node for a reduction in data volume within the communication link itself. Processing overhead is incurred because data translation (compression and decompression) occurs at both ends of the channel. Unless the message contains many repetitive portions and can be compressed significantly, there may be little or no net increase in data transfer rate. Data compression might be justified if either the cost or delay of communication is exceptionally high, compared with processing overhead. As processing devices have become less expensive, data compression has become more common. Rather than using general-purpose CPU cycles, many data communication devices now employ special-purpose processors to perform compression and decompression.

CHANNEL ORGANIZATION AND TRANSMISSION MODES

Another set of variables that affects the rate and efficiency of data transfer includes the ways in which channels can be organized and the pattern of transmission (or transmission mode). For transmission through wires, the basic form of channel organization is shown in Figure 7.2. Note that a single communication line actually requires two wires: a *signal wire*, which carries data, and a *return wire*, which completes an electrical circuit between the sending device and the receiving device.

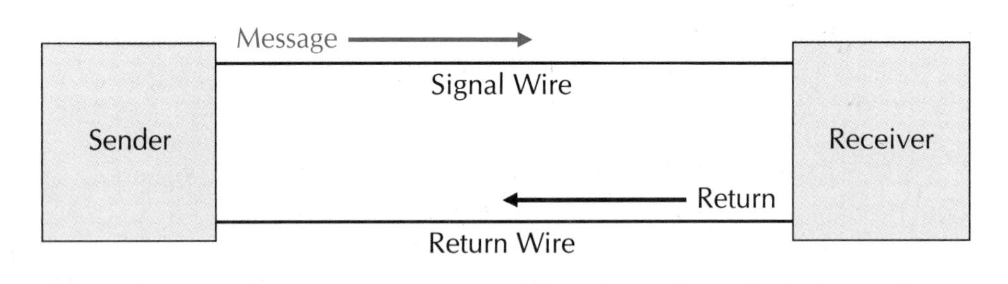

FIGURE 7.2 **The basic configuration of a two-conductor circuit. A sender and receiver are linked by a signal wire and a return wire.**

In some transmission modes, it might be necessary to provide two or more communication paths between sender and receiver. This type of channel organization is shown in Figure 7.3. In such a multichannel configuration, multiple signal wires share a single return wire, or common wire.

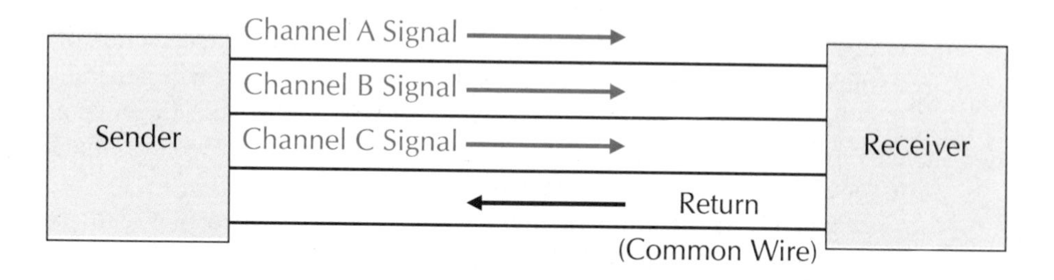

FIGURE 7.3 **Multichannel communication can be implemented with two or more signal wires and a single, shared common wire.**

These alternative methods of channel organization are required to support various tradeoffs among communication speed, capacity, and efficiency. Related alternatives include

- Transmission modes
- I/O modes
- Modulation schemes

Transmission Mode

The pattern of transmission (i.e., transmission mode) relates to the number of channels used and the direction of data transfer within those channels. Possible modes of data transmission include

- Simplex transmission
- Half-duplex transmission
- Full-duplex transmission

Tradeoffs among these alternatives come down to cost, speed, and reliability. In general, the fewer the number of channels, the lower the cost of communication. However, multichannel communication permits higher degrees of error detection and correction, which enhances the reliability of data transfer. Within a given channel, regardless of its organization, there is usually a tradeoff between speed and reliability. That is, reliability can be increased by performing error checking, but at some sacrifice of speed.

Data transfer in one direction only can be achieved in *simplex mode*, as shown in Figure 7.4a. A single channel is used, and transmissions flow from sender to receiver. Transmission in the opposite direction, from receiver to sender, is not possible in simplex mode. A typical use of simplex mode is to send file updates or system status messages from a host processor to distributed data storage devices within a network. In such cases, the same message is transmitted to all devices on the network simultaneously, or in *broadcast mode*.

In simplex mode, error correction capabilities are limited. As discussed later in this chapter, a variety of methods can be used to allow the detection of errors in a transmitted message. If an error is detected, the receiver will generally request that the sender retransmit the message. However, if errors are encountered in simplex mode communication, there is no way for the receiver to notify the sender or to request retransmission over the basic circuit.

Channel organization for *half-duplex mode* is shown in Figure 7-4b. The communication circuit is the same as in simplex mode. However, once a message unit has been transferred from sender to receiver, the direction of communication is reversed, or line turnaround is performed. At this point, the receiver becomes the sender, and vice versa, in a simplex transmission back to the message source.

Half-duplex transmission mode allows the receiver to request retransmission of a message segment in which errors were detected. For example, in character-oriented ASCII communication, the receiver might perform a line turnaround and transmit a status signal after each group of characters is received. A *negative acknowledge (NAK)* control character would be sent if errors were detected, and an *acknowledge (ACK)* control character would be transmitted if no errors were detected. In half-duplex mode, receipt of NAK at the sender causes it to retransmit the preceding message segment after turning the line around again.

The cost of the communication lines is essentially the same in simplex and half-duplex modes. However, the added reliability of half-duplex mode is achieved at a sacrifice in overall data transfer rate. If an error is encountered, the receiver must wait until the entire message segment has been transferred before the line can be turned around and retransmission requested. If the error occurs near the beginning of a message segment, the transmission time used to send the remainder of the message segment is wasted. Also, because errors typically occur in bursts, many retransmissions of the same segment could be required before the entire message is received correctly. Consequently, if a channel is persistently noisy, half-duplex transmission will be relatively inefficient and slow.

The inefficiencies of half-duplex mode can be removed by reserving separate communication lines for transmission in both directions concurrently, as shown in Figure 7.4c. This two-channel organization permits *full-duplex*, or concurrently bidirectional, communication. In full-duplex mode, the receiver can communicate with the sender at any time by using the second channel while the data transmission is under way in the first channel. If an error is sensed, the receiver can notify the sender, which can halt the data transmission immediately and retransmit. Thus, the sender will not transmit needlessly redundant message segments.

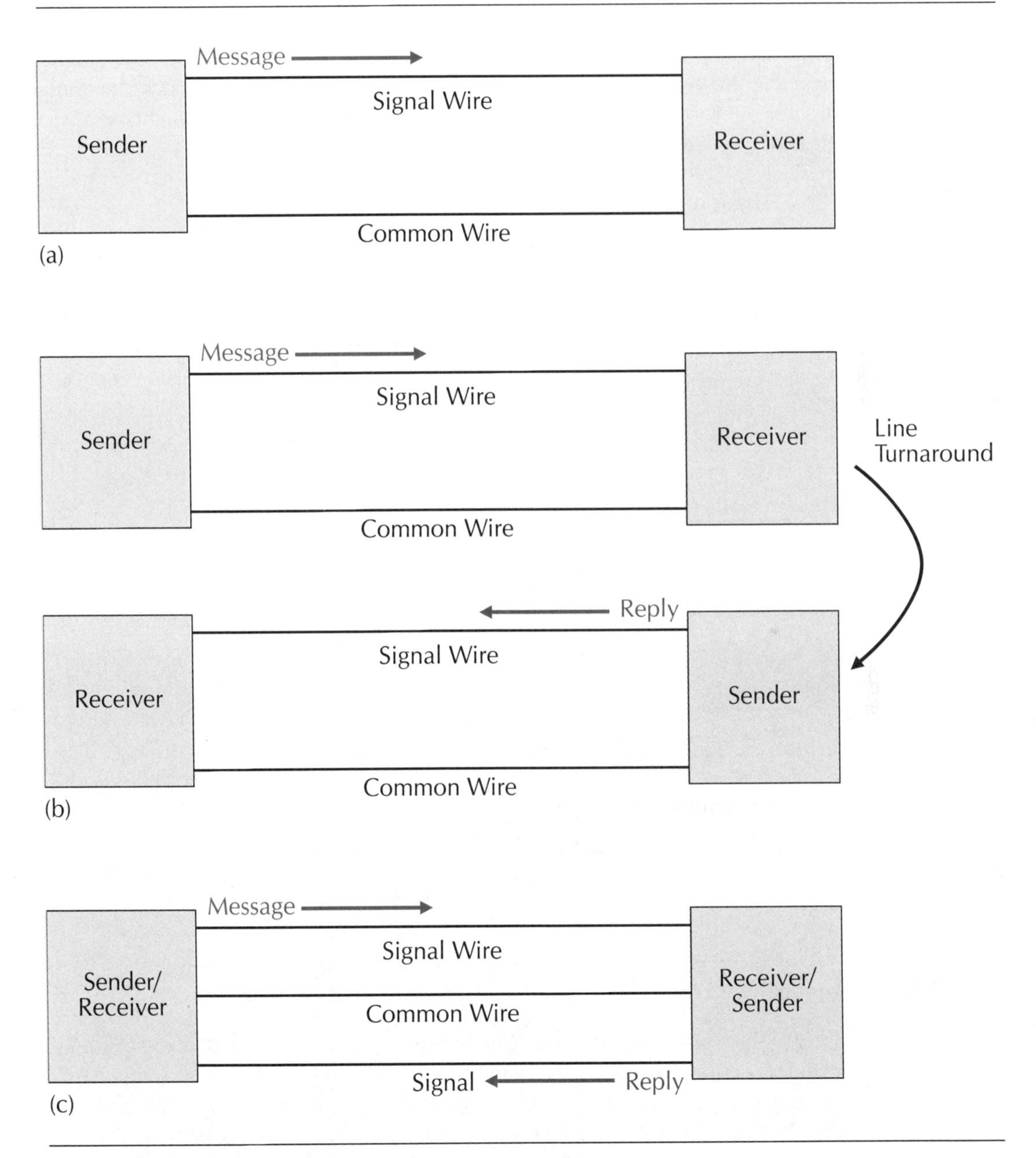

FIGURE 7.4 **Configurations for (a) simplex, (b) half-duplex, and (c) full-duplex transmission modes.**

The speed of full-duplex transmissions can be relatively high, even if noise is present within the channel. Error bursts can be corrected promptly, with minimal disruption to message flow. Compared with single channel modes, the tradeoff for the speed and reliability of full-duplex mode is the additional cost of the second channel.

I/O Mode

The I/O mode of data transfer can be parallel or serial. *Parallel transmission* requires a separate data line for each bit position, as shown in Figure 7.5. The width, or number of lines, of a parallel link typically is one byte or one word, plus a common line. *Serial transmission* requires only a single channel, but multichannel organizations can be used for error checking and communication control.

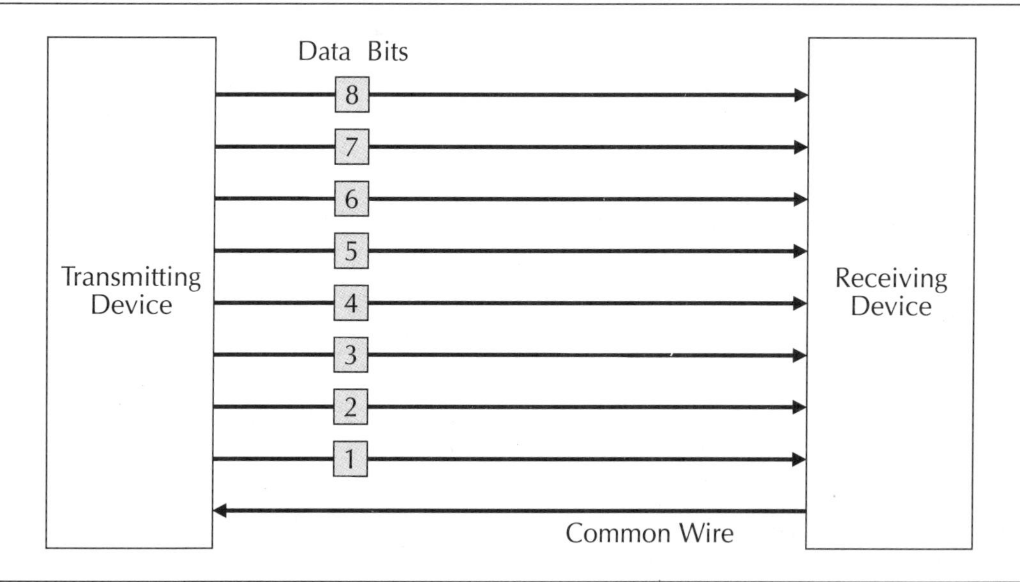

FIGURE 7.5 A configuration that supports parallel transmission of full data bytes.

Because of the number of channels required, parallel communication is comparatively expensive. The maximum distance over which data can be sent reliably is limited by an effect called *skew*. Due to differences among parallel channels, data bits could arrive at the receiver at slightly different times. The timing difference, or skew, between bits increases with distance. At some point, usually within 100 feet or so of the sending device, skew is significant enough to cause errors in signal interpretation. Skew can be corrected by placing repeater equipment at intervals along the communication route, but this equipment adds to the expense of implementing parallel channels.

In serial mode, data is sent one bit after another, within a single channel, as shown in Figure 7.6. Digital communication within networks, and especially in telecommunication media, are handled in serial mode to make the most efficient use of available lines.

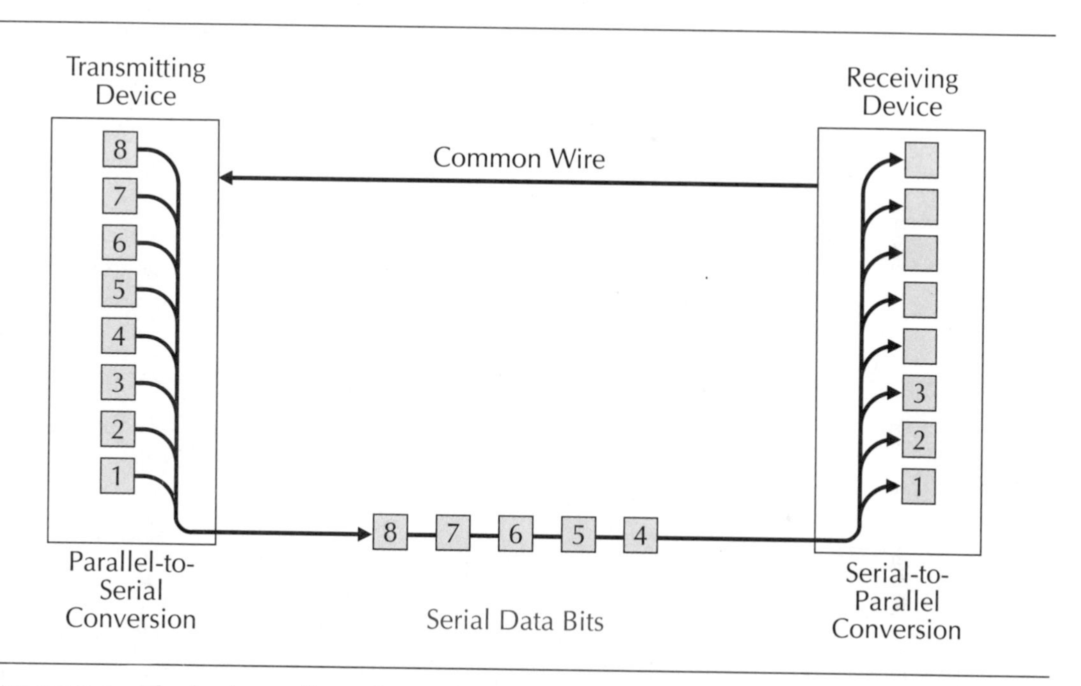

FIGURE 7.6 The basic configuration and operation of a serial transmission channel.

Because parallel transmissions provide a separate line for each bit position, the interpretation of each byte or character received is straightforward. However, in serial mode, the interpreting device must have some reference by which to form meaningful groups of bits from a serial bit stream. The reference is needed to separate the last bit of one character from the first bit of the next.

Techniques for this purpose, referred to as *character framing,* include

- Asynchronous mode
- Synchronous mode

Note that, in this context, character framing does not necessarily refer to character-oriented transmission. Instead, it refers to the grouping of bits into small blocks for purposes of data transmission and error detection. Thus, character framing might be used to interpret a serial binary transmission as a sequence of bytes (eight bits). In this case, the bytes are not interpreted as characters, but simply as eight-bit groupings of a binary (transparent mode) transmission.

If serial communication is performed in *asynchronous mode*, the timing of transmission for each character within the data message can vary. From the standpoint of the receiver, the pattern of transmission is intermittent, or apparently random. For example, data that is input through a keyboard would typically be sent asynchronously to an I/O port of the host CPU. As each key is pressed, a character, composed of a sequence of bits, is sent. From the standpoint of the I/O port, the timing of keystrokes is unpredictable (i.e., the time interval between characters will vary).

For an asynchronous bit stream to be interpreted correctly, the same predetermined character-framing scheme must be applied by both sending and receiving devices. The purposes of character framing are for the sender to indicate and for the receiver to detect character boundaries within a bit stream. As shown in Figure 7.7, character framing in asynchronous serial mode uses a *start bit* and a *stop bit* to indicate these boundaries. Accordingly, asynchronous mode also is called start/stop communication.

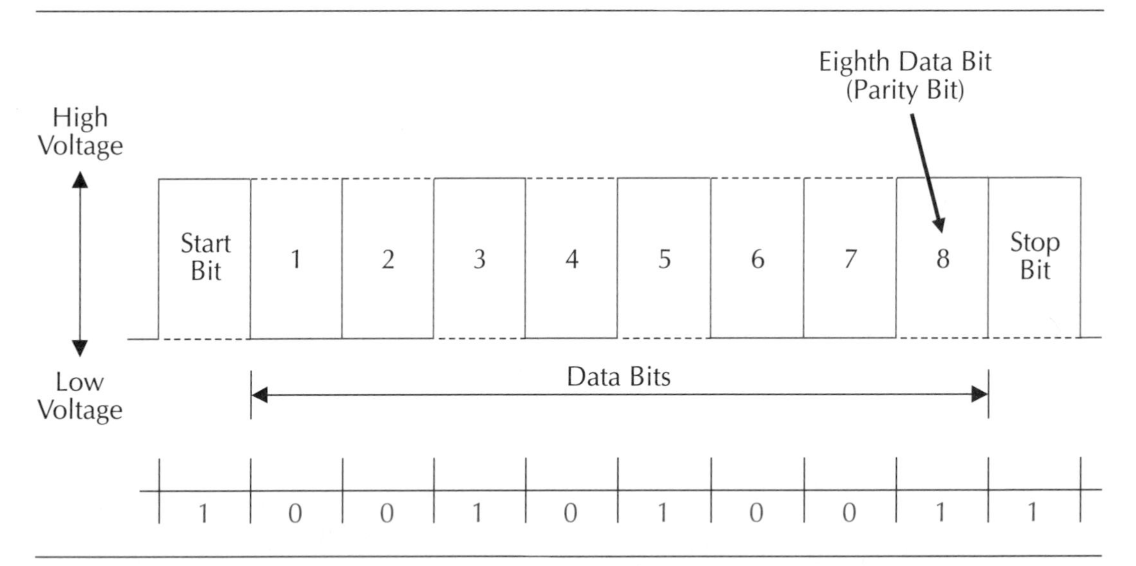

FIGURE 7.7 Asynchronous character framing for serial transmission, including the addition of start and stop bits used for character recognition.

The number of data bits framed by the start/stop bits depends on the coding method. If characters are coded in ASCII-7 with an error check bit, a frame will contain eight data bits. Clearly, the same coding method and number of data bits must be used by both the sender and the receiver. Upon sensing a start bit, the receiving device interprets the next bit received as the first of a character or byte. When a stop bit is received, the transmission of the character or byte is complete, and the communication line is returned to idle (or neutral) state, awaiting the next start bit.

For each bit to be sensed properly, both devices must be set to the same speed. The speed of data transfer is fixed by setting the internal clock of each device to a predetermined bit time. For example, if the devices are set to communicate at 1200 bps, their respective clocks must generate timing (reference) pulses at the rate of one pulse each 0.0008 second. Within the detection logic of the receiver, the incoming signal is sampled (or tested) on each reference pulse. If the signal strength is high—for example, high positive voltage or an intense light pulse—a 1 bit is detected. If the signal strength is low—for instance, low or negative voltage—the bit is interpreted as 0.

In asynchronous mode, the internal clocks of sender and receiver operate independently. Start/stop bits are the means by which these independent clocks are synchronized. Asynchronous serial communication is typically used for low-speed, interactive data transfers between remote terminals and a central host processor, as well as among microcomputers.

If serial transmission is in *synchronous mode,* the clocks of sending and receiving devices operate in precise coordination. Messages are exchanged within predetermined time intervals. Framing of transmissions is by groups of bytes (or blocks) rather than by separate characters, as shown in Figure 7.8. The boundaries of each block are marked by time intervals between blocks (or gaps), during which clock synchronization is checked and readjusted, if necessary. In general, synchronous mode is used to transfer large batches of data at high speed from one processor to another.

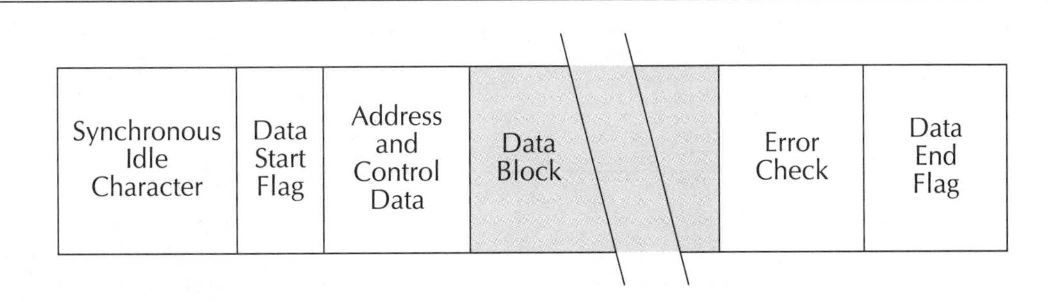

FIGURE 7.8 A typical format for messages to be transmitted using synchronous character-framing techniques.

The framing of synchronous serial transmissions may be either character-oriented or bit-oriented. If the message is interpreted in character mode, each byte within a block is recognized according to a predefined character coding method such as ASCII or EBCDIC. If framing is bit-oriented, byte-length groups within the data stream are handled as message units for purposes of error checking and transfer, but are not interpreted as characters. For example, a file containing an executable program (machine code) might be transferred in a bit-oriented, synchronous transmission.

Modulation Scheme

Digital signals have been used within computers since the appearance of the earliest von Neumann machines. However, the development of data communication has followed a different path. Early data communication methods used existing telephone lines, which carried analog signals exclusively. Today, much of the public service telephone network is digital, although analog circuits are still found in links from residences and businesses to local and regional telephone exchanges. Accordingly, data communication based on analog signalling is still an important and useful technology.

Characteristics of analog signals, or waveforms, are shown in Figure 7.9. These characteristics include

- Frequency
- Amplitude
- Phase

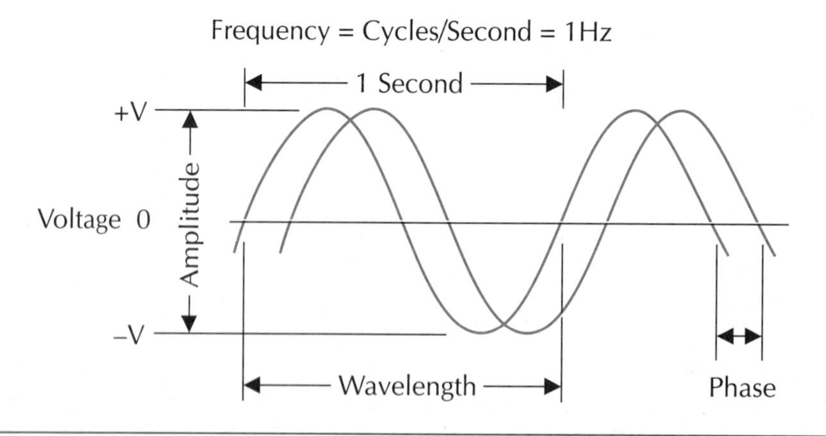

FIGURE 7.9 A typical analog waveform pattern for a telephone circuit.

Analog waveforms also are called *sine waves*, after the mathematical function that describes their shape. Analog signals can be transmitted as electrical currents through wires or as electromagnetic waves (radio frequency energy) through space.

An analog waveform varies continuously between positive and negative levels, at regular intervals, or in cycles. In transmission through wires, these levels correspond to electrical voltage. In radio frequency transmissions, levels correspond to high- and low-energy states of radio waves. The number of cycles completed each second is the frequency of the waveform. Frequency is equivalent to pitch when a wave is converted to sound, as in a stereo speaker. The magnitude of wave peaks is the *amplitude* of the analog waveform. Amplitude is equivalent to volume (or intensity) when a wave is converted into sound. Within the same channel, the timing relationship of one analog waveform to another is called *phase* difference.

In data communication, digital information is encoded in analog signals by varying, or *modulating*, an analog reference signal (or subcarrier). Schemes for modulating analog subcarrier signals include

- Frequency modulation (FM)
- Amplitude modulation (AM)
- Phase-shift modulation

The modulation scheme selected will determine how efficiently the bandwidth of a given channel can be used. Therefore, the modulation scheme potentially affects the data transfer rate.

Frequency modulation. The amplitude of *frequency-modulated (FM)* analog signals is held constant, while the frequency is varied to encode binary values. Usually, a high-frequency signal denotes a binary value of 1, and a low-frequency signal is interpreted as 0 (zero).

FM signals require sufficient bandwidth in the channel to accommodate the variation between high- and low-frequency states. Accordingly, FM transmission is often called *wideband communication* because it uses relatively more bandwidth than other modulation schemes. FM transmission is particularly effective in noisy channels, because signal amplitude can be held at a relatively high and constant level in relation to noise. Thus, even during a noise burst, the S/N ratio in the channel could remain within acceptable limits.

Modems (an abbreviation of MOdulator /DEModulator), which link computers to conventional analog telephone lines, typically encode signals by frequency modulation, as shown in Figure 7.10. Within a conventional telephone voice channel, the frequency of the signal corresponds to audio pitch.

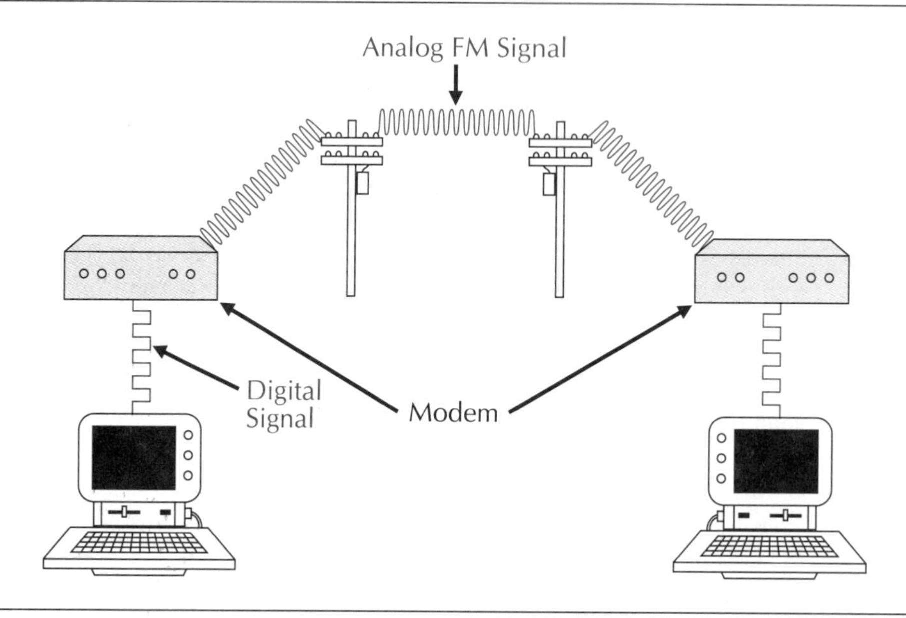

FIGURE 7.10 The placement of modems for signal conversion for computer communication over public telephone networks.

The frequencies assigned for low-speed (300 bps) communication over analog telephone lines are given in Table 7.1. Note that full-duplex mode is made possible under this scheme by assigning separate pairs of frequencies to transmit and receive functions. Transmit-mode frequencies are used by the modem that originates the connection (the one that dialed the phone number). Receive-mode frequencies are used by the answering modem.

TABLE 7.1 A typical pattern of frequency assignments for the transmission of binary data over 300-baud analog telephone lines.

Mode	Signal Frequency (Hz)	Binary Value
Transmit	1070	0
Transmit	1270	1
Receive	2025	0
Receive	2225	1

Amplitude modulation. If analog signals are varied according to *amplitude modulation (AM)*, frequency is held constant, and signal level (or amplitude) changes. Thus, AM transmissions require relatively narrow bandwidth. However, the reliability of AM transmission can be affected significantly by noise in the channel. At low-amplitude signal levels, noise bursts can result in an unacceptably low S/N ratio. In general, the amplitude level corresponding to the off (low) state should be fixed at a level considerably higher than the expected amplitude of background noise in the channel.

Phase-shift modulation. Two analog signals of constant amplitude are used in *phase-shift modulation.* Binary values are encoded as varying intervals by which one waveform leads or lags the other. The phase difference between the waveforms is modulated, or shifted, by momentarily varying the frequency of one of the signals. Phase-shift modulation is typically used if there is a possibility that AM or FM signals at the same frequencies might interfere with the transmission. For example, mobile telephone service is provided over cellular radio, which uses phase-shift modulation to avoid interference with other types of radio.

Multiplexing

Techniques for multiplexing permit the concurrent sharing of a single communication line or channel by multiple users or transmissions. Multiplexing technology has grown from the need to use relatively expensive communication lines efficiently. These methods were originally developed to support high-volume, long-distance telecommunication in order to use existing channels to near capacity. Also, managers of computer centers have used multiplexing to minimize the direct costs of leasing dedicated lines from telecommunication utilities.

Techniques for sharing communication lines through multiplexing include the following:

- Time-division multiplexing (TDM)
- Frequency-division multiplexing (FDM)

Each of these techniques represents a tradeoff between the bandwidth of the common channel (and, therefore, its cost) and total message-carrying capacity. In addition, each method can be used with either analog signalling methods (as discussed above) or digital signalling methods (as discussed later in this chapter).

Time-division multiplexing (TDM). Under *time-division multiplexing (TDM)*, a single, high-speed (usually narrowband) channel can be subdivided into a series of separate time intervals, as shown in Figure 7.11. Messages entering the channel are subdivided into pieces called *packets*. The length of each packet corresponds to its time-interval assignment. Packets from many different messages can be transmitted in apparently scrambled order to maximize the utilization of the channel. Packets also contain information that allows the receiver to reassemble the original messages. Instructions for routing messages to different receiving nodes or users can also be contained in the header.

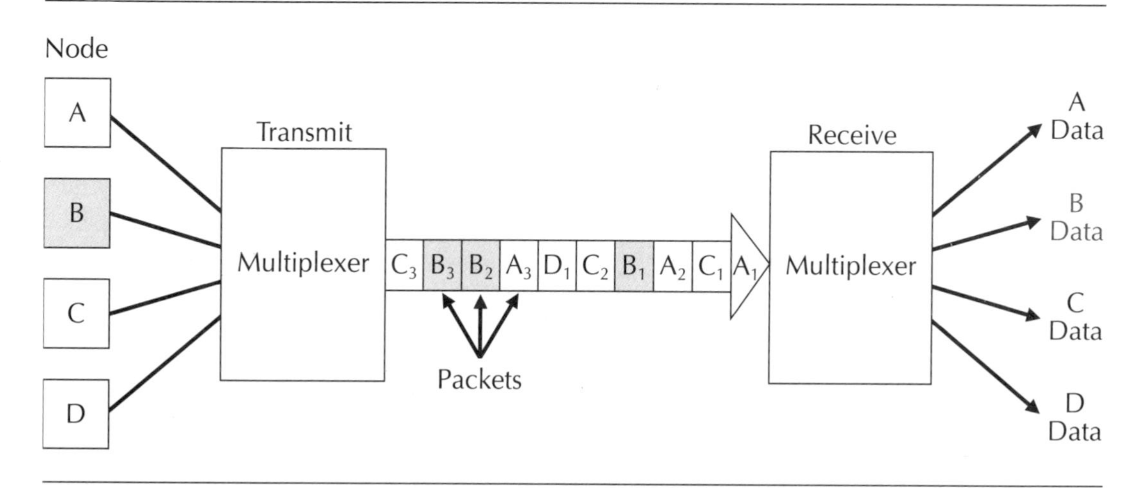

FIGURE 7.11 **The operations that occur in implementing time-division multiplexing under packet-switching techniques.**

TDM and packet switching are used within some types of local-area networks (LANs) to route commands and data to specific devices or nodes. Such networks are called *baseband* local-area networks. If a node within a baseband LAN detects its own address within the header of a given packet, the node acts on the message contained in the packet. (LANs are fully discussed in Chapter 14.)

Frequency-division multiplexing (FDM). Another method by which communication channels can be shared is *frequency-division multiplexing (FDM),* as diagrammed in Figure 7.12. Under FDM, a single *broadband* channel is partitioned into multiple subchannels. Each subchannel represents a frequency range, or *band,* within the overall bandwidth of the channel. Signals are transmitted within each subchannel at fixed frequencies. Digital information is encoded through amplitude modulation. Broadband LANs and cable television (CATV) are examples of systems that use FDM.

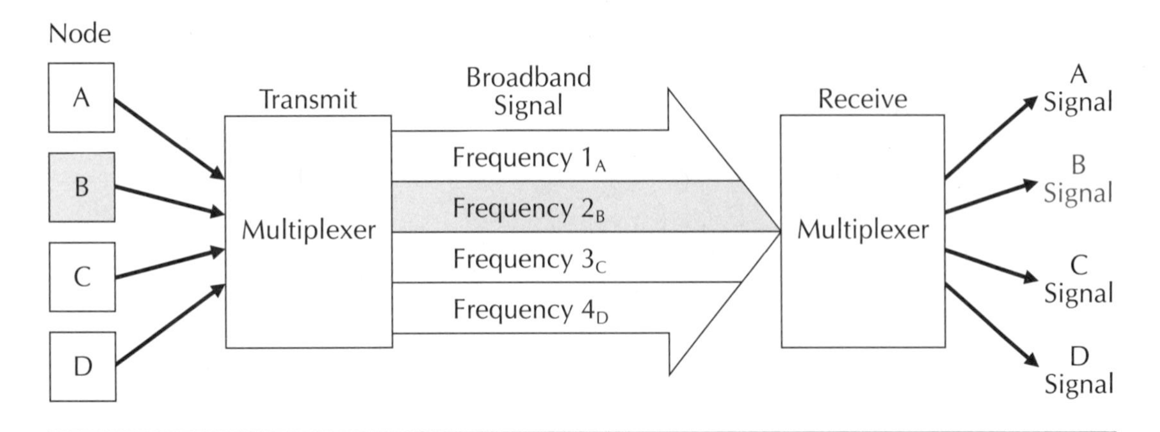

FIGURE 7.12 The operations that occur in implementing frequency-division multiplexing.

DIGITAL COMMUNICATION

Increasingly, the telecommunication industry is moving toward all-digital service, for which no analog signal component is required. As discussed in Chapter 4, digital signals are inherently less subject to noise than are their analog counterparts. In general, this resistance to noise-related problems comes at the expense of message-carrying capacity. However, modern methods of digital signal transmission mitigate this disadvantage to a great extent. In addition, digital signals can take advantage of digital switching technology (e.g., for routing messages over a network), which is inherently faster and more reliable than analog switching.

Many methods can be used to generate and interpret digital signals. Examples of digital signaling methods include

- Transistor-to-transistor logic (TTL)
- Zero-crossing signals
- Pulse-code modulation

As with most other choices in data transmission, these methods represent various tradeoffs among signal efficiency, cost, and susceptibility to transmission errors.

Digital signals within computer hardware can implement *transistor-to-transistor logic (TTL)*. As shown in Figure 7.13, TTL signals are discrete pulses of electrical direct current (DC). A DC signal has no negative component. Digital

TTL signals typically vary between a high-voltage state, such as +5 volts DC, and a low-voltage state, such as 0 volts DC. Transitions between states are abrupt and virtually instantaneous, rather than proceeding in continuous cycles, as analog waves do. Accordingly, digital waveforms are called *square waves.*

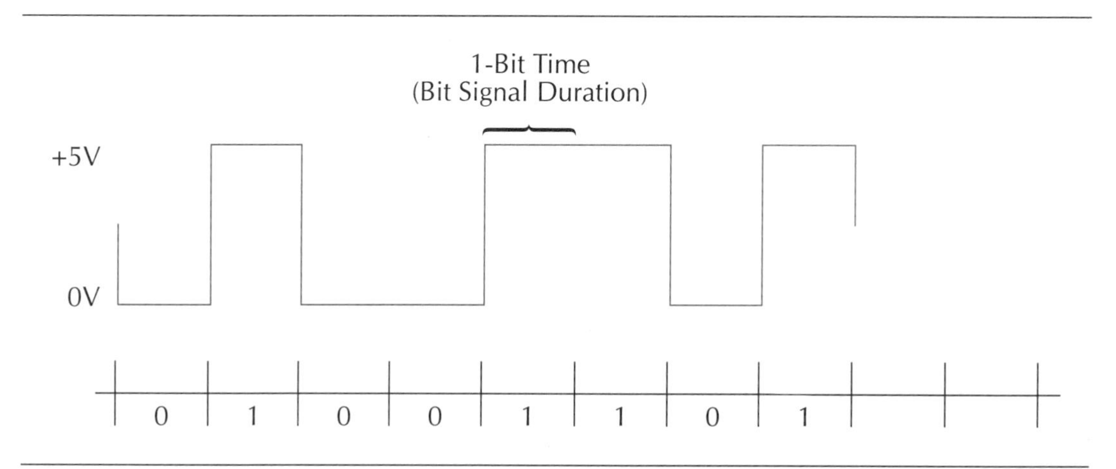

FIGURE 7.13 A signal pattern for TTL transmission, using high- and low-voltage states of a square wave.

Typically, digital signals used in telecommunication have both positive and negative states. An example of such a *zero-crossing signal* is shown in Figure 7.14. A zero-crossing digital signal varies between a high-positive voltage, such as +12 V, and a low-negative voltage, such as –12V. In the zero-crossing scheme shown, the negative state represents a bit value of one, and a positive state represents zero. Zero-crossing signals can be transmitted reliably over relatively longer distances than can TTL pulses. A device that transforms TTL signals within a computer to zero-crossing signals for digital telecommunication is called a *coupler.*

Another type of zero-crossing digital signal is shown in Figure 7.15. The signalling method is called *Manchester coding,* which represents a bit value of one as a positive half bit followed by a negative half bit. Zero values are represented with opposite patterns, or as negative-to-positive transitions. Because each bit value has both positive and negative components, the coding scheme is termed *biphase*. Biphase code pulses are readily distinguishable from noise. Furthermore, biphase signals are *self-clocking,* because the transition within each bit always occurs at the interval of one bit time. For these reasons, Manchester biphase coding is considered an exceptionally reliable method of transmitting digital signals within telecommunication networks.

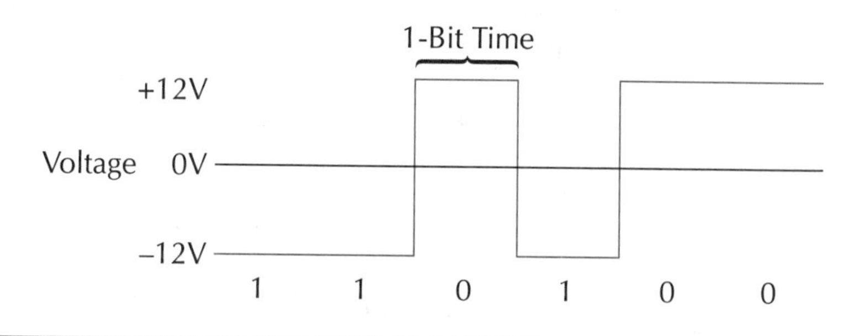

FIGURE 7.14 **The pattern of zero-crossing digital signal transmission, in which positive voltages represent bit values of 1 and negative voltages represent bit values of 0.**

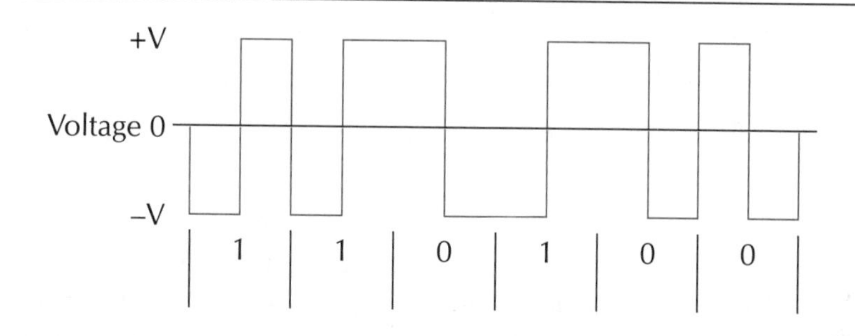

FIGURE 7.15 **Manchester biphase coding. Each bit position contains both positive and negative values implemented through zero-crossing signals.**

Within fiber optic communication channels, signals are coded using *pulse-code modulation (PCM)*. Under PCM, bit values correspond to bursts of light. If an optical fiber carries only one signal, the modulation scheme is called *single-mode PCM*. However, several transmissions can be carried within the same fiber at different frequencies. This approach, called *multimode PCM*, is a form of frequency-division multiplexing.[2] Fiber optic technology can provide considerably more bandwidth than can copper wire channels. Combined with the low-noise characteristics of optical transmission, fiber optic bandwidths can support exceptionally high data transfer rates.

[2] Frequency of light is interpreted as color by the human eye. Thus, multichannel optical transmission uses multiple colors of light simultaneously.

Useful ranges of speed vary widely among transmission media and signalling methods. For example, the maximum reliable rate within a conventional analog telephone circuit is 19,200 bps, asynchronous. Data compression is often used to increase these transmission rates. Within a copper wire local-area network that uses zero-crossing digital signals, typical transmission speed might be 1 megabit per second (Mbps). By contrast, a fiber optic communication line can support 525 Mbps under current technology, with a potential for gigabit rates.

Digital Communication Protocols

For digital devices to communicate successfully, all details of communication transactions must be predetermined, except for the actual content of the data message. As discussed above, critical determinations include coding method, transmission speed, transmission mode, and signal modulation. However, even complete specification of these factors for both devices will be insufficient to achieve data transfer. Specifications for communication transactions also must include

- Predetermined types of communication transactions
- Valid message formats
- Sequence of transactions
- Error-handling measures

In telecommunication terminology, a set of specifications for these requirements is called a *communication protocol.* Other factors that can affect communication between devices are called *conventions,* which might not be included within formal protocols. Often, conventions refer to methods by which potential incompatibilities between specific devices can be resolved.

Protocols and conventions for digital communication include methods for dealing with

- Error detection and correction
- Handshaking
- Synchronous communication
- Efficient use of channels

ERROR DETECTION AND CORRECTION

A crucial requirement for any protocol is the measure by which errors in data reception or interpretation can be detected and corrected. Under most protocols, the correction action is straightforward. Upon sensing an error, the receiver transmits a negative acknowledge (NAK) signal to the sender. Upon sensing NAK from the receiver, the sender retransmits the preceding message unit.

Variations in data format, transmission mode, and transaction sequence will determine

- Time lag between error detection and NAK transmission
- Length of message unit to be retransmitted
- Redundancy of message retransmissions

Although error-correction measures can be similar among protocols, major differences exist in the implementation of error detection. Error-detection measures include

- Parity checking
- Cyclic redundancy checking (CRC)

Error detection for character data can be implemented through *parity checking*. In a character-oriented transmission, one bit (usually the eighth) in each character is designated the *parity bit,* or check bit. The value of this bit is set to correspond with a count of other bits within the character. Depending on the implementation, one of two approaches can be used:

- Odd parity
- Even parity

To indicate *odd parity*, the sending device sets the value of the parity bit to 0 (zero) if the number of 1-valued data bits within the character is odd. Conversely, if the count is even, the parity bit is set to 1. Thus, if odd parity is specified, the count of 1 bits in each byte received should always be an odd number. If the receiving device counts an even number of 1 bits in the byte, it is assumed that bit values have been altered in transmission, indicating an error. Under *even parity*, the sending device sets the parity bit to 0 (zero) if the count of 1 bits is even. If an odd number of 1 bits is found, the parity bit is set to 1. If no errors occur, the receiving device should always find an even number of 1 bits within each byte.

Parity checking provides some degree of assurance that errors will be detected, but this technique is not absolutely reliable. For example, it would be possible for the error to go undetected if several bits within a single character have been altered in multiples of two. That is, a pair of altered bits can act as *compensating errors* on one another. So, even though the received character contains errors, the value of the parity bit will be correct.

Parity checking is a type of *vertical redundancy checking (VRC)*. VRC techniques are based on bit counts within individual characters. In asynchronous, character-oriented transmissions, VRC might be the only practical means of error detection. Because characters are transmitted separately and at unpredictable times, it can be difficult to perform checks on groups of characters.

Parity checking on groups of characters, or blocks, can be performed under *longitudinal redundancy checking (LRC)*. To implement LRC, the sending device counts the number of 1 bits at each bit position within a block. After the block is sent, the sending device derives a *block check character (BCC)* from these counts and transmits it to the receiver.

As shown in Figure 7.16, each bit within the BCC is a parity bit that is set to correspond with the number of 1 bits at the same position within the preceding characters. Either odd or even parity can be used. The receiving device derives a separate BCC after receiving the block of characters; then it compares the BCC developed locally with the BCC that has come from the sender. If the values are not the same, an error is indicated. NAK is returned to the sender, and the entire block is retransmitted. Like VRC techniques, LRC is vulnerable to compensating errors. A higher degree of assurance can be provided by applying both VRC and LRC within the same protocol. However, even with this "double-check" approach, some compensating errors might go undetected.

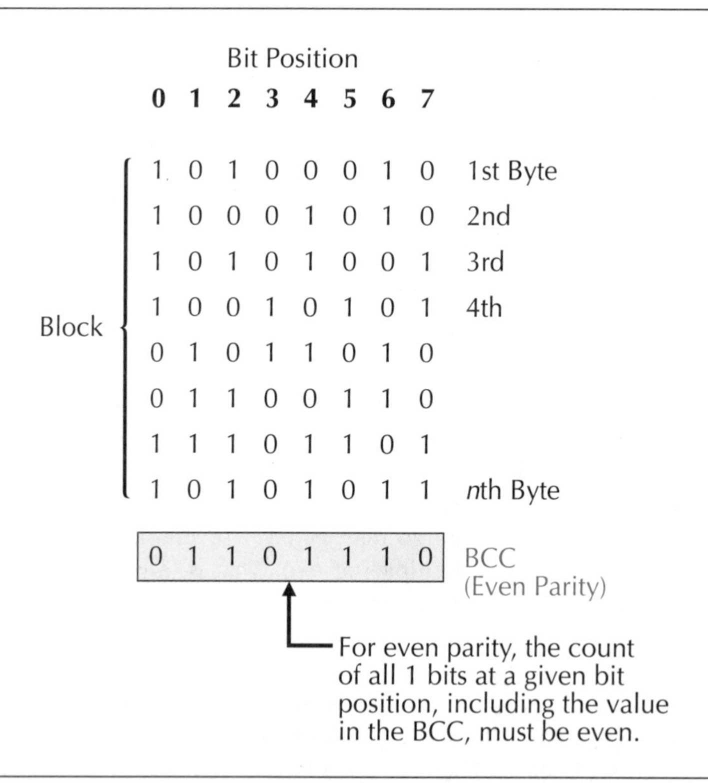

FIGURE 7.16 **This coding pattern shows the use of a block check character (BCC) to implement a longitudinal redundancy check (LRC).**

Another error-detection technique is *cyclic redundancy checking (CRC)*, which applies a mathematical algorithm to a block of characters to generate a BCC. CRC uses a common check (bit) string known by both the sender and receiver and a separate block check bit string that is generated by the sender and appended to the end of each data block. The combined block (data block and appended block check string) can be interpreted as a single large integer. The check string appended to the data block is calculated by the receiver to make the large integer evenly divisible by the common check string. The receiver verifies the data block by dividing the large integer by the common check string.

In general, the accuracy of a particular method of CRC depends on the length of a transmitted block, the length of the block check string appended to the transmitted block, and the value of the common check string used by the sender and receiver. Transmitted blocks generally consist of several hundred to a few thousand data bits. The length of the appended block check string is typically between 16 and 64 bits, depending on the length of the data block. Mathematical algorithms can be used to select a common check string (the divisor) that minimizes the chance of undetected transmission errors.

Like LRC, CRC is commonly used in synchronous character-oriented protocols. With CRC and a sufficiently large check string, the assurance of error detection is substantially higher than with other redundancy checking methods. The tradeoff for such extensive error detection is extra processing overhead. The calculation of the block check string is much more complex than the calculation of an LRC check string. To minimize processing delay, CRC is often implemented in dedicated electronic circuits (i.e., firmware) within sending and receiving devices.

HANDSHAKING

Under a protocol, the timing and sequence of communication transactions are included in specifications for handshaking between devices. In general, the term *handshaking* refers to methods by which a sender and receiver synchronize their communication activities. Such methods are needed to address incompatibilities among devices in processing speed, data transfer rate, or both.

Communication Buffer

Data transfers between devices of dissimilar speeds and capacities can be performed by interposing a *communication buffer*, or temporary storage area. A buffer permits sending and receiving devices to communicate at their own maximum transmission speed. Figure 7.17 illustrates the concept of a communication buffer. The sending device transfers data destined for the receiving device into the buffer. The receiving device removes input data from the buffer.

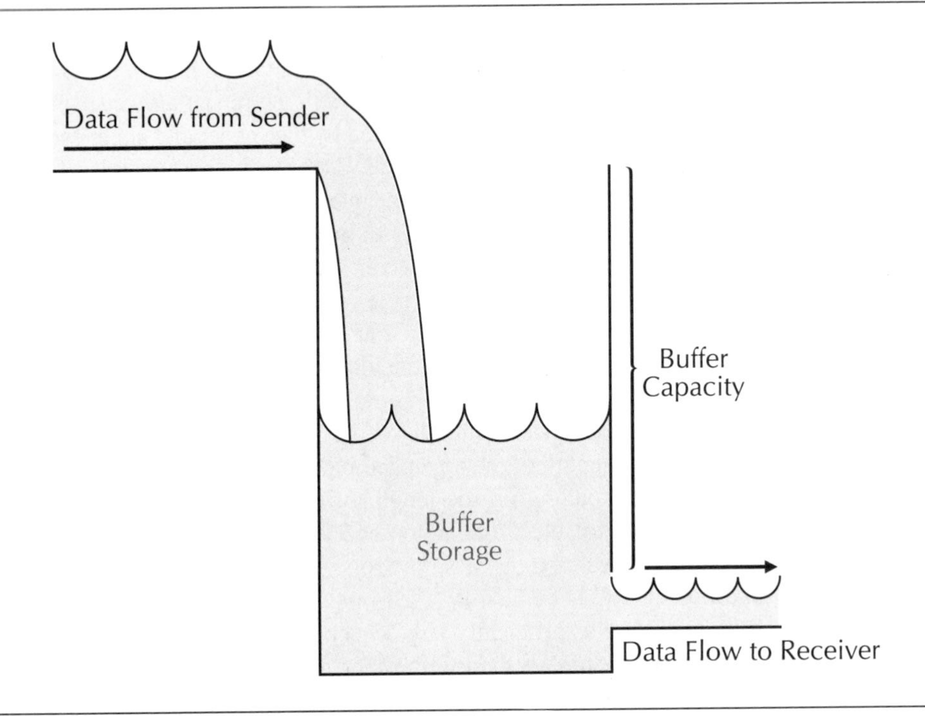

FIGURE 7.17 **The concept and operating principles of an I/O buffer.**

In general, the buffer is implemented within the slower of the two devices. Actual transmission of data occurs at the common speed of the communication channel. However, the buffer in the slower device allows it to process the data at a slower rate than it is being transmitted. For example, a dot-matrix printer typically contains a buffer of several thousand kilobytes. Data transmitted by a personal computer or workstation is first stored in the buffer, then is transferred to the control mechanism of the dot-matrix print head, typically at much slower speeds due to the physical limitations of the printing mechanism.

No problems are generally encountered when the receiving device removes data from the buffer at a faster rate than the sending device sends it. The receiving device simply waits for some data to be transmitted into the buffer. However, the reverse situation requires active intervention on the part of the receiver. Because a buffer has limited storage capacity, handshaking must be used to signal the sender to stop transmission before the buffer overflows. Such handshaking is generally implemented by a pair of signals from receiver to sender.

Overflow conditions can be prevented in a communication buffer by performing transfers in half-duplex mode and by the use of an *XON/XOFF* signalling scheme. This protocol is usually implemented in identical communication control programs that reside in each device. When the buffer is full, the communication line is turned around, and a control character (which represents XOFF) is transmitted to the sending device. On receiving XOFF, the sender stops its transmission, giving the receiver time to read the contents of the buffer. When the control program in the receiver senses that the buffer is empty, it causes an XON character to be transmitted to the sender. The sending device then resumes data transmission into the buffer.

A buffer is often used in conjunction with longitudinal or cyclical redundancy checking. The buffer provides a temporary storage area for incoming data while the receiving device recalculates the block check character and performs comparison operations. If an error is detected, the contents of the buffer must be erased before retransmission is attempted.

Transmission Synchronization

In addition to buffer control, a handshaking protocol also addresses the timing of a message transmission. Timing of individual bits in a message is determined by the common transmission speed of the communication channel, which is not formally a part of the handshaking protocol. Identifying the start and end of a message transmission is a part of the handshaking protocol, and is handled differently for asynchronous and synchronous transmission. By definition, asynchronous communication requires no synchronization between characters (or message bytes). Thus, the start and end of a message must be identified by the sender. Synchronization of asynchronous transmission is implemented through the use of start and stop bits, as described earlier. A start bit is the mechanism by which sender and receiver synchronize their internal clocks.

Some protocols for synchronous communication use *synchronous idle* handshaking to set the internal clocks within sending and receiving devices. Initially, the sender transmits the synchronous idle character, a control code, which the receiver uses to set its own clock. This control code is also transmitted during gaps between blocks so that synchronization is maintained. This and other techniques for handling synchronous communication are discussed in the following section.

SYNCHRONOUS COMMUNICATION

To coordinate devices in synchronous mode, considerable communication overhead is required. Communication overhead encompasses portions of the message that contain control codes rather than data, as well as functions within both devices for processing these codes. As a result, synchronous protocols typically include precise formats for messages. These formats determine the type and sequence of control codes and data, as well as the manner in which the data will be grouped and interpreted. Different synchronous protocols can be categorized by data formats that include the following:

- Character-oriented format
- Multileaving format
- Byte count-oriented format
- Bit-oriented format

Character-Oriented Format

One of the earliest protocols for character-oriented, synchronous communication is *binary synchronous communication (BSC),* which often is referred to by the abbreviated term, *bisync.* In BSC, data is transmitted in blocks of characters. The character-coding method used can be either ASCII-7 or EBCDIC, with the addition of a parity bit. Unlike asynchronous character framing, no start/stop bits are required in BSC format. BSC (with EBCDIC character coding) is widely used in IBM mainframe computers.

As shown in Figure 7.18a, the order and format of a BSC transmission include the following:

- Synchronous idle characters
- Header
- Data block
- Block check character

The header contains communication control and message-routing instructions for the receiving device. The header can include identification of sending and receiving devices, message priority, security level, routing, and description of message content. Other codes, such as *start of header (SOH)*, *start of text (STX)*, and *end of text (ETX)*, are markers, or delimiters, that indicate boundaries of fields within a message unit, or frame. Depending on the implementation, the BCC can be generated with VRC, LRC, or CRC methods.

Portions of BSC transmissions that are interpreted according to ASCII or EBCDIC character codes are said to be *nontransparent,* because the result is a message composed of printable characters. The header, for example, is always sent in nontransparent mode so that it is readable by operators at the receiving site. The data block, however, can be interpreted in either character mode or binary mode. If the message content is in binary format, the transmission mode is *transparent,* but still is said to be character-oriented, because each data byte contains a parity bit. That is, data bytes are transmitted in character mode but are interpreted in bit-oriented mode. Within the frame format, a transition into or out of transparent mode is signalled by the *data link escape (DLE)* control code, as shown in Figure 7.18b.

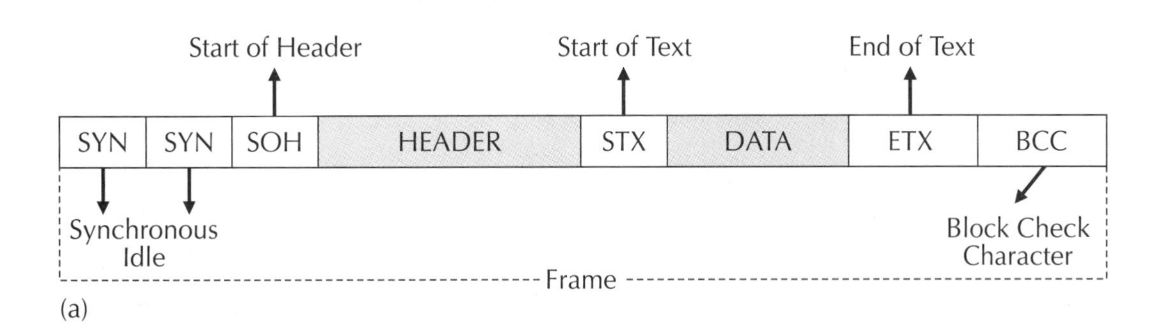

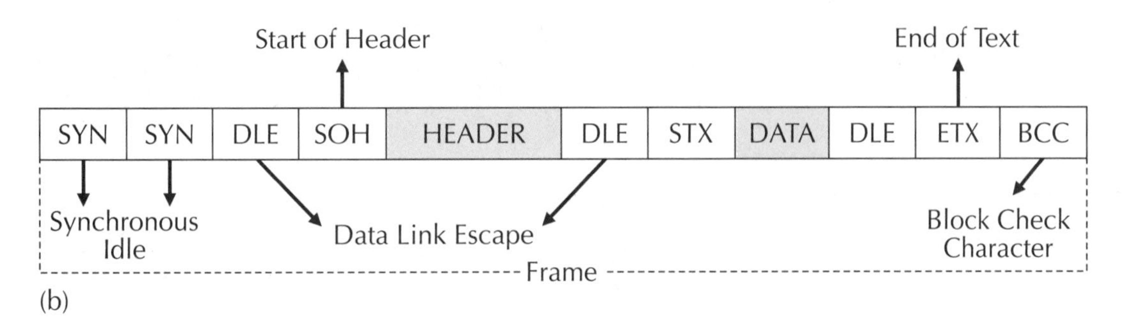

FIGURE 7.18 **These message format diagrams show binary synchronous communication (BSC) in (a) character-oriented and (b) bit-oriented modes.**

Typically, BSC transmissions are performed in half-duplex mode. Line turnaround is required for signalling back to the sender if the receiving device finds no match between the BCC in the message and the BCC it derives locally. If there is no match, NAK is returned to the sender, which then retransmits the entire preceding frame. For this reason, BSC can be highly efficient as long as noise bursts on the line are infrequent. However, any noise will cause the overall data transfer rate to decline significantly, because retransmissions of entire blocks will be largely redundant. Typically, BSC is used in batch transfers between mainframe computer sites.

Multileaving Format

In bisync mode, some of the inherent inefficiencies encountered in noisy channels can be overcome through multileaving. *Multileaving* is a technique by which the receiver might be able to turn a half-duplex line around at any time. Thus, if an error is detected at the character level (incorrect parity), the receiver need not wait to receive the rest of the block or to calculate a BCC before requesting retransmission. In effect, multileaving permits the receiver to interrupt the sender for a priority transmission in the opposite direction. The result is "pseudo-simultaneous" bidirectional transmission, or concurrent use of a single line by sender and receiver. Thus, multileaving approaches the effectiveness of full-duplex operation, with the economies of half-duplex mode.

Byte Count-Oriented Format

The general format of *byte count-oriented protocols* is shown in Figure 7.19. In this synchronous framing format, the data block interpreted in the header is the *byte count* of the data block. The receiver uses this byte count to partition the block into bytes or characters. In the format shown, a two-byte CRC is generated. The first byte is a BCC on the header, and the second byte is a BCC on the data. An example of this protocol is the *digital data communication message protocol (DDCMP).* This protocol has been implemented extensively for transfers among minicomputer facilities.

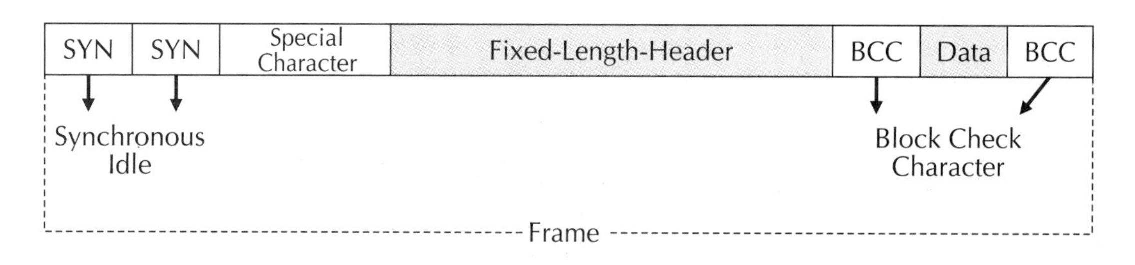

FIGURE 7.19 **A message format diagram that demonstrates the byte-count orientation of the digital data communication message protocol (DDCMP).**

Bit-Oriented Format

Under synchronous protocols that are bit-oriented, data fields are interpreted only in transparent, or binary, mode. If the message contains text, this processing must be performed in subsequent steps by the receiving node. Examples of bit-oriented synchronous protocols are *synchronous data link control (SDLC)* and *high-level data link control (HDLC).* These protocols, which are quite similar, are being used increasingly within large-scale computer networks. The main difference is that HDLC includes communication control instructions for digital packet-switching systems. Because packet switching is used extensively in microwave and satellite networks, HDLC has become an international standard for data exchange over telecommunication carriers. General formats for SDLC/HDLC are shown in Figure 7.20. The data message is encoded as a bit stream within the information field, which can vary in length up to the buffer capacity of the receiver.

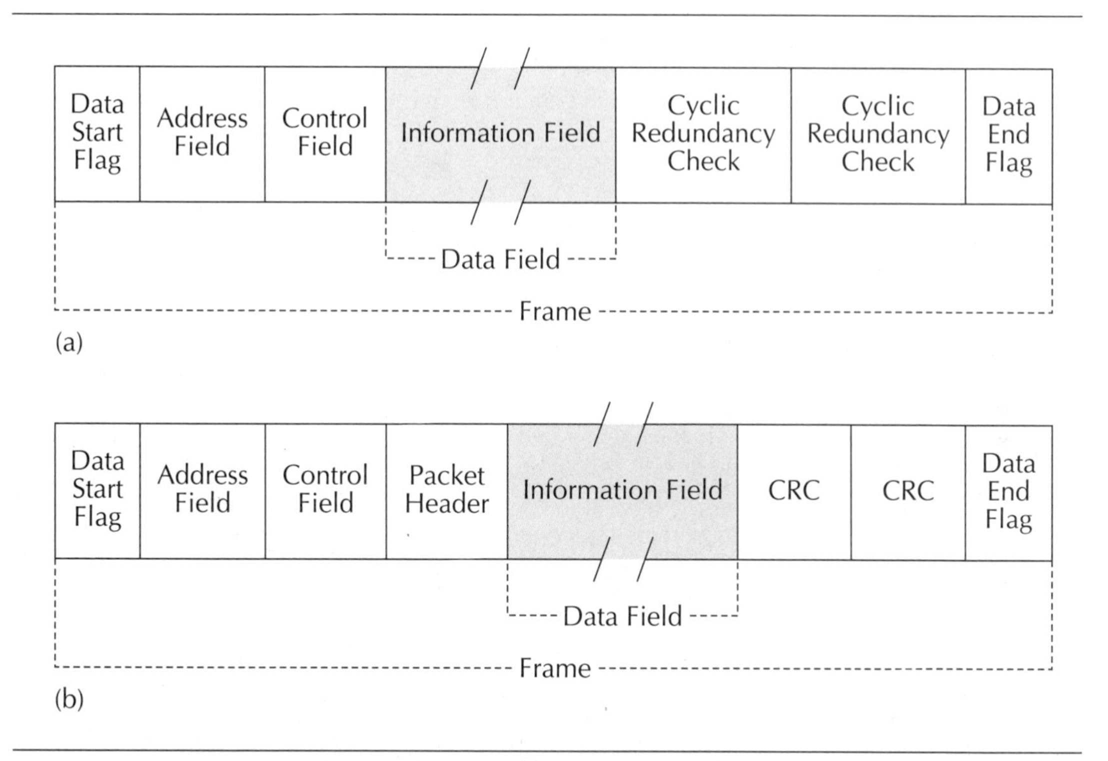

FIGURE 7.20 **Message format diagrams for bit-oriented communication protocols implemented under (a) SDLC and (b) HDLC methods.**

Under these protocols, the method of synchronization is different from that described for BSC. Each SDLC/HDLC frame is bounded by a *data start flag* and a *data end flag*. Each flag contains the bit pattern 01111110, which serves both as a message delimiter and as a reference for clock synchronization. Synchronization is performed at the start of each frame, rather than within pauses between frames, as in BSC. If a pause is sensed at the receiver, a framing error is indicated, and retransmission of the preceding frame is requested. Thus, no pauses are permitted between frames, and any lag must be filled with *continuous high-idle signals*, or a repetitive pattern of data start flags.

The address field, which follows the data start flag, contains a byte that indicates the location of operands or data stores to be used in processing the message. The next byte, the control field, contains a command or specified operation. As shown in Figure 7.20b, HDLC frames also include a *packet header,* which provides information for the reassembly of multiple-packet messages by the receiver. In both SDLC and HDLC, a two-byte CRC is derived from the combination of the address, control, and information fields. If the CRC developed by the receiver does not match the CRC in the message, an error is indicated, and the entire frame is retransmitted.

In SDLC/HDLC, the relationship of sender to receiver is "master to slave." That is, one and only one sending device can control one or more receiving nodes. For example, a mainframe host might transmit in HDLC to multiple remote job entry computers.

SUMMARY

Communication between computers and between computer system components can be implemented using various types of signals and transmission media. Digital or analog signals can be used. Common transmission media include copper wire, fiber optic cable, and radio frequency energy propagated through space. In all cases, messages are encoded in signals and conducted or propagated through the medium.

The communication capacity of a communication channel depends on a number of factors, including transmission medium and signal characteristics, encoding and decoding schemes, and error-detection schemes. The maximum transmission capaity of a signal over a transmission medium is dependent on the physical characteristics of the signal and the medium. These include the susceptibility of the signal and the medium to noise and distortion, as well as the range of signals (bandwidth) that can be carried on the medium. The data transfer rate is a measure of communication capacity that accounts for less than maximal transmission capacity due to encoding/decoding and error detection/correction.

The transmission mode of a communication method refers to the number and use of channels. Simplex transmission uses a single channel. It is capable of transmission in only one direction. Half-duplex transmission allows bidirectional communication on a single channel. The sender and receiver periodically reverse roles and, thus, the direction of signal flow. Full-duplex transmission uses two separate channels, one for messages flowing in each direction.

The I/O mode of data transfer can be either serial or parallel. Serial transmission uses a single channel and transmits bits one at a time. Parallel transmission uses multiple channels to transmit several bits simultaneously. Parallel transmission is more efficient but requires more hardware and is much more limited in maximum channel length.

Character-framing methods are the means by which individual message elements (characters) within the signal are recognized. Asynchronous character framing uses a separate signal value to indicate the start and/or stop of an individual character in the signal. Synchronous character framing is accomplished by explicit signal timing and the synchronization of clocks within the sender and receiver. It is more efficient but requires a substantially more complex communication protocol and message format.

Message elements can be encoded within an analog signal by varying (modulating) the frequency, amplitude, or phase of signal waves. Frequency modulation (FM) encodes bit values as high or low frequencies in a wave of constant amplitude (volume). FM requires a transmission medium capable of carrying a wide range of frequencies (i.e., high bandwidth). Amplitude modulation (AM) encodes bit values as high or low amplitude within a signal of constant frequency. AM is more susceptible to noise than FM. Phase-shift modulation encodes bits as the timing (phase) difference between two waves.

Digital signals can also be used for communication. Options include transistor-to-transistor logic (TTL), zero-crossing signals, and pulse-code modulation (PCM). TTL communication uses direct current electrical signals. Bits are encoded as high- or low-voltage states within a square wave. Zero-crossing signals are similar to TTL signals except that bits are encoded as positive or negative voltage states. PCM transmissions are used in fiber optic communication channels. Bits are encoded as bursts of light (or the absence thereof).

Most communication channels incorporate some form of error detection and correction method. Error detection is commonly implemented by redundancy checks. Methods include vertical, longitudinal, and cyclical redundancy checking. Vertical redundancy checks, also called parity checks, can be further classified into odd and even methods. Error correction is generally implemented by a signal from receiver to sender to retransmit a message or portion thereof.

In parity checking, the sender counts the number of 1-valued bits in a character and sends a bit value corresponding to an odd or even number of such bits. The receiver also counts the 1-valued bits and compares the count to the parity bit. If the parity bit does not match the receiver's count, a transmission error is assumed to have occurred. Redundancy checks are similar to parity checks except that counting is performed on larger groups of data, and more counting bits are used.

Handshaking refers to the mechanisms by which the rate of data flow is controlled. Handshaking is necessary when the sender and receiver can communicate at incompatible speeds. Many handshaking methods use a buffer, which is a temporary storage area that can hold all or part of a message. The sender writes into the buffer and the receiver reads from the buffer. Signals are used to notify the sender when the buffer is full or empty.

Various methods exist for expanding the utilization or capacity of communication channels. The most common method for addressing capacity limitations is data compression, which is often implemented by compressing repeated characters or message fragments into coded transmissions of a shorter length. Data compression increases transmission capacity at the expense of processing overhead by the sender (compression) and the receiver (decompression).

The most common method for addressing channel utilization is the sharing of a single channel among multiple senders or receivers. The technical term for channel sharing is multiplexing. Time-division multiplexing divides access to the channel into time fragments and allocates them to different message senders. This method is frequently used in local-area networks and is called baseband transmission in that context. Frequency-division multiplexing divides the frequency range of a broadband channel into several narrow bands. Each band carries a separate stream of messages.

Key Terms

acknowledge (ACK)
amplitude
amplitude modulation (AM)
asynchronous mode
band
bandwidth
baseband
baud rate
binary signal event duration
binary synchronous communication (BSC), or bisync
biphase
bit time
block
block check character (BCC)
broadcast mode
broadband
byte count-oriented protocol
character framing
character-oriented transmission
communication buffer
communication channel
communication protocol
compensating error
continuous high idle
convention
coupler
cycles per second
cyclical redundancy check (CRC)
data compression
data link escape (DLE)
data transfer rate
data end flag
data start flag
digital data communication message protocol (DDCMP)
distortion
electromagnetic interference (EMI)
even parity
frequency
frequency-division multiplexing (FDM)
frequency modulation (FM)
frequency range
full-duplex mode
half-duplex mode
handshaking
Hertz (Hz)
high-level data link control (HDLC)
longitudinal redundancy check (LRC)
Manchester coding
message
modem
modulation
multileaving
multimode PCM
negative acknowledge (NAK)
noise
nontransparency
odd parity
packet
packet header
parallel transmission
parity bit
parity check
phase
phase-shift modulation
pulse-code modulation (PCM)
radio frequency (RF)
return wire
serial transmission
signal-to-noise (S/N) ratio
signal wire
simplex mode
sine wave
single-mode PCM
skew
square wave
start bit
stop bit

subcarrier
synchronous data link control (SDLC)
synchronous idle
synchronous mode
time-division multiplexing (TDM)
transistor-to-transistor logic (TTL)
transmission medium
transparent (binary) mode transmission
vertical redundancy check (VRC)
XON/XOFF
zero-crossing signal

Vocabulary Exercises

1. _____ transmission sends bits one at a time using a single transmission line.
2. _____ encodes data by varying the distance between wave peaks within an analog signal.
3. A _____ converts a digital signal to an analog signal (or vice versa) to allow data transmission over analog phone lines.
4. A _____ can be used to resolve transmission or processing speed differences between sender and receiver.
5. In synchronous data transmission, a _____ signal is transmitted during periods when no data is being transmitted.
6. Digital transmission using _____ is achieved with the use of square waves.
7. The _____ of an analog signal is measured in _____, or cycles per second.
8. The amount of channel capacity required to transmit a message can be reduced via a _____ algorithm.
9. _____ checking is a form of longitudinal redundancy checking in which bits of a single character or byte are used to derive a check digit.
10. _____ transmission can carry multiple messages simultaneously; _____ transmission carries only one message at a time.
11. Simultaneous transmission of multiple messages on a single channel can be accomplished by _____ or _____ multiplexing.
12. _____ is the time interval by which identical waveforms precede or lag one another.

13. In asynchronous transmission, _____ is implemented through the use of start and stop bits.

14. The term _____ describes the encoding of data as variations in one or more physical parameters of a signal.

15. In _____, blocks or characters arrive at unpredictable times. In _____, the timing of character or block data transfers between sender and receiver is precisely coordinated.

16. _____ transmission implements two-way transmission with two separate communication channels; _____ transmission implements two-way transmission with only one communication channel.

17. _____ encodes data by varying the magnitude of wave peaks within an analog signal.

18. _____ transmission uses multiple lines to send multiple bits simultaneously.

19. A _____ extends the range of data transmission by retransmitting a signal.

20. _____ checking generates a single check digit for each bit position within the bytes of a block.

Review Questions

1. What are the components of a communication channel?
2. Why will the actual data transfer rate of a channel usually be less than the theoretical maximum implied by the technology used to implement the channel?
3. In what ways are optical transmission media superior to electrical transmission media?
4. Define the term *signal-to-noise ratio.*
5. Define the term *bandwidth.*
6. What is data compression? What are its costs and benefits?
7. Describe the differences among simplex, half-duplex, and full-duplex transmission.
8. Describe the difference between serial and parallel data transmission. What are the comparative advantages and disadvantages of each?

9. What is character framing? Why is it generally not an issue in parallel data transmission?
10. What are the differences between synchronous and asynchronous data transmission?
11. Define the terms *frequency, amplitude, phase,* and *modulation.*
12. What are the comparative advantages and disadvantages of frequency modulation to amplitude modulation?
13. How is data transmitted on optical media?
14. Describe the differences between even and odd parity checking.
15. What is a block check character? How is it derived and used?
16. What problems are encountered when a sender and receiver operate at different speeds? By what method(s) can these problems be overcome?
17. Compare and contrast frequency- and time-division multiplexing. What physical characteristics of the communication channel are required by each type? Which provides greater data transmission capacity?

Research Problems

1. As discussed in Chapter 4, ASCII is a character-coding method that also includes codes to control various aspects of data communication and I/O devices. These codes comprise the first 32 numbers in the ASCII sequence. Investigate the use of these codes for implementing handshaking within communication channels. Which codes are designed for asynchronous transmission and how are they used? Which codes are designed for synchronous transmission and how are they used? Are these codes sufficient for modern data communications?
2. Investigate the definition of the RS232C serial communication standard. Describe the physical characteristics of the transmission lines and connectors. Which signals are assigned to which lines? How are data bits encoded? What mechanism(s) are used to implement handshaking and error detection?
3. Kermit is widely used communication protocol (and software package). It is owned by Columbia University and distributed freely. Investigate the mechanisms Kermit uses to transfer files between computers. How is data organized for transmission? What error-checking mechanism(s) are provided? What are the details of the file-transfer protocol?

8 Mass Storage and Input/Output Technology

Chapter Goals

- *Describe the configuration and control of I/O channels.*
- *Describe factors that affect I/O performance.*
- *Describe the physical and logical organization of secondary storage.*
- *Describe the characteristics and implementation technology of common I/O devices.*

INPUT/OUTPUT CHANNELS AND CONTROL

Chapter 6 discussed processor and memory architecture in detail. Chapter 8 extends that discussion by describing the architecture of mass storage and I/O devices. These two hardware subsystems are covered together due to a number of similarities between them with respect to data communication and device control.

An overview of the hardware architecture of a computer system was presented in earlier chapters and is reproduced here in Figure 8.1. Note that a typical computer system has many I/O devices. Actual numbers range from several for a microcomputer (a printer, keyboard, video display, and a mouse) to several hundred for a mainframe computer. Although secondary storage is shown as a single unit, larger computer systems and many smaller ones typically contain multiple secondary storage devices.

Logical and Physical I/O

All secondary storage and I/O devices must be connected directly or indirectly to an *I/O port* on the system bus. Although the system bus provides for some variation in device control procedures—for instance, through the use of various command and status codes—the flow of data between the CPU and all devices is handled in a similar manner. Yet there are major differences in many important characteristics of these devices. These differences include capacity, data transfer rate, data coding methods, and many others.

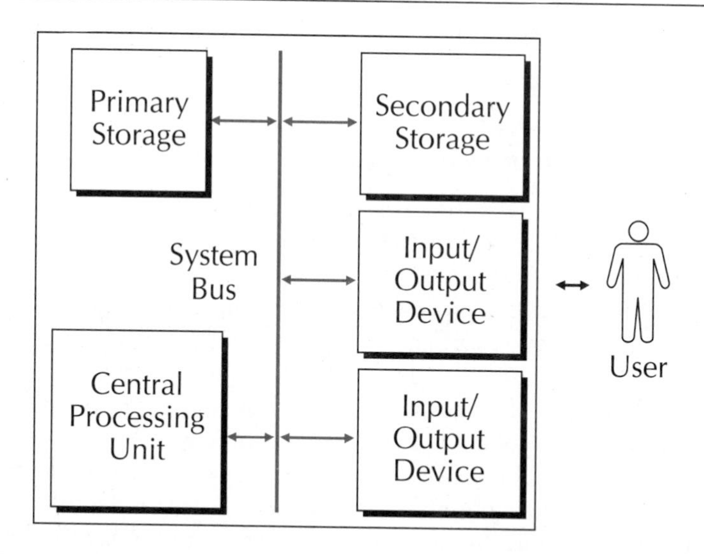

FIGURE 8.1 **The relationship between the CPU and storage devices in a computer system.**

The design of the CPU instruction set and the system bus would be much more complex if all possible differences between I/O and storage devices were accounted for. Physical device details such as how a disk read/write head is positioned or how a certain color is displayed on a video display would require explicit processor instructions and/or bus control signals. Accounting for these complexities would not only increase the cost of the bus and processor, it would limit the types of secondary storage and I/O devices that could be included in a computer system. Only those devices that were "designed in" could be used.

The alternative (and common) approach is to treat each device similarly. That is, provide a small set of generic capabilities for communication with and control of all devices. These generic commands must be simple and few in number so as to simplify processor design, bus design, and I/O protocol. However, they must also provide sufficient power to allow processor control over a wide range of present and future storage and I/O devices.

The use of a generic set of commands and status signals for communication with a storage or I/O device is termed a *logical access*. In contrast, a *physical access* refers to the physical actions of the device that are carried out to meet the request (e.g., the physical actions required to read a sector of a disk or print a character on a printer).

Logical access commands. Generic I/O capabilities can be provided with two simple commands and three status signals. The commands are these:

read <device-id> <address>

write <device-id> <address>

and the status signals are:

ready <device-id>

busy <device-id>

result <device-id> <condition-code>

The parameter <device-id> is a unique numeric code identifying one of the devices attached to the system bus. The parameter <address> is a number identifying a unique location within a device. The parameter <condition-code> is a number that indicates the result of processing. In its simplest form it has only two values: one to indicate success (e.g., 0) and the other to indicate failure (e.g., 1). More elaborate coding schemes can be used to differentiate among various types of failure.

A processor executes I/O commands by placing a command code and device identifier on the control lines of the bus. An I/O or storage device detects these signals and responds by issuing a ready or busy status signal. When the device completes its processing, it places a result status code on the bus. The exact nature of data transfer varies from one computer system to another. In simple architectures, the CPU can use the data lines at the time a command is issued. This method is very inefficient for large transfers, because the CPU must transfer data in small increments via multiple commands. For example, to store 128 bytes of data over a bus with 16 data lines, 64 separate write commands would be required.

Modern computer architectures usually provide for block transfers. A write command issued by the CPU can include the starting address of the data in memory (using the bus address lines) as well as the number of bytes to be transferred (using the bus data lines). Once the command is acknowledged, the CPU exercises no further control over the transfer. This method normally requires that a portion of main memory be dedicated to I/O for each individual I/O and storage device.

Logical device views. The storage locations of a secondary storage device are assumed (by the CPU) to be numbered sequentially starting at location zero and extending to the maximum capacity of the device. This organization is referred to as a *linear address space*. The actual organization of storage locations and the method of accessing those locations will vary between secondary storage devices. Disk devices are normally organized into blocks (or sectors) of 512 bytes to 4 Kbytes. The <address> component of a read or write command identifies one of these blocks.

Translating logical disk addresses into physical addresses requires that the address parameter of a read or write command be converted into a specific disk location expressed in terms of a platter, track, and cylinder. With tape drives, the translation between logical and physical addresses is much more straightforward than for disk drives. This is because the physical organization of blocks on a tape is a linear sequence. An access to a logical block is translated into commands to position the tape head over the appropriate portion of the tape before a read or write operation.

Some I/O devices communicate only in terms of sequential streams of characters. With these devices, the concept of an address or location is irrelevant. Other I/O devices do have storage locations in the traditional sense. For example, the individual character positions of a video display can be considered individual storage locations. With such devices, it is often desirable to write to individual display locations without disturbing data already being displayed in other locations (i.e., to overwrite only a portion of the display). This form of output requires that individual display locations be assigned addresses and that logical addresses in read and write commands be translated into physical (row and column) locations.

Distribution of translation processing. The translation between logical and physical accesses can occur in various places within a computer system:

- In software (e.g., the operating system)
- Within the device
- In a device controller

Each of these approaches has relative advantages and disadvantages, as described below.

Knowledge of the physical organization and access methods of storage and I/O devices can be utilized within systems software. For example, information about the organization of a disk, such as number of platters, tracks, and sectors per track, can be stored within the operating system. Service requests for disk access can be translated by the operating system kernel into explicit commands to the disk drive to access a specific platter, track, and sector. The commands necessary to instruct the device can be numerically encoded and communicated via the system bus in one or more transfers. Thus, all of the intelligence necessary to translate logical accesses into physical accesses would reside within the operating system.

The primary advantage to this approach is flexibility. Changes in device characteristics—for example, replacing a smaller disk with a larger one—are incorporated into the system configuration by updating the corresponding kernel programs. The addition of new devices is also dealt with in this manner. Simple translation would require a relatively simple program; more complex translations would require correspondingly more complex programs. The primary disadvantage of a software-based approach is inefficiency. Recall that the motivation for using logical accesses is to offload complexity and work from the CPU. Although a software-based translation does simplify CPU and bus hardware, it still requires CPU resources to execute the translation programs. For this reason, complete logical/physical translation within software is rarely used.

Translation intelligence can also be embedded within a storage or I/O device. That is, each device implements both its normal physical actions and the processing actions necessary to perform logical-to-physical translation. The primary advantage to this approach is efficiency. Processing overhead associated with storage and I/O accesses is moved to the devices. Other than for logical command and bus control, no CPU cycles are consumed for I/O processing.

The primary disadvantage of device-based translation is cost and redundancy. Processing power must be included in every storage and I/O device. This cost is replicated over all similar devices in a computer system. Consider, for example, a mainframe computer system with 20 identical disk drives. Logical/physical translation within the operating system would use a single processor (the CPU) and store the translation program once. Translation within each disk drive would require 20 different (although relatively simple) processors, each with a separate copy of the translation program.

The most common approach to logical/physical translation is a distributed approach. Small portions of the translation process are implemented within systems software and within device hardware. The majority of the translation process (and associated control procedures) are implemented within auxiliary hardware devices, as described below.

Device Controllers

It is rare to connect storage or I/O devices directly to I/O ports. Instead, such devices are normally connected indirectly through *device controllers*, as shown in Figure 8.2. Device controllers perform several functions, including the following:

- Implement the bus interface and access protocols
- Translate logical accesses into physical accesses
- Allow several devices to share access to a single I/O port

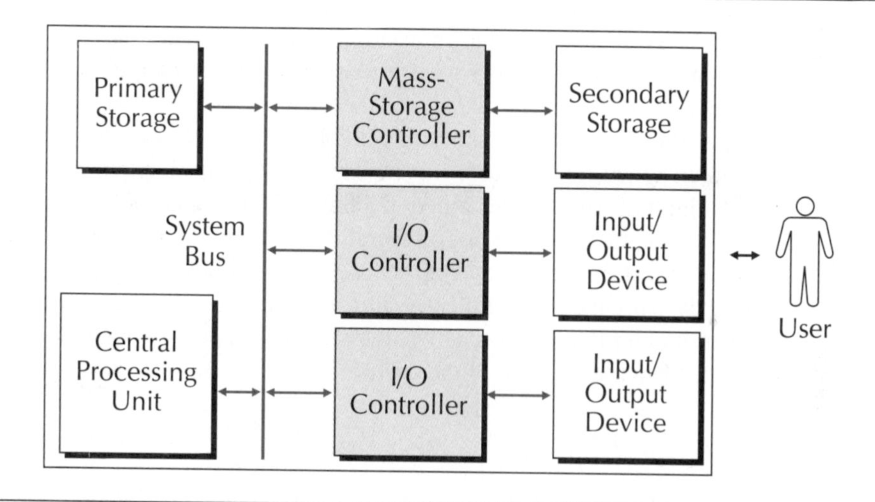

FIGURE 8.2 **The connection of mass storage and I/O devices using device controllers as interfaces to I/O ports.**

Because they are attached directly to I/O ports on the system bus, device controllers are responsible for implementing the bus protocol. The controller is responsible for monitoring the bus control and address lines for signals to attached devices. When detected, these signals must be translated into the appropriate commands to the storage or I/O device. Similarly, data and status signals from the device must be translated into the appropriate bus signals and placed on the bus. All protocols regarding timing, interrupts, signal encoding, and handshaking are thus implemented by the controller.

The translation of signals between the bus and devices is a portion of the conversion between logical accesses and physical accesses. Device controllers must "know" the physical details of the attached devices and must issue very specific instructions to the device. For example, a request for a certain sector of data from a disk must be converted into a command to read from a specific track, platter, and sector. The CPU (or the program executing on it) sees the disk as a linear address space. The device controller provides the translation between that logical view of the disk drive and the physical realities of platters, read/write heads, and sectors.

Device controllers can serve as mechanisms for sharing I/O ports among several storage or I/O devices. This is important because the number of physical I/O ports on the system bus is usually limited (16 is a typical maximum). A device controller can allocate bus access in either dedicated or multiplexed mode. In *dedicated mode*, all lines are used to transmit a burst of data between a single device and the CPU. In essence, the controller acts as a switch or selector, allocating access to the I/O port and translation facilities to only one device at a time. Such controllers are typically used for devices capable of high-speed data transmission, such as disk drives.

In *multiplex mode*, transmissions involving multiple devices are interleaved, or handled concurrently. Such transmission is typically implemented using time-division multiplexing, as described in Chapter 7. The controller combines or separates the communication among several devices and the I/O port simultaneously. Such a controller is typically used with slow-speed devices such as modems and keyboards. This method efficiently allocates the relatively large communication bandwidth of an I/O port to many slower devices, none of which can individually match the data transfer rate of the port.

Mainframe Channels

The use of device controllers greatly reduces the amount of CPU time dedicated to I/O and storage accesses. Within many mainframe computers, the concept of device control is taken one step further. That is, a dedicated (special-purpose) computer is attached to each I/O port. This computer is normally called a *channel*[1] or a *channel processor.*

The difference between a channel and a device controller is not clear-cut. It is a function of power and capability in several key areas, including:

- Number of devices that can be controlled
- Variability in type and capability of attached devices
- Maximum communication capacity

[1] The term *channel* was originally coined by IBM to describe a specific component of its 300 Series (and later 3000 Series) mainframe computers. Due to the predominance of IBM mainframe computers, the term has since gained the generic meaning described in this section. Vendors of other mainframe computer systems often use other terms (e.g., *peripheral processing unit* and *front-end processor*) to describe functionally similar components.

A typical secondary storage controller might be capable of controlling up to eight devices such as magnetic disk drives, optical disk drives, and magnetic tape drives. A typical I/O device controller can control up to 32 I/O devices such as modems and terminals. In general, a device controller can control only devices of similar type and capacity. Its maximum communication capacity (with the CPU) is typically less than 20 Mbytes per second.

A mainframe secondary storage channel will typically control up to a few dozen secondary storage devices. These devices can be of mixed types and capacities (e.g., four tape drives and 16 disk drives). A mainframe I/O channel may control as many as 256 terminals. The maximum communication capacity of a mainframe I/O channel is generally measured in hundreds of megabytes per second. A sample channel configuration for a mainframe computer is shown in Figure 8.3.

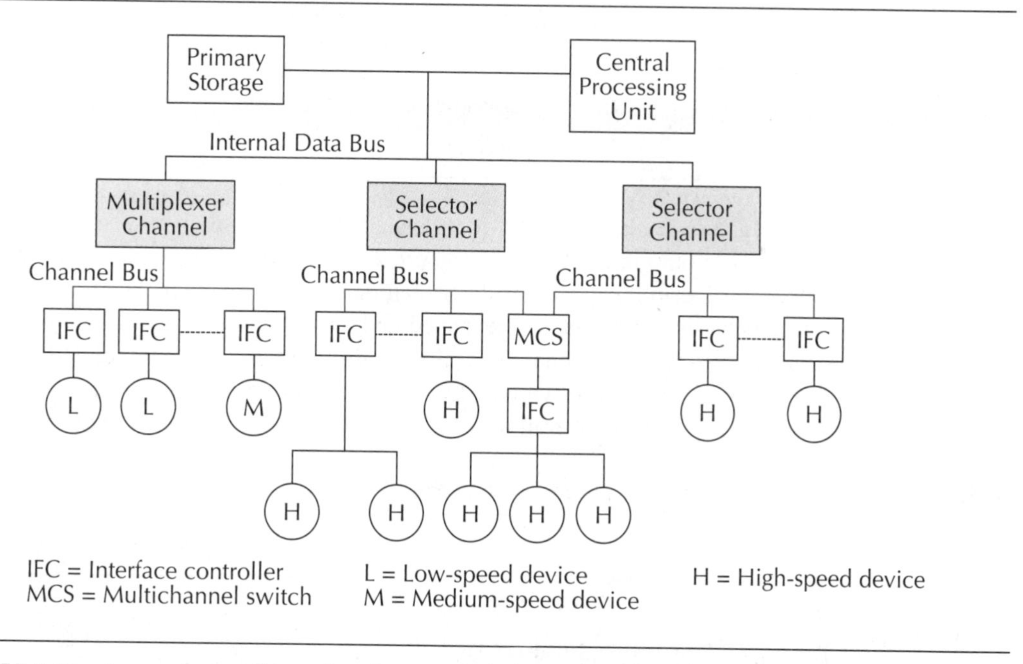

FIGURE 8.3 **A sample configuration for a mainframe computer using three channels.**

PERFORMANCE ISSUES

Differences in the physical operating characteristics of computer devices cause differences in the speeds at which data can be transferred among them. From these differences arise incompatibilities in timing relationships among the devices within a system. Even if other incompatibilities did not exist, these timing differences would still present a major barrier to data transfer and to the integration of computer systems components.

Relative speed and timing differences among storage devices result from differences in performance among levels of the memory/storage hierarchy. These performance differences are summarized in Table 8.1. Recall that a key issue in computer systems configuration is the selection of an appropriate mix of these types of storage. Incompatible data transfer speeds are an inevitable consequence of configuration choices.

TABLE 8.1 Comparison of performance differences among typical memory and storage devices.

Storage Device	Data Transfer Rate	Mean Access Time
CMOS RAM	15 Mbytes/second	70 ns
Hard Disk	250 Kbytes/second	15 ms
Floppy Disk	25 Kbytes/second	200 ms
CD ROM	150 Kbytes/second	200 ms
½-inch Tape	50 Kbytes/second	—

Transfer speeds for I/O devices depend strongly on their respective speeds of input and/or output, as shown in Table 8.2. For example, the operational speed of a dot-matrix print head is far slower than that of a line printer. The timing of data transfers to such devices is determined and limited by the physical actions of the character-producing mechanisms. Speed differentials are compounded by other factors, including delays caused by detecting and correcting errors in transmission. The result is the overall data transfer rate and efficiency of the device.

TABLE 8.2 Comparison of typical speeds for a range of I/O devices.

Device	Speed
VDT	500–2000 CPS
Dot-Matrix Printer	80–400 CPS
Line Printer	300–3000 LPM
Ink-Jet Printer	30–120 LPM
Laser Printer	4–120 PPM

Interrupt processing is the primary means of resolving timing differences among CPU cycles and I/O events. From the standpoint of the CPU, the timing of I/O events is unpredictable. Because of differences in data transfer rate and timing, I/O operations must be performed asynchronously with respect to CPU cycles. Interrupt handling permits the current process in the CPU to be suspended and then resumed after the interrupt has been processed. When interrupts are used to signal I/O events, the only disruption to CPU operation is the actual time interval required to perform the action specified by the interrupt.

Buffered I/O

As shown in Figure 8.4, a *buffer* can be used to compensate for data transfer rate differences between sending and receiving devices. An I/O buffer can be implemented as a reserved area of system RAM, or an intelligent peripheral or controller can contain a local buffer. I/O devices can also incorporate a small area of semiconductor RAM that is a dedicated buffer for communication or I/O.

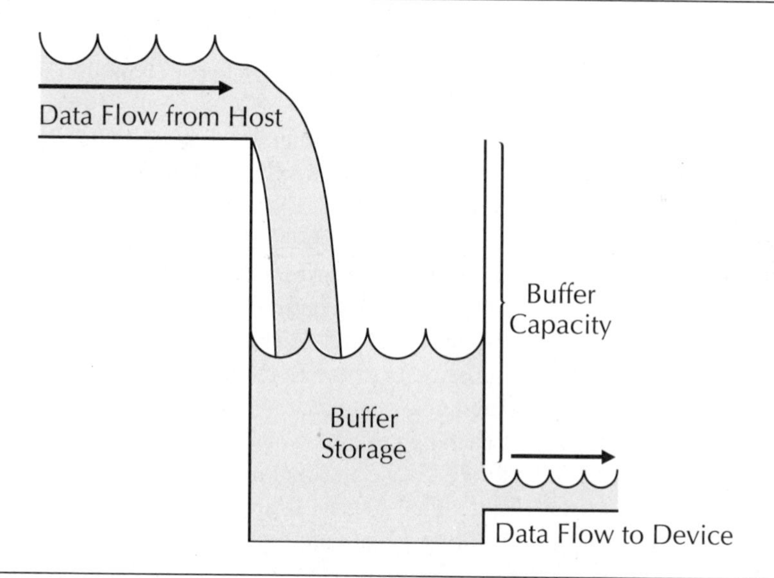

FIGURE 8.4 The concept and operating principles of an I/O buffer.

The purpose of an I/O buffer is to permit each device to communicate at its own maximum data transfer rate. For example, in the diagram in Figure 8.4, the left side of an I/O buffer exists in a device controller connected to one of the system I/O ports. The RAM devices in the buffer are relatively fast, so the CPU can write data to the buffer at the maximum rate of the system bus. When the buffer is filled, the device controller sends an interrupt back to the CPU to signal that the transaction is complete.

Meanwhile, over the data path on the right side of the buffer, an I/O or storage device reads data from the buffer at the maximum data transfer rate of the device. When nearly all the data in the buffer has been read, the device controller generates another interrupt to signal the operating system to initiate the next unit of data transfer. Ideally, the timing of write and read operations is coordinated by the device controller so that the I/O or storage device is kept continually busy.

From a design standpoint, it also is desirable to provide buffer capacity of at least the size of a device's unit of data transfer. For example, a buffer within a disk device or controller can typically hold the content of one sector or track. This figure is typically between 256 bytes and 4 kilobytes.

Cached I/O

A common method of improving I/O performance, especially to and from disk drives, is the use of an *I/O cache*. Recall from Chapter 6 that a cache is a small area of fast storage used to improve the performance of a larger amount of slower storage.[2] The discussion in Chapter 6 focused on the use of a high-speed cache between the CPU and main memory, but the concept can also be applied to mass storage and I/O accesses.

I/O caching can be implemented either at the level of the device controller or by systems software. When implemented within the device controller, RAM is included within the controller and used much as a local buffer. The basic idea behind I/O caching is simple. Accesses to data contained within the cache can occur much more quickly than accesses to the physical I/O device or storage medium. The speed difference is due entirely to the faster access speed of RAM, compared to the access speed of the I/O or storage device. The drawbacks in using cached I/O lie in the costs of the RAM needed to implement the cache and the complexity in managing its contents.

Successful use of caching requires a sufficiently large cache and the "intelligence" to use it effectively. Cache size and intelligent management of its contents are the factors that differentiate a cache from a buffer. The size of a buffer is typically equal to the size of a block of data transferred to or from a device. The size of a cache is typically large enough to hold tens or hundreds of data transfer blocks. When used for output from the CPU or memory, the cache serves primarily as a large buffer. Its unique advantages are realized in transfers from a device to memory or the CPU.

A *cache controller* is a processor that manages the contents of a cache. It attempts to fill the cache with data that it expects the CPU to need. Thus, the cache controller guesses what the CPU will request next, and reads that data from

[2] The reader might want to review the section on caching in Chapter 6 before preceding with the remainder of this section.

the device into the cache before it is actually requested. The mechanisms by which "guesses" are made range from the very simple to the very complex. For disks, a simple strategy is to assume that whatever was recently read will soon be rewritten. A more complex strategy is to assume that a read of a record in a file will soon be followed by a read of the next sequential record.

Cache management can also be implemented within the operating system or by a combination of operating system and device controller processing. The operating system might provide information about file access modes that enable more intelligent cache management. For example, if a file is opened for read-only access, it makes little sense to keep recently read records in the cache. If a file is opened for sequential reading, a lookahead strategy for subsequent records is likely to be advantageous. Strategies for cache management are similar to those used in virtual memory management, as described in a later chapter.

Direct Memory Access

Recall from Chapter 6 that *direct memory access (DMA)* refers to the movement of data between an I/O or storage device and main memory, without the intervention of the CPU. DMA can be implemented in several ways, including a multimaster system bus, an auxiliary DMA controller, and/or auxiliary DMA channels between peripheral devices and main memory. The use of some form of DMA is now nearly universal in computer systems.

The primary purpose of DMA is to reduce processor overhead and improve I/O performance. This is accomplished by eliminating the need for CPU involvement in communication between peripheral devices and main memory, which allows the CPU to perform other tasks while transfers to and from memory are taking place.

The implementation of DMA has a direct influence on the design and implementation of device controllers. In the case of a multimaster bus, the device controller must implement the necessary bus protocol to become a bus master. This implies more complex design and processing within the device controller. Little or no complexity is required in device controllers when auxiliary DMA controllers and channels are used. This method of DMA implementation normally "fools" the device controller into thinking that the DMA controller is, in fact, the CPU.

Data Compression

As first discussed in Chapter 7, *data compression* can reduce the total number of bits used to encode a set of data items (e.g., a file or a network packet). Data compression requires a *compression algorithm* (a program) and a processor to implement the algorithm. Depending on the type of data compressed, a compression algorithm can reduce total data volume from two to 10 times or more. With a 4:1 *compression ratio*, for example, a 100-Kbyte data set would be reduced to 25 Kbytes ($100 \div 4 = 25$).

Data compression is commonly used to reduce secondary storage requirements, as shown in Figure 8.5a. Data sent to the storage device is reduced in volume by a compression algorithm before actual storage. Data received from the storage device is returned to its normal state by applying an inverse compression (decompression) algorithm. Data compression can also be used to expand the capacity of a communication channel, as shown in Figure 8.5b. Data is compressed prior to entering the channel, and is decompressed as it leaves the channel.

The use of data compression always represents a tradeoff between processor resources and communication and/or storage resources. The implementation of the compression and decompression algorithm consumes processor cycles. Algorithms with high compression ratios frequently consume more processing resources than those with low compression ratios. The proper tradeoff between these resources depends on their relative cost, availability, and degree of utilization. Implementing data compression with CPU processing resources is not usually cost-effective. However, implementing data compression through special-purpose processors is often advantageous. Such processors are now commonly available and are frequently used in disk controllers, network interface units, and high-speed modems.

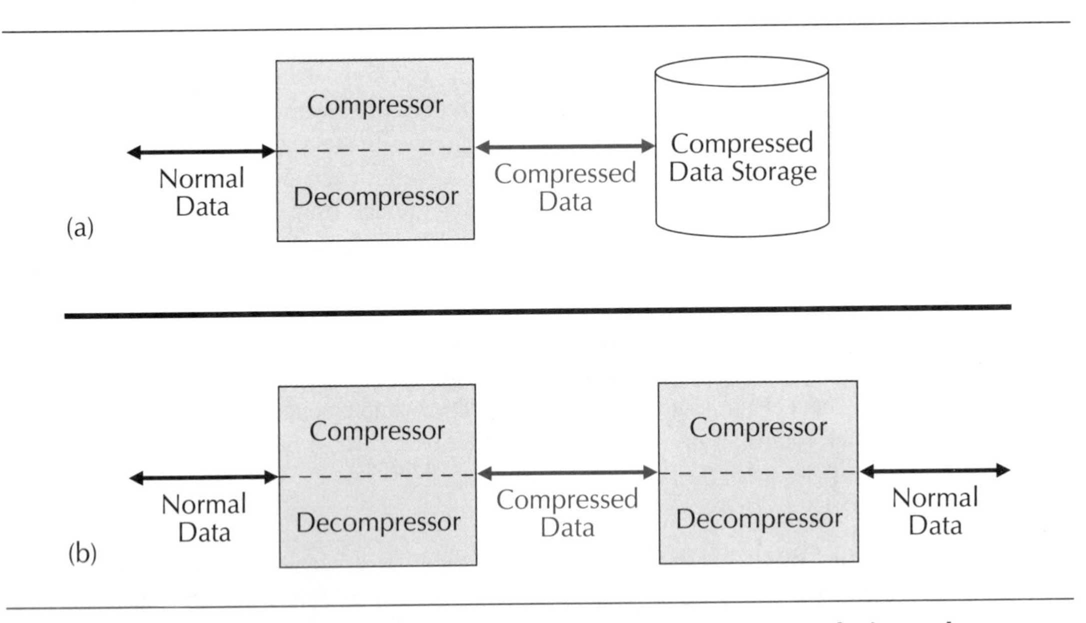

FIGURE 8.5 **The use of data compression with (a) a secondary storage device and (b) a communication channel.**

Principles of magnetic and optical data recording for implementing secondary storage were covered in Chapter 5. To access a secondary storage device such as magnetic tape or disk, the CPU communicates with the corresponding device controller. Signals are passed over the system bus through one of the I/O ports.

The device controller must have sufficient processing capability to direct the physical actions of read or write operations. Once communication has been established with the device controller, the CPU passes it an identifier that corresponds with the desired record. The device controller must search for the record and then perform the requested access operation. To achieve this, logic circuitry within the device controller must be able to

- Read or write at arbitrary positions
- Search for specific positions by using a point of reference
- Locate existing records
- Sense the beginning and end of a record

At the physical level, the device controller must be able to establish a point of reference on the storage medium and to sense the physical organization of data recording. On tapes and some disks, such reference information is contained within a reserved storage area of each volume. The physical organization of data recording is held here as *stored addressing information*, as shown in Figure 8.6. This information is written to the volume when it is initialized, or formatted, under control of the operating system.

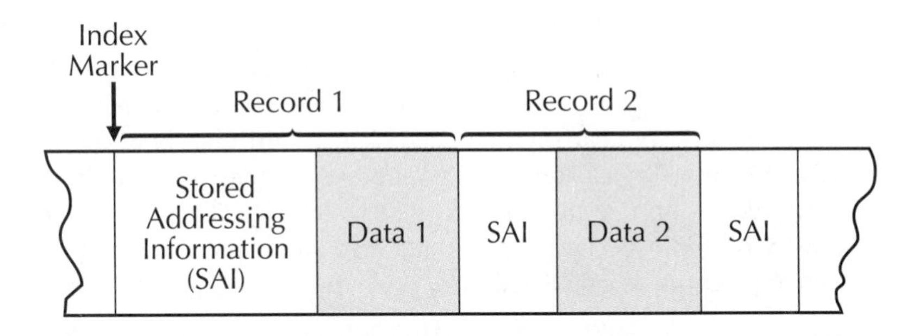

FIGURE 8.6 The recording format for a nine-track tape that provides fields for stored addressing information.

Disk Storage

A disk drive consists of one or more platters of chargeable (or optical) material, a motor to spin the platters, and one or more read/write heads that can be positioned over various tracks of the disk platters. A track is one concentric circle of a platter, or the surface area that passes under a read/write head when its position is fixed. A sector is a fractional portion of a track. The number of sectors per track ranges from eight for low-density floppy disks to the upper teens or lower twenties for high-capacity hard disks. A single sector (or block) is normally the unit of data transfer.

When originally manufactured, all magnetic hard disks (and most floppy disks) are blank. Although the location of tracks within a platter is fixed by the position of read/write heads, the location of sectors within a track is not. One of the operations performed when a disk is formatted is fixing the location of sectors within tracks. This is accomplished by writing synchronization data at the beginning of each sector. This information is the reference by which the controller senses the beginning of a physical record. Synchronization information can also be written within a sector during either formatting or subsequent write operations. The read/write circuitry uses synchronization information to compensate for minor variations in rotation speed and other factors that might disturb the precise timing needed for reliable reading and writing.

Within a sector, data bits are stored sequentially (serially). The read/write head reads and writes these bits one at a time as the sector passes under the head. Communication with the system bus is normally parallel. Thus, serial-to-parallel conversion is required during read operations, and parallel-to-serial conversion is required during write operations. These conversions can be performed within the disk drive itself or within the drive controller. Figure 8.7 illustrates the circuitry required to perform this conversion.

The maximum data transfer rate of a disk drive is a function of the density of bit storage and of platter rotation speed. Once the read/write head has been positioned, it simply waits for individual sectors to pass beneath it and reads or writes data. A typical rotation speed for a hard disk is 3600 revolutions per minute (rpm). Typical bit density ranges from tens of thousands per track for floppy drives up to a million bits or more for high-density hard disks. For a bit density of 100,000 bits per track and a rotation speed of 3600 rpm, maximum data transfer rate is calculated as

$$\frac{3600 \text{ RPM}}{60 \text{ seconds/minute}} \times \frac{100{,}000 \text{ bits}}{8 \text{ bits/byte}} = 750 \text{ Kbytes/second}$$

However, this rate is not sustainable, even for one second. Note that the disk will spin 60 times in one second and, thus, the above calculation assumes that the same track is read or written 60 times.

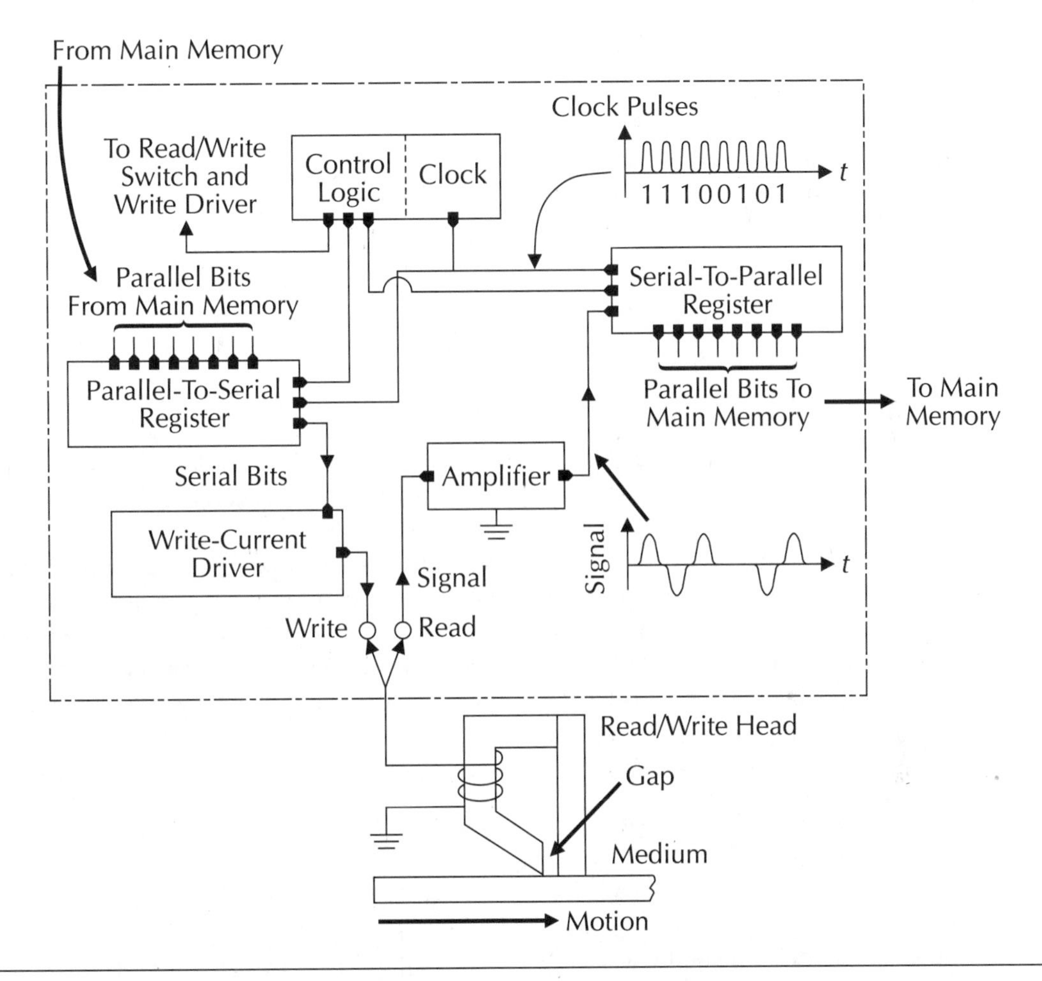

FIGURE 8.7 **The circuit depicted in this schematic implements the logic for serial-to-parallel conversion within a disk drive.**

Files are not normally stored on a disk in sequential sectors and tracks. Rather, the sectors of a file are normally scattered about the disk surface, and sequential access to them requires sequential access to randomly located sectors. Thus, a more realistic estimate of data transfer rate must also account for the time needed to position the head between tracks, called *track-to-track seek time,* and the time spent waiting for a desired sector to rotate under the read/write head, called *rotational delay*. Figure 8.8 illustrates the concept of rotational delay.

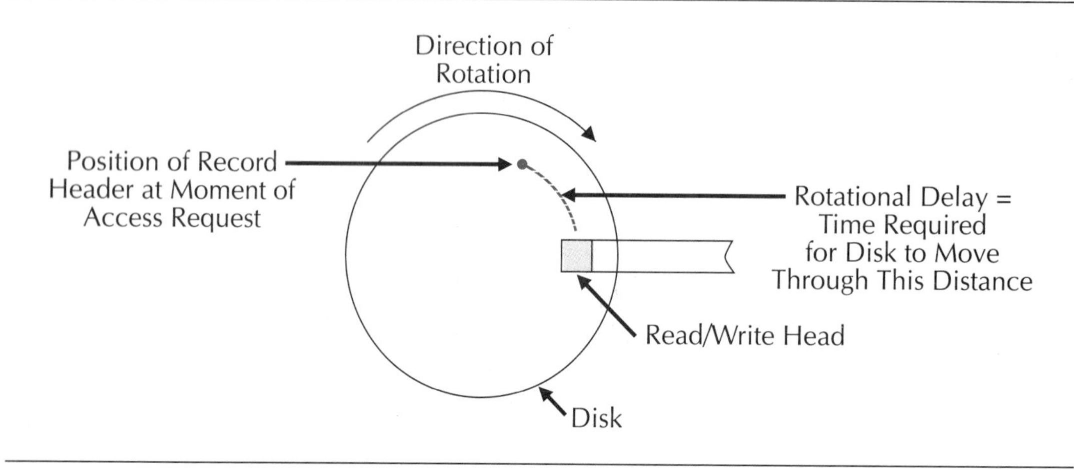

FIGURE 8.8 **The elements that determine rotational delay for disk access operations.**

In some disk drives, a fixed arm is used to hold multiple read/write heads, one for each track, as illustrated in Figure 8.9a. In other drives, a single read/write head is moved back and forth between tracks, as illustrated in Figure 8.9b. On average, the read/write head will need to be moved over one half the number of tracks between each sector read operation. Also, on average, the read/write head

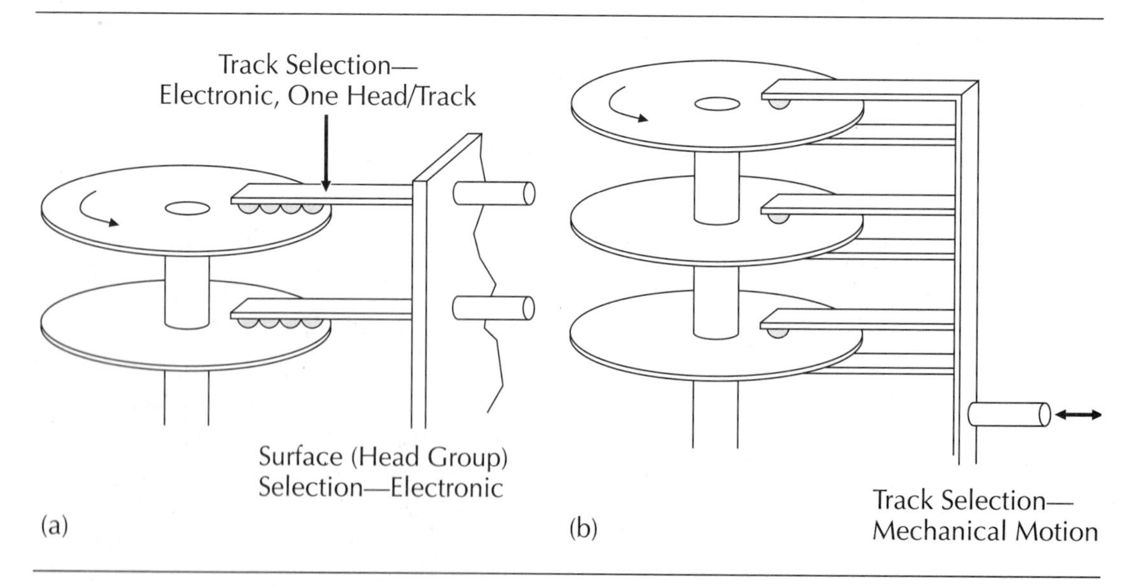

FIGURE 8.9 **Head-positioning techniques for data access on disk drives: (a) a fixed-head mechanism; (b) a movable-arm access.**

will have to wait one half of a platter rotation before the desired sector is positioned beneath the head. Each of these operations incurs a delay and their sum is the *average access time* for a sector. Average access times of between 10 and 30 milliseconds are typical for hard disks.

The translation between logical and physical access is a translation from sequential block number to a physical sector defined in terms of platter, track, and sector within track. This translation is straightforward if the assignment of logical sector numbers is regular with respect to tracks and platters. Typically, the assignment of logical sector numbers proceeds first across sectors within a track, then across platters, and finally across tracks, as illustrated in Figure 8.10. Physical sectors are logically numbered starting with the first platter, track, and sector within track. This organization minimizes the amount of head movement and rotational delay in accesses to sequential logical sectors.

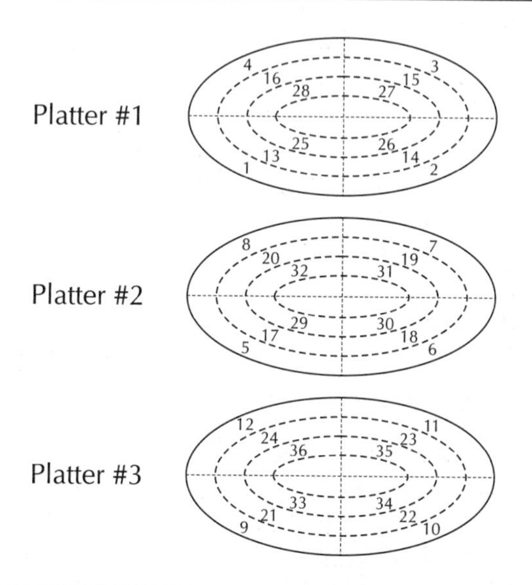

FIGURE 8.10 The assignment of logical sector numbers to physical sectors within a multiple-platter disk drive.

The data transfer rate of a disk device, typically expressed in kilobytes per second (Kbytes/sec), is determined by the net effect of all the factors described above. The number of data lines connecting the device to the controller or I/O port will also be a factor. Remember that data transfer rate describes the speed of error-free data transmission. Error handling within the communication channel also must be considered. Data transfer rates, as well as head-positioning times and average rotational delays, are given for representative IBM disk devices in Table 8.3.

TABLE 8.3 **Head-positioning times and average rotational delays for some IBM disk drives.**

IBM Device Type	Head Positioning (Avg. MS)	Rotational Delay (Avg. MS)	Transfer Rate (Kbyte/Second)
3330	30	8.3	806
3340	25	10.1	885
3344	25	10.1	885
3350	25	8.3	1198
3370	20	10.1	1859
3380	16	8.3	3000

Tape Storage

The beginning of a physical tape record must be positioned beneath the read/write head prior to reading that record. To access the *N*th physical record on a tape, the tape must be wound through the preceding $N - 1$ records. If the position of the desired record precedes the current position of the tape and read/write head, the device controller must rewind the tape to that position. To perform this physical positioning, a tape drive controller must keep track of the current position of the tape (which record is under the read/write head). It must also be able to distinguish the start of each record as it passes beneath the tape head.

A request for a specific logical record on a tape record must be translated to a read operation on a physical record. If logical record length is equivalent to physical record length, the logical record number is equal to the physical record number. If logical records are shorter than physical records, multiple logical records can be stored in a single record. The number of logical records stored in each physical record is referred to as the *blocking factor*. Translation of logical record numbers to physical record numbers utilizes the blocking factor. Thus, with a blocking factor of 20, a read request for logical record #84 must be translated into an access of physical record #4.

Special markers are used to allow the tape drive controller to determine where one physical record ends and another begins. These can include the following:

- Start-of-record markers
- End-of-record markers
- Interrecord gaps

A *start-of-record* marker is simply a special code placed on the tape to indicate the beginning of record. Similarly, an *end-of-record* marker is a special code indicating the end of a record. In many tape drives, only an end-of-record marker is used.

The reliability of read and write operations is improved if a blank space (or *interrecord gap*) is placed between each record. The use of this gap allows the tape to be started and stopped between records. If no gap were provided, the tape-winding mechanism would have to be able to stop instantaneously completely or start and immediately reach normal speed. Because this is not mechanically possible, some interrecord gap is required to allow starts and stops. The space allocated to these gaps depends on record length, but usually is at least 10 percent of the total tape length.

The recording density of a tape is a measure of the amount of linear space allocated to store each bit. Typical recording densities for tape drives range from 200 to 16,000 bytes per inch (Bpi). Physically, recording density can be controlled by controlling either the speed of tape movement or the duration of electrical pulses to the read/write head. The latter approach is the most common because variable speed pulsing is simpler (and more reliable) to implement than variable speed drive motors.

The various recording factors discussed thus far (recording density, use of markers, the blocking factor, and interrecord gaps) are normally stored in a *tape header* record. When a tape reel or cartridge is first mounted, the tape drive controller reads this information from the beginning of the tape. It is used in subsequent accesses to determine exact tape positioning and interpretation of electrical signals to and from the read/write head.

As with disk devices, the overall data transfer rate of a tape device is the net effect of physical device and recording characteristics. Even though tape offers high recording densities, the rate of data transmission from tape devices is slower than from hard disk devices because of generally slower physical access methods.

TECHNOLOGY FOCUS: SCSI. The acronym *SCSI* (pronounced "scuzzy") stands for small computer system interface. It is an ANSI-standard bus designed primarily for mass storage and I/O devices encompassing both physical and logical parameters. Although its name implies that it is used primarily in small computer systems, it is commonly applied in computer systems ranging from personal computers to minicomputers. In its current form, SCSI exists in two primary forms named SCSI-1 and SCSI-2. SCSI-1 was standardized by ANSI in 1986. SCSI-2 is currently in the final stage of ANSI approval (and should be finalized by the time this book is in print). This description will concentrate primarily on SCSI-1.

A SCSI bus implements both a low-level physical I/O protocol and a high-level logical device control protocol. The bus can connect up to eight devices, each of which is assigned a unique device identification number (0 through 7). The bus can be as long as 25 meters. Logically, the bus can be divided into two subsets: the control bus and the data bus. The control bus transmits control and status signals. The data bus is primarily used to transmit data and can also be used to send device identification numbers. In SCSI-1, the data bus is eight bits wide. In SCSI-2, the data bus can be up to 32 bits wide.

A SCSI bus allows multiple devices to be bus masters, thus requiring a mechanism for arbitration when more than one device wants control of the bus. In theory, a SCSI bus could be used as the system bus for an entire computer system. Thus, all computer devices (including one or more CPUs) would be attached to the bus. In practice, the SCSI bus is generally used to connect and control mass-storage devices. A SCSI controller translates signals between the system bus (or a CPU) and the SCSI bus. This architecture is shown in Figure 8.11.

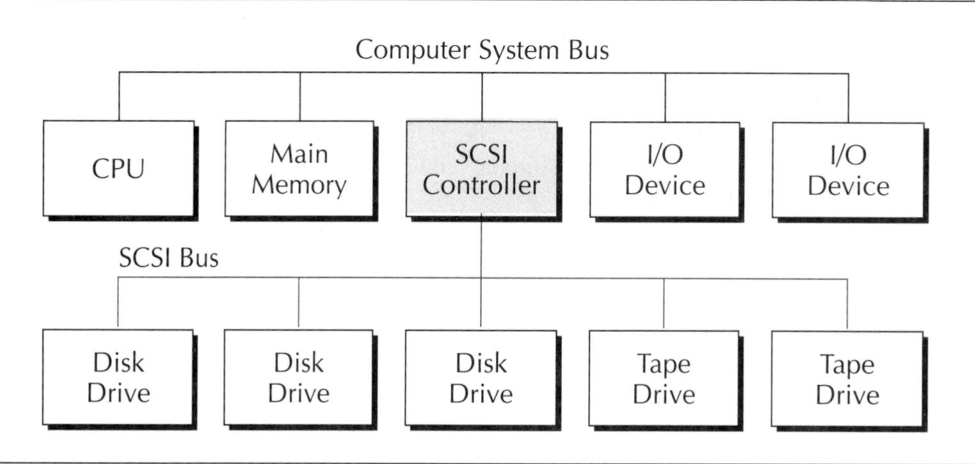

FIGURE 8.11 **The use of a SCSI controller to connect mass-storage devices (on a SCSI bus) to other devices attached to the system bus of a computer system.**

A partial listing of the SCSI-1 control signals is shown in Table 8.4. When a device initiates a transaction, it asks for control of the bus by sending a busy signal and placing its identification number on the data bus. If multiple devices send the busy signal simultaneously, the device with the highest identification number becomes the bus master and is designated as the *initiator*. The initiator then selects a *target* device by sending a select signal and placing the identification number of the target on the data bus.

Once the target device has been selected, the initiator sends it a data structure containing one or more commands. The ability to transfer multiple commands simultaneously is an important feature of the SCSI bus. It allows a series of commands to be sent to a target, with control of the bus released during the time the target processes those commands. The target queues the commands and executes them in sequence. This process avoids much of the control overhead associated with multiple commands. In addition, other devices can utilize the bus while the target is processing commands. Thus, interleaved execution of command sequences between multiple devices is possible.

TABLE 8.4 **A subset of the SCSI-1 control signals.**

Name	Description
Acknowledge	In combination with the Request signal, controls asynchronous data transfer between devices
Attention	Instructs a target to place a message to the initiator on the bus
Busy	Sent continuously by the initiator and/or target while a bus transaction is in progress
Control/Data	Sent by the target to indicate whether the data bus contains data or a control signal
Input/Output	Indicates the direction of data transfer on the bus (target to initiator or initiator to target)
Message	Sent by the target to inform the initiator that a message is being sent on the data bus
Request	In combination with the Acknowledge signal, controls asynchronous data transfer between devices
Select	Used by the initiator to select a particular device as the target; the identification number of the target is simultaneously placed on the data bus

A target can send several status signals to an initiator, including:

- OK
- Busy
- Check condition
- Condition met
- Intermediate

The OK signal is sent to indicate the successful receipt (and intent to satisfy) a command or command sequence. A busy signal is sent if the target is unable to process a request at the current time (perhaps because it is busy processing previously received commands). A check condition signal tells the initiator that an error (for example, an attempt to access a nonexistent storage location) has occurred. The initiator must explicitly ask the target to send a message indicating the nature of the error. The condition-met status signal is sent when a search command is executed successfully. (SCSI defines several such search commands, including searches based on data content.) An intermediate signal is sent to indicate the successful completion of one command in a command sequence.

Data transfer between target and initiator can be implemented either synchronously or asynchronously. In a synchronous transfer, the target sends a request signal simultaneously with the data. The initiator simultaneously sends an acknowledge signal, but the target does not wait for that signal to begin sending the next data item. Synchronous data transfer thus uses the full capacity of the bus

(4 Mbytes per second with SCSI-1) for data transfer. In asynchronous data transfer, the target sends a request signal and data simultaneously. The target does not send the next data item until the initiator explicitly acknowledges the receipt of the previous data item. Thus, asynchronous data transfer can proceed no faster than half the data transfer rate of the bus. (Every other cycle is used by the target to send an acknowledge signal.)

The importance of the SCSI bus in modern computer systems follows from several of its characteristics, including:

- Standardized and nonproprietary definition
- High data transfer rate
- Multimaster capability
- High-level (logical) data access commands
- Multiple command execution
- Interleaved command execution

The standardized definition allows any hardware vendor to manufacture and distribute SCSI-compatible devices. Data transfer rates can be as high as 40 Mbytes per second with SCSI-2. This rate approaches that of mainframe channels. Multimaster capability provides for efficient bus control with a minimum of control overhead. Target devices can become initiators (bus masters) in order to transfer data to a requestor. The ability to combine multiple commands and to interleave command sequences increases the likelihood that most of the bus capacity will be used to transfer data rather than control and status signals.

The access command set is rich and fairly high level. The standard hides many of the complexities of physical device organization from other computer system components. For example, disk devices appear as a linear address space to the CPU and/or operating system. Bad storage blocks are detected within a SCSI disk drive, and the linear address space simply ignores them when processing requests. A wide variety of devices, including magnetic and optical disk drives, conventional and digital tape drives, and data communication devices, can be interfaced to the SCSI bus.

CHARACTER I/O

Keyboard Input

The predominant form of character input to computer systems is a character keyboard. The exact mechanisms by which character data from a keyboard is captured for processing and storage have changed substantially over time. Early computer systems accepted input via punched cards or tape. A stand-alone device such as a keypunch machine was used to generate punched input based on manual keyboard input.

Keypunch devices captured manual keystrokes and converted them into punched holes in a cardboard card or paper tape. A punched card was divided into 80 columns and 12 rows. Each character typed at the keyboard was converted to a sequence of punched holes in one column of the card. Another device called a card reader was attached to the computer system bus. Punched cards were passed over a light source within the card reader. Light shining (or not shining) through each row of a single column was detected and interpreted as an input character. Once interpreted, the character was converted to electrical signals that could be recognized by the CPU.

Modern keyboard devices translate keystrokes directly into electrical signals, eliminating the need for intermediate devices such as punched card or paper tape readers. In keyboards manufactured during the 1960s and early 1970s, keystrokes generated an electrical input to a circuit dedicated to the key pressed. This circuit generated an output bit stream that was routed to the output circuitry of the device.

Modern keyboards that are integrated with video display units usually generate bit streams under control of a microprocessor. Pressing a key sends a coded signal (called a *scan code*) to the microprocessor. The microprocessor generates a corresponding bit stream based on a program or lookup table in memory. With this approach, the output of a key can be changed by altering the microprocessor program or lookup table.

Output signals from a keyboard device can consist of scan codes or can represent coded characters. In microcomputers and workstations, scan codes are normally output. Output from an integrated keyboard/video display device normally consists of characters coded in one of the standardized character coding conventions, such as ASCII. Transmission to a computer system is generally implemented in serial I/O mode. Synchronous character framing is common for devices that are directly connected (i.e., hard-wired) to large computer systems. Asynchronous character framing is common in most other applications.

Keyboards also allow signals to be generated by pressing combinations of keys. Examples include the use of Shift, Control, and Alt functions. The electrical circuitry and/or coding program must recognize multiple input signals and generate a single scan code as output. The use of multiple key combinations expands the number of possible output signals that can be generated from a fixed set of input keys.

Video Display

In older I/O devices, keyboards were often coupled to a paper-printing display device. These devices were typically called *teletypes* or *hard-copy terminals*. In modern I/O devices, a keyboard is usually coupled to a video display device. Some computer systems such as microcomputers consider the keyboard and the video display to be logically distinct devices. Most larger computer systems treat the keyboard and an attached video display device as a single device. Such a device is normally called a *video display terminal (VDT)* or simply a *terminal.*

A VDT consists of five functional components, as shown in Figure 8.12. The physical keyboard is attached to a keyboard driver that generates output signals according to scan codes received from the keyboard. The display generator generates the image actually viewed by a user. Display generators are available in a variety of types and capabilities, as are described later.

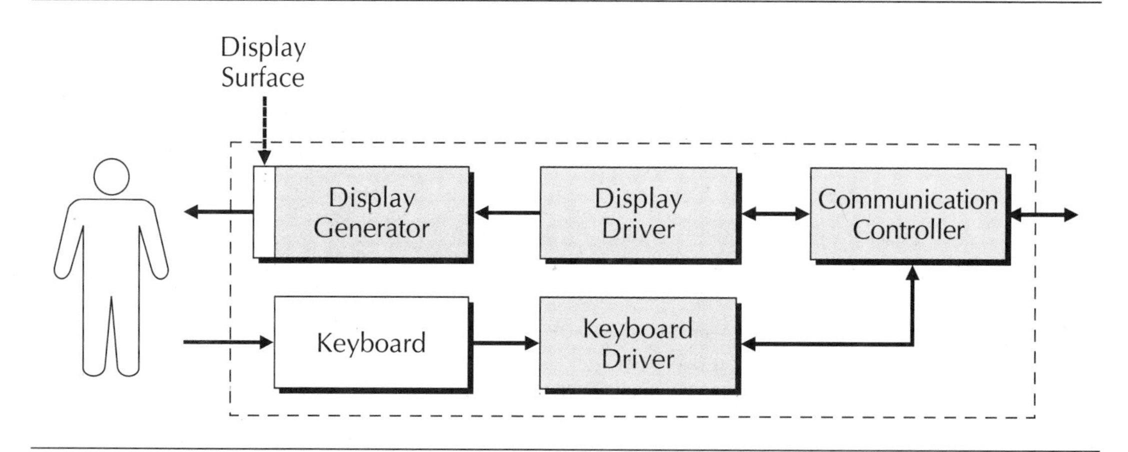

FIGURE 8.12 The functional components of a video display terminal.

Communication between the terminal and an external device is performed through a communication controller. This controller sends and receives coded-character input to and from a computer system. It implements any required aspects of the terminal-to-host communication protocol such as speed conversion, serial-to-parallel conversion, buffering, and handshaking. Signals received by the communication controller are sent to a display driver. The display driver consists of electronic circuitry that translates coded-character input into commands to the display generator. The exact nature of this translation depends on the type and capabilities of the display generator and the display functions supported by the display driver.

The output format for a VDT is illustrated in Figure 8.13. A typical character-oriented VDT can display 80 columns and 24 rows, although larger display dimensions are becoming more commonplace. Characters are normally displayed using a matrix of individually addressable light points. A light point is called a *pixel*, a shortened form of the term picture element. A typical matrix size is 640 horizontal and 400 vertical pixels. For an 80-column, 25-row display, this corresponds to an 8 × 16 matrix of pixels, in which a single character can be displayed, as shown in Figure 8.14.

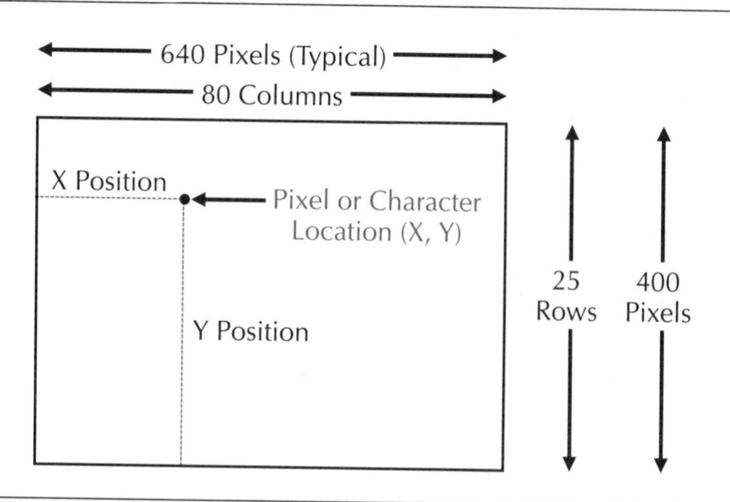

FIGURE 8.13 **The coordinate system used for data representation and reference with two-dimensional displays.**

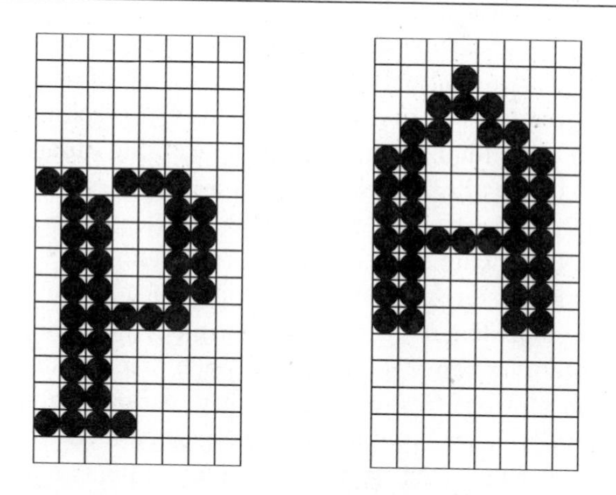

FIGURE 8.14 **The representation of the letters *p* and *A* within an 8 × 16 pixel matrix.**

The pattern of pixels used to display a character can vary from one display device to another or between display modes within the same VDT. A collection of patterns for all printable characters in a common style is called a *font.* Examples of fonts commonly used by video display devices include Courier, Times Roman, and Prestige Elite. Fonts vary not only in appearance but also in size. Variations in size correspond to variations in the dimensions of the pixel matrix used to display a character.

In older video displays, relatively simple electronic circuitry translated coded-character input into the control signals of a corresponding pixel matrix. These pixel control signals were input to a display generator that actually lit individual pixels of the display surface. In modern video display devices, the translation between coded-character input and pixel patterns is controlled by a microprocessor. This allows a great deal of flexibility in character display. Microprocessor-based display capabilities include the ability to alter fonts, sizes, intensity, and other display features.

Display generators that use a single light source for each pixel are inherently monocolored. The color varies, depending on the device that generates the light (e.g., orange for gas plasma, blue for liquid crystal, and red for light-emitting diodes.) Color output can be generated by using three light sources for each pixel. One light source generates red, another green, and the other blue (these colors are the source of the term *RGB* input or display). Thus each pixel on the display is generated by three light inputs, one for each primary color. Various intensities of these three colors can be combined to produce all colors in the visible light spectrum.

Cathode ray tubes (CRTs) have been used for character display almost since the introduction of computers. They are based on the same display technology used in conventional television sets. A CRT is an enclosed vacuum tube. The interior of the display surface is coated with a phosphor that glows when struck by electrons. An electron gun in the rear of the tube generates a stream of electrons that are focused in a narrow beam. The display driver of a CRT controls individual pixel illumination by pulsing (turning on and off) the electron beam. The color of the display is determined by the chemical composition of the phosphor coating. The intensity of individual pixels (or the entire display) can be varied by varying the intensity of the electron beam.

The beam is directed toward individual pixel elements on the interior display surface by magnetic deflection. The magnetic deflection is constantly shifted so that the beam moves continuously between pixel locations in a left-to-right, top-to-bottom motion. Color images are generated using three electron beams directed toward a grid of three different phosphor coatings (red, green, and blue). Different levels of intensity for each color combine to produce a continuous (or nearly continuous) range of color.

The operation of *liquid crystal displays (LCDs)* is based on reflected (rather than generated) light. An LCD panel is a matrix of encapsulated liquid crystals sandwiched between two polarizing panels. Exterior light passes through the top polarizing layer and into the liquid crystal layer. If no power is applied to the liquid crystal, the polarized light reflects back toward the front panel, appearing as a white or light gray spot. If power is applied to the liquid crystal, light passes through the crystal and is absorbed in the rear polarizing panel. Thus, the spot appears dark (usually blue or black).

Another implementation of LCDs, called *backlighting*, generates improved contrast as well as color display. In this approach, a light source is placed behind the display panel. Liquid crystals allow or block the passage of this light through the front of the display, depending on their electrical state. Color output can be generated by backlighting with colored (red, green, and blue) light or by using crystals that mask only a portion of the color spectrum.

Gas plasma displays are constructed using a matrix of electrodes. These electrodes are sandwiched between two transparent plates, and bubbles of neon gas are trapped in the spaces between the electrodes. Applying sufficient voltage to an electrode junction excites the gas into a plasma state, causing it to emit a reddish-orange light. Through proper voltage control, the intensity of the light can be regulated, producing up to 16 different discernible light intensities.

Electroluminescent displays are similar in construction to gas plasma displays. However, instead of a volatile neon gas, these displays use a solid-state phosphor to generate light. Color display can be generated by using a matrix of different-colored phosphors. Current technology has failed to produce "true" RGB solid-state phosphors, but their development is expected in the near future.

Character Printers

The functional components of a character printer are shown in Figure 8.15. Note the similarity to the components of a VDT shown in Figure 8.12. The communication controller and the print driver perform essentially the same functions as their counterparts in a VDT. The exact nature of the translation process performed by the printer driver depends on the type and capabilities of the print generator. Common types of print generators include dot-matrix, daisy wheel, line, and laser print generators.

Control of a *dot-matrix printer* is similar to that of a VDT. Individual characters are represented as a matrix of pin positions. Character codes received by the print generator are transmitted as series of pin control commands to the print head. The print head consists of one or more vertical lines of pins that can be forced outward from the head a fraction of an inch. An inked ribbon is moved in

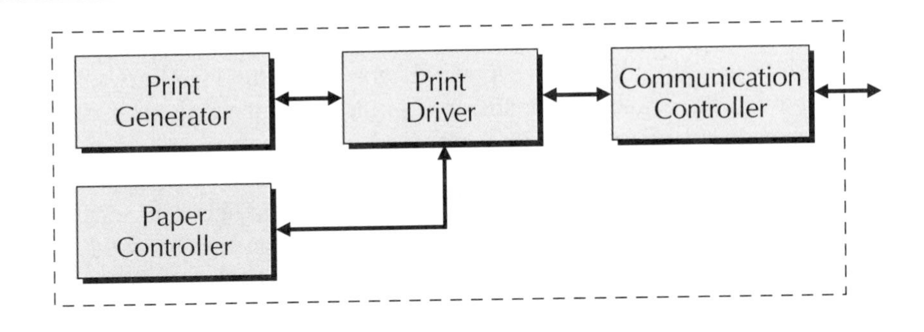

FIGURE 8.15 The functional components of a typical character printer.

front of the pins; the paper is on the opposite side of the ribbon. The print head is physically moved from left to right on each line. Control signals force pins to be pushed out from the print head, impressing a dot of ink from the ribbon onto the paper. A dot-matrix printer thus generates individual characters one or two vertical segments at a time, from left to right.

Paper control is provided by one or more motors controlling rubber rollers and/or sprockets. Paper is normally advanced past the print head one line at a time. The printer thus prints characters on the page in a left-to-right, top-to-bottom order. Advanced control capabilities allow paper movement in increments of less than a single line. Reverse movement of paper and/or print head can also be implemented.

A *daisy wheel printer* uses a print head similar to that found on some typewriters. An entire font of characters is contained on spokes of a small print wheel (resembling the petals of a daisy). Characters are printed one at a time by forcing the topmost spoke of the wheel against an inked ribbon and paper. Characters are selected for printing by rotating the print wheel so that the desired character is at the top of the wheel. Paper control is essentially the same as for dot-matrix printers.

A *line printer* generates an entire line of 80 or 132 characters at one time. A series of print wheels are positioned at each character column, perpendicular to the paper flow. Each wheel contains all of the characters in a font, and a character can be selected by rotating the wheel. Once all wheels are located in their proper positions, they are simultaneously pressed against an inked ribbon and paper. Paper control is implemented through motorized sprockets.

A *laser printer* operates quite differently from other types of printers. No print head is used, nor is any inked ribbon required. The print driver of a laser printer does not transmit individual characters or lines of characters to the print generator. Instead, it generates an internal image of an entire printed page as a matrix of very small pixels (typically between 300 and 600 per inch, both horizontally and vertically). This is necessary because the print generator operates continuously on an entire page. Thus, the print driver must arrange the entire contents of a page before the print generator begins operation. The print driver uses a microprocessor, complex programming, and a memory region to generate and store an image of a page. As characters are received, corresponding pixel patterns are generated and stored in memory at the appropriate location. This process continues until an entire page of characters has been received. At that time, the print driver instructs the print generator to start its cycle.

The print generator operates using an electrical charge and the attraction of ink to that electrical charge, as illustrated in Figure 8.16. A rotating drum is first charged over the width of its surface. A laser is then shone on portions of the drum corresponding to white areas of the page. The laser is pulsed by the print driver according to the stored pixel matrix. The laser light removes charge wherever it shines on the drum. Thus, the drum contains an image of the page where charged areas represent black pixels and uncharged areas represent white pixels. The drum then passes a station where fine particles of toner (a dry ink) are attracted to the charged areas.

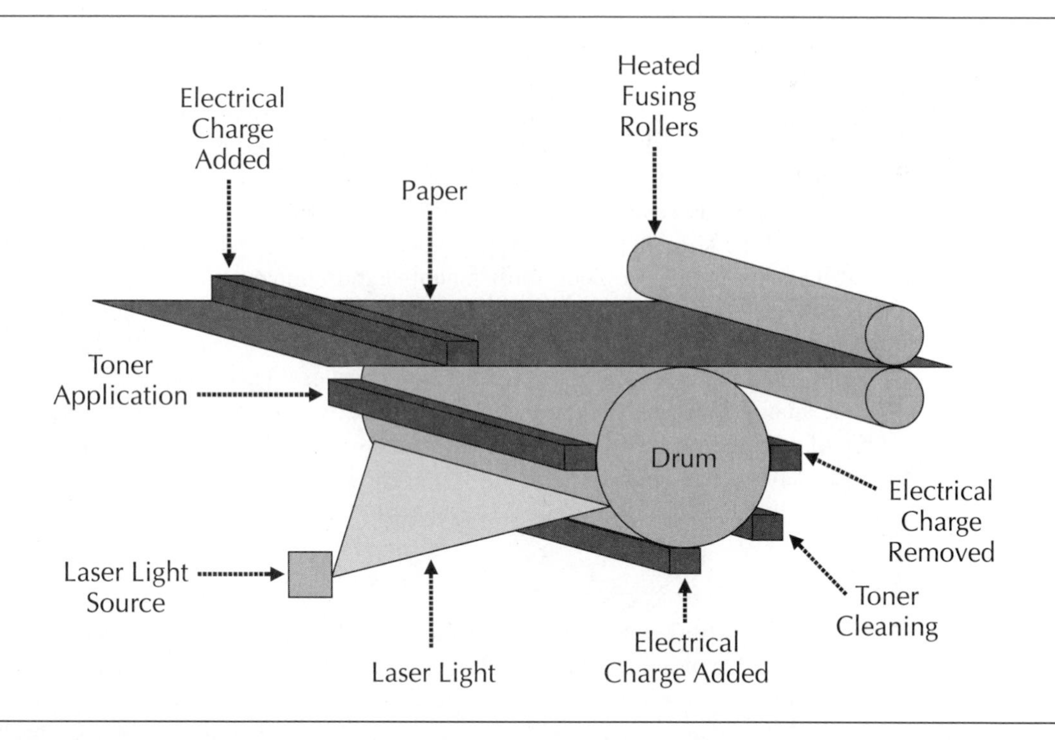

FIGURE 8.16 **The components of a typical laser printer print generator.**

In synchronized motion with the drum, paper is fed through a series of rollers and given a high electric charge. As the paper passes the drum surface, toner on the drum is attracted to the paper due to its higher electrical charge. The paper is then fed through heated rollers, causing the toner to fuse to the paper surface. A light source removes all charge from the drum, and excess toner is removed from the drum by a fine blade and/or vacuum.

Color output can be achieved by using three separate print generators, one for each of the primary colors. This requires precise alignment of the paper as it passes over the three drums. Also, the complexity (and expense) of the printer is substantially increased.

Specialized Device Control

Character-based output devices normally provide a set of specialized display functions that are somewhat dependent on the type of device. For printers, specialized control functions include the ability to change output fonts, font sizes, character intensity such as black or various shades of gray, and character color. Graphic output capability is also usually provided, as we discuss later. Video displays usually incorporate these capabilities as well as formatted output, cursor positioning, and inverse or blinking character display.

As described in an earlier chapter, ASCII and ISO character-coding schemes provide a set of nonprintable *control characters*. These include carriage return (new line), form feed, bell, and a small set of undefined symbols. These undefined symbols can be used by I/O devices to activate (or deactivate) various device control functions. However, the set of undefined characters is not large enough to control all of the specialized display functions available in modern output devices.

To address this problem, most manufacturers implement specialized device control with sequences (strings) of control characters. In many devices, a control character string always begins with an escape character (ASCII 27). Thus, such control strings are sometimes called *escape sequences*. There is considerable variation among devices and manufacturers over the exact format, content, and meaning of escape sequences. Some standardization efforts have been made; they are discussed in a later chapter.

GRAPHIC I/O

Although textual output has dominated human-computer interaction for decades, it is rapidly being replaced or augmented by graphical output. The saying "a picture is worth a thousand words" is a bit dated, but it does convey some of the power of graphic output compared to text output. The human brain is far better equipped to process graphic input than text input. Within the brain, pictures are processed in parallel, but text is processed serially. Parallel processing in the brain is far more efficient than serial processing. Imagine the difficulties in reading this text if all of the diagrams were replaced by equivalent textual descriptions.

Therefore, graphic output is far superior to text for conveying large amounts of information and for showing multidimensional relationships. Consider, for example, the difference between a graph and a spreadsheet for conveying information about trends and comparisons. Graphic output is, however, less well suited to large amounts of detailed information. As a result, graphic output is usually supplemented with text, and the two work together to provide efficient communication.

The use of graphic output is a fairly recent phenomenon, primarily due to the large amount of processing and data transmission required to generate it. Such computer power has only recently become available at a reasonable cost, thus allowing expanded use of graphic output. Given the ever declining cost of computer power, the use of graphic output will likely continue its rapid increase.

Representation of Graphic Images

Devices that produce graphic outputs include graphic VDTs, pen plotters, film recorders, and most types of printers. These devices vary according to the format in which graphic data is presented to them. Formats include

- Bit maps
- Vector lists
- Display lists

Many devices accept graphic input in two or three of these representation formats, in addition to normal character input.

A *bit map* is a numeric representation of an image in memory. Image data is stored as an array of spatial locations, as shown in Figure 8.17. In effect, a bit map is a physical implementation in memory of a row (X-axis) and a column (Y-axis) coordinate system for pixels. Within an array in memory, the numeric value stored for each pixel represents its brightness, or *gray-scale value.* In the case of a black-and-white image, a pixel is either off or on. Thus, only a single bit is required to represent each pixel. Gray-scale images require a greater number of bits (i.e., bit depth) to represent pixel intensities. For example, 4 bits are required to represent 16 gray-scale values (white, black, and 14 shades of gray). For color images, the bit map is often composed of three separate bit planes, one each for the primary colors of red, green, and blue.

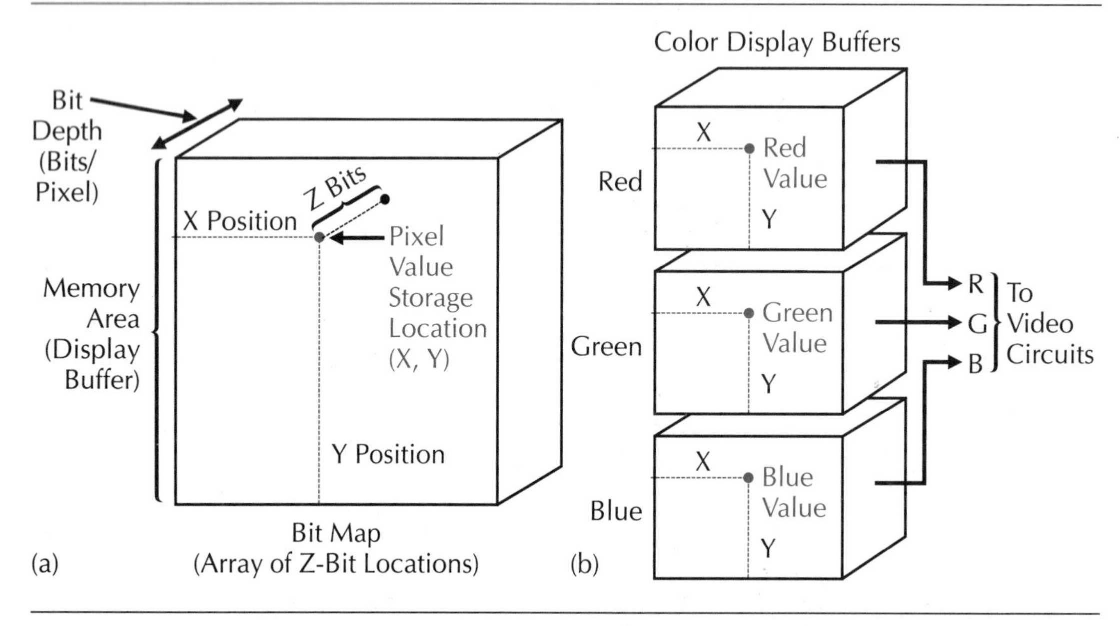

FIGURE 8.17 **Bit-map representation techniques for video displays in (a) monochrome and (b) color.**

One type of display generator used to reproduce bit-mapped images is called a *raster display.* Images are composed of sequences of rasters, or video scan lines, as shown in Figure 8.18. This is essentially the same process used in broadcast television and CRTs. The *spatial resolution,* or sharpness, of a raster display corresponds with the number of pixels along each scan line and the total number of scan lines in the picture. A resolution of 640 × 400 means that the picture has 400 scan lines, each composed of 640 pixels. Spatial resolution can also be expressed in terms of *dots per inch (dpi)* or pixels per inch.

The *chromatic resolution* of the bit map corresponds to the number of bits used to describe the value of each pixel. This value determines the accuracy with which the gray-scale or color value of each pixel can be represented within the bit map. For example, a color display using 8 bits to represent the intensity of each primary color has a chromatic resolution of 16,777,216 ($2^8 \times 2^8 \times 2^8$). Devices such as film recorders that accept bit-mapped inputs are said to be *display driven.*

In mathematical terms, a *vector* is a quantity that has both direction and magnitude. In graphics, a vector is a line segment that has a specific angle and length with respect to a point of origin, as shown in Figure 8.19a. Line drawings can be

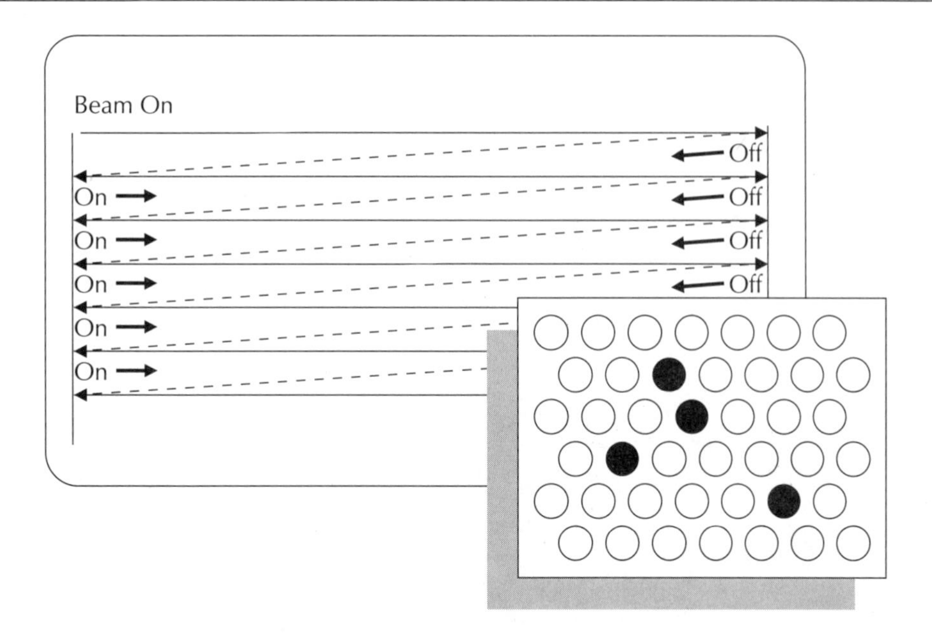

FIGURE 8.18 **Raster-scanning techniques for image generation on a video display.**

described to computer output devices as lists of concatenated, or linked, vectors. Thus, a computer output that is built from a *vector list* resembles a "connect-the-dots" drawing, as shown in Figure 8.19b. Output devices that use vector lists include pen plotters and strokewriter VDTs. In a strokewriter VDT, the CRT beam does not follow a predetermined raster pattern but can be moved around the display screen at random, as directed by vector inputs.

A group of high-level commands for generating graphic output, called a *display list*, describes an image as a collection of *display objects*, or shapes, and *display attributes*, such as color and spatial position. The graphic output device must contain an embedded processor and software for processing the display list to compute the final image and to direct physical output functions.

Unlike bit maps and vector lists, a display list exists independently of the physical requirements or characteristics of the output device. That is, the same display list can be passed to an intelligent raster VDT or to an intelligent plotter. In either case, interface software must translate the display list into hardware control commands for the corresponding device controller. Devices such as film recorders and raster terminals that are capable of generating images from display lists are said to be *data driven.*

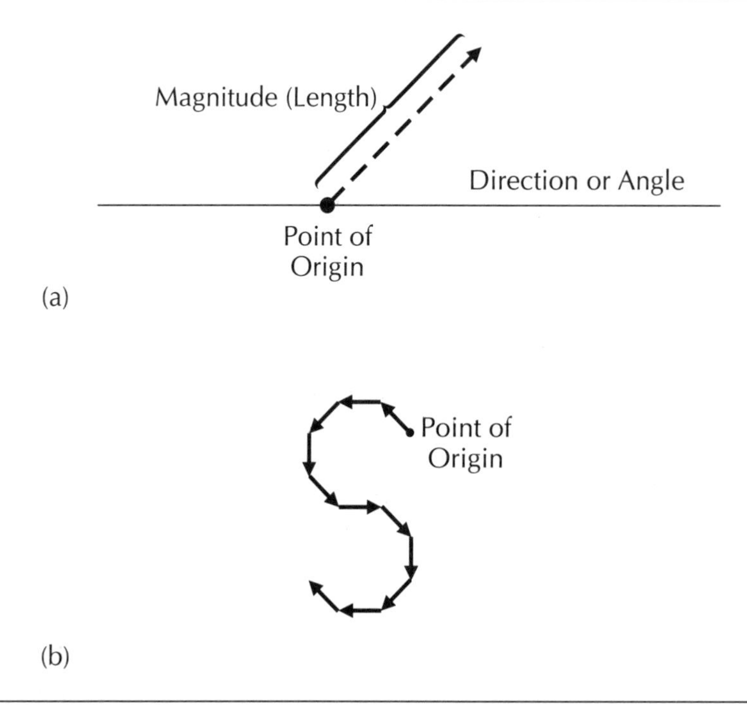

FIGURE 8.19 **Two types of vector list data representations for graphic displays: (a) the elements of a single vector; (b) how a more complex image can be built from concatenated vectors.**

The advantage of graphic output based on vectors and display lists lies in the compactness of the message sent to the display device. Consider a graphic display composed of a square, a circle, and a triangle. The output of such an image to a bit-mapped display must include one or more numbers representing every pixel on the display. A monochrome display with 640 × 480 resolution will require 307,200 bits (38,400 bytes) to be sent to the display device. Gray-scale and color output will require substantially more bits, depending on the number of possible gray scales or colors.

A vector list describing the same three images will be substantially more compact. The square will require sending vector information for four vectors; the triangle will require three. The information for each vector must include origin, length, thickness, and color (or gray scale). This information can be coded in a few bytes per vector. The circle, however, will present a problem because it is not composed of straight lines. To be represented accurately, images containing curved lines and irregular patterns must still be communicated as a bit map.

A display list representation of the same three images is even more compact. If the objects circle, square, and triangle are defined in the display language, the description of those objects can be sent directly. Such descriptions will normally include position or origin, size, line thickness, line color, and fill pattern (if any). A display list representation of a display represents the most economical alternative in terms of communication bandwidth consumed. However, this economy comes at the expense of substantial complexity in the display device itself to process the display image.

The trend toward ever higher resolutions and larger displays has increased the use of display lists for graphics output. The Postscript graphic description has been adopted by many computer and display device manufacturers.

TECHNOLOGY FOCUS: Postscript. Postscript is a display language designed primarily for generating printed documents, although it can also be used to generate video display outputs. Postscript can be considered both a display language and a programming language. In essence, a Postscript display list is a program (a procedure) for producing graphic images. These images can be composed of various graphic objects, including characters, lines, and shapes. Graphic objects can be manipulated in a variety of ways, including rotation, skewing, filling, and coloring. The language is sufficiently robust to represent virtually any imaginable graphic image.

A Postscript display list consists of a set of commands composed of normal ASCII characters. Thus, transfer of a display list between devices (e.g., a computer and a printer) can be performed by the same means as normal character-based communication. Commands can include numeric or textual data, primitive graphic operations, and definitions of procedures. Data is held within a stack. Primitive operations and/or procedures remove data from the stack and use that data as control parameters.

For example, consider the following short program:

```
newpath
400 400 36 0 360
arc
stroke
showpage
```

The first line declares that a new path (straight line, curve, or complex line) is being defined. The second line contains a list of numeric data items that are pushed onto the stack. These are control parameters used by the primitive graphic operation arc (the third line) and are removed from the stack as they are used. The left two parameters of the second line specify the origin (center) of the arc as row and column coordinates. The center parameter specifies the radius of the arc in points (a point is one seventy-second of an inch). The remaining parameters specify the starting and ending points of the arc in degrees. In this case, the starting and ending points represent a circle.

Primitive graphic operations are specified in an imaginary drawing space. Individual locations within this grid are specified as row and column coordinates in points. The first three commands specify a tracing within this space. The fourth command states that this tracing should be stroked (drawn). Because no width is specified, the circle is drawn using the current default line width. The final command instructs the output processor to physically display the contents of the page as defined thus far. Thus, a Postscript printer receiving the above program as input would generate a single page containing one circle.

To improve efficiency and reduce the amount of code in a program, Postscript provides for the definition of procedures. The following program defines a procedure named CIRCLE composed of the primitive commands and parameters used to draw a circle:

```
/circle % stack = x-center, y-center, radius
{ newpath
0 360
arc
stroke } def
```

The first line defines the name of the procedure and includes a comment (the text following the % character) that explains the parameters used by the procedure. The definition of the procedure, which appears between the curly brackets, can be of any length. Note that the procedure defines only the final two parameters needed to draw a circle. The first three must be placed on the stack before the procedure is called. Figure 8.20 shows a program that draws three concentric circles using this procedure.

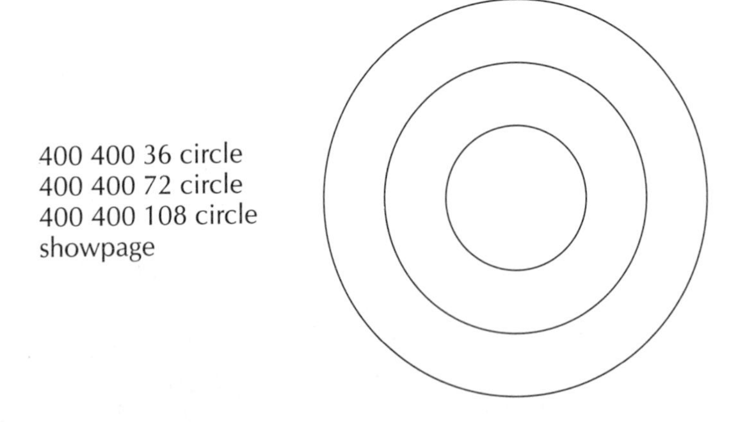
400 400 36 circle
400 400 72 circle
400 400 108 circle
showpage

FIGURE 8.20 A program that draws concentric circles using the previously defined procedure CIRCLE.

Figure 8.21 shows a more complex program for drawing squares. The program includes the definition of a procedure called SQUARE. It draws a square as a sequence of four connected line segments. Because some parameters to the procedure are used more than once, the procedure must duplicate and reorder some of them with the dup (duplicate), exch (exchange), and sub (subtract) operations. These operations manipulate the top one or two elements of the stack, pushing the result (if any) onto the stack. Note also that the procedure controls the width of lines with the SETLINEWIDTH command.

```
/square % stack = x-origin,
y-origin, line-thickness, size
{ newpath
moveto
setlinewidth
dup 0 exch
rlineto                 % up
dup 0

rlineto                 % right
0 exch sub 0 exch
rlineto                 % down
closepath               % left
stroke } def
216 3 180 324 square
144 2 216 360 square
72 1 256 392 square
showpage
```

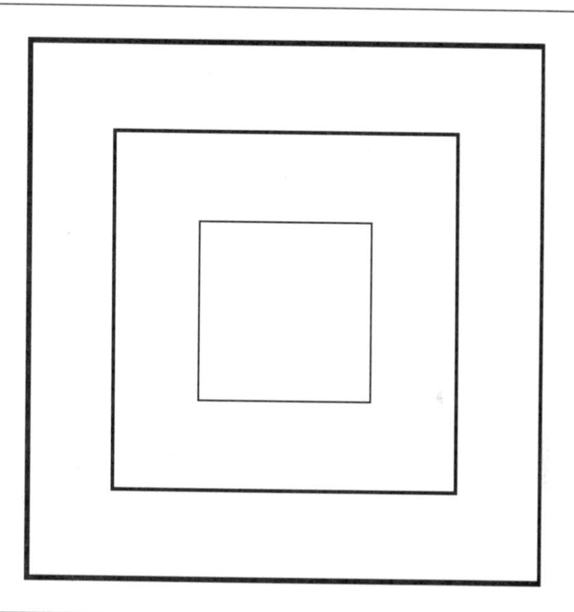

FIGURE 8.21 **A program that defines a procedure for drawing squares and uses it to draw the squares shown.**

As stated earlier, Postscript can generate both video and printed graphic images. Some computer systems such as the NeXT computer use a version of Postscript to generate all video output. Such output is generally quite complex and, thus, so are the Postscript programs that generate it. In such applications, a dedicated processor within the video control interface translates Postscript commands into appropriate commands to control the display hardware. This approach is also generally used within Postscript printers. Figure 8.22 shows a sample Postscript program used to generate a popup menu, such as might be used in a window-based command interface. Note that a video interface would probably encode many of the command sequences in the sample program as predefined procedures.

```
% menu outline
newpath
245 365 moveto
340 365 lineto
340 265 lineto
245 265 lineto
closepath
stroke

% title bar
newpath
245 345 moveto
340 345 lineto
stroke

% text items
/Times-Bold findfont 12 scalefont setfont
250 350 moveto (Applications) show
/Times-Roman findfont 12 scalefont setfont
250 330 moveto (Word Processing) show

250 315 moveto (Spreadsheet) show

250 300 moveto (Database) show
250 285 moveto (Graphics) show
250 270 moveto (Games) show

% mouse pointer
newpath
240 305 moveto
-6 6 rlineto
0 -3 rlineto
-20 0 rlineto
0 -6 rlineto
20 0 rlineto
0 -3 rlineto
closepath
fill

showpage
```

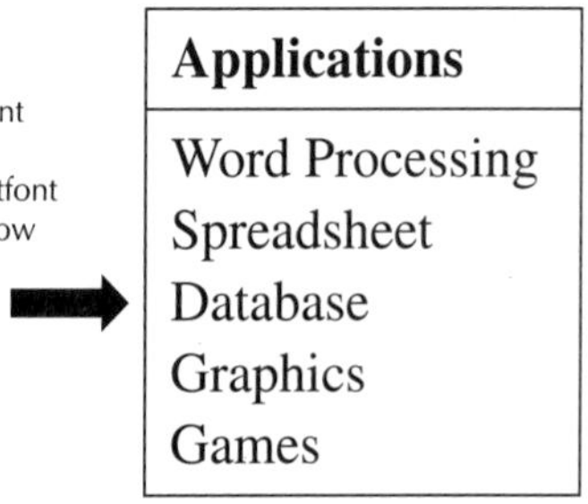

Applications

Word Processing
Spreadsheet
Database
Graphics
Games

FIGURE 8.22 **A sample Postscript program that generates the popup menu shown on the right.**

Graphic Printers

The operation of a graphic printer is quite similar to the operation of pixel map character printers, such as dot-matrix and laser printers, described earlier. The difference lies in the form and interpretation of device input. In a character printer, coded characters are input and translated into properly located pixel patterns. When that same printer is to print graphics, input can be in the form of a bit map, a vector list, or a display list.

In the case of bit-mapped input, the print driver interprets ASCII input as patterns of bits instead of coded characters. These are typically received in the order in which they will be printed on the page. The printer and the sending device must agree on the number of bits per pixel (chromatic resolution) and the output resolution of the printed page. Update of the internal pixel matrix (if used) is relatively straightforward.

In the case of a vector list, the print driver must translate the vector (including origin, direction, length, and thickness) into its pixel-map equivalent. Vector input is typically used only in full-page (e.g., laser) printers that use an internal pixel buffer. Some dot-matrix printers use vector commands to print horizontal lines on a page. As with bit-mapped input, the printer and sending device must agree on various parameters such as page dimensions, output resolution, and chromatic coding and resolution. These parameters are normally established by control sequences sent from computer to printer at the start of communication. These control sequences select printer output modes corresponding to the specified parameters.

Display list input requires the most complex processing within the printer. It is possible only for printers that have sufficient memory for full-page pixel maps. The print generator accepts each element of the display list and computes its corresponding representation within the pixel map in memory. The printer must recognize each element of the display language (e.g., shapes, position, orientation, fill patterns, and colors). Once all display list input has been received and processed, the pixel map is used to generate printing instructions for the print generator. This sequence of operations is typical of laser printers that utilize Postscript or Hewlett-Packard Graphics Language (HPGL) for image description.

Video Display

Graphic display on VDTs is similar to displays using laser printers. Typically the VDT or its controller maintains a memory buffer in which a bit (pixel) map of the display is represented. This map is updated as graphic data is received, and the display generator continuously scans this buffer to generate output images. These two processes (update of the buffer and generation of the image) are performed asynchronously. The display generator typically updates the display based on the buffer contents several thousand times per second. The rate of update is referred to as the *refresh rate* of the display and is generally stated in Hertz (Hz). Typical refresh rates range from 20 to 70 KHz, depending on the complexity, light-generation mechanisms, and output resolution of the display device.

Graphic input can be via bit map, vector list, or display list. Bit-mapped input is the most common, although display list input is increasingly more common. As with printers, vector and display list input require local processing power, thus raising cost and complexity. However, the complexity and data volume of graphic images, coupled with real-time display at high resolutions, virtually require the use of vector and/or display list input.

Pointing Devices

Devices that capture graphic and image data include the mouse, the trackball, the joystick, the light pen, and the digitizer tablet. Essentially, all of these devices perform the same function. They translate the spatial position of a pointer, stylus, or other selection device into numeric values within a system of two-dimensional coordinates. These devices are used to enter drawings into the system or to control the position of a *cursor,* or physical pointer, on a display device. The cursor control function is typically used to indicate selections from visual menu displays presented to the system user on the display.

Digitizers utilize a pen, or stylus, and a digitizing tablet. The tablet is sensitive to the placement of the stylus at any point on its surface. Recognition of stylus position is implemented by pressure-sensitive contacts beneath the surface or by disturbance of a magnetic field in the tablet.

Two uses of digitizers are common. In the first, a digitizer traces the outline of an object or drawing. The drawing is placed over the tablet and the stylus is manually moved around objects or lines on the drawing. The other use of a digitizer is as an alternate form of menu input. This is common in drawing or computer-aided design programs, when the user wants to continuously view a drawing on the video display device while entering commands from the digitizer. A menu or command template is placed over the digitizing tablet and the user selects commands by pointing the stylus to one of the command positions. In either case, the digitizer communicates to a computer the stylus position (or position changes) within the tablet coordinate system. Software programs interpret the meaning of those positions or changes.

Optical Scanning

Devices that convert patterns of light into character or image data are categorized as *optical sensors.* Such devices include video cameras, scanner/digitizers, document readers, mark sensors, bar-code readers, and optical character recognition (OCR) devices. The operating principles behind all of these devices are fairly similar. A light source (usually a laser) is shone onto a printed surface. Light reflected from the surface is detected by a photodetector. High light intensities induce large current outflows from the photodetector. Low light intensities produce little or no current.

Optical scanning devices are differentiated by the following criteria:

- Normal and maximum spatial resolution
- Normal and maximum chromatic resolution
- Formatting requirements
- Amount of local processing

High-resolution devices are capable of spatial resolutions as high a 600 dots per inch. Simpler devices require relatively large marks (no smaller than 5 millimeters) for reliable recognition. Few devices are currently capable of color recognition. Gray-scale recognition typically ranges as high as 256 shades. Certain devices require marks to appear at specific locations on a page or within specific distances of one another. Devices that attempt any form of image recognition or classification—for example, optical character recognition devices—embed sophisticated processing capabilities within devices. Some devices utilize local processing to detect and correct input recognition errors.

Mark sensors and bar-code scanners are examples of relatively simple optical recognition devices. A *mark sensor* scans for light or dark marks at specific locations on a page. These devices (or the input to them) are familiar to students who take standardized multiple-choice tests. Input is typically from a page preprinted with circles or boxes. These locations are marked by filling them with a dark pencil. The mark sensor uses preprinted bars on the edge of the page to establish reference points on the page—for instance, the row of boxes corresponding to the possible answers to question 5. It then searches for dark marks in specific locations with respect to those reference points. Marks must be of a minimum size and intensity to be correctly recognized.

Bar-code scanners operate similarly to mark sensors, but input does not normally have pre-established reference marks. A bar code is a series of vertical bars of varying thickness and spacing. These bars are converted to a numeric code via a standardized algorithm, on which devices from all manufacturers agree. Bar-code readers use one or more *scanning lasers* to detect the vertical bars. A scanning laser sweeps a narrow laser beam back and forth across the bar code. The use of multiple scanners at oblique angles allows for variation in the position and orientation of the bar code. Bar-code readers are typically used to track large numbers of inventory items, such as carried in grocery stores and stored in warehouses.

Optical scanners are designed to generate bit-mapped representations of printed images. Manual scanners require a user to move the scanning surface over a printed page at a steady rate of speed. Automatic scanners use motorized rollers to move paper past a scanning surface. As with bar-code readers, a scanning laser detects light and dark pixels. A full-page scanner will typically scan horizontal rows of pixels across the entire width of the paper. Multiple scans can compensate for scanning errors.

Optical character recognition (OCR) devices combine optical scanning technology with local processing power. Once an image has been scanned, the bit-mapped representation is searched for patterns corresponding to printed characters. In some devices, input is restricted to characters in prepositioned blocks. In these devices, as with mark sensors, preprinted marks provide a reference point for locating individual characters. This greatly simplifies the character recognition process. In more sophisticated OCR devices, the position and orientation of symbols are unrestricted. Such a device faces the dual problem of both finding and correctly classifying printed symbols. Variations in font style, especially among handwritten characters, make the recognition process complex and error prone.

Plotters

A *plotter* is an output device that generates line drawings on sheets or rolls of paper mounted within a control mechanism that can precisely move the paper up or down. One or more pens mounted on a motor-controlled station moves left or right over the paper. Each pen can be raised or lowered to contact (or not contact) the paper.

Images are drawn on paper by moving the pen and/or the paper. Drawing a line from left to right requires the paper to be positioned so that the line position is directly below the pen. The pen is lowered onto the paper at the beginning of the line and moved to the end of the line, then raised. A vertical line can be drawn by correctly positioning the paper and pen, lowering the pen, moving the paper, and then raising the pen. Nonperpendicular lines, curves, and shapes generally require the simultaneous movement of both paper and pen.

The operation of a plotter is an extreme example of digital-to-analog conversion. The initial representation of an image within a computer (and its communication to the plotter) is usually in digital form. The digital representation is translated into analog motor control signals that physically draw the image on paper. Plotters differ primarily by number of pens (colors), size of paper, and the presence (or absence) of local digital-to-analog processing.

If the original image to be printed is a bit map, it must be converted into a series of control signals for the various plotter motors. A sophisticated plotter accepts an image as a bit map, vector list, or display list and performs this conversion process internally. Other plotters require input in the form of direct, digitally encoded motor control signals. For such a plotter, all translation from bit map or other representation formats is performed by software within the computer system.

AUDIO I/O

The use of computers to generate and/or recognize sound is a relatively recent phenomenon. Sound is inherently an analog signal and must be converted to digital form for computer processing or storage. In the process of converting analog sound waves to digital representation, called *sampling*, the content of the audio energy spectrum is analyzed many times per second. For sound reproduction that sounds natural to humans, frequencies between 20 Hz and 20 KHz must be sampled at least 40,000 times per second. A rate of 48,000 samples per second is a common standard. Each individual sample is referred to as a sample point.

Sound varies by frequency (pitch) and intensity (loudness). By various mathematical transformations, a complex sound such as a human voice, consisting of many pitches and intensities, can be converted to a single numeric representation. For reproduction that sounds natural to humans, 16 bits must normally be used to represent each sample point. Thus, one second of sound sampled 48,000 times per second requires 96,000 bytes for digital representation (48,000 × 16 bits × .0125 bytes/bit). A full minute of sound requires more than 5 Mbytes.

Computer generation of sampled sound is relatively simple. The digital information is input to a digital-to-analog converter. The output of the digital-to-analog converter is input to a conventional analog amplifier and then to a speaker. Sampling is also relatively simple. However, the processing and communication power necessary to support both of these operations has only recently become affordable enough to allow widespread application. Compact audio disk players and digital audio tape recorders are two consumer-oriented examples of this technology.

Within a computer system, sound generation and recognition are potentially useful for the following purposes:

- General-purpose sound output (e.g., a warning tone)
- General-purpose sound input (e.g., digital recording)
- Voice command input
- Speech recognition
- Speech generation

General-purpose sound output has existed in computer systems for many years. In early teletypes, a bell was physically rung to signal events that required operator intervention, such as adding paper to a printer. Microcomputers, workstations, and VDTs typically include the ability to sound a single audible frequency at a specified volume for a specified duration of time. This type of sound output is called *monophonic output* because only one sound or note can be output at a time.

As computers have been adapted for multimedia presentation, greater sound-generation power has been added. Many computers now employ *polyphonic* sound-generation hardware with built-in amplification and high-quality speakers. For example, the NeXT workstation has built-in digital signal processing hardware as well as audio output hardware. It has sufficient processing and output power to play digitally encoded music in real time.

Most current research in computer sound processing is focused on voice recognition and generation. Devices resulting from this research include voice command recognition devices and speech generators. For all processing of human speech, problems of word recognition or generation must be addressed.

Speech Recognition

Human speech consists of a series of individual sounds called *phonemes*, roughly corresponding to the sounds of each letter of the alphabet. A spoken word is a series of interconnected phonemes. Continuous speech is also a series of phonemes interspersed with varying amounts of silence. Recognizing individually voiced phonemes is not a difficult computational problem. Sound can be captured in analog form by a microphone, converted to digital form by an analog-to-digital converter, and the resulting digital pattern compared to a library of patterns corresponding to known phonemes, as shown in Figure 8.23.

Although the process is conceptually simple, a number of factors complicate it:

- The need for real-time processing
- Variation among speakers
- Phoneme transitions and combinations

Speech recognition is useful for computer I/O only if it can be performed in real time. This implies that the capture, conversion, and comparison functions must occur fairly quickly. Variation among speakers in the generation of phonemes makes comparison an inexact process. Therefore, comparison to a library of known phonemes cannot be based on an exact match. Rather, the closest approximation must be determined and a decision between multiple possible interpretations must be made. This difficulty pushes the most advanced computer hardware to the limits of its processing power.

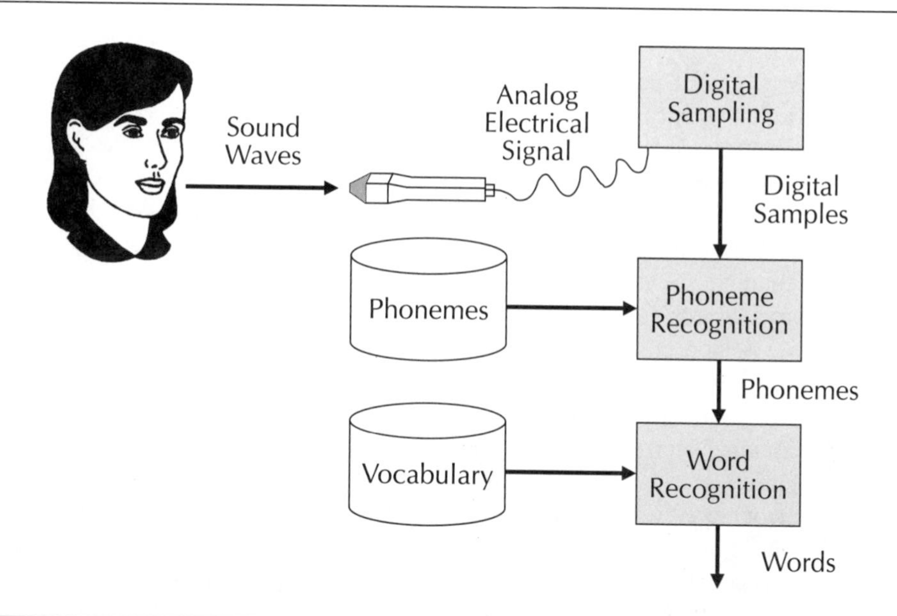

FIGURE 8.23 **The process of speech recognition. Sound waves are converted to an analog signal by a microphone. That signal is sampled to produce a digital equivalent. The digital signal is converted to phonemes and words.**

The most difficult problem in speech recognition arises from the continuous nature of speech. Individual letters sound relatively similar when repeated by the same speaker. However, when combined with other letters in different words, their voicing varies considerably. In addition, a computer must determine where one phoneme ends and another begins, as well as where one word ends and another begins. Once again, the complexity of such processing equals or exceeds the processing power of most currently available computer hardware.

Today most speech recognition systems (hardware and associated software) are not capable of unrestricted speech recognition among multiple speakers in real time. They must generally be "trained" to recognize the sounds of individual speakers. They are also generally restricted to relatively limited vocabularies of a few thousand words. At the extreme of such restrictions, command recognition systems are designed to recognize up to a few hundred spoken words from a single speaker. Such devices are useful when a single user will employ the device and when manual (e.g., keyboard) input is impractical. Such systems have been applied in airplane cockpits, manufacturing control, and input systems for the physically handicapped.

Speech Generation

One type of device that generates spoken messages is an *audio response unit.* Typically, such a device is used to deliver limited amounts of information over conventional telephone instruments. Thus, any telephone becomes a potential output device for the system. Audio response is usually implemented by storage and playback of words or word sequences. Words or messages are stored digitally. Output of the message involves accessing the corresponding storage location and sending it to a device that converts the digitized voice into analog signals that can produce sounds in electronic speakers or earphones.

A more general approach to speech generation is called *speech synthesis.* In speech synthesis, individual vocal sounds, or phonemes, are stored within the system. Character outputs are sent to a processor within the output unit, which assembles corresponding groups of phonemes to generate synthetic speech. The quality of speech output from such units varies considerably. Problems are encountered when combining groups of phonemes to form words. Transitional sounds must be generated between the phonemes, and composite phonemes must be accounted for. However, the difficulties of speech generation are far less formidable than those of speech recognition. Speech generation devices are commonly available, with prices starting as low as a few hundred dollars.

SUMMARY

The CPU communicates with storage and I/O devices through I/O ports on the system bus. To simplify bus and processor design, the commands and status signals used to access these devices are relatively simple. At a minimum, the commands READ and WRITE and the status signals READY, BUSY, and RESULT are all that are necessary to implement communication. An access using these simplified commands and status codes is called a logical access.

The physical processes needed to implement logical accesses are relatively complex and vary substantially from device to device. To implement a logical access, it must first be translated into a specific set of instructions that are understood (and can be implemented) by the receiving device. The instructions and physical processes are referred to as physical access.

The translation between physical and logical access can be implemented in software, within the device, or by a device controller. Implementation within software is highly inefficient because it uses excessive CPU resources to implement each access. Implementation within a device is efficient but entails redundant processing and hardware within multiple devices of the same type. A storage or I/O device controller normally implements the bulk of logical-to-physical translation. In addition to performing this translation, device controllers serve as an interface to the logic and communication protocol of the bus. They also allow multiple devices to be attached to an I/O port. Access by multiple devices through a single device controller can be implemented in either dedicated or multiplex mode.

A mainframe channel is an advanced form of a device controller used in mainframe computers. It differs from a device controller in terms of the number of devices it can control, its ability to communicate with different device types, and its maximum communication capacity. Mainframe channels can communicate with dozens of mass-storage devices or hundreds of terminals simultaneously. They allow a mixture of attached device types and typically have communication capacity measured in hundreds of megabytes per second.

Various measures can be taken to improve communication performance between storage and I/O devices and the CPU. Interrupt processing is a basic means of economizing on CPU resources dedicated to I/O and storage accesses. The use of interrupts allows physical accesses to occur asynchronously with CPU processing. Performance improvements can also be realized through buffered I/O, cached I/O, and direct memory access (DMA), and data compression.

A buffer is a region of memory that holds a single unit of data transferred to or from a device. A buffer allows rapid movement of entire blocks of data. Buffers are generally implemented within the computer system (under operating system control) and within devices or device controllers. An I/O cache is a large buffer implemented within a device controller. When used for input, a cache allows access to input more rapidly than the data transfer rate of the associated device. This is possible if the cache controller can "guess" what data the CPU will request next and load that data into the cache before it is actually requested. Cache implementation is generally complex, requires extensive processing capability within the device controller, and can utilize additional processing by the operating system.

Direct memory access allows large transfers between devices and main memory without CPU intervention. It requires the use of a DMA controller or a multiple master bus and additional processing logic within device controllers. Data compression allows the data stored within a storage device or communicated through a channel to exceed device (or channel) capacity. Data entering the device or channel is reduced in size by a compression algorithm. Data leaving the device or channel is expanded to normal size (and representation) by a decompression algorithm. Data compression trades additional processing resources against reduced resources for data storage and/or communication.

The translation between physical and logical access to a disk drive requires a translation from an access to a logical sector into an access to a specific sector, track, and platter. This translation is normally the responsibility of the device controller. Physical-to-logical access translation for tape drives requires a conversion between a logical record and a specific portion of the tape. The tape drive controller must physically wind or unwind the tape and detect the passage of individual records.

The maximum speed of data transfer for a mass-storage device is limited primarily by the physical motions required to implement an access. In disk drives, these physical motions consist of the rotation of disk platters and the movement of read/write heads. In tape drives, the physical motion consists of the motor control over the movement of the tape. Additional factors affecting the data transfer rates for these devices include error detection and correction as well as device and controller circuitry implementation.

Character I/O devices are designed to implement human/computer communication through printed or displayed characters. Keyboards are the most common form of character input. Physical depression of a key generates a scan code that is sent to a processor. The processor translates the scan code into a coded character that is output to a computer system. Transmission to a computer system is normally implemented as serial I/O with asynchronous character framing.

Video display devices accept coded-character inputs and display them as patterns of light. A video display terminal (VDT) is a video display device combined with a keyboard. Video display of characters is normally implemented by displaying light within a matrix of pixels. Each character input is translated into a matrix of light and dark pixels. A display generator is responsible for lighting individual pixels of the display. Display generators are implemented with several types of devices, including cathode ray tubes, liquid crystal panels, gas plasma panels, and electroluminescent panels.

Character output on paper is implemented by various types of printers. In some printer types, the formation of patterns for printed characters is essentially the same as for video displays. Examples of such printers include dot-matrix and laser printers. Other types of printers operate similar to typewriters. Examples of these printers include daisy wheel and line printers.

Most character display devices allow various forms of specialized display. In printers and video displays, display options include various fonts, font sizes, color control, and print quality. These features are controlled through control code sequences. Control codes sent to an output device by software activate or deactivate specialized display features. Video displays usually include specialized functions for cursor movement.

Graphic I/O allows the display of images in addition to characters. The representation of images in digital form requires the use of bit maps, vector lists, or display lists. A bit map is a pixel-oriented representation of an image. One or more bits represent each pixel in an image. The number of bits needed for each pixel is determined by the chromatic resolution of the image or display device. Color images are represented by a set of three-bit strings, each describing the intensity of one of the primary colors (red, blue, and green).

A vector list describes an image as a series of lines. Each line is described in terms of its origin, length, direction (angle), and thickness. Many types of images can be represented in this manner. However, curves and irregular shapes are difficult to represent as vectors. Vector lists are generally used as a supplement to bit maps or display lists.

A display list describes an object as a collection of shapes and lines. Each shape is described in terms of placement, size, color, and fill pattern. Complex images can be described relatively efficiently using a display list. However, a display list requires extensive processing within the graphic display device (or controller) to decode the image and convert it to bit-map or raster form.

Optical scanning, used for input of graphic images, operates based on reflected light. A laser is shone onto a printed page, and light reflected from the page is detected. Types of scanning devices include mark sensors, bar-code scanners, optical scanners, and optical character recognition devices. Mark sensors detect marks at predetermined positions on a page. Bar-code scanners read input in the form of groups of dark bands of varying thickness and distance from one another. General-purpose optical scanners generate a bit-mapped representation of a printed page. Optical character recognition devices scan printed text, recognize individual printed characters, and generate equivalent coded-character inputs.

Audio I/O capabilities are based on digital representation of sampled analog sound waves. Sound waves can be converted to digital representations and back to sound waves. This capability can be used for simple sound generation (e.g., bells and warning tones), music reproduction and/or alteration, and speech processing. Speech generation requires the conversion of coded characters into sound waves. This requires the reproduction and combination of phonemes and their conversion to analog form. Speech recognition requires analog-to-digital conversion, phoneme recognition, and word recognition. Real-time speech recognition poses special problems that meet or exceed the capabilities of the most powerful computers.

Key Terms

audio response unit
average access time
backlighting
bar-code scanner
bit map
blocking factor
buffer
cache controller
cathode ray tube (CRT)
channel processor
chromatic resolution
compression algorithm
compression ratio
control characters
cursor
daisy wheel printer
data compression
data driven
dedicated mode
device controller
digitizer
direct memory access (DMA)
display attribute
display driven
display list
display object
dot-matrix printer
dots per inch (dpi)
electroluminescent display
end-of-record marker
escape sequences
font
gas plasma display
gray-scale value
initiator
interrecord gap
I/O cache
I/O port
laser printer
line printer
linear address space
liquid crystal display (LCD)
mark sensor
monophonic output
multiplex mode
optical scanner
optical sensor
optical character recognition (OCR)
phoneme
pixel (picture element)
plotter
polyphonic output
raster display
red-green-blue (RGB) display
refresh rate
rotational delay
sampling
scan code
scanning laser
small computer system interface (SCSI)
spatial resolution
speech recognition
speech synthesis
start-of-record marker
stored addressing information
tape header
target
teletype (hard-copy terminal)
track-to-track seek time
vector
vector list
video display terminal (VDT)

Vocabulary Exercises

1. The relative size of a data set before and after data compression is described in terms of the _____ of the _____.
2. A _____ is a special-purpose processor dedicated to managing the contents of a cache.

3. Color display can be achieved by using separate (but closely spaced) elements colored _____, _____, and _____.
4. The display device used in most computer monitors (and televisions) is a _____.
5. _____ are used to control specialized I/O device capabilities using ordinary character-based inputs.
6. A _____ is used to control the location of the ____ or input pointer on a video display.
7. Storage and I/O devices are indirectly connected to a bus (or I/O) port through a _____.
8. A _____ is a primitive component of human speech.
9. The _____ of a tape is the number of logical records contained within a single physical record.
10. The operating system normally views any storage device as a _____, thus ignoring the device's physical storage organization.
11. A _____ display is similar in construction to _____ and _____ displays except that it employs solid-state elements for light generation.
12. With tape storage, a _____ is necessary to allow the drive motor to reach normal operating speed before a physical record is read.
13. A _____ recognizes input in the form of alternating black-and-white vertical lines of varying width.
14. _____ is the process by which analog sound waves are converted to a digital representation.
15. A _____ list differs from a _____ list in that it can describe graphical objects other than lines.
16. Part of the function of a secondary storage device controller is to translate _____ into _____.
17. An individual display element of a video display surface is called a _____.
18. Track-to-track seek time and _____ are the primary limiting factors of _____ in a disk storage device.
19. _____ allows the transfer of data directly between memory and secondary storage devices, without the assistance of the CPU.

20. The output resolution of a graphic display device such as a laser printer is normally measured in terms of _____.
21. A _____ is a high-speed I/O port to which an I/O processor is dedicated.
22. An electronic keyboard generates a _____, which may be translated into ASCII character output by a keyboard controller.

Review Questions

1. How are peripheral devices linked to the CPU at the physical level?
2. What is the difference between physical access and logical access?
3. What functions are performed by a device controller?
4. How are incompatibilities in communication parameters between the CPU and I/O and storage devices resolved?
5. What factors determine the speed at which data on a disk can be accessed?
6. What factors determine the speed at which data on a tape can be accessed?
7. How are graphic images digitally represented?
8. What are the advantages and disadvantages of graphic output compared to text output?
9. What is a display list? What advantages does it have, compared to a bit map?
10. How does the operation of a laser printer differ from that of a dot-matrix printer?
11. What are escape sequences?
12. What is an I/O buffer? Why might one be used?
13. What are the difficulties inherent in speech recognition?

Research Problems

1. Many organizations have begun to store documents as graphic images rather than encoded (e.g., ASCII) textual and numeric data. Examples of applications that use such storage methods include medical records, engineering data, and patents. Identify one or more commercial software products that support this form of data storage. Examine the differences between such products and more traditional data storage approaches. Concentrate on issues of efficient use of storage space, internal data representation, and method(s) of data search and retrieval. What unique capabilities are provided by image-based storage and retrieval? What additional costs and hardware capabilities are required?

2. Many computer scientists are researching the problem of textual data understanding. The problem of visual recognition of character data and its conversion to ASCII or other coded representation is largely solved. However, basic understanding and classification of textual data is still poorly developed. Examples of applications that could be built with such capabilities include automatic indexing, abstracting, and summarization of news articles or other large textual inputs. Several products that address these needs are emerging from research labs into the commercial domain. Identify one or more of these products. What limitations must be placed on the text input to ensure reasonable success in automated recognition? What methods are used to perform classification and summarization? What is the relative performance (speed and error rate) of such systems, compared to humans performing the same tasks? For what traditional information processing applications might this technology be used?

PART 3

Software

9 *Machine-Level Programming*

Chapter Goals

- *Describe the process of assembly language program development.*
- *Describe the structure and content of assembly language programs.*
- *Describe the mechanisms by which assembly language programs are translated into executable code.*
- *Describe the advantages and disadvantages of assembly language programming.*

Chapter 6 examined the close correspondence between processor instruction sets and computer architectures. Recall that an instruction set encompasses the entire group of primitives that can be executed by the processor. These primitives represent the smallest, most fundamental increments by which data transformations and other machine operations can be performed. A given operation is triggered in the machine when the control unit senses the corresponding op code in an instruction. A processor's instruction set contains one unique op code for each type of operation, or primitive, that can be executed within its specific architecture.

At the level of process primitives, or machine level, a program must contain sequences of instructions that specify all required machine actions—and each action must be specified completely. That is, no intermediary software elements such as operating system service routines are interposed between an instruction and the machine actions it specifies. The programmer must anticipate and provide for all input, output, and processing requirements. The programmer must also keep track of data storage locations in memory, contents of registers in the control unit, transfers of data to and from peripheral devices, sequences of data transformations required to follow algorithms, and so on.

Program Translation

Early in the computer era, programmers wrote instructions in raw machine code, or as strings of 1 and 0 notations. As a reference aid, a programmer typically prepared a memory map, a manual table of data items and storage locations. Data items were given symbolic names in these tables so that the equivalent binary values could be found at a glance. That is, the symbolic names were not included in program code. As the size of computer memories increased and programs became more complex, manually keeping track of memory maps became impractical.

Programmers also referred to manual or preprinted tables for the binary equivalents of op codes. It became standard practice to refer to op codes with *mnemonics* such as ADD, MOVE, and XOR. Eventually, translation utilities were developed that permitted programmers to code instructions by using mnemonic and symbolic references. Such programs are called *generative translators* because they direct the computer itself to generate machine code from program statements that contain symbolic references. Such symbolic references are easier for people to remember and to understand.

Assemblers are the simplest type of generative translation program. Machine-level instructions are coded as program statements, or source code. Each program statement is composed of a mnemonic, which represents the op code, and symbolic address names, which represent the addresses of the operands. Source code statements for all instructions in the program are submitted as a batch of inputs to the assembler. The assembler merely replaces each mnemonic with the binary equivalent of its op code and inserts relative address values for each symbolic address name.

An assembly language encompasses *syntax*—a group of rules for coding program statements—and a set of mnemonics. For any assembly language, there will be one mnemonic for each op code in the processor's instruction set. Thus, there is a one-for-one correspondence between the mnemonics in an assembly language and the op codes of a specific processor.

Because instruction sets vary among processing architectures, an assembly language must be tailored to each type of processor. For example, if there are 12 op codes in a processor's instruction set, there will be 12 mnemonics, or instruction types, in its assembly language. In short, an assembly language is machine-specific. With an assembler, the translation process also proceeds in a one-for-one fashion. That is, one program statement must be coded for each instruction that will be generated.

Programs written in assembly language can also contain other statements. These statements are used to control memory allocation and the manner in which the program is built by the assembler. These control statements are covered in depth later in this chapter.

The output of a translator is called its *target language*. Rather than machine code, the target language can be another source code. For example, compilers that translate one high-level language implementation into another are called *cross compilers*. Similarly, assemblers that convert the assembly language of one specific architecture into that of another are called *cross assemblers*.

Linking and Loading

The assembler generates address values as though the program were loaded into memory beginning at address zero. Thus, each address value in the generated code is a measure of the offset from the beginning of the program. Machine code that contains *absolute addresses* can be loaded directly into memory and run. This version of a program is called *executable code*.

The output from an assembler or compiler is *object code*, or machine code in which placeholders have been inserted for absolute addresses. Object code must be *linked* and *loaded* to generate absolute addresses before loading and execution can begin. Thus, another term for object code is *relocatable code*, because its address values are changeable, or not absolute.

The function of a *linker* is to enable multiple machine code modules, or program subroutines, to execute as a single program. To permit concurrent loading of the modules into memory, the linker must allocate separate memory space to each module and must verify that address references are consistent in the resulting program code. In general, linking is performed after code generation and prior to loading.

Absolute addresses must be included in machine code for input to an *absolute loader,* which transfers the instructions into corresponding locations in memory, loads the first instruction into the CPU, and passes control of the computer to the program. Another type of loader is the *relocating loader*. This type of loader can load executable code into memory at different addresses than those used within the executable code. All address references within the executable code are recalculated with respect to a new origin.

ABSTRACTION AND EFFICIENCY IN PROGRAMMING

A key advantage to programming in a high-level language is that concise, symbolic references can represent relatively complex sequences of machine actions. Source code statements can be more "English-like," or more like natural language.

High-level languages typically include predefined subroutines for common functions. To use a subroutine, the programmer need only reference the name of the function in the source code of the program. Because all machine actions for performing the functions are contained in the subroutine, the programmer is relieved of concern for physical processing details. Furthermore, differences among computer architectures can be resolved in language implementation so that differences among machines are not apparent to the programmer. For example, the syntax of the PRINT command in the high-level language FORTRAN is the same for an IBM 3090 mainframe as it is for a DEC 5000 minicomputer, even though the two machines differ significantly in machine-level executions.

The degree to which a language removes the programmer from the details of machine actions is called its *level of abstraction*. High-level languages, as implemented by compilers or interpreters, are relatively more abstract than assembly language. In other words, high-level languages permit the programmer to deal with problems at a logical level, rather than at the level of physical devices.

However, a significant tradeoff accompanies this programming power or abstraction. A programmer using a high-level language has relatively little control over the implementation of instructions in hardware. Indeed, eliminating concern for these details is a primary reason for using the language. The tradeoff is that the programmer also loses control over the efficiency of the machine code that is generated. Machine functions implemented by a compiler or interpreter are coded to be generally useful. Accordingly, the machine code that is produced by the translator could provide relatively sluggish performance in a particular application, for a special problem, or within a specific hardware configuration.

The main advantage of writing programs in assembly language, therefore, is to be able to specify all machine actions fully and completely. For the added complexity and tedium of this detailing, the programmer is able to get the best possible performance from the machine. For this reason, assembly languages have remained important programming tools, even after the emergence of much more sophisticated and powerful high-level languages.

TYPES OF ASSEMBLERS

Implementation options for assembly language include

- Load-and-go assemblers
- Module assemblers

Machine-code generators, called *load-and-go assemblers,* act on a single program that will be loaded to absolute addresses. The translation process involves straightforward conversion of mnemonics and symbolic address references to binary equivalents. A starting address is specified for the program, all addresses are calculated relative to this point, and binary values for the addresses are inserted in the code. No external references, or references to addresses within other modules or programs, are permitted.

An assembly language implementation that is designed for use with linker/loader software is a *module assembler.* Such assemblers permit a single program to be consolidated from multiple modules, or subprograms. Equivalent terms are *routine assembler* or *subprogram assembler.*

Building a program from separate modules typically requires the passing of parameters, data values, or processing results among modules. The degree to which modules depend on one another for their inputs is called *coupling,* and the exchange of data among modules is called *interprocess communication.* Coupling is said to be strong if one module includes external references to addresses or identifiers that are contained within another module.

A key function of a module assembler, then, is to keep track of and to reconcile the external references within each module. The translation process must assure that all external references can be found among the other modules and that a consistent scheme of memory allocation is reflected in the resulting code.

The module assembler correlates external references by stepping through the source code twice. For this reason, these generative translators also are called *two-pass assemblers.* On the first pass, instruction mnemonics are converted to binary equivalents of op codes. Symbolic names for addresses are placed in a *symbol table* that serves as a memory map. Just as manual memory maps were reference aids for early programmers, the symbol table is used by the assembler program to keep a master list of symbolic names that represent numeric addresses. On the second pass, the assembler reconciles all address references and generates an object code file, inserting relative addresses in binary notation for each symbolic reference. After assembly, the resulting code file is ready to be passed to the linker for memory allocation, and then to the loader.

Most assembly language packages today feature two-pass assemblers. Typically, an assembler/linker/loader is a standard item in the software package the computer's manufacturer provides with the operating system. The discussion below covers the functions of a typical two-pass assembler in greater depth, then describes specific statement formats and coding rules.

ASSEMBLY LANGUAGE ARCHITECTURE

The overall function of an assembly language is to enable the programmer to code instructions for a specific processor in symbolic, rather than in binary, form. Specifically, an assembler must:

- Replace symbolic op codes with numeric machine codes
- Replace symbolic names for addresses with numeric addresses
- Reserve memory areas for instructions and data
- Convert constants to binary machine code

Note that the programmer must allocate memory space explicitly. As stated previously, all addresses will be generated relative to the start of the program, or as though the program were loaded in memory beginning with address zero. Once it is generated by the assembler/linker/loader, the executable program will be a sequence of instructions in machine code. Each instruction will contain an op code and the addresses of data items to be used as operands. All instructions in the program will be loaded first, beginning at the lowest available address. At the next available address after the last program instruction, data items are loaded, usually one byte per address. Thus, the address at which a data item is actually stored will depend on the starting address of the program, the number of instructions in the program, and the offset of the data item from the beginning of the data storage area.

Symbolic References

An assembly language generally includes symbolic references for op codes and conventions, or rules for "coining" symbolic names that refer to addresses. Other types of symbolic references are used to allocate memory and to control the assembly process. The main types of symbolic references used in most assembly languages are

- Mnemonics
- Symbolic address names, identifiers, and expressions
- Pseudo-instructions

The term *mnemonics* refers only to the symbolic references that stand for each op code in the instruction set of the processor for which the assembler was designed. Remember that there must be one, and only one, mnemonic in the assembly language for each op code in the instruction set. Because instruction sets vary among processors, a different assembly language must be used for each processor architecture. Thus, each processor will have its own assembly language. By contrast, the same high-level language can be used by many types of processors, with differences handled in the implementation of the compiler or interpreter.

Assembly languages permit some flexibility in the coding of addresses for operands. One approach is to use a symbolic name, which might be an alphabetic string such as DOLLARS or an alphanumeric string such as TOTAL1. Symbolic names are coined by the programmer and typically represent the names of variables or parameters used by the program. The programmer can also use numeric addresses for operands. Numeric address values in the source code are translated as though they were written in decimal notation, although other notation schemes can be used, as discussed below. These numeric values are handled by the assembler as relative addresses, or offsets from the beginning of the program.

Expressions are valid symbolic references that are formulas for calculating an address. Such formulas can include numeric and/or symbolic references that are to be combined under arithmetic operations. For example, DOLLARS + 2 would specify an address that is two bytes higher than the relative address represented by DOLLARS.

A mnemonic that has meaning in an assembly language but that is not included in the target machine's instruction set is called a *pseudo-instruction*. One important use of pseudo-instructions is in *data declarations*. In program text, a declaration is a statement in which a variable or a parameter is defined, or given a symbolic name, and typed, or associated with a specific data type, such as integer, real number, or character.

Pseudo-instructions that give the programmer control over the assembly process are called *directives*. Directives do not appear in the final machine code version of the program being assembled. Rather, directives control the manner of assembly. For example, a typical function of the assembler is to print a listing of the source code. The programmer might insert in the program statements a directive that causes the printer to advance to the next page at that point in the listing. Functions of directives include the following:

- Store code in memory locations out of normal sequence
- Reserve memory areas (blocks) for items that contain multiple words

- Define symbolic addresses and other identifiers
- Pass information to the assembler

Statement Format and Syntax

Most assemblers perform the basic functions described above. However, assembly language implementations vary widely in statement format and syntax. The discussion below summarizes some of the more common features of coding and syntax in assembly language.

The statement format used for instructions in assembly language is shown in Figure 9.1. Note that the format is divided into columns, or *fields*. Each field is identified by a mnemonic of a predetermined size. Proceeding from left to right, fields within an instruction include these:

- Label
- Operation
- Addresses
- Comment

Label	Operation	Addresses	Comments
5-Character Field	10-Character Field	15-Character Field	40-Character Field

FIGURE 9.1 **The format of an assembly language instruction.**

A *label* (also called a location symbol) is simply an identifier. The identifier in the label field corresponds with the starting address of the current instruction in memory. After assembly, address values in the location column correspond with the relative starting address of each instruction, beginning with address zero. The linker and the loader use these locations in allocating memory, recalculating absolute addresses, and relocating the object code.

The *operation field* contains an op-code mnemonic, or *operator*. The set of available mnemonics is entirely predefined. That is, the mnemonic must correspond with either a valid operation from the instruction set or one of the pseudo-instructions available in the assembly language implementation. An example of a pseudo-instruction found in most implementations of assembly language is the mnemonic DWS, which stands for "declare working storage." In a program statement, the operator DWS must be placed in the operation field. A table of commonly encountered pseudo-instructions is shown in Table 9.1.

The *address field* contains symbolic names, identifiers, values, or expressions that are placeholders for the addresses of the operands needed to execute the instruction. In the case of constants, a special syntax indicates that the content of the address field is a data value rather than an address, or location reference. If the instruction is monadic, only one address will be given. For a dyadic instruction, two operands must be identified. In some computer architectures, as in the IBM 370, some instruction formats require three operands.

The *comment field* is a free-form space that can contain character strings. The programmer can use comments to annotate the program text. Liberal use of comments is recommended as a sound programming practice that can enhance the clarity and readability of a program. The comment field is printed in the source code listing but is ignored by the assembler and does not appear in object code.

TABLE 9.1 A typical set of mnemonics that can be used as pseudo-instructions.

	Format			
Name	**Label**	**Mnemonic**	**Address**	**Function**
Origin	Not Used	ORG	Address	Specifies "address" in memory where the code following is to be loaded
Begin	Not Used	BEG	Not Used	Identifies the beginning of the code
End	Not Used	END	Not Used	Identifies the end of the code
Equate	Identifier	EQU	Value	Sets identifier as a synonym for the given "value"
Reserve Memory Bytes	Optional	RMB	Length	Reserve "length" bytes; the optional label is associated with the first byte
Reserve Memory Words	Optional	RMW	Length	Reserve "length" words

```
             .TITLE FIRST
;
;      THIS PROGRAM EVALUATES THE FOLLOWING FORMULA:
;      RESULT = (A+B) * (C+D) + E
;
             .PSECT       DATA,NOEXE,WRT
;
;      INITIALIZE AND DECLARE VARIABLES
;
A:           .LONG        20
B:           .LONG        15
C:           .LONG        10
D:           .LONG        5
E:           .LONG        1
RESULT:      .LONG        0
;
;      MAIN LINE
;
             .PSECT       INSTR,EXE,NOWRT
             .ENTRY       P2,0
             ADDL3        A,B,R1              ; R1 := A + B
             ADDL3        C,D,R2              ; R2 := C + D
             MULL3        R1,R2,RESULT        ; RESULT := R1 + R2
             ADDL         E,RESULT            ; RESULT := RESULT + E
;
             $EXIT_S
             .END         P2
```

FIGURE 9.2 A source code listing of an assembly language program.

Syntax

Assembly language implementations vary considerably in the coding rules, or syntax, used for writing program text. For example, in some implementations, each column of source code must begin at a specific, fixed position. In other implementations, fields must be separated by spaces or by special characters. For example, the semicolon (;) can indicate the end of the executable instruction and must precede the comment field on the same line. Under some versions, the exclamation point (!) means "end of line" (EOL) unless it is contained within a character string. Label and comment fields usually can be left blank. (The assembler will assign a label if none is specified, and the comment field is ignored.) Even if the assembler does not require fixed positions for the columns, good programming practice is to align them for clarity of presentation, as shown in Figure 9.2.

Mnemonics and symbolic names must usually begin with an alphabetic character. Within the symbolic identifier, most valid characters, including all alphanumerics, can be used. Some implementations limit the length of symbolic names, typically to between five and eight characters. In coining new symbolic names, the programmer must avoid duplicating any identifiers that are predefined within the architecture. For example, the registers of some processors have reserved names that cannot be used as address references for other locations.

Besides using symbolic names for addresses, the programmer can give numeric values for absolute addresses. Absolute addresses are normally assumed to be decimal values unless the programmer specifies otherwise. In some implementations, hexadecimal values are indicated by a suffix of H or X, such as:

FFFFH

FFFFX

Binary values usually have a B suffix:

101011100010110B

Octal values generally are followed by O or Q:

100O

100Q

Even though a suffix is not required for decimal values, it is good programming practice to append a *D* to absolute addresses given in decimal notation:

12D

Another kind of reference that can be used in an operand field is a *literal*. The programmer can designate a literal to establish constants for use by the program and to assign storage areas to them. A literal is usually indicated by an equal sign (=), followed by the value of the constant. For example, the literal =679 refers to the decimal value 679. On encountering a literal, the assembler assigns the label, or symbolic reference, to it, and places the literal value at the beginning of the data area in memory, or following the program. Again, if no label is specified, the assembler will assign the value to a placeholder reference.

Another kind of literal is a character or string used as a data constant rather than as a symbolic name. Literal character strings are normally enclosed in single quotation marks. For example, 'TITLE' could be used as a data value for an operand rather than referring to the address of a string value held in memory.

As stated above, an expression is an arithmetic formula for calculating an address and can be used in an operand field to refer to the address. Examples of valid expressions include these:

LOC + 3D

N*2

DOLLARS–6D

Data Declarations

Pseudo-instructions that declare the name and type of data items have the general format shown in Figure 9.3. Fields within such statements can include the following:

- Labels
- Pseudo-instruction mnemonics
- Extent or value fields
- Comments

Label	Pseudo-Instruction Mnemonic	Extent or Value	Comments

FIGURE 9.3 **A typical statement format used for assembly language pseudo-instructions.**

The first column contains a label (symbolic name) that corresponds with a data item or storage area. The mnemonic for the pseudo-instruction or directive is written in the second column. For example, mnemonics used in declarations include WORD, BYTE, or CHAR (character).

Other mnemonics found in many assembly language implementations include the following:

- PROGRAM (name of program)
- BSS (block started by symbol)
- EQU (assignment of value [equal])
- ORG (origin [starting location in memory])
- MACRO (macro declaration [discussed below])
- END (end of program)

If the mnemonic designates a memory allocation, the extent of the storage area in bytes or words is given in the third column. This column can also hold an explicit value that is to be associated with the label, as in assigning a numeric value to a constant. If the directive END is followed by an address value, it is the address at which program execution will begin.

As with other instructions, a comment field can be used if the programmer desires. For example, in a data declaration, the comment field would typically be used to explain the use of the variable in the program.

Generating Executable Code

For programs that are built of multiple modules, or subprograms, the steps involved in producing machine code from a source-code file typically include these:

- Assembly
- Linking
- Macro assembly
- Loading

Assembly

As stated, most assemblers perform translation and code generation in two passes. In the first pass, the assembler steps through each statement in the program. The second field of each statement is converted from mnemonic form into the binary equivalent of the corresponding op code. In each statement that requires allocation of memory, the label in the rightmost column is added to a symbol table maintained by the assembler (Figure 9.4). The symbol table contains a pointer

```
FIRST
Symbol  Table

Symbol          Address

A               00000000    R       D       01
B               00000004    R       D       01
C               00000008    R       D       01
D               0000000C    R       D       01
E               00000010    R       D       01
P2              00000000    RG      D       02
RESULT          00000014    R       D       01
SYS$EXIT        ********    GX              02
```

FIGURE 9.4 A printout of a symbol table maintained by an assembler.

value for each symbolic name. The pointer value is set according to the current value of a location counter, which is incremented each time a byte in memory is allocated. If the computer uses variable-length instructions, the assembler must also keep track of the length of each instruction to determine the proper starting location for the next instruction.

On the first pass, the assembler will not be able to determine the values of external or forward references. An external reference is a symbolic name of another module or a variable within it. Forward references are symbolic names for which values must be calculated or assigned in subsequent portions of the same module. For such a reference, the assembler inserts a placeholder in the symbol table and continues with the first-pass translation. The result of a first-pass assembly operation on a sample program is shown in Figure 9.5.

```
           Statement
Address     Number

  0000        1              .TITLE FIRST
  0000        2    ;
  0000        3    ;    THIS PROGRAM EVALUATES THE FOLLOWING FORMULA:
  0000        4    ;    RESULT = (A+B) * (C+D)+ E
00000000      5    ;
  0000        6              .PSECT     DATA,NOEXE,WRT
  0000        7    ;
  0000        8    ;    INITIALIZE AND DECLARE VARIABLES
  0000        9    ;
  0000       10    A:        .LONG  20
  0004       11    B:        .LONG  15
  0008       12    C:        .LONG  10
  000C       13    D:        .LONG  5
  0010       14    E:        .LONG  1
  0014       15    RESULT:   .LONG      0
  0018       16    ;
  0018       17    ;    MAIN LINE
  0018       18    ;
 00000000    19         .PSECT        INSTR,EXE,NOWRT
  0000       20         .ENTRY        P2,0
  0002       21         ADDL3         A,B,R1            ; R1:=A+B
  000E       22         ADDL3         C,D,R2            ; R2:=C+D
  001A       23         MULL3         R1,R2,RESULT      ;      RESULT:=R1+R2
  0022       24         ADDL          E,RESULT          ;  RESULT:=RESULT+E
  002D       25    ;
  002D       26         $EXIT_S
```

FIGURE 9.5 A printout of the results of first-pass processing through an assembler for the program shown in Figure 9.2.

On the second pass, the assembler again steps through the program. This time, object code is generated. Calculated addresses, or the relative values held in the symbol table, are inserted in the code. Remember that these are relative addresses, or locations in relation to the starting address, which is assumed to be zero. Placeholders will remain in the code for any external references to address values that will be calculated in assembling other modules. The object code generated by the second pass of the same program illustrated in Figure 9.5 is shown in Figure 9.6.

Linking

The purpose of linking is to combine multiple object code modules into a single program. This step is required, for example, in structured programming approaches that build a main program from several subprograms. As an alternative,

```
                                                          Statement
     Generated Object Code              Address           Number

                                           0000                 1
                                           0000                 2
                                           0000                 3
                                           0000                 4
                                     00000000  5
                                           0000                 6
                                           0000                 7
                                           0000                 8
                           00000014     0000  9
                           0000000F     0004 10
                           0000000A     0008 11
                           00000005     000C 12
                           00000001     0010 13
                           00000000     0014 14
                                           0018                15
                                           0018                16
                                           0018                17
                                       00000000                18
                                   0000    0000                19
51    00000004'EF    00000000'EF    C1     0002                20
52    0000000C'EF    00000000'EF    C1     000E                21
      00000014'EF       52 51       C5     001A                22
      00000014'EF    00000010'EF    C0     0022                23
                                           002D                24
```

FIGURE 9.6 A printout of the object code generated during the second pass through an assembler.

a single, long program could be subdivided and coded as separate overlays that will be swapped between memory and secondary storage during execution. In such cases, linking is required to reconcile references within the overlays.

The input to a linker is a set of previously assembled modules. Each module is in object code form. That is, the modules contain no absolute addresses. Address references are either relative addresses (locations in relation to address zero) or placeholders for external references. Alternate terms for the linker include *link editor*, *binder*, or *consolidator*. Its primary function is to account for and reconcile all address references within and among the modules and to replace those references with a single, consistent scheme of relative addresses.

Linking is typically done in two passes. In the first pass, the linker scans the object code to identify all modules, or segments. For each segment, a *header* is created, as shown in Figure 9.7. The header lists all modules referenced within the segment. The header also lists the symbolic names of any references to other modules and any data areas that are shared among modules. The linker compares and correlates the segment headers to reconcile all external references. For each module, all references to other modules, as well as any required subroutines within those modules or in the library, are found. Library routines also are scanned for external references, and these are checked to assure that the required modules are included in the linking process.

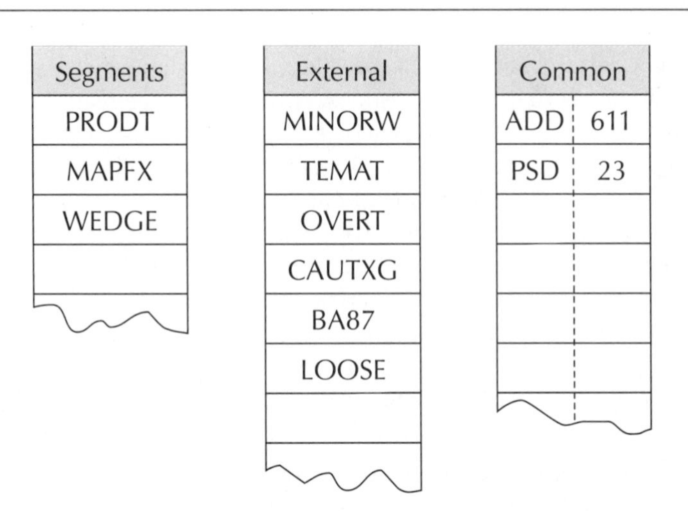

FIGURE 9.7 The format and structure for a header maintained by a linker.

On the second pass, the linker generates a memory map, or memory allocation scheme, for the program. As shown in Figure 9.8, the map is a series of records. Each record contains the name of a module, its object code instructions, and the relative addresses of its data storage allocation. After linking has been performed, all address references among modules are consistent with one another. However, the references still are relative to a hypothetical starting address of zero.

Module Name	Object Code Instructions	Relative Addresses

FIGURE 9.8 **The format a programmer uses to establish and maintain memory-mapping notes.**

The output of the linker is a single program file. This file contains the memory allocation map, headers for each segment, and object code with a consistent set of relative addresses.

Macro Assembly

Higher level assembler/linker/loader implementations also provide programmers a coding shortcut called a *macro*. A macro is a template for a sequence of program statements. The template is a source code version of the sequence, with the required parameters and variables omitted. A macro can be created in the declaration portion of a program or it can be an existing template that is retained in a library in systems software. To generate the macro, the programmer need only code its symbolic name, followed by the required parameters, in the order in which they are used by the macro's instructions.

Translation and insertion of macros is typically done by the linker/loader because macros are typically passed among modules. Upon encountering the name of a previously declared macro, the assembler or linker/loader performs the following actions:

1. Update all segment headers.
2. Fetch the instructions from the macro library.
3. Translate the macro's instructions into object code.
4. Insert the parameters given by the programmer.
5. Insert the macro's object code into the program.

Note particularly that macros differ significantly from subroutines, another technique often used for handling modules. In the case of a macro, the required object code is generated at the time of assembly and is inserted into the resulting program. That is, each time the name of the macro appears in the source code, its instruction sequence, along with relative address values for its parameters, will be inserted into the program.

By contrast, a subroutine is a module of executable code that is called from a library in systems software at the time of execution. That is, the calling program contains only a symbolic name for the subroutine. Upon encountering the symbolic name during a program run, the processor fetches, loads, and executes the code in the library. Thus, a macro delivers on line code, or code that is inserted directly into program object code during the assembly process.

From a practical viewpoint, the output of a program will be the same whether a module is coded as a macro or as a subroutine. The programmer can choose either of these techniques. Each technique brings a different set of tradeoffs. Coding a process as a macro prevents some inefficiencies of branching during program execution. Because code is inserted on line with the main program, no additional processing overhead will be incurred during execution to fetch instructions from secondary storage. Macros also save time in writing program code. A major drawback of coding with macros is the size of the object code file. Because code is inserted on line, extensive use of macros will result in a longer program. If subroutines were used, the impact on the size of the main-program file would be negligible.

A variation in macro assembly is the *conditional macro*. A conditional macro is a sequence of instructions inserted into the object code only if a specified test condition is true. Some programmers also feel that extensive use of macros can sacrifice program clarity. Especially difficult to understand are *recursive macros*, or macros that call other macros.

Loading

The functions of the loader are to

- Relocate the program's object code to specific memory addresses
- Transfer the resulting machine code into memory
- Transfer control of the processor to the first instruction in the program

In some implementations, the functions of linking and loading are combined in a single utility program called a *linking loader*.

A loader must be tailored to the memory allocation scheme of a specific computer. When the loader is installed in the system, it must be given the bounds, or highest and lowest addresses, of the working storage area in memory, the portion of memory not used by systems software. Thus, a loader is related closely to the operating system and memory allocation scheme of a specific configuration of hardware and systems software. An essential function of the loader is to recalculate all addresses according to the map generated by the linker. Depending on the architecture, absolute address values will be calculated during either loading or execution.

If the computer uses absolute addressing, actual addresses are calculated during loading by adding an offset to the relative addresses in the object code. The offset corresponds to the number of storage addresses in low memory that are reserved for systems software and its working storage. Machines that use indirect addressing, such as the IBM 370, use a *base register* architecture. Under this approach, a separate register, the base register in the CPU, holds a value for the memory address at which program loading may begin. The CPU adds the contents of the base register to all address values as they are encountered in fetching or interpreting instructions.

In computers that support multiple users, multitasking, or multiprogramming, indirect addressing permits the operating system to change the value in the base register and to reallocate memory at the time of execution. Thus, reallocation is done without affecting the addresses that the loader has assigned.

System Loaders

A paradox arises if the program to be loaded is the operating system itself. For example, when electrical power to a computer is turned on, no valid instructions reside within either memory or the registers of the CPU. Without a loader program running in the CPU, it will be impossible to load either the rest of the operating system or any application programs.

A straightforward solution on some computers is to require the operator to set the initial value of the instruction register manually. Values in the register are controlled from the front panel of the CPU by a row of toggle switches. The position of each switch corresponds to a value of 0 or 1. By consulting a reference manual, the operator sets the switches to the binary value of the first instruction, which contains the op code for an input operation and the address of the next instruction. By pressing a button, the operator causes the processor to execute this instruction. To do so, it requires the next instruction, which the operator again loads manually on the row of switches. This instruction commands the processor to load the next program instruction. The third and last instruction that is entered manually is to return to the first instruction. Thus, a program loop is created by which the processor is instructed to load one program instruction after another.

The sequence of instructions that are loaded manually, or through the toggle switches, is the equivalent of the following:

```
0 INPUT 1
1 INPUT *
2 GOTO 0
```

The value of the asterisk identifier (*) is incremented with each iteration of the loop. This simple program, with only three statements, is called a *bootstrap loader*, or *boot program*, because it loads itself, pulling itself up "by its own bootstraps," in effect.

A bootstrap loader used to load the operating system is called a *system loader*. The first program to be loaded is the *supervisor*. As it is loading the supervisor, the system loader actually causes itself to be overwritten in memory with the last few instructions of the supervisor. As the last instruction is loaded, control of the system is passed to the first instruction of the supervisor program.

Most computers today do not require the boot to be loaded manually. Rather, boot instructions are retained within the system in firmware. Typically, a read-only memory (ROM) chip containing the instructions is activated when power is turned on. Firmware used for this purpose is called a ROM boot. Instructions in the boot program cause the operating system to be loaded into memory from secondary storage, such as a disk. Then control of the processor is passed to an instruction at a predefined address in the loaded program.

PROGRAM DESIGN CONSIDERATIONS

Considerations in designing assembly language programs include the advantages and disadvantages of coding at the level of the processor's instruction set. Tradeoffs of writing programs in assembly language will be encountered in the following areas:

- Data structure
- Memory allocation
- Processing efficiency
- Program structure
- Portability

Data Structure

The programmer working in most high-level languages such as COBOL or FORTRAN is limited to the data types and structures that are predefined within the language implementation. An advantage of using assembly language is that the programmer can define new data types and structures through pseudo-instructions. For example, the programmer can control the number of bytes that represent floating-point values.

A disadvantage of data typing in many assembly language implementations is a lack of multidimensional data structures such as arrays, tables, records, and files. Such data structures are predefined in most high-level languages. To specify an array in assembly language, the programmer must write an algorithm for calculating addresses within the array.

A two-dimensional array, diagrammed in Figure 9.9a, can be regarded as a set of numbered, horizontal rows and vertical columns. The position of each value in the array is given by the number of its row and the number of its column. However, memory allocation by the assembler is one-dimensional, as shown in Figure 9.9b. Values in the array will be stored in sequence, one row after another. There is no predefined addressing scheme for the array, so an algorithm must be coded for deriving the address from the row number and column number in assembly language.

Memory Allocation

Coding in assembly language gives the programmer considerable control over memory allocation. Pseudo-instructions can be used to declare working storage, and data within each storage area can be organized in bytes, words, characters, and so on. In machines that use absolute addressing, the programmer can make specific allocations anywhere within working RAM. In machines that use indirect addressing, the programmer can control allocation relative to the program's base address. The loader or the operating system will then determine the location at which loading will begin.

Memory allocation schemes become especially crucial if the computer has a limited amount of working storage. Coding in assembly language might be necessary to control the swapping of program overlays into memory from secondary storage.

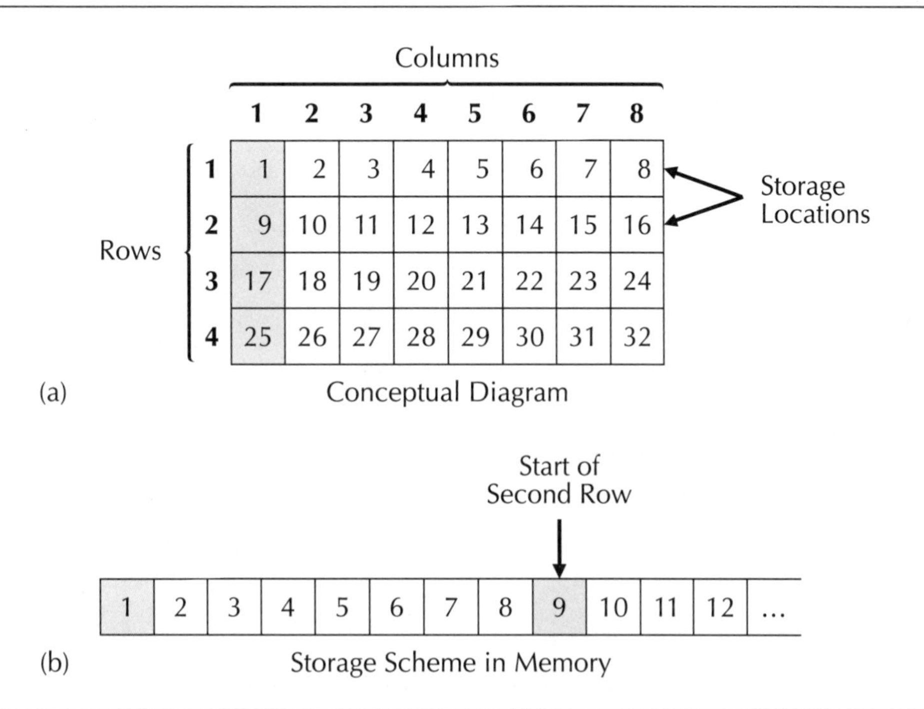

FIGURE 9.9 (a) A conceptual diagram for a two-dimensional array; (b) along with a one-dimensional storage scheme for the same data.

Processing Efficiency

Using assembly language to specify machine actions in small increments has both advantages and disadvantages. The main advantage is that code can be tailored for optimum performance in executing a specific process in a given machine. By contrast, algorithms that are implemented within compilers often sacrifice efficiency for generality.

Program Structure

Some high-level languages, such as Pascal, impose rules of structure on program text. That is, the compiler not only checks the syntax of each statement, it also requires the sequence of statements to follow a specific format. For example, definitions and declarations usually must precede the main body of the program. Program text also can be grouped in *program blocks* to form subroutines and modules.

The underlying structure of a program can be shown in a *structure chart*, as in Figure 9.10. Such charts illustrate and document the logical design of a program, which is independent of the language in which it will be coded. In structured programming, a separate program module would be coded for each box on the structure chart.

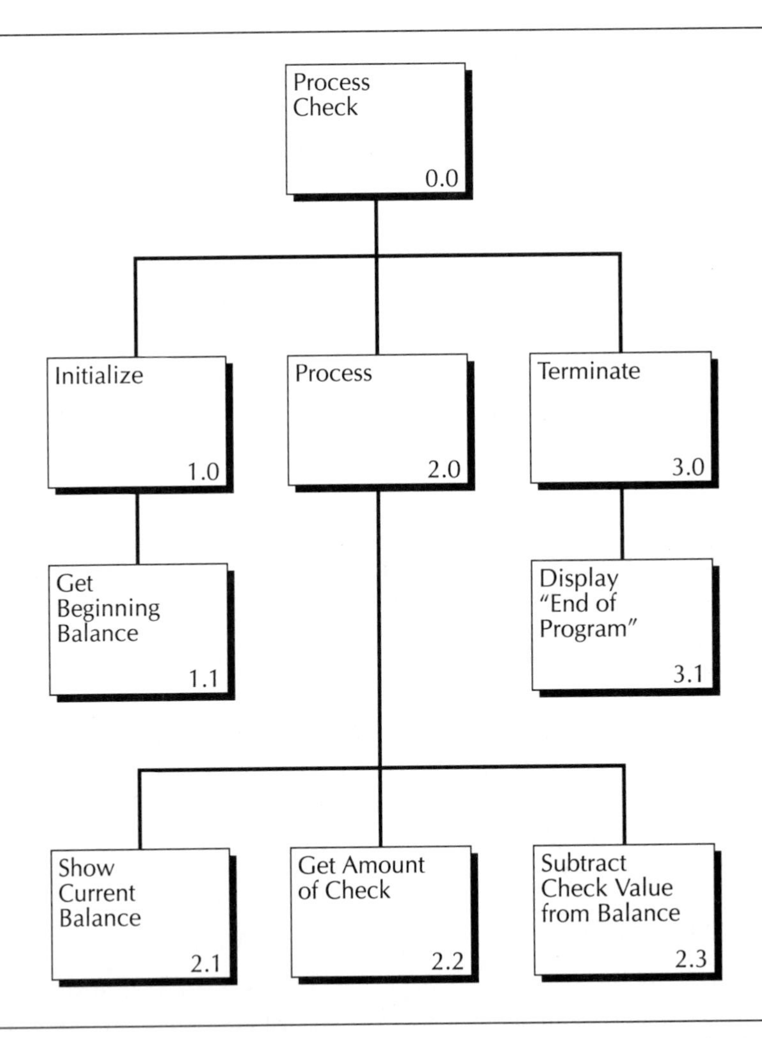

FIGURE 9.10 **A sample of a structure chart used to establish and present a logical overview for the modular structure of a program.**

The purpose of applying structure in programming is to promote understanding of the underlying problem and to enhance the readability and clarity of the program text. To promote clarity further, comments should be used liberally throughout the program text to explain logical functions of the code. A key objective is to make it easy for other programmers and systems professionals to understand program functions.

In itself, assembly language is the least structured of all coding methods. Source code can be written on line, and there are few constraints beyond the required syntax for individual instructions or pseudo-instructions. Furthermore, the mnemonics used in assembly language are usually concise abbreviations and do little to describe the corresponding operations. Because assembly language programs specify machine actions in detail, the application or the business problem being solved might not be readily apparent, especially to readers who are unfamiliar with the machine's instruction set.

Even though the assembler does not enforce structure, it is good programming practice to write the text as though a block structure were required. It also is preferable to code and assemble modules separately, then pass them to the linker.

Portability

A major tradeoff in using assembly language is the sacrifice of portability, or interchangeability of source code among machines. Because the language implementation is unique to each machine's instruction set, a program usually must be rewritten completely for execution within a different architecture. By contrast, high-level languages such as Pascal and C are highly portable, or machine-independent. With such languages, differences in machine architectures are handled in the compiler implementations. The syntax of the source code can be virtually the same from one computer to another. Thus, to move a Pascal program to a different computer, it is usually necessary to only recompile the source code.

Portability is an important consideration in an era of rapid technological change. Each modification or evolutionary step in processor architecture requires a corresponding change in the instruction set or assembler program. Portability also becomes crucial within distributed systems, which can have different processor architectures at each node.

Many systems development projects require both high performance and portability of software. To achieve favorable tradeoffs in both areas, programmers might write code initially in a structured, portable language such as C. Doing early program versions in such a language facilitates communication and understanding among project team members and carries modular design concepts through the coding effort. Program modules are then compiled, linked, and loaded. Software testing includes performance tests in which the execution time of each module is evaluated. Selected modules (typically those with high processing overhead) can then be recoded in assembly language for optimum performance and linked with the remaining modules to form the production version of the program.

Within the scope of current software development practice, assembly language can be regarded as a specialized tool for "fine-tuning" system performance. Significant trends are moving in the opposite direction toward increased involvement of nontechnical end users and higher level languages. Assembly language is presented here primarily as an aid to understanding computer architecture. With this background, the next several chapters examine program development using high-level languages and the operating system. These two tools form a collection of programs that effectively removes users and programmers from concerns about processing details at the machine level.

SUMMARY

Assembly programs consist of instructions that closely match a CPU's instruction set. In general there is one assembly language statement for each different machine instruction. Assembly language source code is translated into machine language (executable) programs in two steps: assembly and linking. An assembler is a generative translator that translates assembler language statements into object code. One or more object code modules are combined into a single executable program by a linker. An executable program is loaded into memory and given control of hardware by a loader.

The primary advantage to assembly language programming is the programmer's ability to exercise precise control over instruction execution and memory utilization. Compilation of higher level language programs often results in less than optimal program efficiency as measured in CPU cycles and memory consumed during execution. Assembler programmers have the opportunity to maximize program efficiency. The primary disadvantages of assembly programming are the labor intensity of the activity and the lack of program portability.

Labor intensity arises from the large size of most assembler programs. Lack of portability is due to the dependence of assembler language on the instruction set of a single CPU.

Assembly language programs are composed of instructions and pseudo-instructions. Instructions consist of symbolic instructions (mnemonics) and data references. Data references can be expressed as memory addresses, symbolic data names, or constant values. Pseudo-instructions allocate memory to variables and control certain other assembler actions. Data declarations are pseudo-instructions that assert the existence of a data item. A data item is defined by its type, length, and symbolic name.

Assembly language programs are normally translated into object code by a two-pass assembler. In the first pass of the assembler, a symbol table is created. The symbol table contains entries for each data item declared within the program. In the second pass, instruction mnemonics are translated into machine language op codes; memory addresses from the symbols table are used as operands.

Link editing combines multiple object code modules into a single executable program. This requires the resolution of external references between individual modules and the organization of program instructions and data within an integrated memory allocation scheme. Object code modules are assumed to occupy sequential memory regions. The linker calculates the address occupied by each instruction and data element and inserts or updates appropriate memory references in machine language instructions.

Key Terms

absolute address
absolute loader
address field
assembler
base register
bootstrap loader (boot program)
comment
conditional macro
coupling
cross assembler
cross compiler
data declaration
directive
executable code
expression
field
generative translator
header
label (location symbol)
level of abstraction
linker (linkage editor)
linking loader
literal
load
load-and-go assembler
loader
macro
module (routine, subprogram) assembler

mnemonics
object code
operator
portability
program block
pseudo-instruction
recursive macro
relocatable code
relocating loader
structure chart
supervisor
symbol table
syntax
system loader
target language
two-pass assembler

Vocabulary Exercises

1. An assembler maintains a _____ to keep track of the memory locations of data items and program segments.
2. A _____ is a short program that loads the operating system into memory after the computer is powered on.
3. A _____ translates assembly language for one processor into assembly language for another.
4. A _____ is a program that translates _____ into machine code.
5. A ____ is a short name for a CPU instruction.
6. An executable program is placed into memory by a _____. A _____ is used if address references within the program do not match the program's placement in memory.
7. A _____ defines data or other program objects.

Review Questions

1. When coding programs at the machine level, what functions must the programmer specify in detail?
2. What is the purpose of a generative translator?
3. How do the functions of linking and loading relate to generative translation?
4. What are some differences and tradeoffs among computer programming languages?

5. What is a two-pass assembler, and how can such tools facilitate modular programming?
6. What are the primary functions of an assembler?
7. How is a relocating loader related to the function of memory allocation?
8. In view of the power and convenience of high-level languages, why is assembly language still in use?
9. What might be the advantages and disadvantages of writing programs in high-level language?

10 *Application Development*

Chapter Goals

- *Describe common methods of application system development and software support for those methods.*
- *Describe software to support the development of system models.*
- *Describe software to support the development and translation of application programs.*

THE PROCESS OF APPLICATION DEVELOPMENT

Recall from Chapter 2 that the primary role of application software is to translate user requests into the machine instructions required to satisfy those requests. The process by which application software is developed can also be considered a translation process. It is, in essence, a translation from the user's statement of his processing needs into application software that is capable of fulfilling those needs. The exact steps required to perform this translation process depend on the particular method used to develop the application program(s).

A principle common to most application development methodologies is the concept of a *system model*. A system model is a statement of information processing requirements. As such, it is an intermediate step in the translation process between the users' statement of processing requirements and the machine instructions that implement those processing requirements. The system model is the basis for developing application programs, which are in turn translated into machine instructions. The entire translation process is depicted in Figure 10.1.

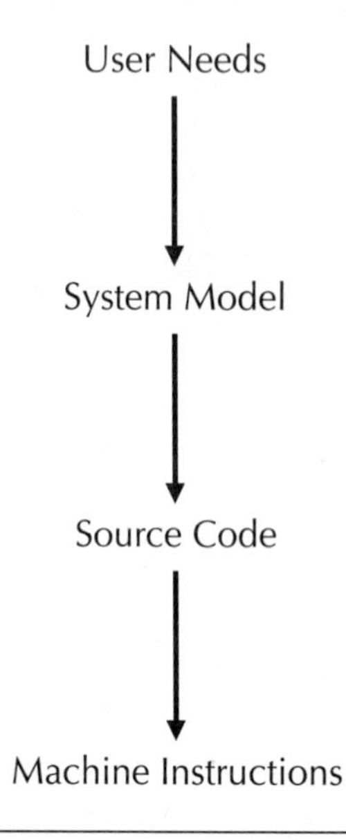

FIGURE 10.1 **The various translation steps needed to develop application programs (source code).**

There are several different methodologies for application development. There are also many variations of each methodology as they are currently practiced. For the purposes of discussing application development software, our discussion of each methodology will be restricted to its most general implementation. Methodologies will be differentiated according to

- The number of system models used
- The characteristics of each model
- Types of tools used to build each model
- Software used to support modelling and translation

At present, the most widely practiced application development methodologies are the *structured system development life cycle (structured SDLC)* and *rapid prototyping*. These were initially discussed in Chapter 1 and are further explained in the following sections. An older methodology (referred to as *classical systems development)* will also be discussed for comparative purposes.

Classical Systems Development

Classical systems development methodologies primarily used system flowcharts as models. A *system flowchart* shows the flow of data between manual processes, automated processes (i.e., programs), and files. Such flowcharts are normally supplemented by written descriptions of the model components. The system flowchart provides a graphical overview of the system components and flow detail. The written descriptions of each component provide a detailed basis for implementing programs, initializing files, designing input and output documents, and documenting user procedures. All of these model components also serve as system documentation for later changes or upgrades to the system.

An example of a system flowchart for a payroll system is shown in Figure 10.2. Each of the rectangles represents a separate application program. A *program specification* is written for each of these programs. An example of such a specification is given in Figure 10.3. Written program specifications can also be supplemented by graphical program descriptions such as *program flowcharts*. Program specifications must describe processing requirements in sufficient detail to allow a programmer to write an application program that implements the process.

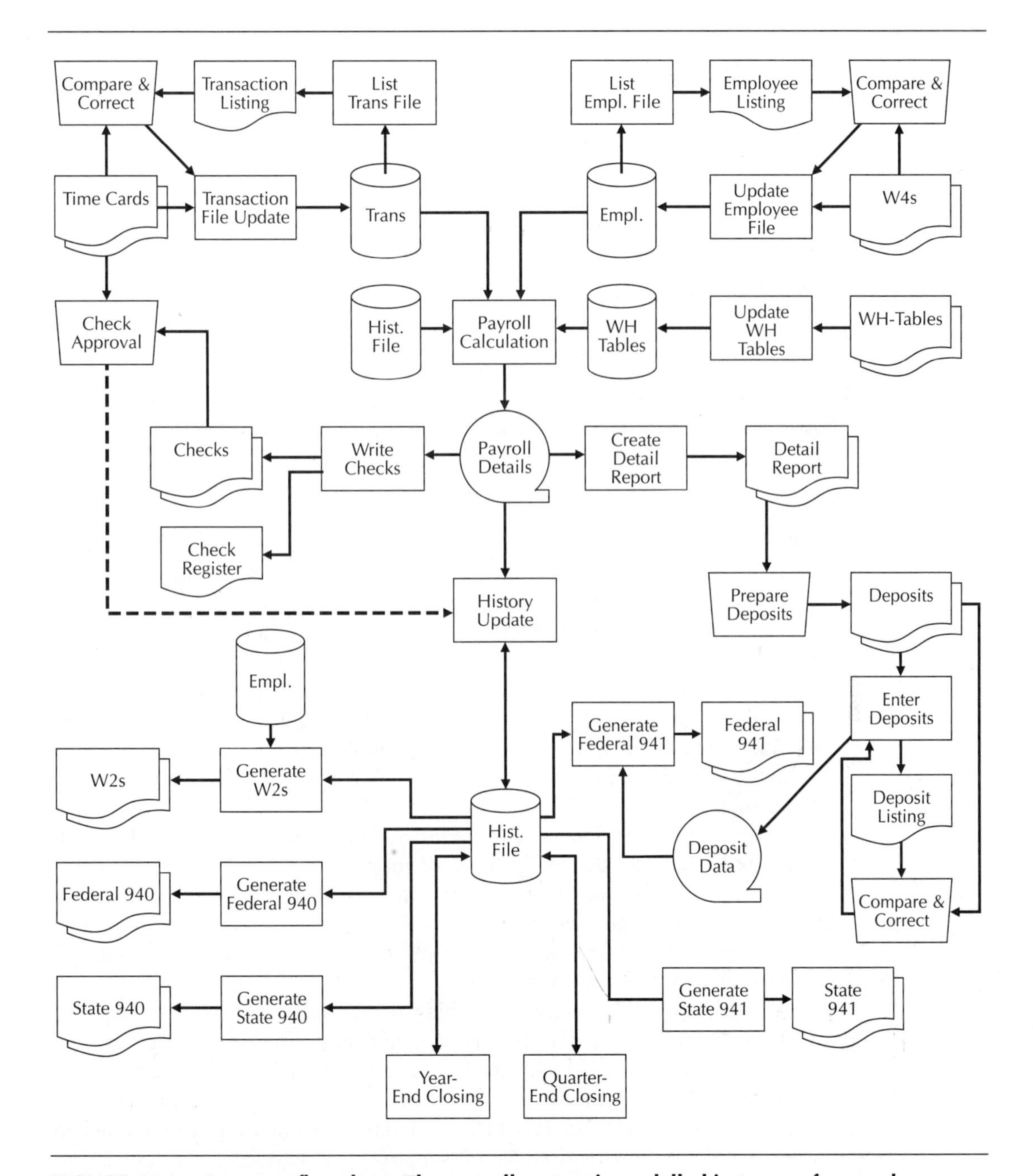

FIGURE 10.2 A system flowchart. The payroll system is modelled in terms of manual processes, automated processes, automated files, and input and output documents.

```
Program Name:        Update History File
Inputs:              Payroll Details File
Outputs:             History File
Description:
     For each record in Payroll Details File do
     read History File record matching
          Social-Security-Number in
          Payroll Details File
     YTD-Gross-Pay = YTD-Gross-Pay + Gross-Pay
     YTD-Federal-WH = YTD-Federal-WH + Federal-WH
```

FIGURE 10.3 A program specification for the Update-History program in the payroll system.

In a system flowchart, files are represented by one of two symbols. In the example shown, the symbol for the payroll history file indicates that it is a permanent file. This same symbol is sometimes used to indicate random access files stored on disk. The symbol for the payroll details file indicates that it is a temporary file. Such files generally exist only until other programs that read them have executed. This same symbol could also be used to indicate a sequential access file or a file stored on tape.

Each file is described by a written *file specification* that shows the composition of the file in terms of records and fields within records. Each field is further described in terms of its data type and length. The organization and/or access methods for the file are also stated. An example of a file specification is shown in Figure 10.4.

Additional written descriptions are provided for inputs, outputs, and manual procedures. Inputs are defined in terms of format and content. Input constraints such as editing and existence criteria can also be included. Outputs are generally defined in terms of form and content, which can include a sample of the actual output (e.g., a screen or report layout). Written procedures are described in a manner similar to program specifications.

A system flowchart (and its supporting written specifications) is an example of an *implementation model.* That is, it is not only a statement of processing requirements but also the specification of a particular method for implementing a system to satisfy those requirements. As such, implementation based on the model is very straightforward. An application program is written for each automated process represented on the flowchart. Each program is developed and tested, usually in a high-level programming language. These programs are then translated into machine instructions by program translators (e.g., interpreters or compilers).

```
File Name:      Employee File
Access Method:  Indexed on Social-Security-Number
Content:        Fixed Length Employee Records

Record Definition:

    Social-Security-Number        String    999-99-9999
    Last-Name                     String    X(30)
    First-Name                    String    X(30)
    Middle-Initial                String    X(1)
    Street Address                String    X(60)
    City                          String    X(60)
    State                         String    X(2)
    Zip-Code                      String    99999
    Salary                        Real      99999.99
    Salary-Code                   String    X
    Filing-Status                 String    X(2)
    Number-Of-Exemptions          Integer   99
    Additional-State-WH           Real      9999.99
    Additional-Federal-WH         Real      9999.99
```

FIGURE 10.4 A file specification for the employee file in the payroll system.

Files are initialized according to the written specifications provided, and application programs directly utilize those files. Each manual input or output is specified in terms of content and format, and these specifications determine implementation requirements for both automated and manual processes. After testing and installation, the programs and files are ready to be used by the end user.

Structured Systems Development

The structured systems development methodology models user-processing requirements by constructing two separate system models. These are the *analysis model* (or *logical model*) and the *implementation model* (or *physical model*). The analysis model is a model of user requirements that is independent of implementation method. A graphical model constructed of these components is called a *data flow diagram (DFD)*. A data flow diagram can also be used for the physical model. However, the more appropriate (and common) format for constructing a physical model is the system flowchart described earlier.

A sample data flow diagram (DFD) for the payroll system described earlier is shown in Figure 10.5. Processing requirements are modelled in terms of processes, data flows, files (or data stores), and external entities. Processes are represented by circles or rectangles with rounded corners. External entities are sources or destinations for data that are outside of the system. They are shown as squares, usually at the edges of the diagram. Files are represented by shallow rectangles that are open on the left or right side. Data flows are represented by named arrows that show the movement of data among processes, files, and external entities.

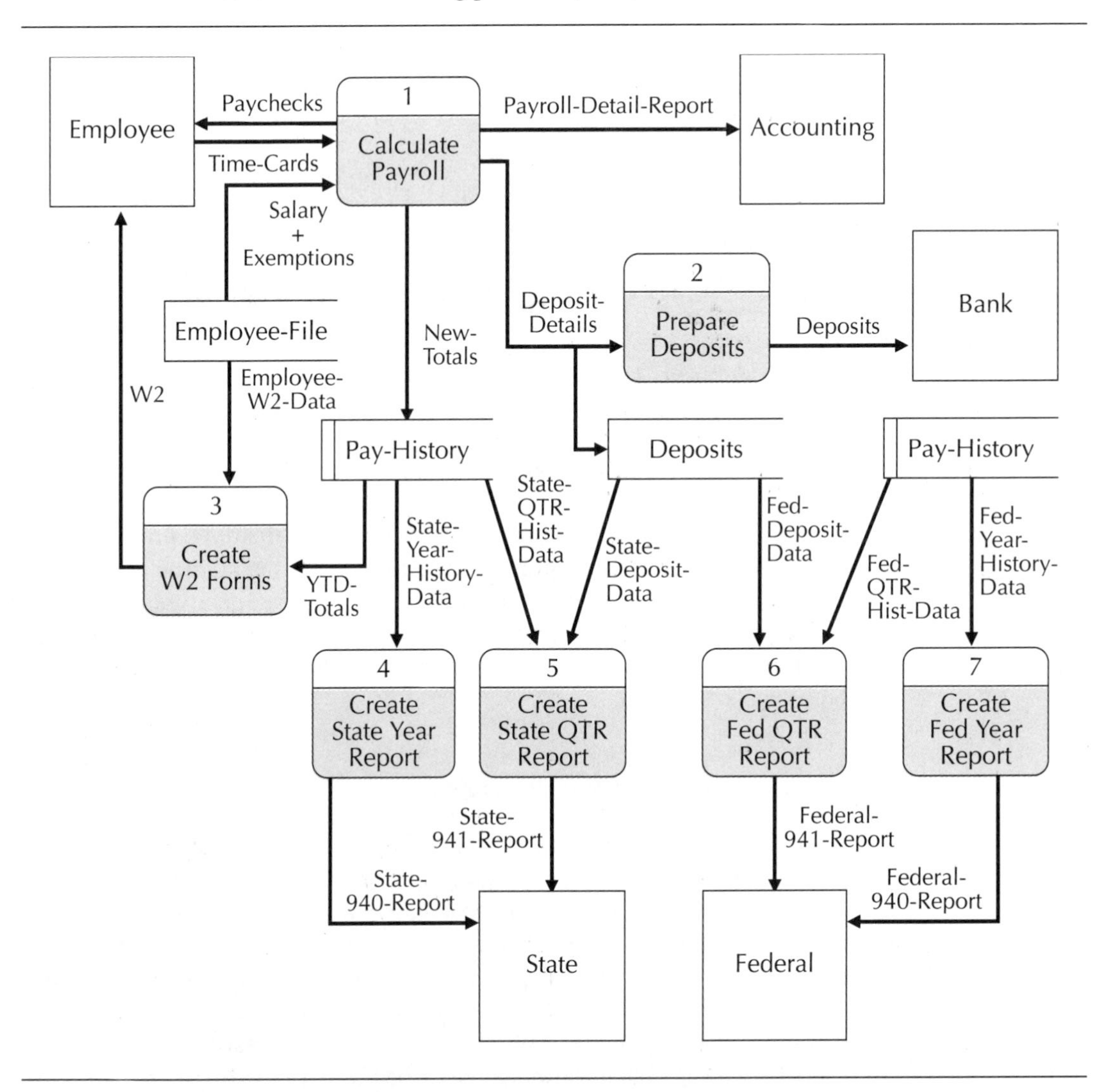

FIGURE 10.5 A data flow diagram. The system is modelled in terms of external entities, processes, data stores, and data flows.

As with system flowcharts, individual components of the data flow diagram are further defined by textual descriptions. Unlike system flowcharts, these written specifications are not designed to be directly used for implementing programs or files. Rather, they describe system details so that they can be read and validated by users. In general, implementation issues such as data (or file) formats and specific processing algorithms are ignored. The model concentrates on the content of data, its movement, and the rules by which data inputs are transformed into data outputs. Only after the user has validated all of the model components will a detailed implementation model be prepared.

Processes on a DFD are further defined by written *process specifications*. These are generally similar to the sample program specification shown earlier. Data flows and data stores are defined in terms of their data content (data elements) and their internal structure. These definitions are stored in a *data dictionary*. Implementation of specific information such as output format, file organization, and method of implementation for processes is purposely omitted. Sample data dictionary entries are shown in Figure 10.6.

```
Employee-File =
    1{ Social-Security-Number + Last-Name + First-
         Name + Middle-Initial + Street-Address +
         City + State + Zip-Code + Salary + Salary-
         Code + Filing-Status + Number-Of-Exemptions +
         Additional-State-WH + Additional-Federal-WH }

Time-Card =
    Social-Security-Number +
    1{ Date + Start-Time + End-Time + Hours-Worked } +
    Total-Hours-Worked + Supervisor-Initials
```

FIGURE 10.6 Sample data dictionary entries for a data flow and a file in the payroll system data flow diagram.

The analysis model is a basis for both describing user processing requirements and evaluating alternative methods of implementing a system to meet those requirements. Implementation options are reviewed after the analysis model has been created and validated. Various alternatives for automated or manual processing, form of input, form of output, file organization, and grouping of processes into application programs are considered. These high-level design decisions are the starting point for establishing a detailed system design (physical model) and a plan for implementation.

The implementation model is a description of the application requirements in terms of specific manual and automated procedures. This model can be represented as a system flowchart, as in classical systems development. Individual components of this model are further defined by textual descriptions that include program specifications, file specifications, and descriptions of input and output documents. The implementation model can also be represented with data flow diagrams and their associated textual descriptions. In this case, processes represent specific programs or manual procedures. Process specifications describe program logic (or algorithms), and data dictionary definitions describe physical as well as logical content.

The entire process of structured systems development is shown in Figure 10.7 as a series of translation steps. The user's description of his processing requirements is first translated into the analysis model. This is the process of *systems analysis*. The analysis model is then translated into an implementation model. This is the process of *systems design*. The implementation model is then translated into one or more high-level language programs. These programs are translated into executable machine instructions by program translators, and the entire translation process is thus completed.

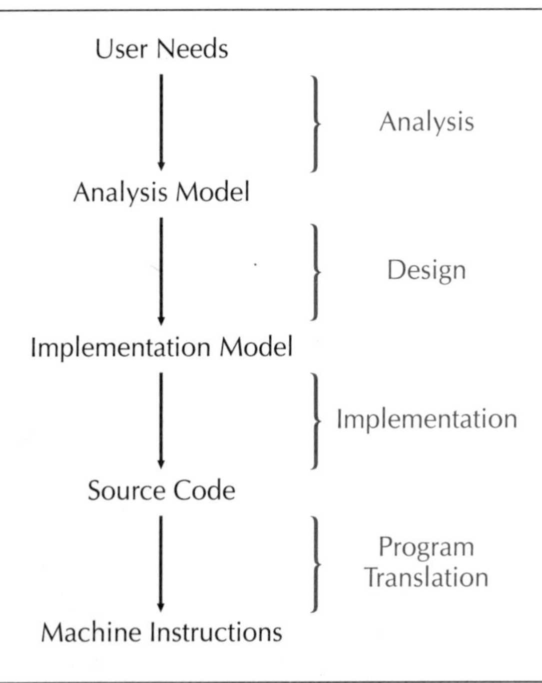

FIGURE 10.7 The translation steps used in the structured system development life cycle.

Rapid Prototyping

A common characteristic of the classical and structured development methodologies is their use of paper models. Be it through a data flow diagram or a system flowchart, each of these methodologies uses an abstract model to work out all of the details of the system *before* any software is actually written. The *rapid prototyping* methodology takes an opposite approach. Rather than concentrating on developing and validating an abstract model, the method focuses on developing, validating, and debugging a prototype system. Thus, the user has an actual working model to see and use.

The steps of rapid prototyping are shown in Figure 10.8. The first step is to develop an initial set of requirements. This can be accomplished through traditional analysis tools such as data flow diagrams or by less formal interaction with the user. Often, these initial requirements involve only a subset of the full system. Once initial requirements are identified, a prototype is created as quickly as possible. This necessitates the use of powerful application development software that allows application programs to be developed rapidly—in a matter of hours or days—and modified easily. Once created, the prototype is given to the user(s) for validation.

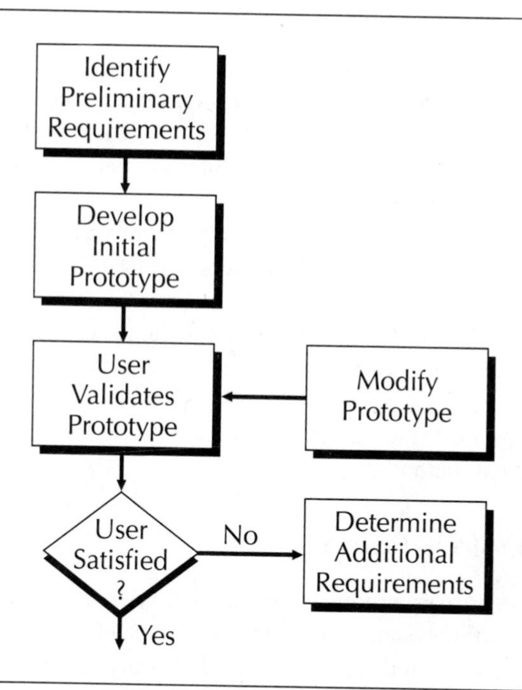

FIGURE 10.8 **The process of prototype development and evaluation.**

During validation, the user(s) identify deficiencies in the prototype. These might include missing functions, poor interface design, or any number of other problems. These problems are the basis for revisions to the prototype. Once revised, the prototype is again given to the user(s) for validation. These steps are repeated many times until all requirements have been incorporated into the prototype and all problems resolved.

For relatively small systems, rapid prototyping could be the development methodology exclusively used (i.e., a "stand-alone" or "full life-cycle" methodology). For larger systems, rapid prototyping is usually combined with various steps of the structured methodology. This combination occurs primarily in two ways:

- Use of a prototype to define user requirements
- Prototyping as an alternate or supplemental methodology for detailed design

When used to define user requirements, rapid prototyping is, in essence, an alternate form of systems analysis. It is especially useful in this role when users are not entirely sure of their needs or desires. In such situations, rapid prototyping provides a mechanism for adding concrete reality to abstract needs and desires. The prototype thus both defines and refines user requirements.

Prototyping can be used as a design and implementation methodology. In classical and structured development, most of the low-level details of an implementation program are specified before any software is written. Examples of these details include structure of menus, screen layouts, and order of data entry and validation. With rapid prototyping, design and development are combined through the iterative development, validation, and refinement of a prototype. The use of rapid prototyping in this manner is very common for the interactive portions of an information system.

It is often the case that the prototype cannot be used as the actual implementation of the system, perhaps due to software or hardware incompatibilities with existing applications, the prototyping tool's inability to implement systems that can deal with anticipated processing volume, or other factors. In this case, the prototype can be considered a tangible (as opposed to abstract paper) model of the users' processing requirements. These requirements can be converted to paper, if desired, and design and implementation can then proceed via structured development methods.

Methodology Comparisons

One of the primary criticisms of classical systems development is the use of a single model (the system flowchart) to represent both the analysis model and the implementation model. There is seldom only one possible way to implement a system. However, by stating requirements in terms of only a single implementation method, many design decisions are made very early in the development process, and other implementation possibilities are precluded from further consideration. One of the most important of these decisions is the choice of an automation boundary (i.e., the portions of the system that will be implemented on a computer). Other important design decisions include methods of data capture, input formats, file formats, output formats, and the allocation of specific processing tasks to specific software and hardware elements.

The lack of a clear distinction between the requirements and implementation statements makes it very difficult for users and system designers to discuss alternative methods of meeting users' processing requirements. No mechanism is provided to state requirements independently of very specific design decisions. This discourages thorough examination of implementation alternatives and their associated costs and benefits.

In contrast, structured systems development provides a very clear transition point between analysis and design. Although the user and the analyst might have preconceptions about implementation alternatives, the analysis process and the model used to state requirements do not convey (or require) any decisions regarding implementation. At the conclusion of the analysis phase, a clear model exists that states *what* the system must do, with no information or bias as to *how* the system will be implemented. This distinction between "what" and "how" is the essential difference between a logical model and a physical model.

In structured systems development, analysis is followed by a thorough examination of important (high-level) design decisions. This examination is made with *all* user requirements fully stated and understood. The costs, benefits, and tradeoffs associated with each design decision can thus be made based on a complete understanding of all user needs. This is especially important with large systems due to the large number of tradeoffs and dependencies among their various components.

A primary criticism of both classical and structured systems development arises from their use of abstract (paper) models. The development and validation of extensively detailed abstract models causes two primary problems: the long time periods required for model development and poor user feedback. The detailed development of analysis models consumes a substantial amount of time and human resources. User needs must be initially assessed and documented as an abstract model. The model must then be validated with the user and refined until all requirements have been correctly specified. In structured systems development, systems analysis typically consumes between one third and one half of the total project schedule.

Implementation models are not generally as time consuming to develop. In structured systems development, systems design consumes approximately 20 percent of the project schedule. In classical systems development, implementation models require substantially more time to develop, because they represent both analysis and design.

During the time allocated to analysis and design, the only feedback provided to the user is contained within paper models. Due to the relatively abstract nature of these models, understanding and validating them can be very difficult for users. Imagine the purchase of an automobile based on only pictures, a list of features, and detailed design and performance specifications. This is similar to what the user is asked to do when structured or classical systems development is used. A "test drive" is a far better method for eliciting user response and determining his or her level of satisfaction with the product.

The reliance of older methodologies on paper models is based on a common underlying assumption: that application software is very costly to develop and difficult to change once implemented. In the analysis phase, this assumption implies that all user requirements should be stated completely and precisely. This requires a substantial level of detail as well as a great deal of user involvement. The extra effort expended on analysis is supposed to ensure that the system that is designed and implemented is exactly what the user wants and needs.

In the design phase, costly development and maintenance implies that a substantial amount of time should be spent working out all possible technical and implementation problems ahead of time. The internal aspects of application programs should be specified in great detail so that the design can be evaluated in terms of ease of initial programming and future changes. With respect to both analysis and design, resources are added to planning processes so as to reduce resources consumed during implementation.

Until relatively recently, the assumption of costly software development and change was well founded in fact. The task of writing, testing, and debugging application programs was time consuming and costly, even when the best of available tools were used. If development is costly, it makes good economic sense to work out the details of the system to the greatest extent possible before programming begins. This results in savings during implementation that more than offset the cost of developing and validating paper models.

Although application software is still costly to develop and change, a number of factors have altered the tradeoff between these and other costs. The most important of these factors is the availability of improved tools for programming and other application development tasks. Newer tools for application program development, such as fourth-generation languages, database management systems, and report generators, have drastically reduced the time needed for programming, testing, and debugging. Costs of program development have been reduced as a result.

Thus, the extra resources used for extensive planning in the classical and structured methodologies no longer lead to substantial savings during implementation, because programming is now cheaper. This is not to say that analysis and design are no longer needed or that programming is cheap. It is simply that the economic balance between these activities and programming has shifted.

Systems Software Support for Application Development

Recall from Chapter 3 that application development software was classified into the following categories:

- Program translators
- Program development tools
- System development tools
- Data manipulation tools
- Input/output tools

With the exception of system development tools, each of these classes of software directly supports the development of application programs. These tools support the creation, testing, and translation of application programs written in high-level languages. System development tools are a relatively new class of software that has been developed to support earlier phases of application development (e.g., analysis and design).

The declining cost of computer hardware and the increasing cost of computer-related labor—analysts, designers, and programmers—has led to a shift in the use of hardware resources. This trend has been accelerated even further by users' demands for increasingly more powerful and sophisticated applications. It is a classic economic tradeoff between labor and capital resources.

In the early days of computers, hardware was so expensive that its use could be justified only for high-value applications. The use of computers to directly support the development of applications was extremely limited. It made more economic sense to use large amounts of labor for application development, because labor was relatively cheap compared to computer hardware. As hardware costs decreased, this economic balance shifted. It was this shift that led to the introduction of systems software designed to support the development of applications.

Early operating systems, assemblers, and programming language compilers are examples of this shift. They are all methods by which computing hardware can be used to directly support the development (as opposed to the execution) of application programs. As the economic balance continued to shift, the number, power, and complexity of these tools continued to grow.

The proliferation of application development tools today is simply a continuation of this trend. Database management systems, fourth-generation programming languages, and CASE tools are all examples of relatively recent tools for application development. Each of them supports application development by increasing the consumption of hardware resources while reducing the consumption of labor resources. If the decline in hardware cost relative to labor cost continues, the proliferation and extension of these tools will also continue.

The role of various tools in the application development process is summarized in Figure 10.9. Most of these tools are discussed in detail in the remainder of this chapter.

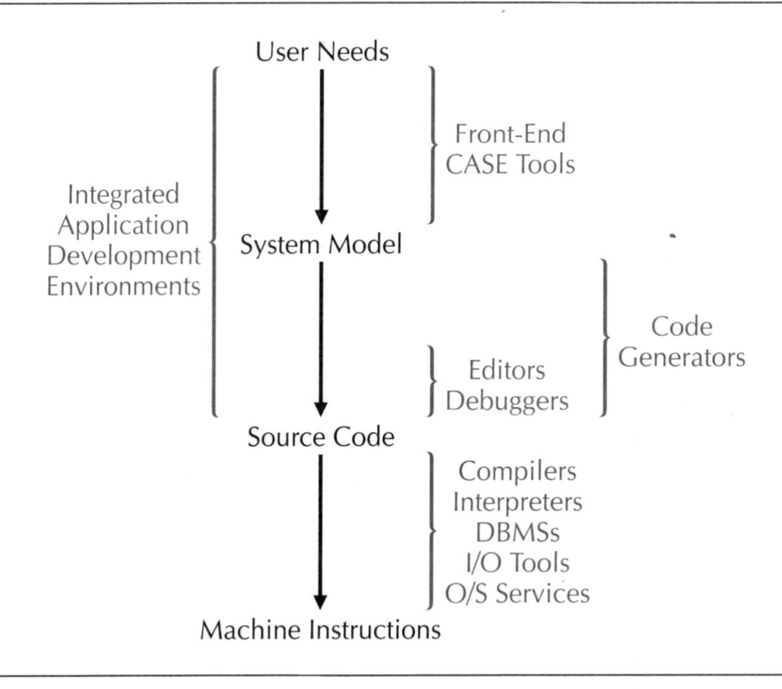

FIGURE 10.9 **The application development translation process and the role of various development tools within that process.**

PROGRAMMING LANGUAGES

The earliest shift of computer hardware resources to a direct support role in application development came with the development of programming languages. Because any programming language other than machine language must be translated, the use of a programming language requires the use of an automated program translator. Any hardware resources consumed by a program translator represent capital resources shifted from executing existing applications to developing new applications.

The history of programming language development has been driven largely by a desire to make application programs easier and easier to develop. To realize this goal, it is necessary to substitute the use of systems software for labor in the translation from implementation model to executable code. The exact distribution of effort (and thus resources) between the programmer and the program translator is dependent on characteristics of the programming language.

Instruction Explosion

Recall from Chapter 3 that statements in a high-level programming language can correspond to many machine instructions. Depending on the type of statement and the programming language, a source code statement can be implemented by anywhere from one to thousands of machine instructions. This one-to-many relationship between high-level programming language statements and the machine instructions needed to implement them is called *instruction explosion*.

Programming languages have various degrees of instruction explosion. An older programming language such as FORTRAN will generally have a relatively low degree of instruction explosion. This degree of explosion can be described as a ratio of source language statements to machine instructions. For example, if 1,000 machine instructions result from translating a 20-statement FORTRAN program, the degree of instruction explosion can be stated as the ratio 50:1. However, such a ratio is only an average. Within the same high-level programming language, various types of statements can have vastly different degrees of instruction explosion. In general, statements that describe mathematical computation typically have relatively low instruction explosion, whereas statements that describe I/O operations typically have relatively high instruction explosion.

The degree of instruction explosion in a programming language has a direct effect on the distribution of translation effort between the human programmer and the compiler or interpreter used to translate the program. The amount of effort required to perform the translation is essentially constant. For a given application processing requirement and a particular machine instruction set, a fixed amount of translation effort is required. However, programming languages distribute this effort differently between human programmers and automated program translators.

Figure 10.10 shows two equivalent programs in a third- and fourth-generation programming language. Note that the number of program statements is much larger in the third-generation language, thus implying a greater degree of effort on the part of a human programmer. However, the translation effort of a compiler or interpreter would be substantially greater for the SQL program compared to the C program. Note that the C program states an explicit procedure for searching the files and matching records based on account number. The SQL program merely states the record-matching criteria (CUSTOMER.ACCT_NUM = TRANSACTION.ACCT_NUM), thus leaving it up to the compiler or interpreter to determine the correct procedure.

3GL Example - C

```
balance_report() {
FILE *cust_file,*trans_file;
int  status,acct_num,a_num,balance,amount;
char *name;
cust_file=fopen("customer","r");
trans_file=fopen("transaction","r");
status=scanf(cust_file,"%d%s%d\n",&acct_num,name,&balance);
while (status != EOF) {
   status=scanf(trans_file,"%d%d\n",&a_num,&amount);
   while (status != EOF) {
      if (acct_num == a_num) {
         balance+=amount;
      }
      status=scanf(trans_file,"%d%d\n",&a_num,&amount);
   }
   printf("%d %s %d",acct_num,name,balance);
   trans_file=freopen("transaction","r");
   status=scanf(cust_file,"%d%s%d\n",&acct_num,name,&balance);
}
close(trans_file);
close(cust_file);
exit(0);
} /* end balance_report */
```

4GL Example - Structured Query Language (SQL)

```
open database banking;
select     customer.acct_num,customer.name,balance+sum(transaction.amount)
from       customer,transaction
where      customer.acct_num=transaction.acct_num
group by   customer.acct_num,customer.name;
```

FIGURE 10.10 Equivalent programs in a fourth-generation programming language (SQL) and a third-generation programming language (C).

Programming languages with a high degree of instruction explosion allow processing instructions to be stated in relatively few source language statements. Thus, the amount of (translation) effort required to produce the source code program is reduced. Such a language is said to be powerful, because processing requirements can be described in few words or statements. However, the amount of effort expended by the compiler or interpreter to produce machine instructions from such a language is increased. This division of labor is represented graphically in Figure 10.11.

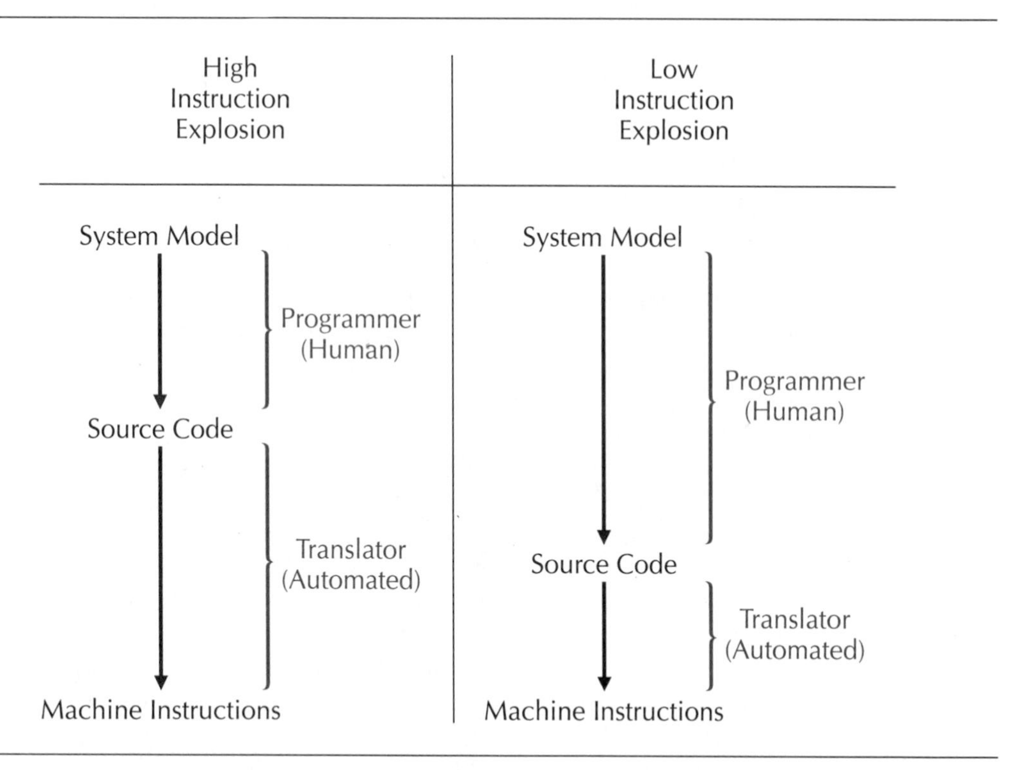

FIGURE 10.11 The distribution of translation effort/intelligence when using programming languages with high and low degrees of instruction explosion.

If a programming language has a low degree of instruction explosion, a considerable amount of effort is required to program in that language. This is because it takes a relatively large number of source language statements to state processing instructions. However, the amount of effort required to translate that program into machine instructions is relatively low. The human programmer assumes more responsibility for the translation effort, and the work required by the program translator is correspondingly reduced. This situation is also represented in Figure 10.11.

High-Level Programming Languages

High-level programming languages are used to develop both application and systems software. Rather then requiring the programmer to specify primitive machine actions, these languages provide the programmer with a set of language constructs or statements that each implement a sequence of primitive machine actions. As a result, a programmer needs to write substantially fewer program instructions than would be necessary if the same function were programmed using assembly or machine language. High-level languages also insulate the programmer from much of the physical detail of the hardware, especially concerning mass-storage and I/O devices.

Most of the high-level languages that have been discussed thus far are termed *third-generation languages* or *3GLs*. The majority of these languages (COBOL, FORTRAN, BASIC, C, Pascal, and so on) were developed between the late 1950s and the early 1970s. The limitations of computer hardware capability during that time period are reflected in the basic capabilities (or lack thereof) in these languages.

COBOL, for example, was developed in the late 1950s. At that time, secondary storage devices were quite primitive and expensive. The computing power and technology to support online interactive systems was virtually nonexistent. Most applications consisted of batch programs that utilized cards or tapes for input and produced punched cards, tapes, or printed output.

COBOL's developers were actually quite forward-thinking in many respects, especially regarding file manipulation. They foresaw the increased use of large disk files in information-processing applications, and embedded many capabilities in the language to manipulate these files. The ability to define and manipulate complex file structures (e.g., indexed files) was built into the programming language. They did not, however, endow the language with capabilities for database processing, full-screen input/output, or interactive execution of programs. Although later adaptations did address some of these areas, they were far from complete.

Shortcomings of Third-Generation Languages

In general, 3GLs suffer from design limitations with respect to modern hardware capabilities and modern requirements in information-processing applications. Two areas are of particular concern: mass-storage management (or file manipulation) and interactive I/O capabilities. The traditional 3GLs have failed to evolve sufficiently in these areas to meet the needs of modern information-processing applications.

Advanced mass-storage management. Until the 1970s, the majority of applications were designed as single programs or as a relatively small set of integrated programs. These systems operated on a common set of files, and the responsibility for those files was localized to the department or organizational unit that owned the application.

Applications and their data were thus viewed as stand-alone entities subject primarily to local control. Although this state of affairs was not problematic when applications were few and localized, it became so as the number of applications and their scope increased. For example, consider data and applications concerned with employees. Thirty years ago, the only automated applications that processed employee data were payroll systems. These systems were typically automated due to their high transaction volume and their connection to general accounting functions.

However, other systems did exist that were concerned with employees and data about them. Examples include recruitment and promotion, training, and insurance. These systems were not generally automated. It was thus feasible (and efficient) to maintain entirely separate sets of data within those systems and not to integrate that data with the automated payroll system.

But as computer technology became cheaper, more and more of these systems were automated. At this point, redundancy in data storage among various systems became a problem. It was quite possible for similar employee data such as name, address, and salary to be stored in the files of many different automated systems. Ensuring consistency of multiple stores of data and providing access to them across organizational and application boundaries was a difficult problem. *Database management systems (DBMSs)* were developed partly to address these issues.

The primary purpose of a database management system is to provide a common (integrated) repository for an organization's data. A DBMS allows data to be shared by many applications using a common interface. It also allows data to be more effectively managed and controlled. As an abstract concept, DBMSs are an excellent idea. To be useful, they must be implemented as a set of tools that can be used to develop application programs. At the time DBMSs were introduced, none of the available programming languages had the capability to manage and interact with a database. Their capabilities were instead directed toward individual files.

Although database manipulation functions could have been implemented by extending the capabilities of programming languages, they were typically implemented as separate (add-on) software components. These were usually composed of general-purpose database definition facilities, data manipulation facilities, and interactive query facilities. The data manipulation facilities were typically implemented as libraries of subroutines that could be linked into application programs written in third-generation languages. The other two components were generally implemented as stand-alone programs or systems. The entire set of facilities was developed and marketed as a package. Examples of these packages include DB2 and IDMS.

There are advantages and disadvantages to extending programming language capabilities with add-on software components. One of the primary advantages is that the add-on software can be used with many different programming languages. With DBMSs, a single set of data manipulation library routines can be used with the compilers for many different languages. Variations among programming languages in data structure definitions must be accounted for, and a mechanism must be provided for moving data between program data structures and the add-on software package.

The primary disadvantage of using add-on software is a lack of standardization, which hampers the portability of application programs. Although widely accepted standards such as ANSI exist for most 3GLs, there are few standards for database manipulation functions. If these functions were implemented within programming languages, they could be incorporated into the standards for those languages. If they are implemented as add-on software components, their developers are free to set their own standards.

A program written entirely in a standardized language such as COBOL is guaranteed to run on many different computers and operating systems. If that program uses an add-on database manipulation library, portability is guaranteed only to software and hardware environments supported by the DBMS vendor.

Advanced input/output. Programming language limitations similar to those described for database manipulation also exist for advanced input and output functions. Older 3GLs such as FORTRAN were designed to accept batch input as streams or blocks of individual characters from devices such as tape drives and card readers. Similarly, these languages were designed to use character-based output to printers and other devices.

Newer third-generation languages such as BASIC and C were designed with some facilities for interactive I/O using terminals. However, these languages still tend to process both input and output as streams or blocks of characters. Thus, they interact with a keyboard in much the same manner as a tape drive, and interact with a video display unit in much the same fashion as a printer. These similarities are also exploited by older 3GLs.

This approach to interactive I/O ignores some of the unique capabilities of modern I/O devices. Full-screen I/O, for example, is poorly handled, if at all. The concept of cursor motion (e.g., using a Tab key to move from one input field to the next) is very difficult to implement with character-based I/O. Similarly, special display attributes such as inverse video and half intensity require that character sequences be assigned to turn these various display modes on and off. Special-purpose keys such as function keys and cursor movement keys also present this problem. The complexity of interactive I/O has been magnified in recent years by the introduction of graphical I/O devices such as graphic display terminals, pointing devices, and digitizing tablets.

The approach to addressing the I/O limitations of older programming languages has been much the same as the approach to implementing advanced data manipulation. Rather than extend the I/O capabilities of these programming languages (or develop entirely new languages), add-on software components have been designed to add these capabilities to existing languages. Examples of these include ADDS-ON-LINE and CICS. These packages typically consist of a stand-alone screen definition facility and a library of input/output subroutines that can be linked into object code. They have basically the same benefits and problems as stand-alone DBMSs.

Fourth-Generation Application Development

As these add-on database and I/O tools grew in popularity, the complexity of application programming increased substantially. Some of this complexity was unavoidable due to the additional demands placed on the application programs. Some of it, however, was a direct result of the number of interfaces required to develop and execute application programs as well as problems of portability and compatibility.

Application programs that interact with add-on database or I/O management systems typically contain specialized code for database and I/O manipulation. This specialized code requires special translators (or preprocessors) and complicated procedures for generating executable code. Portability of applications suffers because a new hardware/software environment must provide both a compiler for the 3GL and the complete set of supporting DBMS and I/O tools.

Partly to address these problems, a new class of application software emerged. The new programming languages are generally referred to as *fourth-generation languages (4GLs)*. Integrated sets of application development tools based on these languages are referred to as *fourth-generation development environments*. Examples of fourth-generation development environments include dBASE IV, Natural, Paradox, and Oracle. These languages and environments address problems of complexity and portability by integrating basic programming language constructs, advanced database capabilities, and advanced I/O capabilities into a single integrated package.

In these packages, a single interpreter and/or compiler that addresses all types of processing is provided. The various subcomponents are designed to work together. This approach has greatly reduced the complexity of application development.

The combination of all functions into a single integrated software package has improved portability because only one set of components (from a single vendor) must be available in a new software/hardware environment.

Fourth-generation languages also simplify application programming by providing a higher degree of instruction explosion. This higher instruction explosion is partly due to the inclusion of more powerful I/O and data manipulation tools and partly due to more powerful statements for representing other types of processing. Third-generation languages can require anywhere from 10 to 1,000 times as many program statements as 4GLs for equivalent processing tasks. However, this reduction in program size (and thus programming labor) comes at the expense of much more complex (and resource-hungry) program translators and somewhat less efficient executable code.

The development of 4GLs is also closely tied to the emergence of rapid prototyping as a viable application development methodology. Smaller programs are not only faster and easier to write, they are faster and easier to change. The rapid development, evaluation, and refinement of an application prototype is simply not possible with 3GLs. With 4GLs, this process can be performed in a matter of minutes, hours, or days, as opposed to days, weeks, or months with 3GLs.

PROGRAM TRANSLATION

Regardless of what generation of programming language is used, all higher level language programs must be translated into executable code. The functions of assemblers, compilers, and interpreters were initially described in Chapter 3. Compilers, interpreters, and some additional program translation tools are further described below.

Compilation

A *compiler* is a program that translates a high-level language program. Each statement of the program is read and appropriate translations are performed. The original program is referred to as *source code*; the translated program is referred to as *object code*.

Source code statements are read into the compiler one at a time, usually from a file. The compiler can take various actions, depending on the type of statement encountered. Source code statements (or statement fragments) can be divided into three classes:

- Data declarations and references
- Data operations
- Control structures

As *data declarations* are encountered in source code, memory locations are set aside to store the declared data. The amount of memory allocated will depend on the type of the data (e.g., integer, real, character array, and so on). The compiler builds an internal table to keep track of the data names, types, and assigned memory addresses.

For example, consider the following portion of a COBOL data division:

```
77  A        PIC 9(6)
77  B        PIC 9(6)
77  C        PIC 9(6)
```

This example declares the existence of three working storage variables (named A, B, and C), each of which is an integer of 6 significant (decimal) digits. On encountering these declarations, the compiler creates an entry for each of these variables in its internal *symbol table*. These entries include the variable name, the fact that the name refers to a data element, the type of data element, and its memory location. The compiler assigns memory locations based on data type and the number of bytes needed to store the variable. Sample entries are shown in Table 10.1.

TABLE 10.1 Sample symbol table entries for data declarations.

Name	Type	Length	Address
A	Integer	4	1000
B	Integer	4	1004
C	Integer	4	1008

As *data operations* are encountered, they are translated into the sequence of machine instructions necessary to implement those operations. In general, these instruction sequences will include primitive data manipulation instructions as well as any necessary data movement instructions. When data must be moved to or from memory, the compiler finds the name of the data item in the symbol table and uses the corresponding address for data movement instructions.

For example, consider the following COBOL source code statement:

```
ADD A TO B GIVING C.
```

Assume that data items A, B, and C are as previously declared and that the CPU is capable of executing computation instructions only on data stored in registers. The compiler translates the above statement into the following sequence of machine instructions:

```
MOV 1000 R1        ; move A to register 1
```

```
MOV 1004 R2        ; move B to register 2
IADD R1 R2 R3      ; add the contents of registers 1 and 2 and
                   ; store the result in register 3
MOV R3 1008        ; move the contents of register 3 to C
```

Various CPU capabilities might imply a different set of primitive instructions. For example, a machine with an instruction set that allows computation directly on data elements in memory would not require the MOV instructions in the above example. The following single instruction would suffice:

```
IADD 1000 1004 1008
```

Other complexities might also be encountered. For example, assume that A and B are integers, but that C is a floating-point number. In this case, the addition would proceed as before, but the contents of register 3 would have to be converted from integer to floating-point representation before being moved to memory.

Control structures are language statements that control the execution of other language statements. Typical examples of these include branches and loops. Consider the following example:

```
IF A IS EQUAL TO 5 THEN ADD A TO B GIVING C.
```

The IF statement is a control structure that governs whether or not the subsequent computation is performed. It can be implemented as a conditional branch, as follows:

```
0100    MOV 100C R2       ; move the constant '5' to register 2
0104    MOV 1000 R1       ; move A to register 1
0108    XOR R1 R2         ; compare the contents of registers 1 and 2
010C    CJMP 011C         ; branch if XOR produced zero result
0110    MOV 1004 R2       ; move B to register 2
0114    IADD R1 R2 R3     ; add the contents of registers 1 and 2 and
                          ; store the result in register 3
0118    MOV R3 1008       ; move the contents of register 3 to C
011C                      ; next instruction following the IF statement
```

In this example, the memory location of each CPU instruction is listed in the left column. The constant 5 and the value of A are moved from memory to registers and an XOR instruction is executed to compare them. The result of the comparison is stored in a status bit in the program status word. This bit is tested by the CPU when executing the CJMP instruction, and a branch occurs if the status bit indicates a non-zero result of the XOR. As before, various machine capabilities (instructions sets, etc.) would result in a different sequence of instructions.

Support Libraries and Linking

From the previous examples, it might appear that object code consists entirely of executable machine instructions and data. This is not the case. Certain language statements and contents are converted directly into executable code. These include certain data declarations, most control structures, and many computational functions.

Other language statements such as file manipulation, I/O operations, and floating-point calculations are typically handled differently. The usual method for translating these types of language statements is via support libraries. These support libraries are of two types: those provided with the compiler (for instance, the *compiler library*) and those provided by the operating system. These libraries consist of previously translated programs that perform specific types of functions. Thus, a library is a collection of object code for specific procedures and is indexed by the names of the procedures. A request to execute one of these functions is referred to as a *library call.*

Library calls. The use of libraries and library calls is not a necessary feature of compiler design and operation. It is possible to construct a compiler that directly translates all programming language statements into executable code. The primary advantage of using libraries is to allow simpler compilers and to provide a measure of portability and flexibility.

As an example, consider the use of floating-point computations in a source language program. Although many application programs can be expected to utilize these functions, not all CPU architectures support them directly. Particularly in smaller computers, such as computers that use the Intel 8086, 80286, or 80386 microprocessor, CPU instructions are often not provided for floating-point operations. These operations must instead be implemented as complicated sequences of integer operations on the whole and fractional portions of real numbers.

It is common to include support for floating-point operations within a separate compiler library and to translate program statements for floating-point computation into calls to these library routines. This approach increases compiler simplicity and flexibility for the following reasons:

- The complex translation of floating-point calculation into equivalent integer operations is moved out of the compiler program and into the library routine. The size of the compiler is thus reduced.
- Alternate libraries can be provided for the same functions. In floating-point operations performed with integer calculations, there is often a tradeoff between computational accuracy and speed. Two separate libraries can be provided: one that uses slow but accurate algorithms and another that uses fast but less accurate algorithms. The programmer can choose the library that best suits his or her needs.

- Alternate libraries can be provided for a variety of CPU architectures. For example, one library can be provided that uses only integer operations; another can be provided to take advantage of a floating-point coprocessor such as the Intel 80387. Multiple libraries could be provided for the use of different coprocessors (e.g., an Intel or a Weitek). The programmer would select the library that corresponds to the CPU architecture of the computer system. This approach allows compiler writers to write a single compiler that allows for some variation in hardware architecture.

When a program statement is encountered that requires the use of a library routine, the compiler inserts a call to that routine in the object code. This call includes the name of the library routine as well as the addresses of any data that must be passed between the program and the routine. For example, consider the earlier COBOL example:

```
ADD A TO B GIVING C.
```

Assume that A, B, and C are all floating-point numbers and that the compiler supports floating-point calculations via a library. A call such as the following would be inserted into the object code:

```
CALL ADD_FP(1000,1004,1008)
```

The name ADD_FP refers to a library routine, and the numbers between the parentheses are the memory addresses of the operands and the result. During program execution, control would be passed to the instructions of the library routine. The library would load data from 1000 and 1004 and store the result of the operation to 1008.

Linking. Note, however, that the above statement (CALL ADD_FP) is not a CPU instruction. It is a reference to a set of executable code that has been previously compiled and stored in a library for later use. It must, therefore, be replaced by the corresponding executable code before the application can be executed. This replacement is not performed by the compiler. It is instead performed by a program called a *linker* or *link editor*.

The function of the linker is to combine separate sets of object code into an integrated set of executable code. Each set of object code contains executable code and can contain calls to external (library) routines. The call to the library routine ADD_FP shown above is an example of such a call. In the terminology of linking, this type of call is known as an *unresolved reference*. It is a request for a procedure or for data by name, where the name is unknown within that set of object code.

The linker performs its function by searching object code for unresolved references. As they are found, it searches the index of one or more support libraries for the names of those references. If it finds those names in a library, it extracts the corresponding executable code and/or data from the library and inserts (links) it to the object code. This process is called resolving the reference. The practical result is that the library call is replaced by its corresponding executable code and/or data.

If all of the unresolved references are resolved by the linker, the result is a program consisting entirely of executable code and data. This program can then be loaded into memory and executed. The process of developing a program (executable code) via compilation and linking is shown in Figure 10.12.

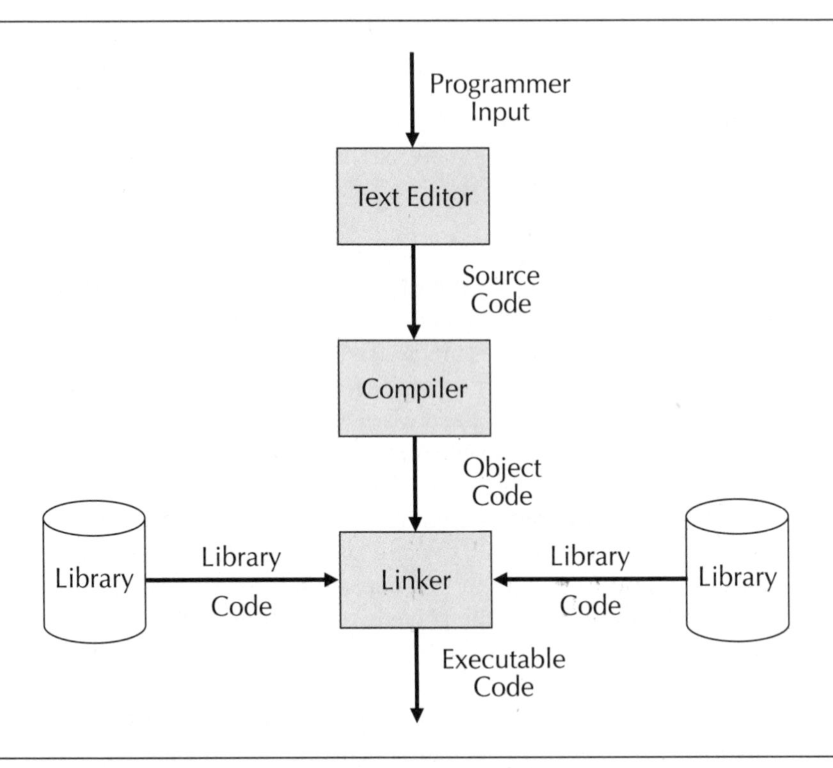

FIGURE 10.12 **The process of developing executable code with a compiler and a linker.**

Operating system library calls. Recall from earlier discussions of operating systems that the service layer consists of a set of routines (service calls) that can be used by application programs. These routines are provided by the operating system as an interface to lower level hardware operations. The majority of these service routines provide basic file manipulation and input/output capabilities. Application programs that need these services request access to them from the operating system. The operating system satisfies these requests and returns results (if any) to the application program for further processing.

The compiler "knows" the set of service routines that are available in the operating system. It also "knows" which statements in the programming language require access to operating system services and which particular service routines are required. When such a statement is encountered within source code, it is translated into a call to the appropriate service routine(s). Whether or not this translation results in executable code depends on the particular operating system, the mechanism by which an application program requests system services, and the compiler. Service requests are generally made via either interrupts or library calls.

When services are requested via library calls, the compiler inserts data into the object code to represent those calls. This data consists of the name of the service routine and the names or addresses of any data elements that are passed between the service routine and the application program. For example, a compiler might insert the following statement into the object code to represent a request to the operating system to open a file:

```
open(0F00,0F50)
```

In this example, the name OPEN refers to the operating system service routine that opens files. The first parameter of the call is the address of the first byte of the filename; the second parameter is the address of an integer used to store a status code generated by the service routine. When the service routine is actually executed, it will use these addresses to find the name and location of the file to open and to store an integer representing the status (success or failure) of the operation.

If O/S service requests are made via interrupts, the translation will usually produce executable code. Typically, the service routine will be called by issuing an interrupt of the appropriate number (or code). The compiler will produce instructions to generate the appropriate interrupt, either by an explicit interrupt (INT) instruction or by loading the interrupt number from memory to the interrupt register.

In interrupt-driven operating systems such as MS-DOS, registers will normally be used to pass data values between the application program and the service routine. For example, when a request is made to open a file, the name and location of the file must be passed to the service routine, and error codes (if any) must be returned to the application program. A service routine to open files might assume that register R1 contains a pointer to the filename and location as stored in memory. Status codes such as SUCCESS, FILE NOT FOUND, and DISK ERROR might be placed in register R2 by the service routine.

In addition to generating code for the proper interrupt, the compiler must also generate code to manipulate the registers used to communicate with the service routine. In the above example, the compiler must generate instructions to load the address of the filename and location into R1 and to process the status codes returned in R2. More complex service calls might require more data and, thus, more complex operations involving registers to support the transfer of that data.

Interpreters

Although both are methods of translating source code into object code, *interpretation* is a fundamentally different process than compilation. When a program is compiled, each statement of the source code is read and converted into object code. The program cannot be executed until all of the statements have been translated and the object code has been linked. In other words, compilation and linking operate on an *entire* program at once.

In contrast, interpretation converts source code into executable code one statement at a time. Each statement of the source program is read, translated into executable code, and immediately executed before the next statement is read. There is no explicit linking process as in compilation. In essence, one program (the interpreter) is executing another (the source code) one statement at a time.

The differences between interpretation and compilation result in both advantages and disadvantages. The primary advantage of interpretation over compilation is a superior ability to detect and report execution errors. The primary disadvantage is a substantial increase in the consumption of memory and CPU resources.

Resource utilization. The differences in resource utilization arise due to differences in how and when source code statements are translated and how and when library references are resolved. Each software component used for compilation resides in memory for only a limited amount of time. The text editor resides in memory while the programmer is preparing source code. When source code preparation is finished the text editor is removed from memory.

The compiler resides in memory only during the compilation phase. Once compilation has finished, the compiler is terminated. The linker resides in memory during link editing and is then terminated. When the executable code is finally loaded and run, all of the software components used to create it are no longer in memory, because they are no longer needed. Thus, the memory requirements of the application program consist only of the memory needed by the executable code.

In contrast, the interpreter must reside in memory during the entire execution of the program. This is because each statement must be translated and executed individually. Interpreters, like compilers, are fairly large programs, and this requirement substantially raises the memory requirements at execution time. Essentially, there are two programs in memory: the interpreter itself and the portion of the application program currently being executed.

Another increase in hardware resource utilization results from multiple translations. Consider the following program loop:

```
open input_file
read A,B
while not eof(input_file) do
    C=A+B
    print A,B,C
    read A,B
end while
close input_file
```

The three statements in the WHILE loop will be executed for as long as the file contains additional records. If the input file contained 1,000 records, these statements would each be executed 1,000 times.

If this program were compiled, those statements would each be translated into executable code exactly once. If this same program is interpreted, those statements are translated 1,000 times. Thus, the interpreter unnecessarily translates those statements 999 times. These extra translation operations waste CPU time and significantly increase program execution time. In reality, most modern interpreters incorporate procedures to eliminate some of this extra work by storing translated statements in memory. Although this eliminates a great deal of the inefficiency, it rarely achieves the level of efficiency of a compiled program.

Inefficiency also arises from executing an application program multiple times. Consider an application program that is run every day (e.g., a batch daily reporting program). If the program is used for a year without changes, execution via an interpreter will translate the entire program 365 times. Compilation and linking will reduce the number of translations to one, thus saving the computer resources necessary to perform the other 364 translations.

An additional inefficiency arises as a result of library calls. Like a compiler, an interpreter uses one or more libraries (referred to collectively as the interpreter *run-time library*). Because the program is being executed and translated one statement at a time, the interpreter must have the *entire* contents of this library available during program execution. This is because the interpreter hasn't "seen" the entire program yet and, therefore, doesn't know which library routines will be needed during its execution.

The run-time library will generally cause increases in either memory or I/O usage. If the run-time library is held in memory, some memory is being wasted to store library routines that won't be used by the program. In contrast, a linker links only those library routines that are actually used into the program, so only those routines actually used are in memory while the program is executing.

An alternative is to store the run-time library in secondary storage and load only those portions that are needed as they are needed. This greatly reduces memory requirements because a typical application program uses only a small portion of the available library routines. However, the overhead of loading and unloading library routines from secondary storage slows program execution drastically. This is especially true if many routines are used or if some routines are used many times.

The resources used to translate and execute a program under compilation and interpretation are summarized in Table 10.2. In general, interpretation requires two to five times as much memory as executing a compiled program. The execution speed of an interpreted program is typically five to 10 times slower than that of a compiled program. This resource inefficiency is incurred each time the program is executed.

TABLE 10.2 Summary of the memory and CPU resources consumed during the execution of an application program.

Resource	Interpretation	Compilation
Memory Contents (during execution):		
Interpreter or Compiler	Yes	No
Source Code	Partial	No
Executable Code	Partial	Yes
Library Routines	All Routines	Only Those Actually Used
CPU Instructions (during execution):		
Translation Operations	Yes	No
Library Linking	Yes	No
Application Program	Yes	Yes

Error detection and correction. Although interpreted programs utilize more CPU and memory resources than do their compiled equivalents, they are substantially easier to develop and debug. This ease arises from the nature of the interpretation process and the simultaneous processing of source and executable code. The process of developing, testing, and debugging source code with both compilers and interpreters is shown in Figure 10.13. Various aspects of these processes are described below.

Once a program has been compiled and linked, the relationship between source and executable code is difficult to establish. Variable names are replaced by memory addresses; symbolic source code instructions are replaced by CPU instructions. If a compiled program generates an error while executing, the error can be reported only in terms of the memory locations of CPU instructions and data items. For example, a divide-by-zero error can be reported to the user only in terms of the location of the particular divide instruction that was being executed when the error occurred.

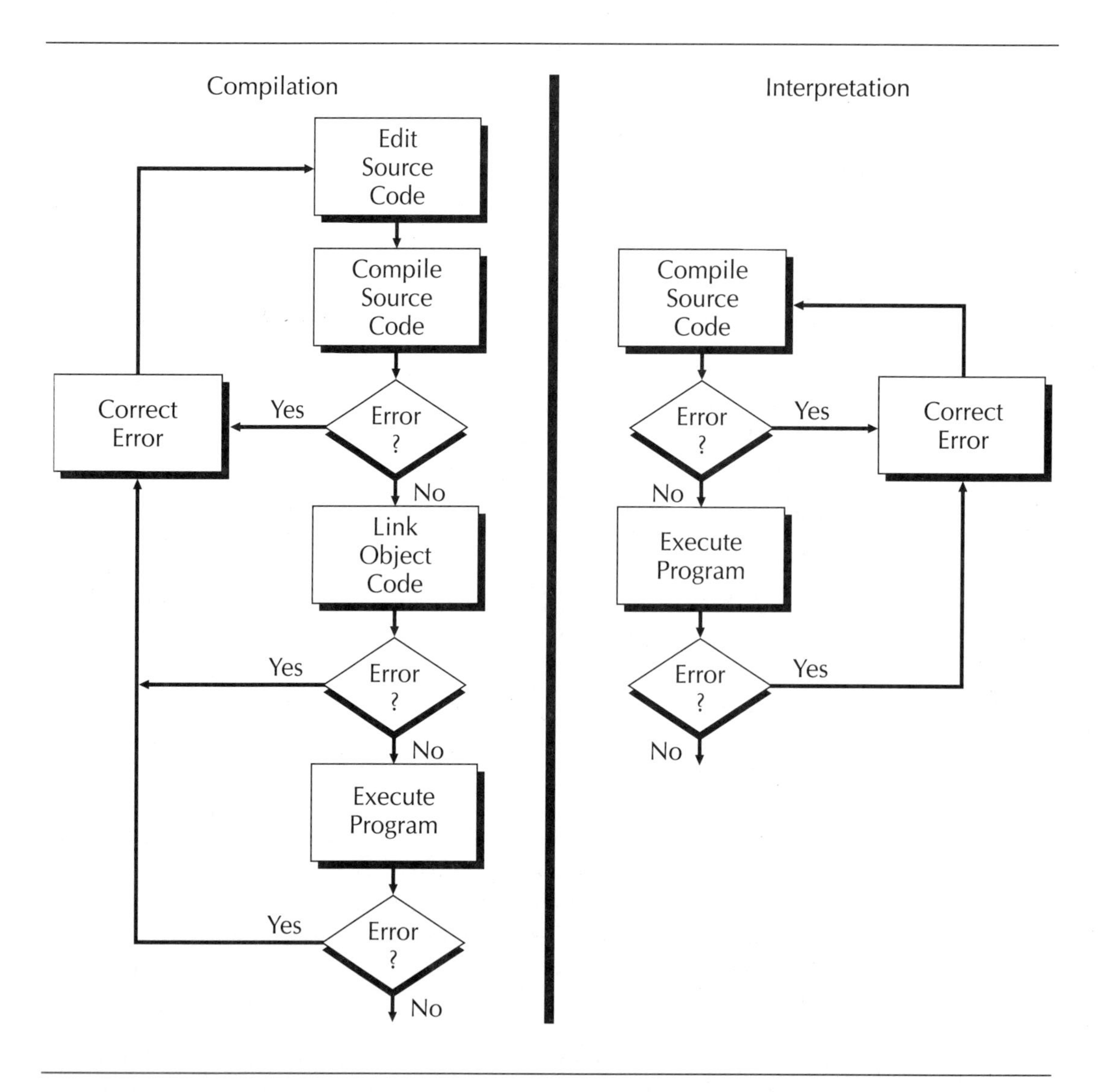

FIGURE 10.13 The process of source code development, testing, and modification.

It is up to the user or programmer to determine which statement in the original source code corresponds to the CPU instruction that generated the error. Similarly, if the address of a data item is reported as part of an error message, it is up to the programmer to determine to which data item (variable) in the source code that address refers. Determining this correspondence is not an easy task. As part of the linking process, a *memory map*, or *link map*, is sometimes produced. This map shows the correspondence between program and library routines and the memory locations in which those routines are stored. Using such a map, it is possible for a programmer to trace an error message containing memory addresses to the corresponding program or library routine. However, tracing those addresses to particular source code statements or data items requires a very detailed memory map and a thorough understanding of machine code and of the compiling and linking processes. Thus, although it is possible to trace run-time errors to particular source code statements, it is a difficult task.

When executing an interpreted program, the correspondence between source code statements and machine instructions is much easier to determine. Because each source code statement is translated and immediately executed, any run-time error must have been caused by the most recently translated source code statement. Thus, the interpreter is able to report a run-time error in terms of the actual source code (e.g., source code statements by line number and data items by variable name), not just its executable equivalent.

Most interpreters include integrated facilities for editing source code. This is in contrast to program development with compilers, where a separate text editor is used to develop and modify source code. Because the interpreter allows source code modification, it is possible to correct errors as they are detected and rerun a program immediately. With a compiler, the programmer must first use the text editor to change the source code, then recompile, relink, and rerun the program. Thus, interpreters make it easier for programmers to modify source code to correct detected errors.

This ease of modification applies not only to run-time errors but also to translation (e.g., syntax) errors. When a program is compiled, such errors are normally detected by the compiler, and a printed error message is generated, indicating the line number of the incorrect source code. As before, the programmer must first change the program with a text editor and then recompile. With an interpreter, translation errors are detected while the program is being executed. These errors can be corrected immediately with the built-in editor and execution can resume quickly.

A final advantage of interpreters for source code development and debugging arises from the way in which variable names and source code lines are tracked. Recall that a compiler builds a symbol table that is used to track references to data and procedure names. As references to those names are encountered in the source code, they are looked up in the symbol table and corresponding addresses are inserted into the object code.

An interpreter also builds and maintains a symbol table, but this symbol table is held in memory during program execution. This allows the programmer a great deal of flexibility in testing and debugging source code. It is possible to stop a program in mid-execution and examine the contents of one or more data items by variable name. New values can be placed into these variables if desired. The program can then be restarted with the new values in place. It is also possible to start execution at a point other than the beginning of the program. This is useful when testing individual segments of the code in isolation.

Many of the testing and debugging features described thus far can also be obtained when using a compiler. This requires the use of an interactive program called a *symbolic debugger*. Its use requires that the compiler produce additional information in the object code besides instructions and library calls. Specifically, the contents of the symbol table must also be stored within the object code. This is normally accomplished by a specific option or instruction to the compiler.

With the symbol table stored in the object code (and ultimately the executable code), it is possible to automatically trace memory locations and instructions to their corresponding statements and data names in the source code. The debugger operates much like an interpreter in that it executes the compiled and linked program one instruction at a time. It also provides facilities for starting and stopping execution, tracing changes to variables, and changing the contents of variables. However, unlike an interpreter, it does not provide a source code editor and it performs no translation functions. Thus, errors must still be corrected through text editing, recompilation, and relinking.

ADVANCED SUPPORT FOR APPLICATION DEVELOPMENT

The previous sections have discussed the translation of high-level language application programs into executable code. Although the languages and their capabilities have changed substantially over time, the labor-intensive activities that precede program development—for example, analysis and design—have changed much less. The purpose of this section is to describe more recent tools that support or fundamentally change these labor-intensive activities.

CASE Tools

As you learned earlier, the term CASE is an acronym for *computer-assisted software engineering*. In essence, any systems software that aids in the development of executable application programs is a *CASE tool*. This would include not only aids normally considered CASE tools, but software such as compilers, interpreters, and link editors. The ambiguity implied by the literal interpretation of CASE has led to a great deal of confusion as to exactly what is and is not a CASE tool.

For the purposes of this discussion, we exclude any tools that are used to turn high-level language programs into executable code (e.g., compilers). We also exclude tools that directly support program development, such as text editors and source code debuggers. We restrict our attention to those tools that support the definition of user requirements, the generation of analysis and implementation models, and the generation of source code based on those models. Tools that support requirements definitions and model generation are sometimes called *front-end CASE tools*. Tools that generate source code based on system models are sometimes called *back-end CASE tools* or *code generators*.

The advent of front-end CASE tools can be traced to the late 1960s and early 1970s. They were designed to address problems and complexities associated with developing and documenting implementation models for large systems of application programs. Examples of these tools included drawing tools for developing system and program flowcharts and data dictionary facilities (the latter were often associated with database management systems). Such tools allowed models and documentation to be quickly developed and analyzed. In addition, because they made development easier, they also made changes easier. Thus, alternative implementations could be developed *and refined* on paper.

As introduced, these front-end CASE tools were little more than automated drawing and text editing programs. Some of their features—standard flowcharting symbols, templates for file definitions—were specifically geared toward developing system models. In practice, they were often used to document systems *after* they were built rather than to support the evaluation of system models *before* the system was built.

Although providing automated assistance for drawing and text editing does improve the productivity of analysts and designers, it is a very limited improvement. Further productivity increases require a tool that "understands" what is being modelled and that can assist the analyst or designer to refine a model and evaluate its quality. Certain quality checks are universally applicable to the evaluation of system models. Examples of these include checks for completeness (e.g., does a textual specification exist for each program on a system flowchart?) and checks of consistency (e.g., is it possible to generate the specified program outputs based on the specified program inputs?).

Both completeness and consistency are relatively simple to verify, but the processes are extremely tedious if performed manually. The need to perform such checks became even greater in the 1970s and 1980s as structured systems development became widely practiced. However, the drawing and text editing tools of the time were simply not capable of performing these checks. Although these verification processes are simple, they require some knowledge of the objects that are being modelled. That knowledge is simply not present in a general-purpose drawing or text editing tool.

Modern front-end CASE tools incorporate the editing features (textual and graphic) of older tools as well as the specialized knowledge required to manipulate the objects being modelled. Thus, a CASE tool "knows" that a rectangle on a system flowchart is a program and that it should have a corresponding program specification. It "knows" that an arrow flowing from a file symbol to a rectangle implies that the program reads data from that file. This knowledge of what is being modelled makes it possible to perform error and quality checking as well as sophisticated query processing and report generation.

For example, a CASE tool is capable of comparing the definitions of a program's outputs to the definitions of its inputs and listing those data items in the outputs that are not read into the program. It is also capable of answering queries such as LIST ALL PROGRAMS THAT READ OR WRITE TO FILE XYZ. Facilities such as these can be used to find errors in the model, to generate initial data declarations for a program, or for numerous other purposes. All of these things lead to an improvement of the quality of system models.

Front-end CASE tools make it possible to consider more model alternatives and to produce better models. The rapid query and display facilities allow models to be viewed and modified with ease. Their error-checking facilities allow hardware resources to be substituted for labor in many "straightforward but tedious" model validation procedures. All of these things lead to a reduction in labor input to the analysis and design phases of application development. In other words, the productivity of analysts and designers is increased.

Code Generation

Although front-end CASE tools can increase the productivity of analysts and designers, they do little to address the productivity of programmers. High-quality models provide a better basis for writing application programs, but they don't directly support that activity. The translation from implementation model to a set of files and application programs still requires a substantial labor input. A code generator (or back-end CASE tool) is a tool that generates application programs and/or file (or database) definitions from all or part of an implementation model. Although code generators have existed for decades, the technology is only now becoming widely used. This is partly due to the trend toward cheaper hardware resources and partly due to the increased use of sophisticated modelling tools.

The input needed for code generation must be complete, specific, and in a standardized format. Thus, for example, the automated generation of a report-writing program requires complete specification of the inputs, outputs, and rules that describe the transformation between them. This implies a substantial amount of detail in the implementation model regarding file content, format, and content of outputs. It also requires rigorous specification of program logic (data transformation rules).

Code for many programming languages can be generated by a code generator. Although it is possible to directly produce executable machine code (i.e., a code generator combined with an interpreter or compiler), it is more common to generate code in a traditional third- or fourth-generation programming language. This makes it much easier for a human programmer to validate and modify the generated programs (if necessary). The generated code is then translated with a conventional interpreter or compiler.

For code generation to be successful, the implementation model must be complete and in a standardized format. Because of this requirement, code generators are usually designed to work with a specific front-end CASE tool. Front-end CASE tools usually employ standardized formats for descriptions of processes, files, reports, and the like. This standardization simplifies the interpretation of the model for purposes of code generation. In addition, the consistency and completeness checking procedures of the CASE tool can be used to validate the input prior to code generation. Problems can be brought to the attention of the designer for correction to ensure that valid programs are generated.

Integrated Application Development Environments

An integrated application development environment is a complete set of tools to support all phases of application development. Although their exact composition varies considerably, they typically consist of a front-end CASE tool, a code generator, and fourth-generation language facilities. Such environments provide support for analysis, design, and implementation and do so with tools that are designed to work together.

SUMMARY

Systems of application programs are developed using a systems development methodology or life cycle. Current methodologies include structured systems development and rapid prototyping. A principle common to all systems development methods is the use of a system model, which is a representation of user processing requirements or of a system designed to satisfy those requirements.

Structured systems development uses two separate models: the analysis (or logical) model and the physical (or implementation) model. The analysis model consists of a data flow diagram and a written specification for each of its components. An analysis model is designed to convey requirements for processing, data flow, and data storage. Information about possible methods of implementing a system to meet those requirements are purposely omitted from the model.

The implementation model normally consists of a system flowchart and a written or graphic specification of each of its components. An implementation model describes the movement of data between system components, including application programs, files, and manual processes. The implementation model represents a specific plan for building a system to satisfy user requirements. The development of application programs does not normally begin until the implementation model has been completely specified.

Rapid prototyping is a methodology for the rapid development and evaluation of application programs. Initial requirements are determined either using structured systems analysis or by less formal means. These initial requirements are used to quickly create a prototype system. The prototype is given to the user for evaluation. Changes or additions requested by the user are incorporated into the prototype. This process proceeds interactively until all user requirements are satisfied.

Each development methodology has relative strengths and weaknesses as well as specific software support requirements. The structured methodology ensures that all requirements and design decisions are fully specified before program development begins. This generally results in rapid program development with a minimum of problems and changes. Software support for the structured methods includes computer-assisted software engineering (CASE) tools and program development tools. CASE tools are used to develop and refine system models. Program development tools are used to create application programs and translate them into executable machine code.

Rapid prototyping requires the use of CASE tools if an analysis model is developed. Rapid prototyping requires the use of the most powerful program development tools. These are needed because prototypes must be produced and refined quickly. Rapid prototyping generally produces superior user feedback due to the concrete nature of the system model (the prototype). However, the need for powerful program development software and many iterations of prototype refinement leads to high hardware resource requirements.

Modern programming languages are generally classified as either third-generation (3GL) or fourth-generation (4GL) languages. Third-generation languages are characterized by hardware capabilities and limitations at the time of their development (late 1950s to early 1970s). In general, they are poorly equipped to handle some modern application processing tasks, including database management and complex interactive I/O.

A number of add-on software systems have been developed to augment the capabilities of traditional 3GLs. These include database management systems (DBMSs) and I/O management systems. These software packages extend the capability of 3GLs to handle the complexity and demands of modern applications. However, they do so at the expense of complexity in application development and portability of application programs.

Fourth-generation programming languages incorporate database management, complex interactive I/O, and other advanced functions. They are more powerful than 3GLs due to these added capabilities and due to a higher level of instruction explosion (the number of machine instructions generated for each source code instruction). These features make them especially well suited to rapid prototyping.

All high-level language programs must be translated into machine instructions before execution. There are two basic approaches to this translation: compilation and interpretation. A compiler is a program that translates source code into object code. Object code consists of machine instructions and calls to previously defined machine language programs or library routines. These programs are provided either by the compiler or the operating system.

A link editor is a program that converts object code into executable code. It searches object code for calls to library routines. It then searches one or more libraries for the corresponding machine language programs. The original object code and the library routines are combined to create an executable program.

An interpreter performs program translation, linking, and execution one statement at a time. A source code statement is read from a file, translated, linked (if necessary), and immediately executed. Interpretation provides superior error reporting and correction capabilities because errors can be reported in terms of source code statements and variables. However, interpretation consumes substantially more memory and CPU time during program execution than does the execution of compiled and linked programs.

A code generator or (back-end CASE tool) is a program to create source or executable code directly from a system model. Code generators require that the system model be complete and detailed. Code generators are generally designed to work with a specific front-end CASE tool. They are normally used to generate source code in a 3GL or a 4GL. This allows human programmers to examine and modify source code prior to compilation and linking.

Key Terms

analysis (logical) model
back-end CASE tool (code generator)
CASE tool
classical systems development
compiler
compiler library
computer-assisted software engineering (CASE)
control structure
database management system (DBMS)
data declaration
data dictionary
data flow diagram (DFD)
executable code
file specification
fourth-generation development environment
fourth-generation language (4GL)
front-end CASE tool
instruction explosion

interpretation
interpreter
library call
linker (link editor)
memory (link) map
object code
physical (implementation) model
process specification
program flowchart
program specification
rapid prototyping
run-time library
source code
structured system development life cycle
symbol table
symbolic debugger
systems analysis
systems design
system flowchart
system model
third-generation language (3GL)
unresolved reference

Vocabulary Exercises

1. A compiler allocates storage space and makes an entry in the symbol table when a _____ is encountered in source code.
2. A _____ is produced as output of the analysis phase of the structured SDLC.
3. A link editor searches object code for _____.
4. The structured SDLC produces a _____ as output of the design phase.
5. A fourth-generation programming language has a greater degree of _____ than does a third-generation programming language.
6. _____ contains executable code and _____ calls.
7. A _____ translates an entire program before linking and execution; a _____ interleaves translation and execution.
8. A _____ describes the content of a data flow or file within a _____.
9. The result of compilation is _____; the result of link editing is _____.
10. A _____ generates source code based on a _____.
11. Unlike _____, _____ proceeds under the assumption that a working model of the system is easy to develop and modify.
12. A compiler library is similar in function to an interpreter's _____.

Review Questions

1. What types of system model(s) are used in the classical, structured, and prototyping approaches to application system development?
2. What characteristics of application development software are needed to support rapid prototyping?
3. What is instruction explosion? What are the advantages and disadvantages of using a programming language with a low degree of instruction explosion?
4. With respect to the requirements of modern applications, what are the shortcomings of third-generation programming languages in the areas of data manipulation and user interface?
5. What is a compiler? What is an interpreter?
6. How do source code, object code, and executable code differ?
7. Compare and contrast the execution of compiled programs to interpreted programs in terms of CPU and memory utilization.
8. What is a linker? What is a library? What are the advantages of using them?
9. Compare and contrast the error detection and correction facilities of interpreters and compilers.
10. What is a CASE tool? What is the relationship between a CASE tool and an application development methodology?
11. What is a code generator? What is required to efficiently use a CASE tool and code generator to support application development?

Research Problems

1. Investigate a modern CASE tool such as IEF or ADW. What types of system models can be built with the tool? How is an analysis model translated into an implementation model? What programming languages, operating systems, and database management systems are supported by the ba ck-end CASE tool? Can the tool(s) be used to support system development via rapid prototyping?
2. Investigate a modern implementation of BASIC such as QuickBASIC or Visual BASIC. In what ways is the translation program similar to an interpreter? In what ways is it similar to a compiler? What program editing tools are provided? What tools are available to support run-time debugging?

11 *Operating Systems*

Chapter Goals

- *Describe the primary components and functions of an operating system.*
- *Describe the evolution of operating system functions and capabilities.*
- *Describe the resource allocation functions of an operating system.*
- *Describe the mechanisms by which an operating system manages programs and processes.*
- *Describe the mechanisms by which an operating system manages memory.*

OVERVIEW OF OPERATING SYSTEM FUNCTIONS

The operating system is the most important systems software component in an information system. It plays a dual role as service provider and hardware manager. In its service role, it has a critical impact on the types of capabilities that can be provided to users through application programs. In its hardware management role, it directly controls the allocation of hardware resources to user tasks, as well as the efficiency of hardware utilization. Thus, the selection and configuration of an operating system largely determine the types of services available to end users and the efficiency with which those services are delivered.

The primary purpose and functions of an operating system can be summarized as follows:

- Provide an interface for users and application programs to low-level hardware functions
- Efficiently allocate hardware resources to users and their application programs
- Provide facilities for loading and executing application programs
- Provide facilities for managing secondary storage
- Provide controls over access to hardware devices, programs, and data

A functional view of an operating system is provided in Figure 11.1. These functions can be loosely divided between those directed toward hardware resources (CPU control, memory control, mass-storage control, and I/O control) and those directed toward users, their processes, and their files. Note that control of mass-storage hardware and of user files is combined within the file management subsystem.

Operating system functions can be fulfilled in many ways. Differences in the set of functions that are implemented, the method and efficiency of implementation, and the computer system being controlled account for the wide variety of operating systems available. Some basic criteria upon which to distinguish operating systems include

- Single-tasking versus multitasking
- Single-user versus multi-user
- Batch versus timesharing

Single-tasking operating systems are designed to execute only one process at a time. These operating systems are generally restricted to microcomputers and certain technical applications. *Multitasking* operating systems are designed to execute multiple programs concurrently. Note the use of the term *concurrent.*

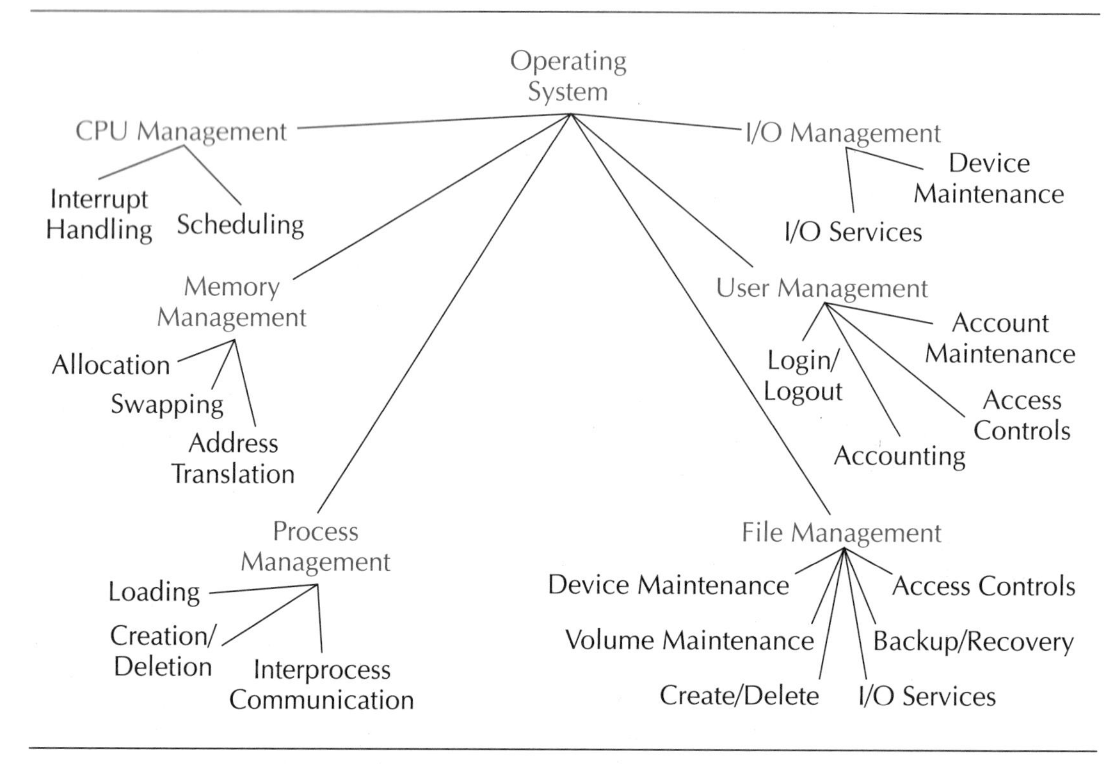

FIGURE 11.1 A functional view of an operating system.

A von Neumann computer with a single CPU is incapable of *simultaneous* execution of more than one program, because a single CPU cannot execute more than one instruction at a time. Execution of multiple programs is performed by dividing CPU time among them. This usually results in *interleaved execution*, in which the CPU suspends execution of instructions of one process, executes some instructions of other process(es), and then returns to execute the next instruction of the suspended process.

Multi-user operating systems are designed to be used by more than one user at a time. They are by necessity multitasking, because each user will execute one or more processes. Multi-user operating systems normally employ fairly elaborate mechanisms to account for resource use by individual users (e.g., logins, logouts, and related accounting procedures). They also normally implement ownership of and access controls over both programs and data.

Batch operating systems are designed to execute programs that do not require active user intervention; these are called *batch processes*. Such processes normally use non-interactive I/O devices, such as document scanners, or secondary storage for input and return results to those same devices. They are unusual today due to the widespread use of online processing. Such systems are still in use, however, especially in large batch transaction processing environments—for example, in nightly processing of checks in a bank or preparation of credit card statements—or as components of a distributed network of computers.

Operating System Layers

The operating system can be described as a layered set of software. This layering is illustrated in Figure 11.2. The *kernel* is the operating system layer that provides the most basic functions. These include loading and executing processes as well as allocating hardware resources (access to disk drives, CPU time, memory, and so on) to individual processes. Most of the interaction between hardware and software is localized to this layer.

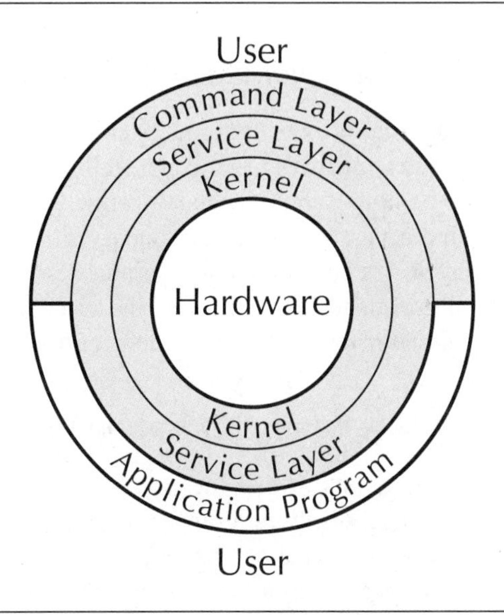

FIGURE 11.2 **A model of an operating system as software layers.**

The *service layer* consists of a set of programs used by application programs or by the command layer. These programs provide the following types of services:

- Low-level access to I/O devices (e.g., the movement of data from an application program to a printer or terminal)
- Low-level access to mass-storage devices (e.g., the movement of data from a tape drive to an application program)

- File manipulation (e.g., opening and closing files, reading and writing records)
- Basic process and memory control functions (e.g., sending signals between application programs, allocating additional memory to a program during execution)

In addition to these common functions, other services—such as window management, access to communication networks, and basic database services—can be provided. The entire service layer can be considered a set of reusable programs that are available to application programs on demand.

The *command layer* provides end users a direct interface to the service layer. This allows the user to request many service functions without using an application program. This layer also provides the user a means of managing hardware and files independently of application programs. It also provides some basic management and access controls through facilities such as user accounts and file ownership.

Basic Process Management and Resource Allocation

The management of processes (programs) and the allocation of hardware resources to those programs is primarily the responsibility of the kernel. Processes must be provided with memory to hold data and instructions, access to the CPU to execute instructions, and access to I/O and secondary storage devices. The operating system is responsible for managing these various resources to meet the needs of each process and to guarantee efficient use of the machine. This aspect of the operating system is discussed in detail in later sections of this chapter.

The Service Layer and Application Interface

The service layer exists to provide access to complex hardware functions for application programs, utility programs, and the command layer. It is essentially a set of utility subroutines that can be called by other programs. A request for service from one of these subroutines is referred to as a *service call.* The functions of these programs can be broadly classified into the following categories:

- I/O
- Process and memory management
- File manipulation

I/O service calls provide capabilities to send or receive data from the various I/O devices attached to the computer system. The primary purpose of these service calls is to hide the complexity associated with hardware level I/O instructions from application programs. Low-level control of I/O devices requires extremely complex programming, and its specifics vary from one I/O device to another. Providing a generic set of I/O services through the service layer centralizes all of this complexity within the programs that implement the service calls. It also tends to increase the portability and flexibility of application programs.

The devices that can be accessed through I/O service calls include terminals, graphic display devices, printers, modems, and network interface units. Varying sets of system calls can exist for each type of device, or the operating system can combine system calls to similar types of devices. The system calls dedicated to each type of I/O device are programmed to meet the specific capabilities of that device. The number of I/O service calls in modern operating systems is steadily increasing in response to an increase in the types of devices and their complexity.

Process management service calls and memory management service calls provide the following capabilities:

- Allocation of memory to a process
- Creation of subprocesses
- Communication between executing processes

A program is normally allocated one or more regions of memory at the time the program is loaded and execution commences. Some programs require additional memory to be allocated (or deallocated) during their execution. A text editor, for example, normally requests additional memory if the user wants to edit a file that is too large to fit in its current memory allocation. Interpreters also use this facility to load additional library routines as they are needed. One or more service calls exist to implement this function.

Executing processes could also need to create other processes and to communicate with those processes during their execution. The creation of a process by another process is sometimes referred to as *spawning*. The original process is called the *parent*; the newly created process is called the *child*. This facility is often used when the operating system provides a general-purpose program (other than a system service call) that is frequently used by other programs. A text editor, for example, can allow a user to sort the contents of all or part of the file currently being edited.

Rather than implement instructions for sorting within the editor program itself, the editor might request that the sort utility program be loaded and executed. This request is made via a service call for loading and executing child processes. Interprocess communication facilities are required for the parent (editor) process to pass unsorted data to the child (sort) process and to receive the sorted data from it. The parent process would issue a service request to establish the necessary communication link(s) between itself and the child process. This type of interprocess communication is called *piping*.

Some application programs are purposely implemented as a set of cooperating but independent programs. This is particularly useful when an application performs multiple functions that can be executed simultaneously and when the computer system contains multiple processors. Coordination between cooperating processes is usually accomplished by allowing processes to send signals to other processes. Operating systems that allow this type of communication must provide system service calls to establish the necessary communication links and to pass signals between processes.

Support services for I/O devices, process management, memory management, and interprocess communication are fully described in Chapter 13.

Secondary Storage Access and Control

Numerous facilities are normally provided for the management of secondary storage devices as well as the management of programs and data stored on those devices. The management of storage devices is provided primarily by a set of utility programs in the command layer. The utilities provide a mechanism for adding or deleting secondary storage devices and device controllers. They also provide for the initialization of such devices through facilities such as formatting and partitioning.

The management of programs and data stored in files and file systems is also provided by the command layer. Utility programs are provided for manipulating individual files as well as for managing groups of files. Utilities for creating, copying, and deleting individual files are normally provided. These are supplemented by utilities to manipulate directories of files or entire file systems.

Command-layer facilities for file and directory manipulation are provided for use directly by end users. Many of these functions are also required by application programs. Access to these and other file-related services is provided by the service layer. File manipulation service calls provide capabilities to create and modify files, manipulate directory structures, and access (read or write) data within files.

The structure of the file system and its implementation through service calls are needed to provide an organizational framework for data and programs stored on secondary storage. Organizing data on secondary storage into directories, files, and records provides structure to the relatively large content of secondary storage. This structure allows a program to find and access the relatively small portion of stored data that it needs.

File manipulation service calls are also needed to hide the complexity associated with accessing physical locations on specific secondary storage devices. Converting a request for access to a specific record or file into the corresponding hardware instructions to secondary storage devices and controllers is an extremely complex task. The specific location of the data or file (e.g., which disk, platter, track, sector, and byte(s) contains the record) must be determined, and the appropriate instructions to the storage controller must be issued to access that location.

A complete discussion of secondary storage management, file systems, and file manipulation is provided in Chapter 12.

THE EVOLUTION OF OPERATING SYSTEMS: RESOURCE ALLOCATION AND PROCESS CONTROL

Systems software has evolved in parallel with computer hardware technology. At times, operating systems have been designed to push existing hardware capabilities to practical limits. In other instances, software approaches had to be modified to take advantage of innovations in hardware technology. In most cases, the underlying motivation has been to maximize the performance and capacity of the system for each of its users, while attempting to minimize cost. Milestones in the evolution of systems software include:

- Single-job processing
- Overlapped processing
- Stacked job processing
- Dynamic processing
- Multiprogramming
- Virtual memory
- Multiprocessing

Each of these is discussed in detail in the following sections.

Single-Job Processing

In early computers, application programs were not supported by programs within the service layer. All machine actions had to be specified fully and completely within each program. Programs were loaded manually into the system, one at a time, typically as stacks of punched cards. Application programs had to control allocation and use of memory as well as all other hardware devices. Data also was input on cards. A program and its data inputs comprised a *job*. In *single-job processing*, one data record passed through the system completely before the next record was input. As shown in Figure 11.3, the computer's processor was idle, or unused, during input and output operations, or most of the time.

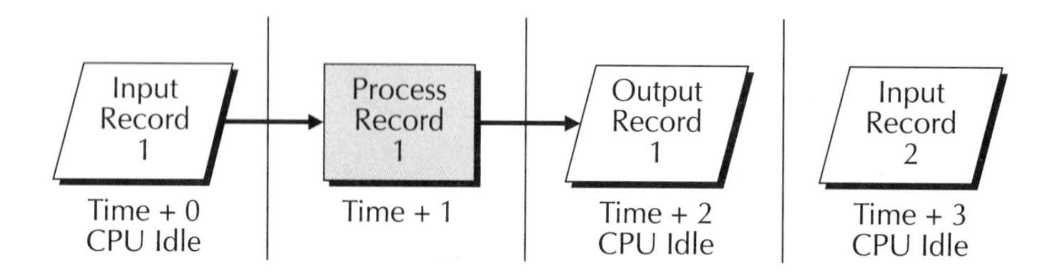

FIGURE 11.3 **Early computers processed jobs in single streams and in stages, with the computer handling one function at a time. This approach led to high levels of idle time for computing equipment.**

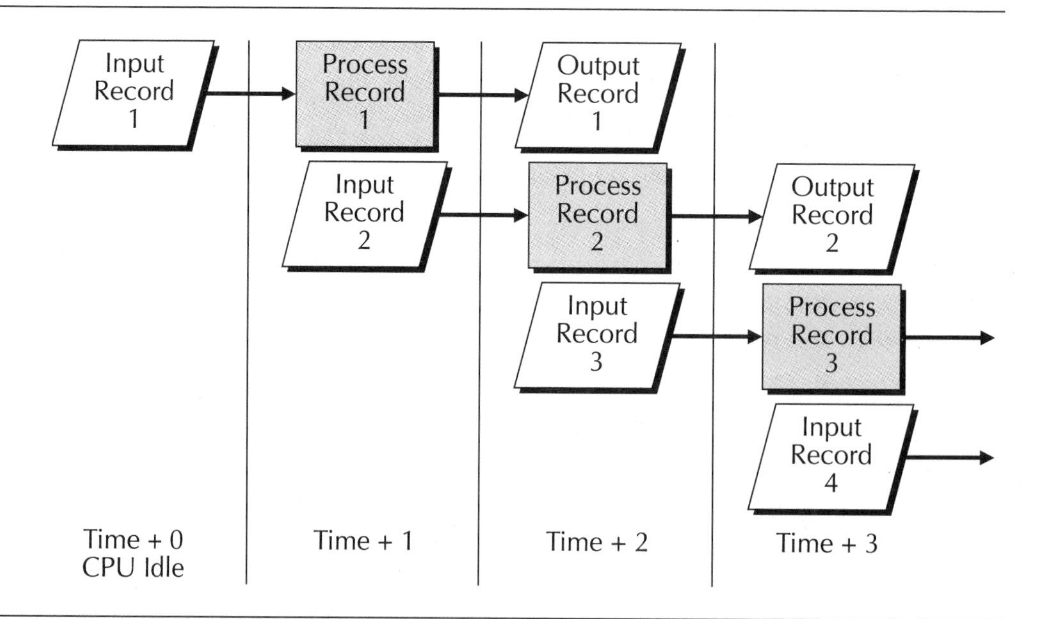

FIGURE 11.4 **The overlapped processing approach filled in gaps of idle computing time by staggering execution of record processing. However, the computer still operated on a single job stream.**

The inefficiencies of single-job processing were partly addressed by techniques for *overlapped processing*. As shown in Figure 11.4, the next data input record in the job was loaded as the current record entered the processor. Subsequent records were processed while the preceding record was being output. This pattern of job loading and processing was made possible by new hardware and software techniques. With these techniques, input and output devices could function independent of the CPU.

ɟf the next enhancement of systems software was to stack multi-
ate the delays of setting up each job separately. This technique,
processing, permitted a sequence of jobs awaiting execution to
b queue, as shown in Figure 11.5. Each program or portion of a
e was called a *job step.*

in systems software for handling each job step were activated
lled *job control commands*. These commands automated much
ng that formerly had been performed manually by computer
commands were provided for identifying users, translating pro-
le code, loading and executing job steps, identifying and locat-
pecifying output devices, and so on. The set of commands
ontrol is referred to as a *job control language (JCL).*

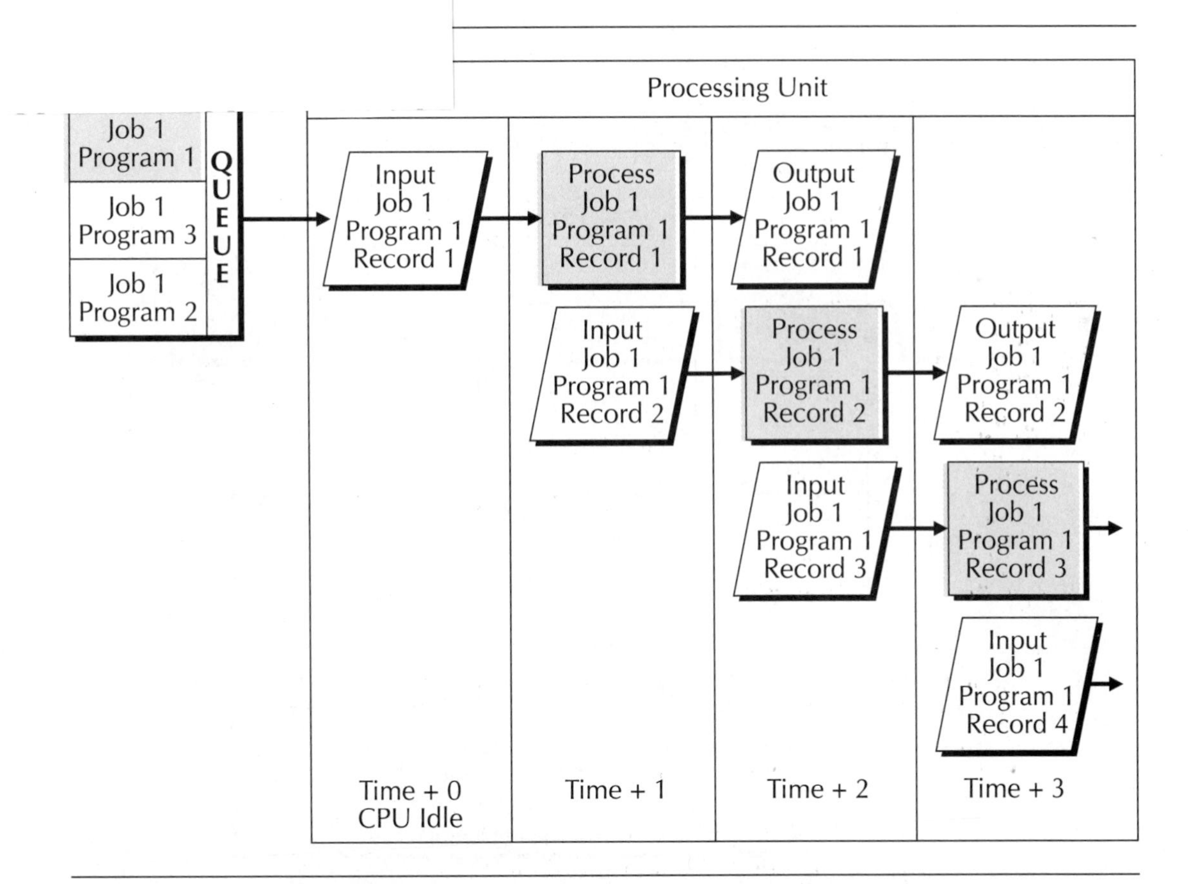

FIGURE 11.5 Implementation of stacked-job processing techniques made it possible to queue jobs for continuous execution. This approach maintained the advantages of overlapped processing and eliminated downtime for job setup.

With stacked job processing, systems software began to provide some significant processing efficiencies. However, capabilities were still limited. The CPU remained idle during set-up and take-down procedures for each job queue. Also, there was no provision for modifying the order of the queue during processing, a necessary capability for interrupting the job flow in case priorities changed.

Time Sharing and Multiprocessing

The technique of *dynamic processing* was developed to overcome some of the limitations of previous approaches to job queuing. Systems software for dynamic processing retained the concept of stacking jobs in queues, with the added capability of moving portions of jobs in and out of the CPU. As shown in Figure 11.6,

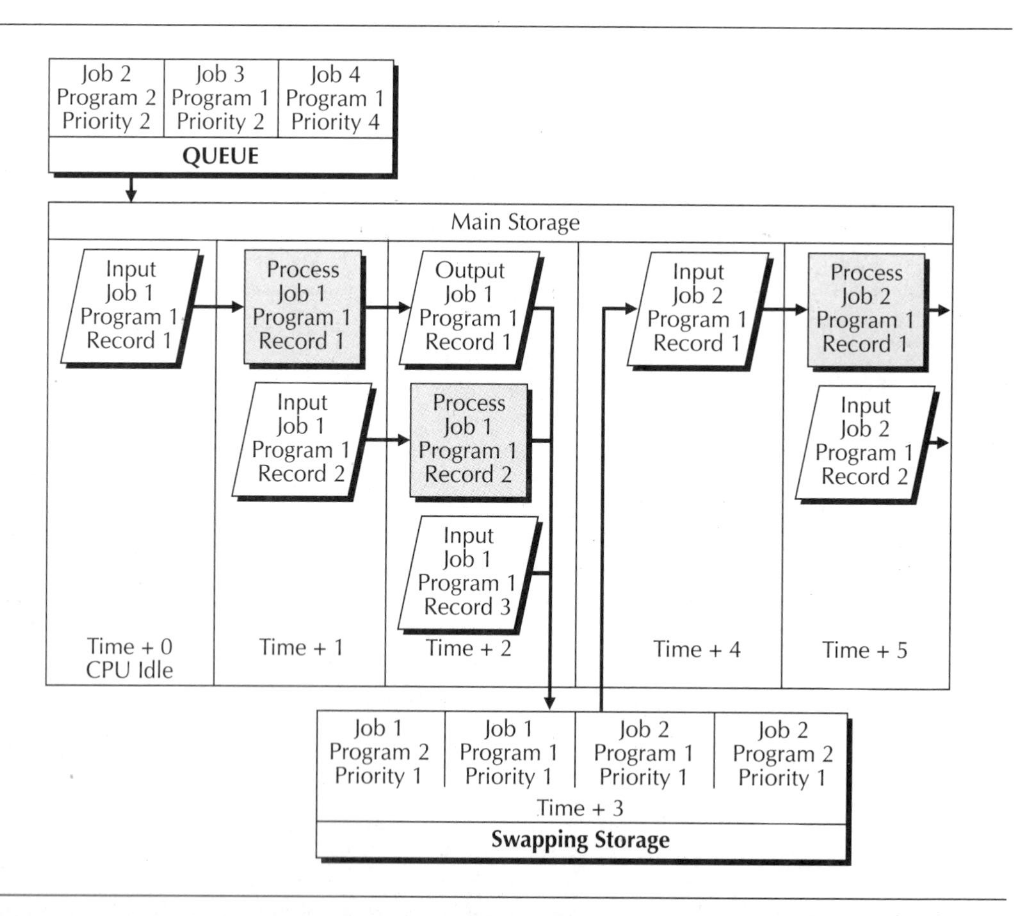

FIGURE 11.6 **Dynamic processing software made possible the interleaving of multiple jobs by partitioning memory to support multiple jobs and by swapping pages of jobs between memory and secondary storage.**

this swapping was performed between main memory and secondary storage, such as magnetic disk. Under this approach, each job step was assigned a processing priority. Time intervals of CPU usage were divided into discrete time slices, or small fractions of a second. Based on priority, systems software allocated time slices, or CPU usage, among job steps.

The number of time slices allocated to a job step determined the duration of its access to the CPU. After this time, the remaining instructions in the job step were written from main memory to disk, and another job step was fetched. Processing proceeded continuously in this manner until all job steps were executed. The operating system coordinated swapping so that high-priority job steps were allocated more CPU time and, hence, tended to be completed first. Because priority job steps took precedence over other job steps in gaining access to the CPU, this approach was also called priority interrupt processing.

A technique that is related to dynamic processing is *spooling,* an acronym for *shared peripheral operation online*. In spooling operations, system inputs or outputs are written to disk to compensate for the typically slower speeds of I/O devices such as card readers or printers. Thus, spooling permits data to be moved to or from the CPU at relatively high speeds, avoiding the delays that would be incurred in transferring data directly to slower I/O devices. Within a system that implements spooling, input and output operations are performed offline, leaving the CPU free to handle ongoing processing tasks.

Although dynamic processing improved system efficiency, its inherent limitations eventually pointed to the need for different approaches. Potentially productive processing time was consumed by the swapping operations themselves. Swapping requires moving both programs and data between main memory and secondary storage. At some point, it becomes impractical to increase the loading of job steps. A more efficient approach was developed that permitted multiple programs to be held in memory. This technique, called *multiprogramming*, is shown in Figure 11.7.

Under multiprogramming operating systems, main memory was divided into separate partitions. Each partition held a different program. The CPU alternated among these programs at high speed, interleaving various sequences of execution. Although instructions were processed individually and sequentially, all of the interleaved programs were handled during the same time as far as users were concerned. Alternate terms for multiprogramming include *concurrent processing* and multitasking, the latter being the term most commonly used today.

A drawback of multiprogramming was that the operating system handled all programs as though they had similar resource requirements. In practice, however, different types of programs place varying loads on system functions. For example, programs that are *I/O bound* might require large volumes of data to be input from a keyboard, written to disk, or printed. By contrast, programs that are *compute bound* (or *CPU bound*) contain extensive calculations and place heavy demand on the processor.

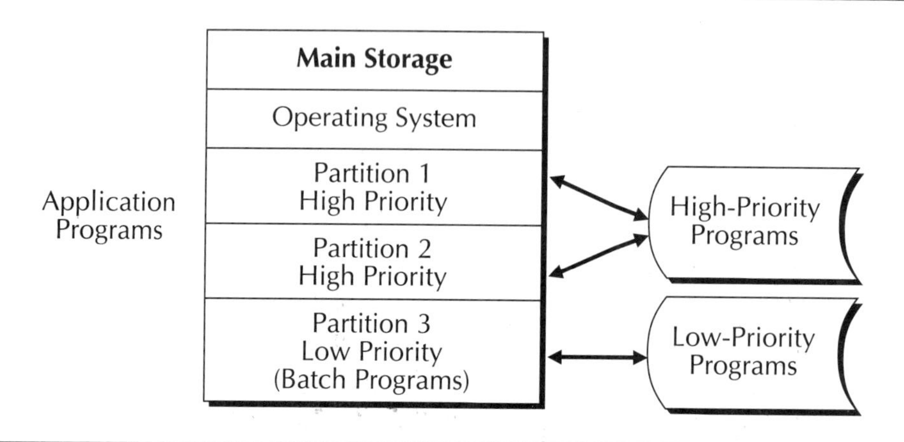

FIGURE 11.7 **Multiprogramming software made possible a capability for assigning and dynamic shifting of priorities for queued jobs. This capacity led to continuous, increasingly efficient use of high-speed processing time.**

Online and Interactive Processing

For almost 20 years after computers first were used in business, computer systems and related facilities represented major investments that could be justified by only the largest organizations. A company that had multiple office sites typically would have only one computer, located at headquarters. Providing computing services to field offices ultimately became a priority. This situation also stimulated the emergence of service bureaus, providers of computer information processing as an outside service to organizations that did not have access to computers.

A typical application was payroll processing. Initially, dealing with central office and service bureau operations involved transporting source documents (such as employee time cards) to the computer facility and returning (usually the next day) to pick up printed checks and payroll reports. To save time and expense, client organizations were encouraged to perform data capture operations internally and to submit jobs in machine-readable form, such as on punched cards.

An important advance involved eliminating the physical transport of inputs and outputs between the remote location and the data center. Newly developed data communication devices such as modems made it possible to perform this transfer over conventional telephone lines. Client organizations were equipped with *remote job entry (RJE)* terminals, which included a card reader/card punch and a modem. Stacks of cards into which source documents had been captured were loaded into the RJE terminal for transmission to the processing facility. Similarly, outputs could be directed to a card punch or to a printer.

Support of RJE terminals required communication control software at the host computer. These programs were used to set up communication links, monitor transmissions, detect errors, and request retransmissions, as necessary. Data inputs were transmitted and run as separate jobs, or in *batch mode*. Techniques that had been used in dynamic processing and multiprogramming eventually made it possible for access to a single computer system to be shared concurrently among multiple users. *Time-sharing* operating systems relied on previously developed capabilities for dividing CPU usage into separate time slices, allocating memory through partitioning, and running multiple programs concurrently.

Time-sharing operating systems also included communication control programs for supporting users at terminals in remote locations. The interleaving of different programs and the allocation of memory and other hardware resources gave each user the impression of personal control of the entire computer. Furthermore, the technique of priority interrupt processing made it possible for these users to control processing interactively, or during execution. Remote terminals connected to the host for interactive use are said to operate online.

Virtual Memory

Up to this point, the capacity of a computer's main memory was a limitation on the size of application programs and on the number of applications and users that could be supported concurrently. Under previous approaches, an entire program had to be loaded into memory from secondary storage, and it resided there throughout execution. However, because the CPU accesses only one instruction in each processing cycle, it is not necessary for the entire program to be present continually.

Under virtual memory, programs are subdivided into smaller portions called *pages,* as shown in Figure 11.8. Under control of the operating system, pages of a program are swapped into main memory from secondary storage as needed to support ongoing processing. Thus, within a virtual memory system, secondary storage units can be regarded as extensions of main memory. This approach permits the use of relatively large programs with minimal demands for physical memory space.

At the time of its development, virtual memory represented a breakthrough because secondary storage devices were relatively less costly than main memory and could be expanded more readily. As a further benefit, more users and applications could be supported concurrently, and each user had the impression of controlling a relatively large area in primary storage.

The advent of virtual memory systems increased time-sharing capabilities further. Each user could have exclusive access to a relatively large virtual memory capacity composed of a small partition within main memory and a larger area of relatively inexpensive disk space. In effect, each user has control of a "computer within a computer." The set of resources apparent to a user of a virtual memory time-sharing system is called a *virtual machine.*

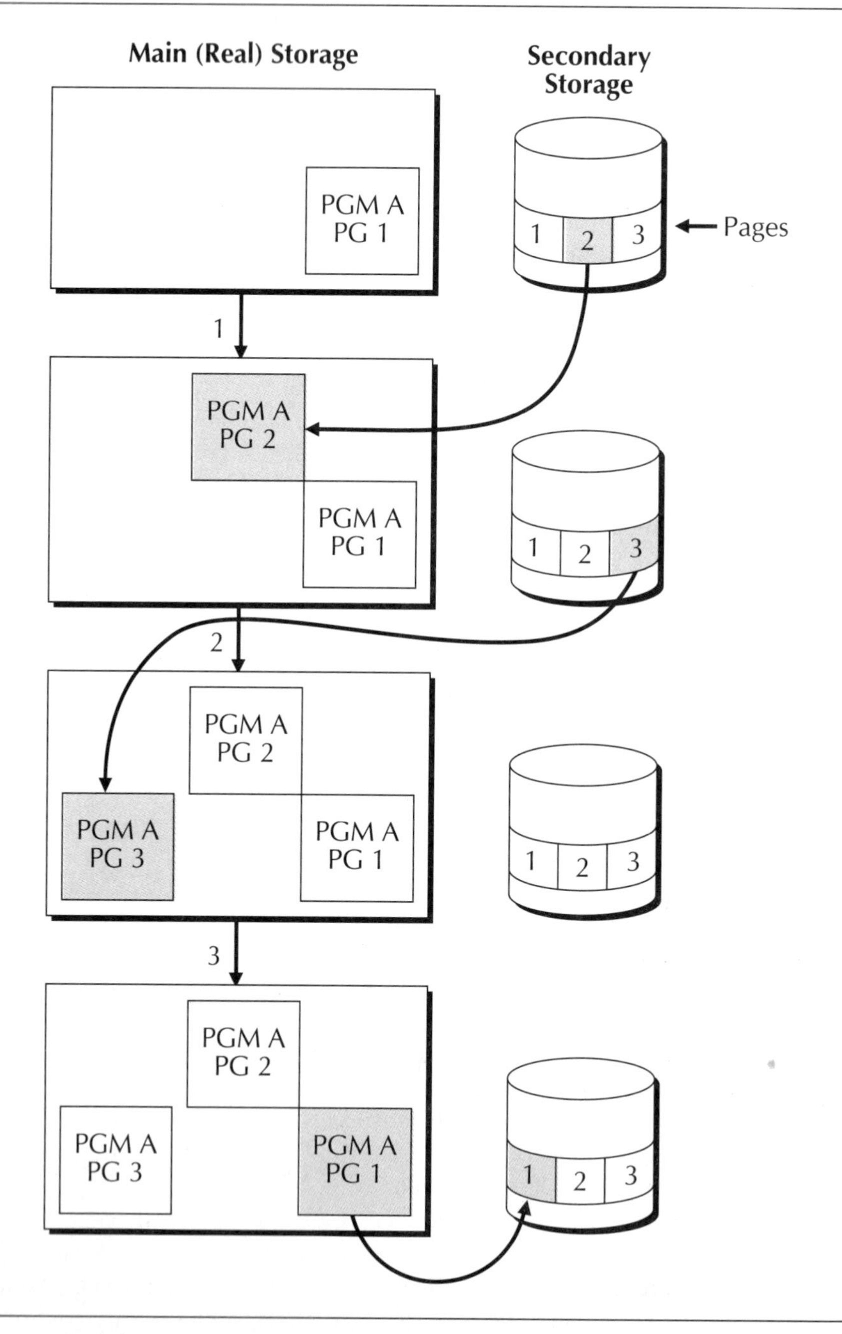

FIGURE 11.8 **Virtual memory systems like the one diagrammed here treat random-access storage devices as extensions of main memory. Pages of data or programs are brought into memory as needed to support job execution.**

Multiprocessing

Another step in the evolution of systems software was necessary to keep pace with the growing complexity of hardware configurations. These new configurations linked multiple processors to achieve increased processing speed and capacity. Efficiency was also increased by permitting the linked processors to share input, output, and storage devices. Operating systems had to be developed to allocate workloads among these shared elements.

A common multiprocessing approach was to offload processing tasks from an overburdened host processor to a smaller auxiliary processor (or slave processor), as shown in Figure 11.9. In some configurations, the slave served as a front-end processor, handling virtually all I/O operations and leaving the host free to perform calculation. In other multiprocessing configurations, the slave handled file storage and retrieval operations. In such cases, the auxiliary unit is called a back-end processor.

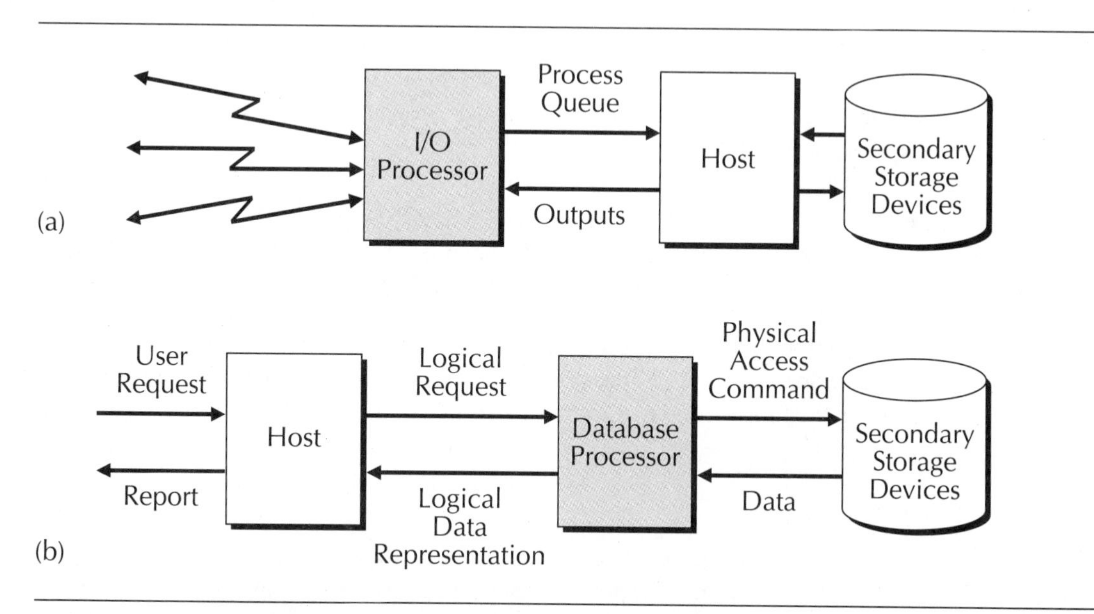

FIGURE 11.9 Multiprocessing made possible simultaneous processing of multiple commands through the coordinated use of two or more computer systems: (a) direct linking of two processors; (b) linking of a high-capacity processor and a specialized database device.

The use of two or more processing units within a computer system makes simultaneous processing possible. That is, multiple instructions can be executed at the same time (one by each processor). This approach represented an improvement in performance over multiprogramming techniques, which permitted concurrent, or interleaved, handling of multiple processes within a single CPU.

RESOURCE ALLOCATION

The efficient utilization of a computer has traditionally focused on the management of two key resources: space and time. That is, storage space and processor instruction cycle time must be used as productively as possible. In the early years of computers, memory devices and processor hardware were so costly that considerations of machine efficiency were paramount. Elaborate schemes, spanning several generations of systems software, were developed for sharing a single, high-capacity machine among many users and applications. Of course, within a von Neumann architecture, even these large multi-user machines were capable of processing just one instruction at a time.

Since then, hardware costs have plummeted while software development costs have continued to rise. One effect has been that large computers have grown in capacity. In addition, personal-sized machines that fit comfortably in a briefcase have become feasible and economical. These changes have had considerable impact on the rationale for resource sharing. On a per-user basis, memory space and processor time are comparatively less costly today. Therefore, resource-sharing approaches increasingly are considering other factors, such as staff hour cost of highly paid user/managers, individual and departmental productivity, organizational effectiveness, and so on.

A set of actual, physical capabilities (devices and associated systems software) within a computer system is called a *real resource*. As allocated by the operating system, a set of capabilities that is apparent to a user is a *virtual resource*. As a result of the evolution of systems software toward multi-user and multiprogramming systems, techniques have been developed by which a set of virtual resources can appear to the user to exceed the available capacity of real resources.

To allocate memory space and processing time, the operating system must manage the following real resources:

- Data storage
- The Processor

Allocations of space for data storage can be from either primary or secondary storage devices in the computer system. Typically these consist of main memory (primary storage) and various secondary storage devices including fixed disks, removable disks, and tape storage.

Because its quantity is usually quite limited, the allocation of main memory is a critical task. Processes require main memory to hold instructions and data for access by the processor. The operating system must usually allocate memory to multiple processes. In cases where demand for memory by processes exceeds physical memory, the operating system can employ techniques such as virtual memory. Such techniques require the operating system to coordinate the loading and swapping of memory pages to and from secondary storage.

The allocation of secondary storage is usually an easier task. The operating system is responsible for allocating secondary storage locations and access to the devices that control those locations. The allocation capabilities are normally centered within the file management portion of the operating system. Files can be created, moved, or deleted by applications or directly by users through the command layer. File operations require the operating system to allocate (or modify the allocation) of secondary storage space to specific files and/or users.

Operations on files can require the operating system to perform allocation, access, or both. The input of existing data from secondary storage requires the operating system to allocate access to the appropriate storage devices to a process or user. Output to a file also involves access operations and can require the operating system to allocate additional space (to hold new data) on a secondary storage device. Both input and output require the operating system to translate file-oriented requests by applications (e.g., read next record) into the low-level operations necessary to implement them.

Appropriate control information (data structures) for the secondary storage device and file system must be updated as a result of such operations. The operating system must keep track of free and allocated space on each storage device. Changes must be made to structures such as directories and file allocation tables each time space is allocated or freed. The operating system must also coordinate operations between secondary and primary storage. When main memory is used to buffer I/O for secondary storage, the operating system must coordinate the allocation of both types of storage as well as the movement of data from one to another.

Processor time is allocated to processes by loading the address of the next pending instruction into the program status word (PSW) register or instruction pointer. The instruction that is loaded into the control unit from this address potentially has control of the entire system, including any of its functions. However, after any execution cycle, the operating system can preempt or postpone further execution of the current instruction sequence and swap a different address into the instruction counter. This swapping is normally initiated through the interrupt processing facilities of the CPU.

A basic objective in allocating processor time is to maximize the number of instructions executed per unit of clock time. That is, the processor must be kept waiting as little as possible for the completion of external events. In time-sharing systems, the duration of a processor access operation is limited to a fixed number of clock cycles, or *time slices*, for each pending process. If a process does not terminate within the allotted time, it is swapped out and must wait for another time-slice allocation.

Despite declining costs of memory space and processor time, most of today's operating systems are designed with machine efficiency as a key objective. However, definitions of efficiency vary. Machine architectures and operating system implementations can represent widely different policies that govern resource sharing. Differences in policy stem from the tradeoffs between implementing an operating system within a machine architecture and a specific set of assumptions about service requirements.

A basic problem of computer resource allocation is that users demand system resources in differing proportions. Furthermore, for each user, demand can fluctuate. The volume of demand might depend on the time of day or the season of the year, or it might be unpredictable. For some users, peak demand can be much higher than average demand. The total number of users and the total number of transactions per hour can also vary, particularly within interactive systems.

An important function of the operating system, therefore, is to maintain expected levels of service despite relatively unpredictable changes in the pattern and volume of demand for system resources. Sufficient resources must be allocated to a user or program so that useful results can be produced. However, if a user were allocated a disproportionate amount of resources, the cost of computer service could easily exceed the benefits. Also, for reasons of security, one user or program should not be permitted to consume more than its reasonable share of the resources available.

Thus, the operating system must allocate resources so as to achieve a balance of computer performance and cost for each user. Particularly for interactive or multi-user systems, allocation must be dynamic (performed continually during execution) to keep pace with fluctuating demand.

Single-Process Resource Allocation

Allocation of resources within a single-tasking operating system is a relatively simple matter. This simplicity is due to coexistence of only two executing programs within the computer: the operating system itself and a single application program. Resource allocations to the application program are made on demand if such resources exist within the computer system.

In single-tasking operating systems, there is rarely any contention for resources between the operating system and the application program. When an application program is loaded and its execution begins, control of the hardware is turned over to that program. The program retains this control until it terminates, an error occurs, or it needs service from the operating system.

Errors and service calls are normally processed through interrupts. When an interrupt is detected, control of the hardware is passed back to the operating system. If an error has occurred, the operating system could attempt to correct it so that the program can continue. If this is not possible or if the operating system is not programmed to correct errors, the application program will be terminated and the operating system will retain control of the hardware.

The MS-DOS operating system is an example of a single-tasking operating system with only rudimentary error-correction abilities. It attempts error correction for relatively few types of errors. For example, some failures during access to secondary storage produce the message "Not ready error reading drive A - Abort, Retry, Ignore?" If the error resulted from a user-correctable condition, such as a floppy drive bay door that was not closed, the user can correct the condition and request a retry of the access. However, most errors during program execution will cause the operating system to terminate the program and return control of the hardware to the command layer.

During the execution of a system call, the operating system has control over all system resources that are not allocated exclusively to the application program. Normally, memory is the only resource that can be allocated exclusively to a program, so the operating system is free to use any other system resource, such as access to I/O devices, as necessary.

Multiple-Process Resource Allocation

Although single-tasking operating systems are still in use in some situations—for instance, in simple microcomputers and dedicated real-time processors—multitasking operating systems are the norm today. This is true even for single-user systems, because a single user might often want to have several processes running simultaneously.

Multitasking capabilities also provide a great deal of flexibility in the construction of systems software. Network interface units, for example, typically have a dedicated operating system process that continually monitors the status of the network interface unit to detect incoming data traffic. Multitasking allows this process to be continually active, saving the time necessary for loading and initializing the process each time network traffic is detected. Similar strategies are often used for managing and monitoring other I/O devices such as printers and terminals.

The allocation of resources within a multitasking operating system is substantially more complex than for a single-tasking operating system. The operating system must allocate resources to multiple processes and must ensure that processes do not interfere with one another. CPU time, storage (primary and secondary), and access to I/O devices must all be managed such that each process receives the resources it needs. The operating system must perform this task in an efficient manner so that excessive hardware resources are not consumed by the management and allocation functions themselves.

Process States

Even within a relatively straightforward von Neumann computer, many types of processes can be handled concurrently. As shown in Figure 11.10, in any serial machine, a process can be in one, and only one, of the following states of operation:

- Ready
- Running
- Blocked

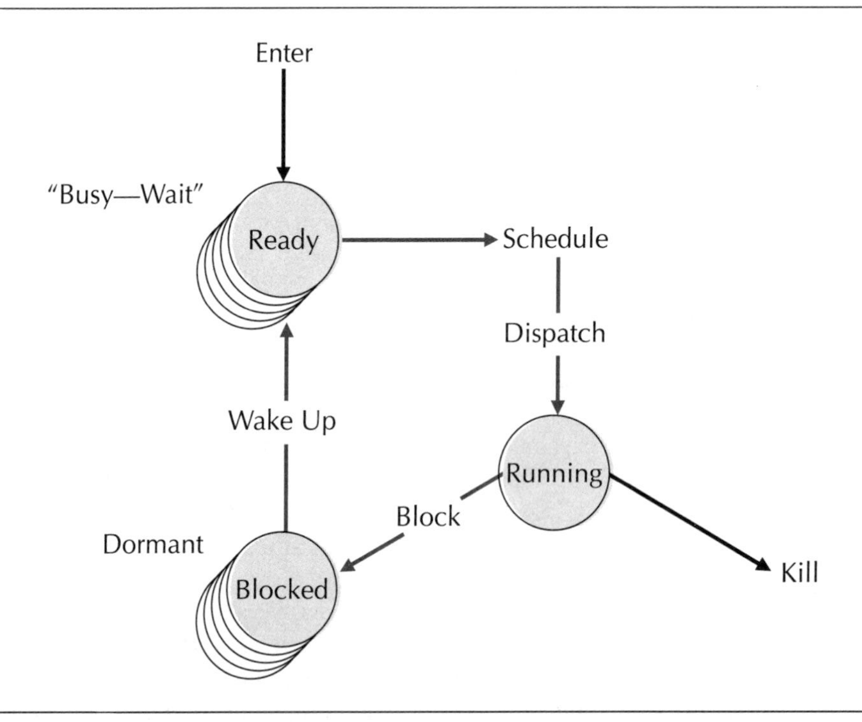

FIGURE 11.10 **The relationships among process states.**

Processes in the *ready state* are waiting for access to the processor. That is, all prerequisites for process execution have been completed. Once it is activated, a process is entered into a queue of pending processes. As the processor becomes available, the next process in the queue is dispatched by the operating system. Multiple processes can be in the ready state concurrently.

In a single-CPU computer, only one process can be in the *running state* at any given time. This process, having been dispatched by the operating system, has access to the instruction pointer and has control of the system. Once it is initiated, a process can be removed from the running state if an interrupt is sensed by the control unit. If the process is halted and not resumed, it is said to be killed. Or the process can be suspended temporarily.

A process that has control of the processor can leave the run state for the following reasons:

- The process terminates (halts) normally. Control of the processor is returned to the operating system, which dispatches the next process.
- The process is suspended because it must wait for I/O, the correction of an error condition, or some other event before it can continue.
- The process is suspended because it has used its current allocation of CPU time or to allow a higher priority process to access the CPU.

A process in the *blocked state* is suspended until a specified event occurs. For example, a process that has been blocked by a priority interrupt must be suspended until the interrupt-handling process terminates. Or a process that is blocked during an I/O operation must be suspended until the data transfer is complete.

A process in the blocked state is not eligible for resource allocation. Once the specified event is sensed, the process is made to "wake up," or pass into the ready state. In other words, a blocked process must pass through the ready state before processing resources can be allocated to it. Any number of processes can be in the blocked state at a given time.

PROCESSOR MANAGEMENT AND PROCESS CONTROL

To initially enter the ready state, a process can be recognized by the operating system and have an initial set of resources allocated to it. Recognition occurs as a result of a request for process creation received either from the command layer (in response to user input) or via a process creation service call from another active process.

In either case, the task of activating a process (bringing it into the ready state) consists of the following steps:

- Locate the executable process image (usually stored on disk)
- Allocate a region of memory to hold the process
- Copy the executable image into memory
- Create a new process control block for the entry

The exact ordering of these steps (particularly the third and fourth steps) and the exact mechanisms by which each step is implemented will vary from one operating system to other. The following discussion will describe these steps in general terms, applicable to most multitasking operating systems.

Typically, a request to the kernel for process loading will identify the exact location of that process's executable image on secondary storage. The command layer will often implement a searching function to determine that location before passing the request to the kernel. This usually involves a search procedure through one or more directories or file systems for an executable file matching the user's request (e.g., via the PATH environment variable in MS-DOS or Unix).

The kernel will verify that this file exists and will attempt to determine its size. Based on the file size, an initial allocation of memory will be made to hold the program. The kernel will then execute the necessary CPU instructions to load the contents of the file into the allocated memory region. This makes the process's instructions and data available for access by the CPU.

In order to manage the process during its lifetime, the kernel needs to keep track of certain information about it. The kernel maintains a set of data structures, each of which is referred to as a *process control block (PCB)*. The information contained in this data structure can include

- The process state (ready, running, or blocked)
- Events for which the process is waiting (reason it is blocked)
- Allocated memory locations
- Files currently being used by the process
- Hardware devices to which the process has exclusive access (if any)
- Accounting information (e.g., CPU time consumed thus far)
- Ownership and/or access privileges
- Priority or other scheduling information

Many operating systems might not store all of this information, because functions that would utilize it are not implemented. The latter three items, in particular, are generally present only in sophisticated multi-user/multitasking operating systems.

Process control blocks for each active process are stored in a reserved area of memory, usually in either a sequential or linked list. They are sometimes collectively referred to as the *run queue*. The operating system records information in a process's PCB as process states change, files are opened, and device access is requested.

Scheduling and Dispatching

Once a process has been loaded into memory and a PCB created, it is now in the ready state. All processes in the ready state compete for access to the CPU. The process of assigning the CPU to a particular process is called *dispatching*. The term *scheduling* refers to the acts of creating or terminating a process or to the decision to give CPU access to some process other than the one currently running.

Operating systems vary widely in their approach to scheduling processes. Scheduling and dispatching procedures are based on any of the following methods:

- Priority-based methods
- Time-sharing methods
- Real-time methods

Many operating systems combine these methods; the exact mechanisms by which they are implemented vary widely.

Priority-based and preemptive scheduling. Priority-based dispatching methods assume that some processes should receive greater access to the CPU than others. These are often used in environments where online and batch processes compete simultaneously for access to the CPU. Because it is usually desirable to minimize response time for online processes, they will normally be given a higher priority than batch processes.

An almost universal prioritization scheme is to give higher priority to ready processes than to blocked processes. Not to do so would result in wasting CPU time while the process in control of the CPU is blocked. Removing control of the CPU from a blocked process is an example of *preemptive scheduling*. Simply stated, preemptive scheduling allows control of the CPU to be taken away from a process before it terminates normally. In contrast, *nonpreemptive scheduling* gives control of the CPU to a process, and the process does not release it until it terminates normally or an unrecoverable error occurs.

Preemptive scheduling and interrupt processing are very closely related.[1] Interrupts are the normal mechanism by which processes are preempted and by which they change state. Consider, for example, the processing of a service call for input from a file. The process that needs the input will normally execute a service call by generating a software interrupt. In order to process the interrupt, a portion of the operating system must be executed. This implies that the running process must be preempted to allow operating system access to the CPU in order to process the interrupt.

The portion of the operating system that is first executed in response to an interrupt is sometimes called the *supervisor*. The supervisor's task is to determine the cause of the interrupt and pass control to the appropriate interrupt handler. When the supervisor first takes control, it will check the interrupt register or program status word to determine the type of interrupt. The left side of Figure 11.11 shows a sample sequence of steps for processing an interrupt generated by an application program's request for input from an I/O device.

[1] The reader might want to review the discussion of interrupts and interrupt processing in Chapter 6.

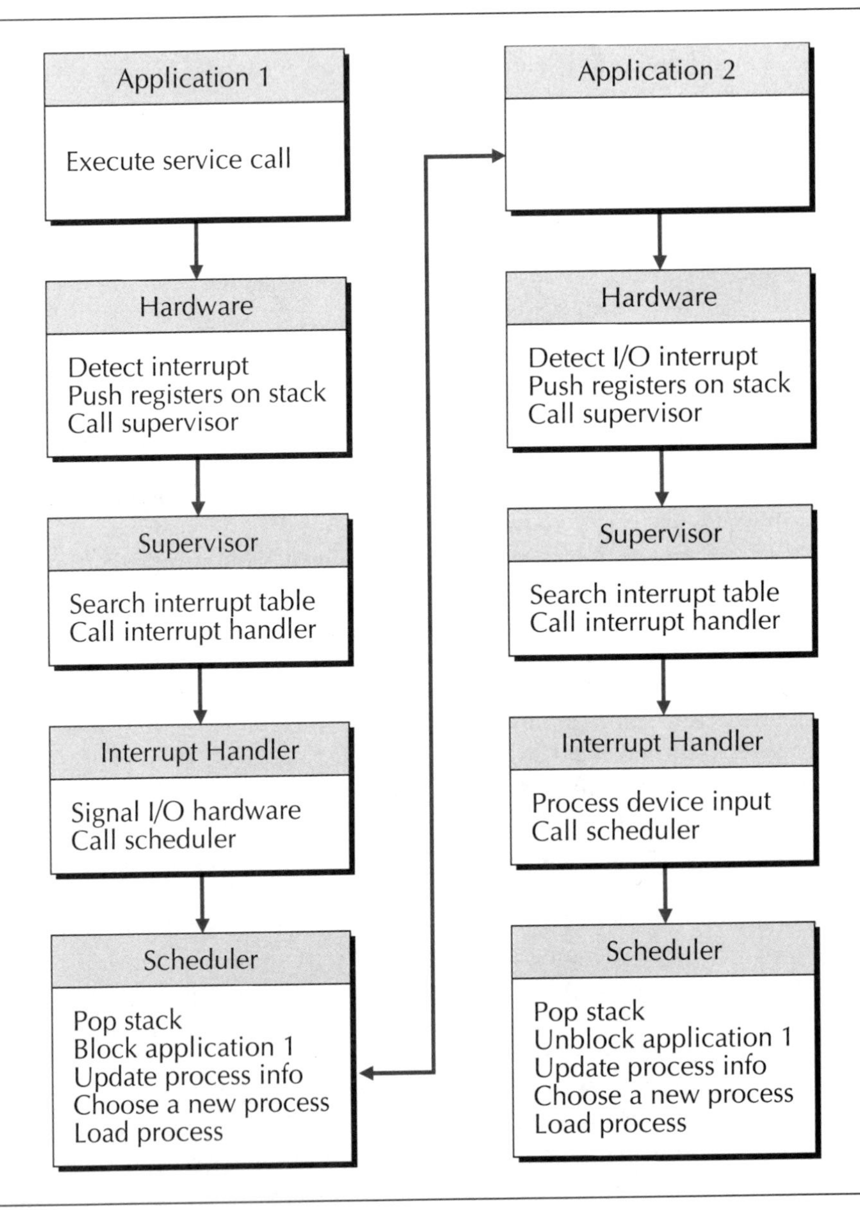

FIGURE 11.11 **The steps in processing an interrupt. The steps on the left result from an I/O service call by application 1. The steps on the right result from the completion of the I/O task by the I/O device.**

The supervisor will search an interrupt table to determine the appropriate interrupt handler to call. This table is indexed by interrupt number and contains the starting address (in memory) of each interrupt handler. When the interrupt handler terminates, control passes to the scheduler, which determines the next process to be dispatched. In the case of an input service call, the process that requested the service will not be immediately eligible for dispatch until the input operation is complete. Completion of the input request must normally await the receipt of another interrupt from the I/O device or controller that is servicing the request. The scheduler will place the requesting process in the blocked state, pending completion of the I/O request. It will then choose and load a new running process.

When an interrupt is received from the I/O device, hardware will detect the interrupt, and control will pass to the scheduler and then to the appropriate interrupt handler. If the input operation was successful, the process that made the original input request no longer needs to be blocked. The scheduler will unblock the process and update the appropriate information in the process control block. The process will then become eligible for dispatch to the CPU.

Scheduling methods. Various methods are available to determine which ready process should be dispatched to the CPU. These include the following:

- First-come, first-served
- Shortest time remaining
- Explicit prioritization

An operating system can implement any or all of these methods. The exact methods for implementing the latter two options vary widely.

The simplest scheduling method is *first-come, first-served (FCFS)*. The implementation of this method requires only that the scheduler keep track of when each job most recently entered the ready state. This can be accomplished by recording the time of the state change in the process control block or by maintaining a queue of ready processes in order of arrival. An ordered queue is normally implemented as a linked list of process control blocks.

FCFS makes no allowance for process prioritization or for preemption of a running process for anything other than an I/O interrupt or error. *Explicit prioritization* utilizes a priority system whereby each process is assigned a priority. This priority can be used in several ways. The two most common are

- Always dispatch the process of highest priority when the CPU is available
- Assign larger time slices to processes of higher priority

The first strategy ensures that high-priority processes will always be dispatched first. Some operating systems combine this strategy with a preemptive strategy, whereby the creation of a high-priority process (or a change in its state from blocked to ready) will cause a running process of lower priority to be suspended. Other operating systems wait until the running process is suspended for some other reason (e.g., an I/O interrupt). In either case, there must be some mechanism by which priorities are initially assigned to processes. This may be based on the default priority of the process owner (user), a priority stated explicitly in a JCL command, or some other method.

Priority-based processing can result in extremely long wait times for long jobs of low priority. This is because they are constantly moved to the "back of the queue" and are unable to obtain sufficient CPU time to complete. Many operating systems automatically increase the priority level of "old" processes to compensate for this effect.

Shortest time remaining scheduling chooses the next process to be dispatched based on the expected amount of CPU needed to complete the process. This can be accomplished directly by dispatching processes based on information about time remaining to completion. It can be accomplished indirectly by increasing the explicit priority of a process as it nears completion. In either case, the scheduler must know how much CPU time will be required for the process to execute completely and how much time has already been used. This implies that the required CPU time was provided to the scheduler when the process was created, and that this information was stored in the process control block. It also implies that the amount of CPU time a process uses is stored in its process control block and is updated each time the process leaves the ready state.

Time sharing and real-time scheduling. Time-sharing systems utilize interrupts generated by a hardware clock. In such systems, clock interrupts cause the current process to be automatically suspended and the next ready process to be dispatched. Priority scheduling can be combined with this method by always dispatching the next process of higher priority or by allocating multiple time slices to higher priority processes.

A *real-time scheduling* algorithm guarantees certain minimum levels of resource allocation to processes, if requested. Real-time scheduling is employed when one or more processes must be allocated enough resources to complete their function within a specified time interval. Examples of processing environments that require real-time scheduling include some transaction processing applications, data acquisition, and automated process control. In the case of a transaction processing application, a real-time processing requirement might specify a minimum response time for a query against a database. For example, an automated bank teller machine will typically expect a response to a customer's balance query within a specified time interval.

Data acquisition and process control are similar in that data arrives at a constant rate from one or more hardware devices. In a data acquisition process, data inputs are simply copied to a storage device such as tape or disk. Data analysis is performed by a separate process at a later time. This type of processing is typical in many scientific environments such as radio astronomy. A process control program analyzes the data as it arrives. For example, a chemical manufacturer could utilize a process control program that receives a constant flow of data from sensors within pipelines. This information might represent temperature, flow rate, or other data of significance for controlling the manufacturing process. Typically, the data inputs from these sensors are buffered, and the process control program extracts input from the buffers. The real-time processing requirement arises from the need to extract and process the data inputs quickly enough to keep the input buffer(s) from overflowing (and thus losing data). Thus, the process control program must be allocated sufficient resources to ensure that it can consume incoming data at least as quickly as it arrives.

In general, real-time processing requirements state a minimum time interval (elapsed or "wall-clock" time) for completing one *program cycle*. A program cycle can involve the processing of a single transaction, the retrieval and storage of data from an I/O device, or many other possible actions. The operating system must be informed of both the CPU time required to complete one program cycle and the maximum allowed elapsed time. The scheduler must track the progress of the process by updating CPU time used in the process control block. Furthermore, it must regularly check the status of real-time processes to determine whether or not they need immediate dispatching to meet their response requirements. This usually necessitates a check of the process control block every time a clock interrupt is generated. The exact implementation of this method can be extremely complex and demand a great deal of processor time (i.e., overhead) to implement.

Interrupt Processing

To rephrase an earlier definition of the term, an *interrupt* is a signal by which priorities and instruction scheduling sequences can be controlled. Interrupts function at the lowest, or machine level of the system by triggering the swapping of addresses in the instruction counter. That is, the interrupt signal is the machine function by which the operating system manages the CPU.

With respect to the CPU, an interrupt can be either external or internal. External interrupts are caused by events that are asynchronous with the current process. For example, an interrupt generated by the interval timer within a time-sharing system is external to the timing relationships within the current process. Similarly, the completion of a data transfer in an I/O operation would cause an external interrupt. By contrast, an internal interrupt is synchronous with the current process. For example, the sensing of arithmetic overflow would generate an error condition interrupt that is a direct consequence of executing the current instruction.

To implement process scheduling, interrupt signals are coded by priority. A typical operating system handles interrupts in the following priority, from high, or most urgent, to low:

- Peripheral requests
- Service requests
- Errors
- Clock signals
- Operator commands

This set of priorities, in itself, reflects deliberate design decisions. Peripheral requests are assigned the highest priority so that I/O instructions are processed first. The rationale is that peripherals, typically the slowest elements of the system, must be kept busy to promote overall system throughput. All other processing is performed in the intervals between I/O interrupts. Service requests occupy the next priority position for similar reasons. Because many service requests result in I/O processing, they must be processed as soon as possible to keep I/O devices busy.

If an interrupt has low priority, the scheduler must determine the appropriate action. A low-priority interrupt can be either enabled or masked. If the interrupt is *enabled*, it is permitted to set the value of the instruction counter. If it is *masked*, it is either held as a pending process (blocked) or discarded (killed). Thus, priority interrupts sometimes are called *nonmaskable interrupts.* A service request always generates a priority interrupt, because the process generating the request typically cannot proceed until the request is satisfied.

The generation of interrupts on execution errors helps prevent the current process from terminating abnormally. Without such a mechanism, the system might not be able to recover from the error, effectively killing the current process or causing all processing to halt entirely. Examples of execution errors that generate interrupts include invalid op code, invalid operand, and arithmetic overflow.

Most modern operating systems process interrupts through the use of a stack, as described in Chapter 6. An attractive feature of a stack is that this method of handling interrupts and suspended processes simplifies the task of managing process priorities. If a process is interrupted, its next instruction is pushed onto the top of the stack. Other processes can be pushed onto the stack as they are interrupted, each being added to the top. As the processor becomes available, the next process is popped, or removed from the top. Accordingly, the process on top of the stack is always the most recently interrupted.

The order of the stack maintains process priorities, because instructions at the bottom are blocked until higher priority instructions have been removed from the stack. Thus, in a stack computer, the operating system does not need to tell the control unit where to find the next instruction. It is always on the top of the stack. Scheduling is determined largely by the scheme of interrupt priorities.

MEMORY MANAGEMENT

Memory management and allocation is one of the primary functions of the operating system kernel. The complexity of this task depends on the number of applications that are supported and on the type of memory addressing and allocation used. For a modern multitasking operating system, the goals of the memory allocation and management function are as follows:

- Allow as many processes as possible to be active
- Respond to changes in memory demand by individual processes
- Prevent unauthorized changes to a process's memory region(s)
- Implement memory allocation and addressing as efficiently as possible

Unfortunately, these goals are somewhat in conflict and, thus, achieving them requires a balanced and carefully designed approach to memory management.

Single-Tasking Memory Allocation

As with the allocation of other resources, memory allocation in a single-tasking operating system is relatively simple. A memory map for a single-tasking operating system is shown in Figure 11.12. The bulk of the operating system normally occupies the lower addresses within memory, and the application program is loaded immediately above it. Addressing individual instructions and data items within the application program is normally accomplished by some form of offset addressing, as discussed in Chapter 6.

Because many operating systems would consume an excessive amount of memory if entirely loaded into memory, it is common for some operating system routines to be loaded into memory only upon demand. If such routines must be loaded during the execution of an application program—for example, to answer a service call—they will normally be loaded in the highest regions of memory, as depicted in Figure 11.12. This allows free space between the top of the application program and these service routines to be easily allocated to the application program if needed.

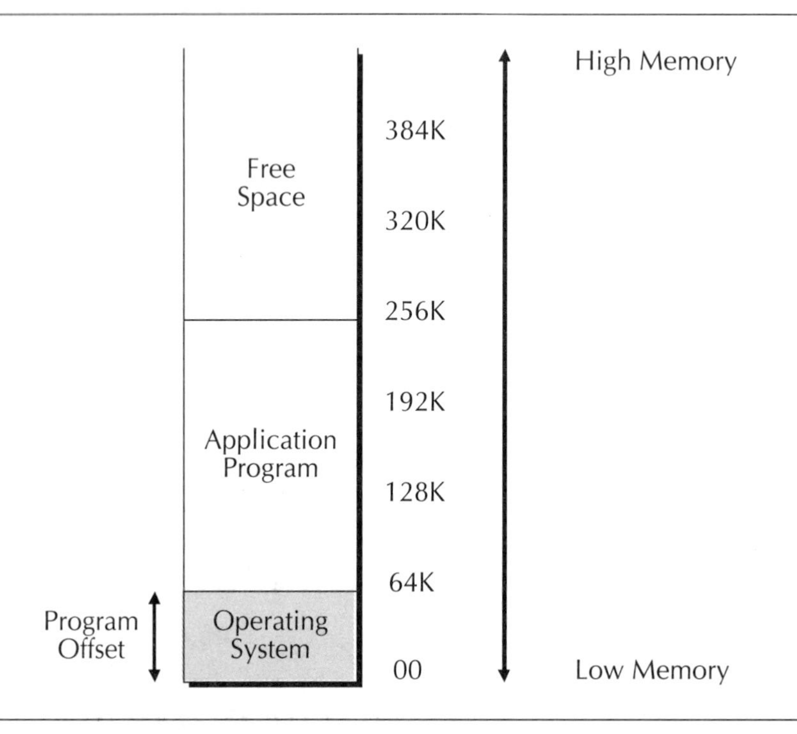

FIGURE 11.12 **The allocation of memory between the operating system and a single application program.**

For the application program in Figure 11.12, memory allocation is contiguous. That is, all portions of the program are loaded into sequential locations within memory. In contrast, the memory allocation for the operating system shown in Figure 11.13 is noncontiguous, or fragmented. Although most of the operating system is stored contiguously in low memory, a portion is stored in high memory, with intervening space that is not allocated to the operating system.

As discussed in Chapter 6, most CPUs utilize some form of offset addressing. Offset addressing allows programs to be located within physical memory in locations other than those assumed at the time of compilation and linking. Normally, programs are compiled and linked, assuming that the first instruction will be located at the first address (zero) in physical memory. If offset addressing is to be used, such programs can be referred to as relocatable, because they can be placed anywhere within physical memory.

A memory reference within a program occurs when a program instruction uses a memory address as an operand. Examples of instructions that contain memory references include load, store, and branch. The process of determining the physical memory address that corresponds to an address reference within a program is called *address mapping* or *address resolution*. The resolution of addresses used within a program is quite simple when memory is allocated contiguously. When offset addressing is used, each address reference within a program can be mapped to its equivalent address in memory by adding the offset value. For the application program shown in Figure 11.12, all address references can be mapped to physical memory addresses by adding the value 65536 (64K + 1) to the reference.

Just as the operating system might require the allocation of additional memory while it is executing, so too might an application program. As stated earlier, the placement of additional operating system routines in high memory leaves the largest possible amount of contiguous free space above an application program. Thus, if the program requests additional memory, it is simply allocated from the empty region immediately above it in memory.

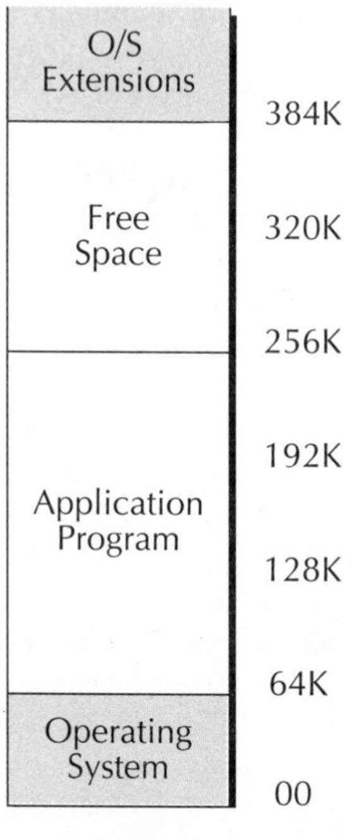

FIGURE 11.13 **The allocation of memory between the base operating system, a single application program, and extended operating system routines.**

Because the region above the application can contain additional operating system routines, the operating system must keep track of memory allocated for this purpose so that it is not allocated to the application program. Thus, each time high memory is allocated to the operating system, that fact must be recorded so that a request for additional memory by the application program can be checked against that boundary. If no operating system routines are present in high memory, application program requests must still be checked to ensure that the requested memory is available between the current top of the program and the last address in physical memory.

Contiguous Memory Allocation for Multitasking

Memory allocation is much more complex when multiple programs are allowed to reside in memory simultaneously. Memory must be partitioned for multiple programs that do not coordinate their activities or their demands for memory. Thus, the operating system is responsible for finding free memory regions to load new programs, loading those programs into those regions, and reclaiming those regions when application programs terminate. Multitasking capabilities require a more sophisticated approach to both the procedures for allocating memory to programs and the mechanisms by which free and allocated memory are managed.

Essential to multitasking memory management is the concept of *partitioned memory*. In its simplest sense, this refers to the division of memory into a number of regions, each of which may hold all or part of an application program. Figure 11.14 depicts memory divided into several fixed-size partitions. Each partition can hold an operating system fragment, an application program fragment, or nothing at all (i.e., free space).

Figure 11.15 shows three programs and the operating system loaded into fixed-size memory partitions. Note that all programs do not fit exactly into a single memory region. Program 2 only occupies a portion of its allocated region. The unused portion of this region is wasted space, because it cannot be allocated to another program if fixed-size regions are used. Program 3 is too large to fit in a single region. It is allocated three entire regions, although it uses only a small portion of its third region. As with program 2, the unused portion of the last region is wasted space.

A multitasking operating system using partitioned memory must keep track of the status of each partition. This requires a data structure in which to store the status of each memory partition.

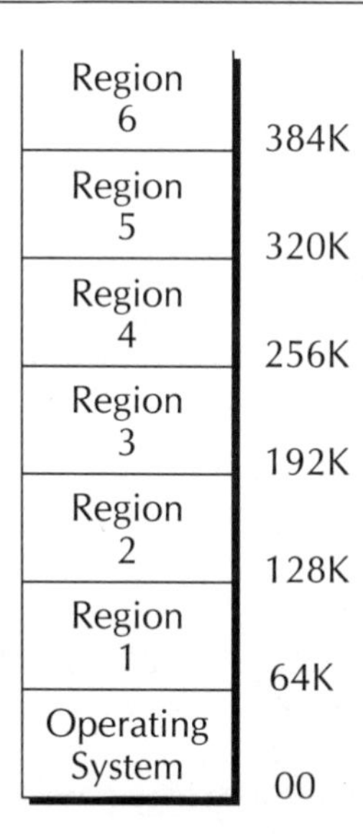

FIGURE 11.14 The partitioning of main memory into several fixed-size regions.

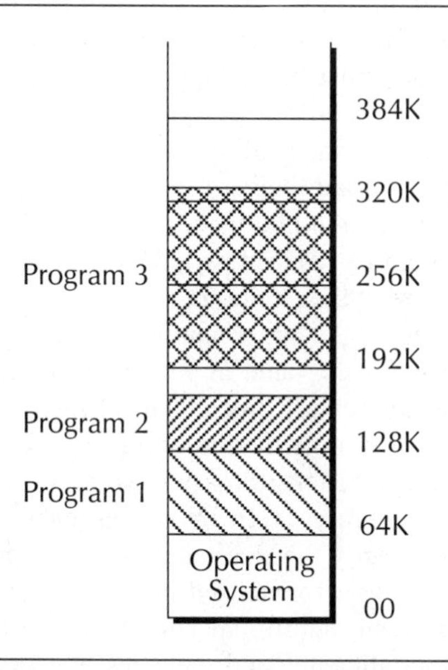

FIGURE 11.15 Several programs and the operating system loaded into memory that has been partitioned into segments of fixed size (64 kilobytes).

An example of such a data structure is provided in Table 11.1. The operating system must track each memory partition in terms of its location, status (allocated or free), and the program to which it is allocated (if applicable).

TABLE 11.1 **Memory partition data maintained by the operating system.**

Partition	Starting Address	Status	Allocated To
1	0	Allocated	O/S
2	64K	Allocated	Program 1
3	128K	Allocated	Program 2
4	192K	Allocated	Program 3
5	256K	Allocated	Program 3
6	320K	Allocated	Program 3
7	384K	Free	

When a program is ready to be loaded for execution, the operating system must search this table to find a sufficient number of contiguous free partitions to hold the program. If they are found, the table is updated with the appropriate data, and the program is loaded into the free partition(s). When a program terminates, the table must be updated to show that its memory partition(s) are now free.

Contiguous program loading, coupled with fixed-size memory partitions, usually results in wasted memory space. As with programs 2 and 3, the unused portion of the last region allocated to a program is wasted. The total amount of wasted space could be reduced by reducing the size of the partitions. In general, the smaller the size of the partitions, the less wasted space will result. This benefit is offset, however, by the larger number of partitions that must be managed. This increases the size of the table in which partition allocation data is stored, and increases the time necessary to search that table.

One way to overcome the limitations of fixed-size memory partitioning is to allow memory partitions to vary in size. An example of this type of allocation is shown in Figure 11.16. Note that each program occupies a memory partition that is exactly the size required by the program. As a result, no memory is initially wasted for unused portions of fixed-size partitions. Variable-size partitions create additional complexity in the procedures by which the operating system allocates and manages memory. The data structure used to track partitions and their allocation is more complex, as shown in Table 11.2. In addition, the procedures for maintaining this table are substantially more complex.

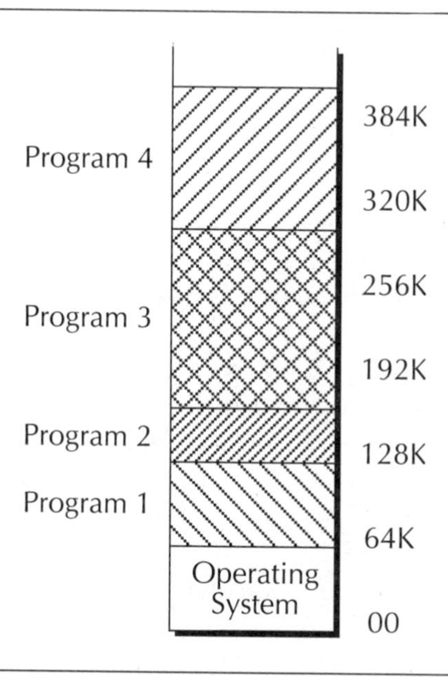

FIGURE 11.16 The allocation of memory to programs using variable-size partitions.

To search for free space for a program, the table must be searched for a free partition of sufficient size. If an exact match is found, the entry in the table is simply updated to indicate the allocation. If the partition is larger than needed, the corresponding entry in the table must be split. One entry is made for the portion of the partition allocated to the program; another entry is made for the unallocated (free) portion. In general, the size of the table can grow or shrink as adjacent free partitions are split or combined. In contrast, fixed-size partitions guarantee a table with a fixed number of entries. Updating a table of fixed size is substantially easier than updating a table of variable size.

TABLE 11.2 Memory partition data maintained by the operating system.

Partition	Starting Address	Size	Status	Allocated To
1	0	64K	Allocated	O/S
2	64K	64K	Allocated	Program 1
3	128K	42K	Allocated	Program 2
4	170K	144K	Allocated	Program 3
5	314K	96K	Allocated	Program 4
6	410K	38K	Free	

Both fixed- and variable-size memory partitioning memories suffer from problems of memory fragmentation. As programs are created, executed, and terminated, the allocation of memory partitions changes accordingly. In general, the continual allocation and deallocation of memory partitions leads to an increasing number of small free partitions and/or free partitions that are separated by allocated partitions.

This process is depicted in Figure 11.17. The memory allocation in Figure 11.17a is essentially the same as shown in Figure 11.16. Figure 11.17b depicts the allocation of memory after program 3 has terminated and program 5 has been allocated a portion of the newly freed partition. Now, because program 5 is not as large as program 3, a new partition of free space has been created. In Figure 11.17c, program 2 has terminated, and its former partition is shown as free.

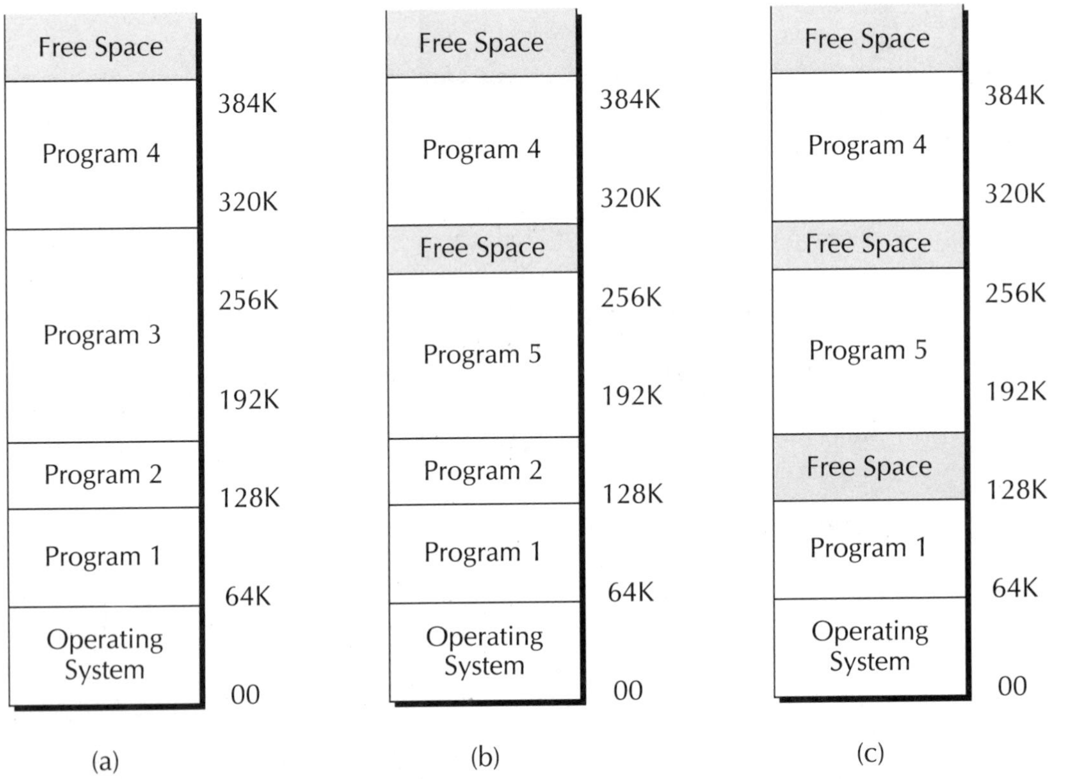

Free Space
384K
Program 4
320K
Program 3
256K
192K
Program 2
128K
Program 1
64K
Operating System
00

(a)

Free Space
384K
Program 4
320K
Free Space
256K
Program 5
192K
Program 2
128K
Program 1
64K
Operating System
00

(b)

Free Space
384K
Program 4
320K
Free Space
256K
Program 5
192K
Free Space
128K
Program 1
64K
Operating System
00

(c)

FIGURE 11.17 **This series of diagrams shows changes over time in the allocation of variable-size memory partitions as programs execute and terminate.**

There are now three small partitions of noncontiguous free space. The largest of these is 64K in size, and the total free space is slightly more than 128K. Although other jobs waiting in the run queue might require less than 128K of memory, only those jobs that require 64K or less can be run due to the fragmentation of available free space. In general, large programs will have difficulty in obtaining required memory allocations due to the necessity to store the entire program in a contiguous block of memory.

One way to address the problem of fragmented free space is to periodically relocate all programs in memory. This process, referred to as *compaction*, results in all free space being collected into a single partition in high memory. If compaction were performed on the memory allocation in Figure 11.17c, program 5 would be relocated just above program 1, and program 4 would then be relocated just above program 5. All of the available free space would then be contained in a single partition starting after program 4.

The process of compaction is extremely expensive. The entire contents of programs must be moved within memory, and many entries in the partition allocation table must be updated. The amount of work entailed by these operations is generally larger than the overhead required to implement a more common strategy—noncontiguous memory allocation to programs, usually coupled with virtual memory techniques.

Virtual Memory Management

In general, the only portions of a program that must be in memory at any given point during execution are the next instruction to be fetched and any data required to support that instruction, such as an operand referenced by a memory address. Thus, at most, only a few bytes of any program must reside in memory at any one time. Although no operating system carries memory allocation strategies to this logical extreme, several techniques attempt to minimize to relatively small portions the amount of program code and data that is stored in memory. These techniques free large quantities of memory for use by other programs and greatly increase the number of processes that can execute concurrently. However, these techniques also require a considerable amount of operating system overhead to manage memory allocation and the swapping of program fragments between primary and secondary storage.

Virtual memory management divides a program into segments called *pages*. Each page is a relatively small portion of a program (normally between 1 and 4 kilobytes), and page size is a constant. Main memory is also divided into pages of the same size as program pages. Each page of main memory is sometimes referred to as a *frame*. During program execution, one or more pages are held in memory; the remainder are held in secondary storage. As pages in secondary storage are needed for current processing, they are copied into main memory by the operating system. If necessary, existing pages in main memory can be swapped out to disk to make room for pages being swapped in. Each memory reference made by a program must be checked to see if it refers to a page in memory. References to pages not contained in memory are referred to as *page faults*.

As in simpler memory addressing and allocation techniques, tables are used to store information about the allocation of program segments to pages and the location of those pages in primary or secondary storage. These tables are called *page tables*. Each active process will have a page table or portion of a page table dedicated to storing its page information. The information in the table includes the page numbers, a status field indicating whether or not the page is currently held in memory, and the page frame number in main memory or on disk. An example of a page table is shown in Table 11.3.

TABLE 11.3 A portion of a program's page table, which is used when resolving memory references made by the program under virtual memory management.

Page Number	Memory Status	Frame Number	Modification Status
1	In Memory	214	no
2	On Disk	101	n/a
3	On Disk	44	n/a
4	In Memory	110	yes
5	On Disk	252	n/a

Because page size is fixed, memory references relative to address zero can easily be converted to the corresponding page number and offset within the page. The page number can be determined by an integer divide of the memory address and by the page size. The remainder of the division is the offset into that page. For example, if page size is 1K, a reference to address 1500 is equivalent to an offset of 476 (1500 – 1024) into page number 2 (1500/1024 + 1). If the table is stored sequentially, the corresponding entry in the process's page table can be quickly computed as an offset into the table. If the reference is to an address in a page held in memory, the corresponding memory address is an offset into the memory page indicated in the table. Using Table 11.3 as an example, a reference to address 700 (offset of 700 into page 1) would translate into an offset of 700 into memory page 214. Once again, fixed page size allows us to quickly calculate the corresponding memory address as 219836 (217 * 1024 + 700).[2]

[2] Note that all of these calculations are similar to those used for addressing array contents, as described in Chapter 4.

An area of secondary storage is normally reserved exclusively for the task of holding pages not held in (or swapped from) main memory. This space is normally referred to as the *swap space*. This space is divided into pages in the same manner as main memory. A memory reference to a page held in the swap space results in that page being loaded into a page frame in memory. As with the resolution of addresses in main memory, the location of the page within the swap space can be quickly computed by multiplying the page number by the page size.

If the system is fully loaded (i.e., all page frames are currently in use), a page fault in the currently executing process implies not only a page swapped in from memory, but an existing page swapped out of memory. The selection of the page to be swapped out (sometimes called the *victim*) can be made on any number of criteria. Some common methods for selecting the victim are

- Least recently used
- Least frequently used

Each of these strategies requires the operating system to maintain information about the utilization of each page in memory. Searching and updating this information is a part of the system overhead associated with virtual memory management.

When a victim has been selected, it might or might not be copied back to the swap space. Some operating systems maintain a copy of all process pages in the swap space. In this case, the contents of the page in memory are simply copied to the corresponding page in the swap space. The sample data in Table 11.3 shows an entry that indicates whether a page has been modified since it was swapped into memory. If a page has not been modified, the copy held in the swap space is identical to the copy in memory. This implies that the contents of the memory page do not have to be copied to the swap space if that page has been selected for replacement.

Memory Protection

Memory protection refers to the protection of memory allocated to one program from unauthorized changes by another program. This can apply to interference between programs in a multitasking environment or interference between a program and the operating system in either a single- or multitasking environment. A lack of memory protection allows errors in one program to generate errors in another. If the program interfered with the operating system, this could result in the computer "locking up."[3]

[3] This phenomenon is a frequent occurrence with simple microcomputer operating systems such as MS-DOS. Normally, the only way to recover from such an error is to reboot the computer.

In its simplest form, memory protection requires that each write to a memory location be checked to ensure that the address being written has been allocated to the program performing the write operation. Complicating factors include the use of various forms of indirect addressing, virtual memory, and cooperating processes (i.e., two or more processes that "want" to share a memory region). Regardless of the form of addressing used, memory protection adds overhead to each write operation to check ownership and other protection information.

Memory Management Hardware Trends

Early implementations of multitasking, protected memory writes, and virtual memory were based exclusively on modifications to systems software. However, implementation through software imposes severe performance penalties on the system as a whole. Consider, for example, the overhead required to map program memory references in a virtual memory environment. Each reference requires the operating system to search one or more tables to locate the appropriate page and to determine the corresponding memory or disk location of that page. Thus, a memory reference that should consume only one or a few CPU cycles consumes many additional cycles for the paging, swapping, and address-mapping functions.

The benefits of advanced memory addressing and allocation schemes are quickly offset when they must be implemented in software. Because of this, one trend in the design of computer systems in general and of the CPU in particular has been to incorporate support for these techniques directly in the hardware. This hardware support has appeared in mainframe systems for many years, and has recently appeared in microcomputers.

As an example, consider the performance difference between the Intel 80286 and 80386 processors. The former is a 16-bit processor without virtual memory support; the latter is a 32-bit processor with virtual memory support (paged memory and related addressing functions). When running applications in a real (not virtual) memory operating system such as MS-DOS, the difference in performance can be explained almost entirely by their differing word sizes. When running applications under a virtual memory operating system such as Unix, the 80386 can be as much as 20 times faster than the 80286. The bulk of this difference is due to the virtual memory hardware support functions present in the 80386.

This trend is present not only in memory allocation and addressing, but also for multiple-processor systems and applications that share memory. The latest generation of hardware is being implemented with direct support for multiple CPUs and the sharing of memory between those CPUs. These advances promise substantial performance improvements over the software-based allocation and control procedures used with more traditional hardware.

SUMMARY

An operating system is the most complex component of systems software. Its primary purpose is to manage hardware resources and to provide support services to users and application programs. In functional terms, the operating system is responsible for managing the CPU, memory, secondary storage, and I/O devices. It provides service functions by allocating these resources through file, process, and user management services.

From an architectural viewpoint, the operating system consists of the kernel, the service layer, and the command layer. The kernel performs all hardware management and access functions. The service layer consists of a set of programs that provide basic system services to users or their application programs. These can be classified as process, memory, file, and I/O services. The command layer is the user interface to the operating system. It is used to control the execution of application and utility programs and for direct user control over hardware resources and files.

The resource allocation and process control functions of operating systems have evolved in parallel with hardware capability. Early operating systems provided no system services and were restricted to executing single application programs in batch mode. Later milestones in the ability of operating systems to manage application programs include overlapped processing, stacked job processing, dynamic processing, multiprogramming, and virtual memory. Each of these represented successively more sophisticated capabilities for process control as well as more efficient utilization of hardware resources.

The operating system is responsible for allocating hardware resources to individual user processes on demand. This is a simple matter in single-tasking operating systems, because only one application program can be active at a time. Multitasking operating systems must implement substantially more complex procedures for resource allocation. The operating system must mediate contention by multiple processes for access to the same resources while ensuring that each process receives the resources it needs. This mediation is performed primarily through operating system procedures for scheduling, dispatching, and interrupt processing.

An active process is always in one of three states: ready, running, or blocked. In a single-CPU machine, only one process can be in the running state at any one time. Ready processes are waiting only for access to the CPU. Blocked processes are waiting for some event to occur. Such events are normally the completion of a service request or the correction of an error condition. Various methods exist for determining the order and duration of process access to the CPU. Methods for determining order of dispatch include first-come, first-served; shortest time remaining; and explicit prioritization. Methods for determining the duration of a process's control of the CPU include priority interrupt and time sharing. Operating systems often implement dispatching and scheduling by a combination of these methods.

Memory is normally allocated through some form of partitioning scheme. A memory partition is a fixed or variable portion of available main memory. Processes are allocated to one or more memory partitions to store instructions and data. The operating system maintains tables of information concerning the allocation of specific memory partitions to specific processes. Memory references to a program are mapped to its memory partitions through table lookups and address calculations.

Modern operating systems implement a form of memory allocation and management called virtual memory. Under virtual memory, portions of processes are held in small partitions of memory called pages. Some pages might be held in main memory; some might be held in the swap area of secondary storage. Process pages are swapped between main memory and secondary storage as dictated by the needs of individual processes and the availability of main memory.

Complex memory management procedures require a substantial amount of operating system overhead. Techniques such as partitioned, virtual, and protected memory require address calculations and the maintenance of tables of memory allocation information. To reduce the amount of processor time allocated to such tasks, modern CPUs implement many of these functions within the CPU. This speeds such processing considerably and leaves more CPU cycles free to execute user application processes.

Key Terms

address mapping (resolution)
batch mode
batch processing
blocked state
child process
command layer
compaction
compute (CPU) bound
concurrency
concurrent processing
contiguous allocation
dispatching
dynamic processing
explicit prioritization
first-come, first-served (FCFS)
frame
interleaved execution
interrupts enabled
interrupts masked
I/O bound
job
job control command
job control language (JCL)
job queue
job step
kernel
multiprogramming
multitasking
noncontiguous (fragmented) allocation
nonmaskable interrupt
nonpreemptive scheduling
overlapped processing
page
page fault
page table
parent process
partitioned memory
piping

preemptive scheduling
process control block (PCB)
program cycle
ready state
real resource
real-time scheduling
remote job entry (RJE)
run queue
running state
scheduling
service call
service layer
shortest time remaining
single tasking
single-job processing
spawning
spooling
stacked job processing
supervisor
swap space
time sharing
time slice
victim
virtual machine
virtual resource

Vocabulary Exercises

1. A type of processing in which the operating system supports multiple active processes is called _____ processing or _____.
2. Under virtual memory management, the location of a memory page is determined by searching a _____.
3. A _____ occurs when a program requires a memory page not held in memory.
4. A process in the _____ requires only access to the CPU to continue execution.
5. In the _____ scheduling method, processes are dispatched in order of their arrival.
6. Under virtual memory management, the location of a memory page is determined by searching a _____.
7. _____ scheduling refers to any type of scheduling in which a running process could lose control of the CPU to another process.
8. The act of selecting a running process and loading its register contents is called _____ and is performed by the _____.
9. To achieve efficient use of memory and a large number of concurrently executing processes, most operating systems utilize _____ memory management.
10. A _____ application program requires no user interaction during its execution.
11. When a process makes an I/O service request, it is placed in the _____ until processing of the request is completed.

12. Memory pages not held in primary storage are held in the _____ of a secondary storage device.
13. On a single-processor computer, multitasking is achieved through _____ execution of processes.
14. A _____ is apparent to a process or user, although it might not physically exist.
15. Under a _____ memory allocation scheme, portions of a single process can be physically located in scattered segments of main memory.
16. A _____ process spawns a _____ process by means of a process creation service request.
17. Under virtual memory management, memory references by a process must be converted to an offset within a _____.
18. Information about a process's execution state (e.g., register values, status, etc.) is held in a _____ for use by the _____.
19. Under the _____ scheduling method, processes requiring the least CPU time are dispatched first.
20. The detection of a _____ normally causes the currently executing process to be _____ and control to be passed to the _____.

Review Questions

1. Describe the functions of the kernel, service, and command layers of the operating system.
2. How/why does single-job processing make inefficient use of computer resources?
3. What common motivation underlies many of the advances in operating system capability (e.g., dynamic processing, multiprogramming, and virtual memory)?
4. What is the difference between a real resource and a virtual resource?
5. What hardware resources are allocated to processes by the operating system?
6. How/why does a process move from the ready state to the running state? How/why does a process move from the running state to the blocked state? How/why does a process move from the blocked state to the ready state?

7. What is process control block and for what is it used?
8. What alternatives are available for determining the movement of processes between the ready and running states? Briefly describe the operation of each.
9. What scheduling complexities are introduced by real-time processing requirements?
10. What are the comparative advantages and disadvantages of memory allocation using fixed- and variable-size memory regions?
11. Describe the operation of virtual memory management. Under what assumptions is it more efficient than other methods of memory management?
12. What is memory protection? What capabilities does it imply in the operating system and/or hardware?

Research Problems

1. As stated in the chapter, the Intel 80386 (and later processors in this family) incorporate hardware support for virtual memory management. Investigate the architecture of the 80386 processor, with particular emphasis on memory addressing and virtual memory management. What functions are provided to the operating system to support virtual memory management? What functions must still be performed by the operating system itself?

2. Versions of Microsoft Windows (up to and including version 3.1) add multitasking capabilities to the MS-DOS operating system. However, MS-DOS itself is not a multitasking operating system. Investigate the process and memory management facilities of Microsoft Windows. How is multitasking supported? What scheduling method(s) are used? Is virtual memory management supported? How efficient are the methods used for multitasking and virtual memory?

12 Mass-Storage Access and Management

Chapter Goals

- *Describe the components and operation of a file management system.*
- *Describe the logical and physical organization of files and directories.*
- *Describe the operations that manipulate data, files, and directories.*
- *Describe the mechanisms used to share and protect files.*

FILE MANAGEMENT OVERVIEW

In most operating systems, a file is the fundamental unit of storage for both data or programs. Files are typically stored on secondary storage devices, including optical disks, magnetic disks, and magnetic tape. Because of the number and importance of files, their management is a critical function of systems software. For purposes of clarity, we will refer to the entire collection of systems software programs that perform file management and access functions as a *file management system*. In most systems, the bulk of these programs and functions is provided by the operating system. However, some systems utilize additional software—for instance, a database management system—to extend the file management capabilities of the operating system.

File Content and Type

To restate the definition given in Chapter 4, a *file* is a collection of related data items. Typically, a file has a unique identifier such as a symbolic name, and can be manipulated in various ways as a single entity by copying, deleting, and so on. Many types of data can be stored in a file:

- Character, numeric, and other types of data used by application programs
- Symbolic instructions
- Machine instructions

Data used by application programs can be of many types. For example, a customer file used by a billing program could contain account numbers, names, addresses, account balances, and other data items. Each of these items can be of a different data type or structure. For example, the account number might be encoded as an integer, the name and address as character arrays, and the account balance as a floating-point number.

Symbolic instructions are normally stored as sequences of characters. Thus the source code of a COBOL or FORTRAN program is typically stored as a sequence of characters comprising the program statements. The sequence of characters in the file would normally follow the sequence of the characters as read or written on a printed page. In contrast, machine instructions are generally stored in a format that can be directly manipulated by the CPU. Thus, an executable program stored in a file consists of a sequence of instructions encoded in the instruction format of the CPU. Individual components of the file might be encoded as integers (e.g., the op code), addresses (e.g., operands of a load or store instruction), and other primitive data formats.

Due to the wide variety of file contents and methods of accessing those contents, files may require varying methods of storage and manipulation. Because file manipulation is such a pervasive aspect of information processing, it is desirable to implement all file manipulation functions within systems software (e.g., as utility programs and/or as service-layer functions). However, it is difficult to design a file management system that accounts for all possible variations in file content and organization. Such a system would be excessively complex and have a very large number of file-oriented commands and service routines.

In most file management systems, a compromise is struck in which a limited number of predefined *file types* are provided. For example, a simple file management system can define the following types of files:

- Executable programs
- Operating system commands
- Data

Generally, when file types are defined, every file must be of one (and only one) type. Among other things, the type of file determines

- The physical organization of data items and structures within secondary storage
- The methods of I/O that can be used with the file (e.g., sequential, random, or indexed access)
- The operations that can (or cannot) be performed on the file
- Restrictions on the symbolic name of the file

The file management system will normally use differing physical organizations for differing types of files. Thus, the physical storage format of an executable program (a file of machine instructions) will normally differ from the physical structure of a file that contains data. These differences are usually implemented for the sake of efficiency. For example, files containing machine language programs are normally organized to minimize the complexity of loading their contents into memory for execution. Operating systems that implement virtual memory management can utilize a physical organization that facilitates paging and swapping operations.

Some file management systems further differentiate physical structure within file types. For example, a variety of physical storage formats can be used for data files, depending on whether access to those files will be sequential or random. In this case, an application program would typically declare the type of access when interacting with a file through service-layer functions. The operating system will normally provide a different set of file I/O service calls for each type of access. A programming language, such as BASIC, or an application development tool, such as a database management system, could impose additional physical structure upon files containing data.

Within many file management systems, the type of file determines the operations that can be performed on it. For example, it should not be possible for a user to load and execute any file that does not contain a machine language program. Similarly, it should not be possible to use an ordinary word processor or text editor to edit a file containing a machine language program. Implementing such restrictions on access and appropriate use requires a mechanism by which systems software can determine the content or structure of a file.

File type is normally declared at the time a file is created. In some systems, file type is merely stored as part of a file's directory entry. The file type can be viewed when a user examines a directory, but it is not otherwise obvious. In other systems, the file type can be indicated by a naming convention for the file's symbolic name. A common approach is to require that the filename extension, or appending identifier, match the intended use. The file management system uses these identifiers to match file type to allowable operations.

For example, consider the naming conventions for executable files used by MS-DOS. Filenames with the extensions .COM and .EXE are assumed to contain executable machine language programs, each stored in a prespecified format. Filenames with the extension .BAT are assumed to contain the text (ASCII characters) of one or more (batch) commands to be interpreted by the operating system command layer. Many programs that execute under MS-DOS make additional assumptions about file-naming syntax and extensions. Examples include .ASM (macro assembler source code file), .DBF (dBASE IV database file), .WKS (Lotus 1-2-3 spreadsheet file), and many others.

A tradeoff for the protection and efficiency provided by file typing is the complexity and size of the operating system. Additional routines are required to generate, test for, and act appropriately on each file type. A further disadvantage can be that programmers and users will be limited to the set of predefined file types, none of which might be entirely suitable for a specific application.

File Management System Functions and Components

The primary purpose of the file management system is to provide I/O capabilities to and from files. Thus, many of its functions directly implement file I/O operations. However, because the number of files in a typical computer system is large, the file management system must also provide facilities for managing and administering groups of files and secondary storage devices. All of the functions can be loosely grouped into the following categories:

- Creation and manipulation of files
- File I/O operations
- Management of secondary storage devices

- File security
- File backup and recovery

These functions are normally implemented within the operating system. However, some systems might use additional systems software (e.g., a database management system) to extend the native capabilities of the operating system.

A file management system is implemented in layers similar to those of the operating system as a whole. These layers form a hierarchy that spans from the lowest (or device) level to the highest (or user) level. From high to low, layers within a file management system include these:

- Application and command
- File control
- Storage I/O control
- Secondary storage devices

These levels of the file management hierarchy are illustrated in Figure 12.1 and compared to the more general representation of operating systems software layers.

At the hardware level, a secondary storage device must be capable of sensing specific tracks or locations on data-recording media and of controlling the device so that a desired storage location is positioned beneath a read/write head. This low-level function is generally implemented by a special-purpose processor within the device or device controller. As discussed in Chapter 8, the device controller presents a logical view of the storage device or media to systems software. This logical view is usually a linear sequence of storage locations (or blocks).

Software modules within the *storage I/O control layer* include device drivers and interrupt handlers. These modules generally reside within the kernel of the operating system. Their function is to provide for data transfer between main memory and secondary storage devices. Such transfers are typically handled in blocks, with error checking performed on each block. Accesses are made via the system bus to an I/O port corresponding to a secondary storage controller. Access requests from the CPU generate signals on the control, address, and data lines of the system bus.

Within the *file control layer*, basic file manipulation and I/O capabilities are implemented through service routines. Also within this layer, operations on files are correlated (mapped) to specific storage locations on secondary storage devices. This is the level at which the directory structure of the file management system is implemented. The directory structure is an integrated and consistent map that correlates all files in the system with secondary storage locations that hold those files. Accordingly, physical allocation of storage space is handled at this level.

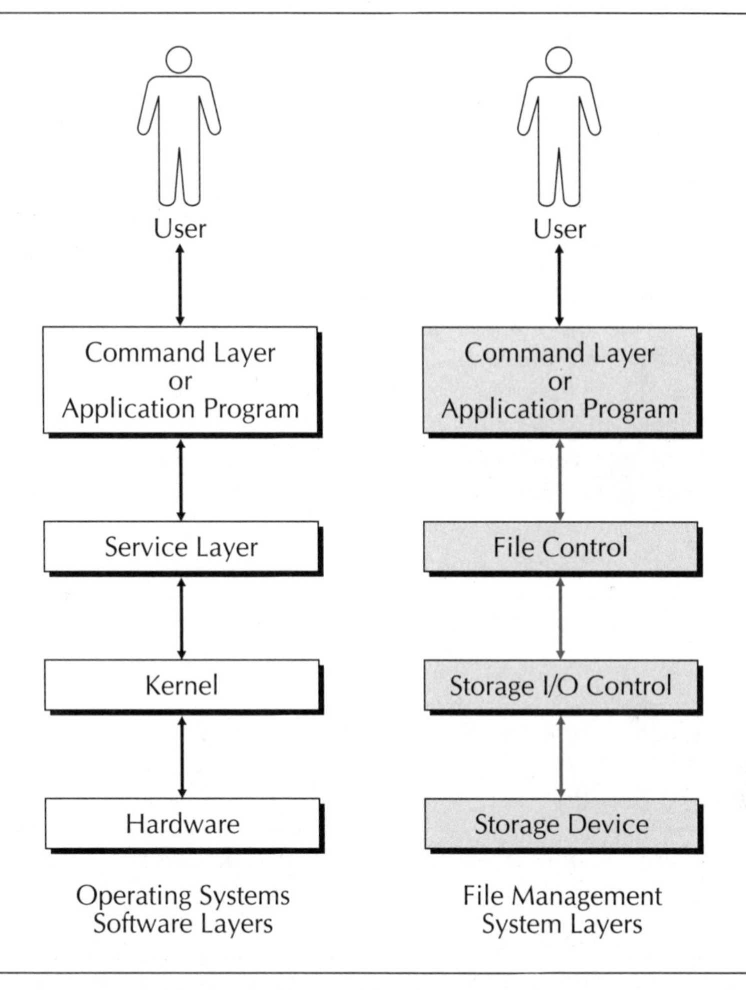

FIGURE 12.1 **Layers of the file management system and their relationship to the more general concept of operating system layers.**

FILE AND DEVICE ORGANIZATION

The file control layer forms a bridge between the physical file management system and the logical file management system. Within this layer, directory entries and storage locations of specific devices are correlated with symbolic names by which files are referenced at higher levels. Users and application programs can perform operations on files and directories by referencing these names. These operations are translated into the equivalent low-level commands needed to physically implement these operations. As with other aspects of layered software, this approach insulates users and application programmers from the details of low-level device manipulation.

Logical File Structure

The *logical file view*, or *logical file structure*, of file content is the user's perspective. That is, the logical structure of a file corresponds with presentation of its data under a format that is meaningful to users. As shown in Figure 12.2, the logical structure of a typical data file is subdivided into multiple *records*. A record typically corresponds to a single person (e.g., a customer or employee), thing (a product held in inventory), or event (a transaction). Each record is composed of multiple *fields*. A field usually contains a single data item that describes the subject of the record. For example, a record might contain information about a customer and be composed of individual fields for account number, customer name, customer address, and so on.

Records	Fields: 1	2	3	4	5	6	
1	99635	Smith	George	R	1414 Oak Road	Los Angeles	...
2	04667	Jones	Alice	L	12 Main Street	Buffalo	...
3	52046	Torres	Maria	H	9823 7th Ave.	Chicago	...
4	34421	Miller	Fritz	S	451 Quincy Rd.	Atlanta	...
5	63752	Chang	Wendy	B	310 G Street	Fort Worth	...

FIGURE 12.2 **The logical structure of a file as a group of related records, each of which is in turn a group of related fields.**

The logical structure of a file exists independent of the physical devices on which it might be stored. In essence, the logical structure of a file corresponds to the user's view of its contents. The *physical file structure* corresponds to the manner in which individual bits and bytes of data are represented and organized on physical storage media. These two structures are rarely equivalent. A number of characteristics of physical file structures are simplified or ignored in their corresponding logical file structures, including

- Physical storage allocation
- Method(s) of physical data access
- Data encoding

Allocation considerations include the placement of fields and records within a file and the distribution of file components across storage locations on specific devices. Physical data access considerations include the use of directories (and related structures) and the use of indices, links, and other forms of nonsequential storage allocation. Issues of data encoding include the data structures and coding methods that represent individual fields. Related issues include data encryption and data compression. These issues are fully discussed in later sections of this chapter. In sum, the logical view of a file is a sequential set of records and fields that are immediately and exclusively available to a single user or application program. Physical storage and access considerations are simply not relevant to this view.

Storage Allocation

Storage allocation refers to the methods by which individual storage locations are correlated with individual files. The unit of storage allocation generally corresponds to the unit of data transfer to or from secondary storage devices (i.e., a block). Block sizes typically range from 512 bytes to 4 Kbytes, in multiples of 512 bytes. Storage allocation is generally simplest for systems that use a single unit size for all secondary storage devices. Thus, if a system supports disk drives with block sizes ranging from 512 bytes to 4 Kbytes, the largest common multiple (4 Kbytes in this case) might be the common unit of storage allocation. However, some file management systems do use differing allocation unit sizes for devices with differing block sizes. This tends to utilize the storage space of each device most efficiently at the expense of more complex storage allocation procedures.

For any storage device, a record of the allocation of storage locations to specific files must be kept. This record (or map) is sometimes called a *file allocation table*. The format and content of this table depend on the basic method of storage allocation. In general, allocation can be either contiguous or noncontiguous.[1] As with the allocation of other types of storage such as memory, contiguous allocation is relatively simple and requires little management overhead. Noncontiguous allocation is comparatively complex, but more flexible. With either method, there must be some degree of coordination between the file allocation table and the data structures used to store directory information.

Contiguous storage allocation. With contiguous allocation, most storage allocation information can be stored with a file's other directory information. This includes the identifier (number) of the first block allocated to a file and the number of blocks allocated to a file. Figure 12.3 shows a set of storage blocks allocated to three files. The blocks allocated to each file are similarly shaded. Note that all blocks allocated to an individual file are in a contiguous sequence. Table 12.1 is a simplified directory table that includes storage allocation information for these files. Note that a fourth file, SYSFREE, has been added to the table to represent the unused storage blocks.

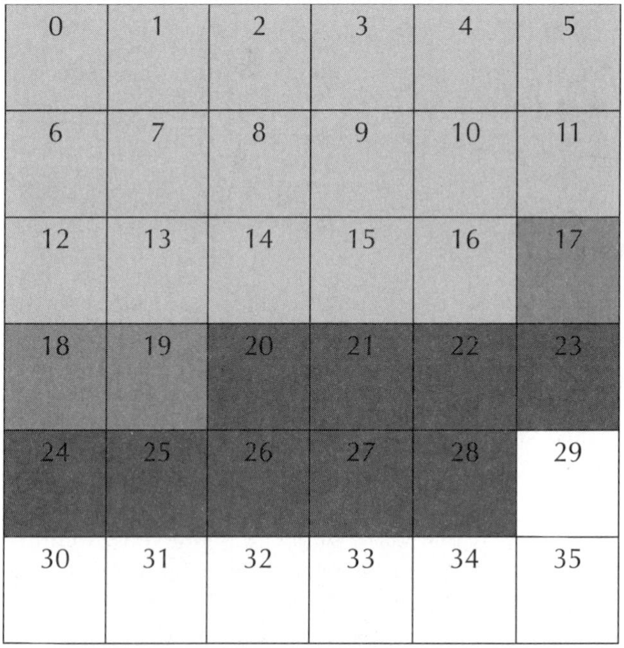

0	1	2	3	4	5
6	7	8	9	10	11
12	13	14	15	16	17
18	19	20	21	22	23
24	25	26	27	28	29
30	31	32	33	34	35

FIGURE 12.3 The contiguous allocation of storage blocks to three files.

[1] The reader may want to review the discussion of memory allocation in Chapter 11 prior to reading this material.

TABLE 12.1 Simplified directory entries for the files depicted in Figure 12.3. The entries include storage allocation information.

File Name	Owner Name	First Block	Length (In Blocks)
FILE1	SMITH	0	17
FILE2	JONES	17	3
FILE3	SMITH	20	9
SYSFREE	SYSTEM	29	7

The most significant problem with contiguous storage allocation is the fragmentation of storage units that occurs after creating and deleting many files. Figure 12.4 shows the storage map from Figure 12.3 after the deletion of FILE1 and the creation of a new (but smaller) file. Note that there are now two separate regions of unallocated space on the device. Thus, the simple use of the dummy file SYSFREE is no longer adequate to connect all unallocated storage blocks.

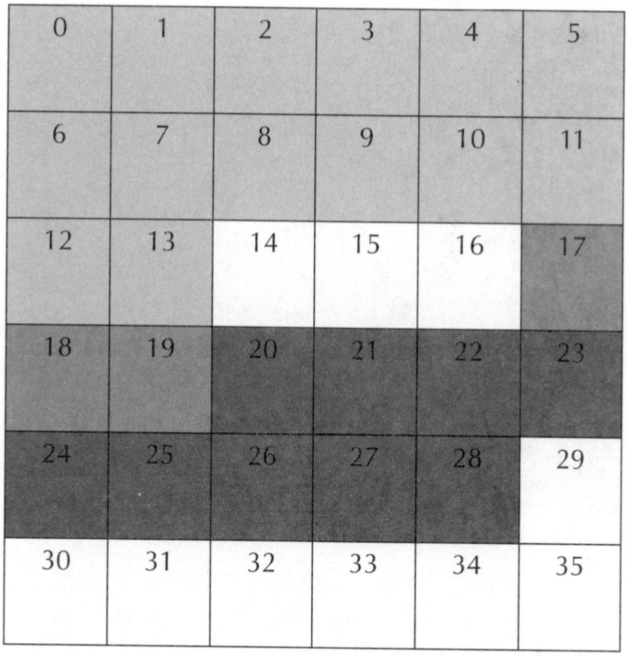

0 1 2 3 4 5
6 7 8 9 10 11
12 13 14 15 16 17
18 19 20 21 22 23
24 25 26 27 28 29
30 31 32 33 34 35

FIGURE 12.4 The allocation of storage blocks after the deletion of FILE1 and the creation of a new (smaller) file.

Multiple files are required (one for each contiguous region of unallocated blocks). Over a long period of time, the number of these regions tends to grow, and the size of individual regions tends to shrink. Because blocks allocated to a file must be contiguous, the maximum size of a newly created file is limited to the size of the largest region of unallocated blocks.

One method of counteracting the problem of many small unallocated storage regions is *compaction*, which refers to the collection of multiple groups of unallocated storage blocks into a single contiguous region. It is performed by moving all files to the lowest numbered blocks possible. This has the effect of "squeezing out" small unallocated regions and moving them to higher numbered storage blocks. Figure 12.5 shows the results of compacting the files shown in Figure 12.4. The process of compaction is relatively expensive because it requires the movement of large numbers of blocks. In some file management systems, compaction is automatically performed when needed—for instance, when a user attempts to create a file that exceeds the size of the largest available group of unallocated blocks. In most file management systems, compaction is performed by a special utility only when explicitly requested by the system operator or administrator.

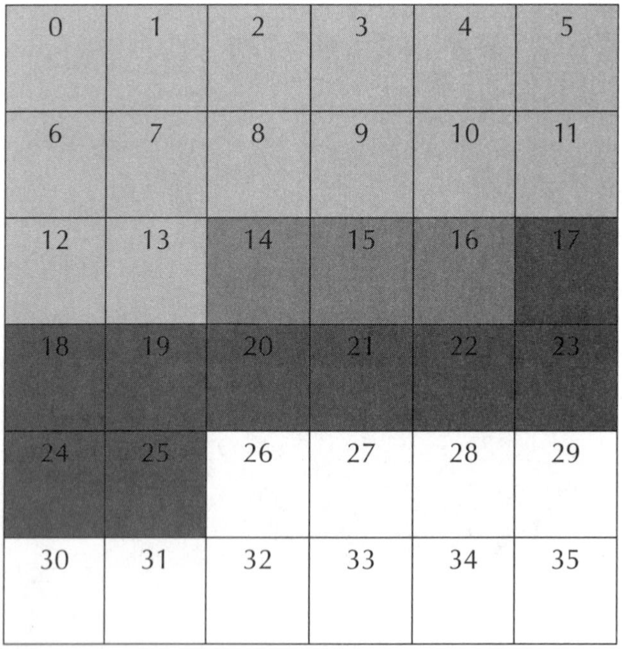

0	1	2	3	4	5
6	7	8	9	10	11
12	13	14	15	16	17
18	19	20	21	22	23
24	25	26	27	28	29
30	31	32	33	34	35

FIGURE 12.5 The results of compacting the files in Figure 12.4.

Noncontiguous storage allocation. Because of the method's relatively low degree of flexibility, few file management systems utilize contiguous storage allocation. Instead, some form of noncontiguous allocation is used. However, noncontiguous allocation methods require substantially more complex procedures for recording the allocation of storage blocks to individual files. Keeping track of unallocated blocks can also be more complex. Both types of complexity add additional processing overhead to any file I/O operation (e.g., appending new records to an existing file) that requires storage blocks to be allocated or deallocated.

Figure 12.6 shows a noncontiguous allocation of storage blocks to the files shown in Figure 12.5. Simplified directory entries for these files are shown in Table 12.2. Note that these entries are identical in structure to those in Table 12.1. Although the storage allocation information stored in the directory is the same under contiguous allocation, it cannot be used in the same manner. This is because the second, third, and subsequent blocks of a file can no longer be assumed to immediately follow the first block. Because blocks allocated to a file can be widely dispersed throughout a storage device, some method other than the assumption of contiguous allocation must be used to correlate allocated blocks.

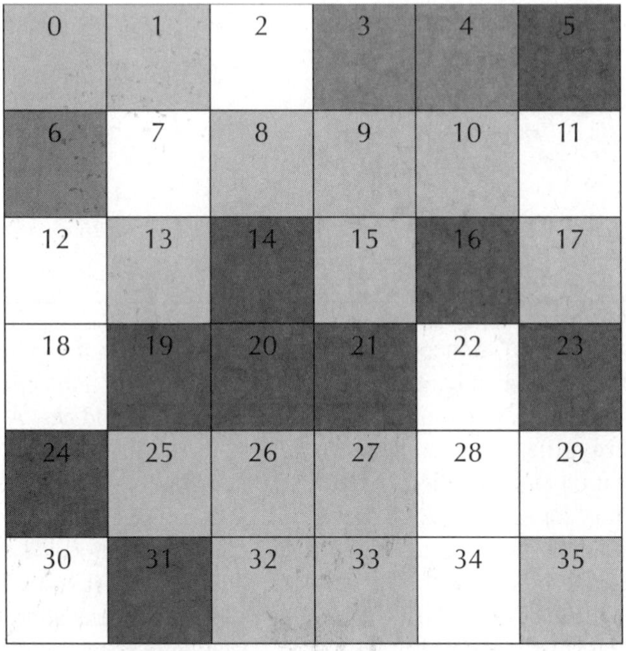

FIGURE 12.6 **The noncontiguous allocation of storage blocks to three files.**

TABLE 12.2 **Simplified directory entries for the files depicted in Figure 12-6. The entries include storage allocation information.**

File Name	Owner Name	First Block	Length (In Blocks)
FILE4	SMITH	0	14
FILE2	JONES	3	3
FILE3	SMITH	5	9
SYSFREE	SYSTEM	2	10

TABLE 12.3 **Storage allocation information for a system that uses noncontiguous storage allocation.**

Block	Entry	Block	Entry	Block	Entry	Block	Entry
0	01	9	10	18	22	27	32
1	08	10	13	19	20	28	29
2	07	11	12	20	21	29	30
3	04	12	18	21	23	30	34
4	06	13	15	22	28	31	End
5	14	14	16	23	24	32	33
6	End	15	17	24	31	33	35
7	11	16	19	25	26	34	End
8	09	17	25	26	27	35	End

In general, noncontiguous allocation requires a separate data structure to record the allocation of blocks to individual files. A number of data structures can be used, including arrays (tables), linked lists, and indices. A common structure is an array (file allocation table) containing linked lists. Table 12.3 shows such a file allocation table for the storage device depicted in Figure 12.6. The table contains an entry for each block on the storage device. The entry for each block is a pointer to the next block. Thus, block 0 (the first block allocated to FILE4) contains a pointer to block 1 (the second block allocated to FILE4). These pointers form a linked list (or chain) that ties together all of the blocks allocated to a specific file. The last block allocated to a file must contain a special code to indicate that it is the final block. Note that all unallocated blocks are also connected through table entries into a single chain. This simplifies the task of finding free blocks to allocate to new or expanded files.

Blocking and Buffering

For storage allocation purposes, files are structured at two levels:

- Logical records
- Physical records

At the level of application programs, files are composed of *logical records*. A logical record is the collection of data items, or fields, that is presented to an application program in a single read operation. A *physical record* is the unit of data storage that is accessed by the CPU or storage device controller in a single operation. For disks and other devices that use fixed-size units of data transfer, a physical record is equivalent to a block. For storage devices with variable-size data transfer units (e.g., tape drives), block size might differ from physical record size, or it might be undefined.

If logical record size is less than physical record size, multiple logical records can be contained within a single physical record. If logical record size is larger than physical record size, multiple physical records are required to hold a single logical record. The grouping of logical records within physical record storage areas is called *blocking*. If a physical record contains just one logical record, the grouping is said to be *unblocked*. However, logical records seldom fit neatly into predefined physical areas. If the size of a logical record is smaller than the physical record area and only one logical record is contained therein, the result is an internal fragment, or wasted space. The objective of blocking is to increase the efficiency of storage allocation by storing two or more logical records within a single physical record, as shown in Figure 12.7.

Within the file management system, a *buffer* is a region of main memory that is dedicated to file I/O operations. The user generally has no knowledge of or control over this area, which is managed exclusively by systems software.

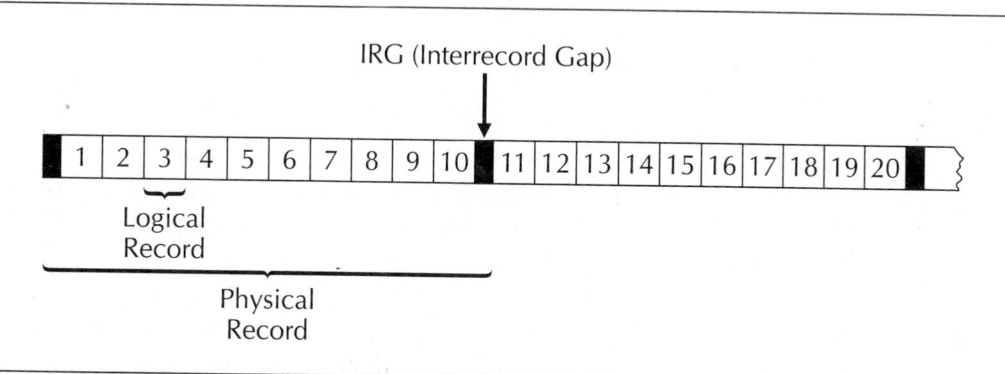

FIGURE 12.7 The blocking technique for magnetic tape, under which multiple logical records are included within one physical record.

The user's working storage area, or program data area, is a separate portion of main memory that is under the control of the application program. *Buffering* is the process of interposing a buffer area between the program data area and files that reside in secondary storage, as shown in Figure 12.8. The objective is to enhance the efficiency of system I/O by using main memory to aid the conversion between logical and physical record formats.

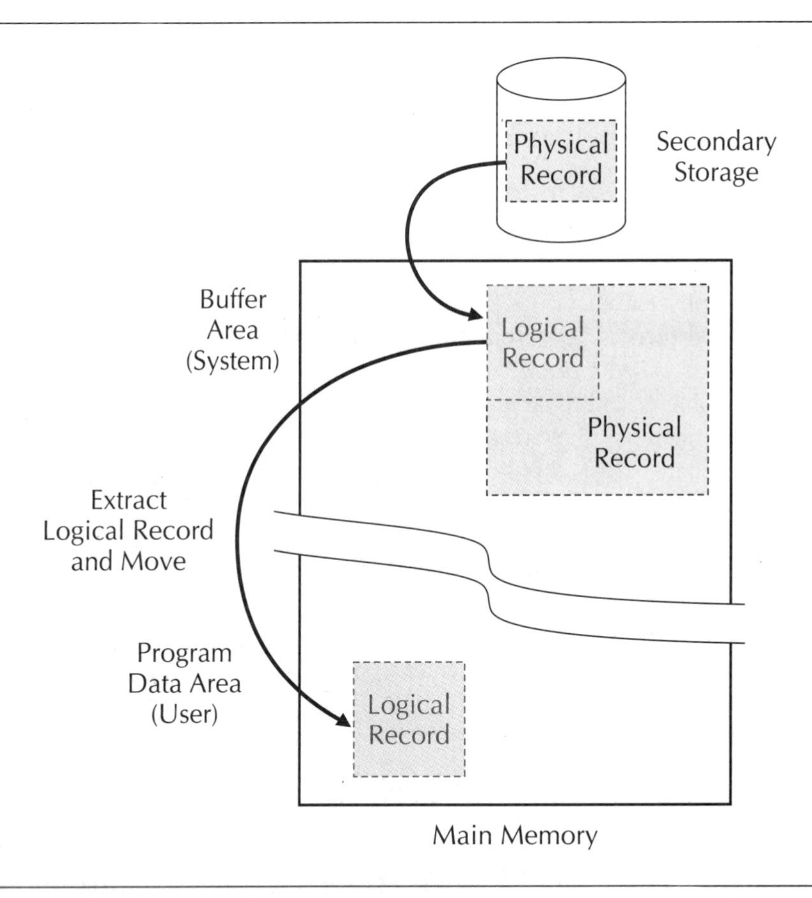

FIGURE 12.8 A buffering technique under which the file system loads a physical record from disk into a buffer area in memory, then extracts the desired logical record and moves it to the program data area.

As stated above, at the level of application programs, logical records are the unit of file I/O. However, data is moved between memory and secondary storage in units of physical records. When an application program requests access to a logical record, the file management system instructs the device controller to read the physical record in which the logical record is stored. When the storage device controller has accessed the physical record, it is moved into the buffer. Again, this physical record can contain two or more logical records, only one of which was requested by the application program.

The file management system extracts the desired logical record from the buffer, moves it to the program data area, and returns system control to the application program. If the application program issues another read request for the next (logical) record, the file management system simply retrieves that record from the buffer, without a corresponding read operation to the storage device. Thus, the use of a sufficiently large buffer can reduce the number of physical read operations from a storage device. Because accesses to secondary storage are typically much slower than access to RAM (i.e., the buffer), performance could be substantially improved.

If the logical record is large and is stored in multiple physical blocks, the file management system must combine the physical records to reconstruct the required logical record. Typically, the file management system will use a buffer that is large enough to hold all of the physical records corresponding to a single logical record. If the size of the logical record exceeds the capacity of the buffer area, the logical record must be moved to the program data area in a series of physical read and buffer copying operations.

File Organization Methods

File organization refers to the placement of records within the storage space allocated to a file. Note that this is an entirely separate issue from the allocation of storage blocks to a file. That is, the blocks allocated to a file should appear as linear sequence for purposes of accessing file contents. The physical position of those blocks on a storage device should be unknown (and irrelevant) to file manipulation procedures. This division of knowledge is implemented through the layers of the file management system, as illustrated in Figure 12.1. In particular, issues of file access and organization are dealt with entirely within the file control layer. The mapping of logical file blocks to physical storage blocks is implemented entirely within the storage I/O control layer. In essence, the storage I/O control layer "fools" the file control layer into thinking that each file is sequentially stored in blocks, starting at the beginning of the storage device.

Several methods of file organization, including sequential, direct, and indexed, are possible. Also, hybrid methods are sometimes implemented (e.g., indexed sequential). *Sequential file organization* is the simplest method, because it makes no assumptions about file content and requires no special data structures to implement file access methods. The records of the file are simply stored in adjacent logical storage locations. Although data or records can be sorted or otherwise ordered within this sequential space, that fact is unknown by (and irrelevant to) the file management system.

Direct file organization assumes that the storage location of a record depends on the value of some field within that record. In general, it requires the use of a key field that can be mapped into a logical record number within the file. For example, consider a customer file in which each record is identified by a unique field called ACCT#. Assume further that account numbers are always six-digit integers, and that the lowest numbered account is 100000. Direct access to records of this file can be implemented by correlating logical record numbers to account numbers via the formula

ACCT# – 100000 = logical_record_number

Thus, for example, the customer record corresponding to account number 100423 is stored in logical record 423 (100423 – 100000).

Direct file organization works well under the following assumptions:

- The key field is unique and is easily converted into a logical or physical storage address.
- Allowable key values fall within a relatively narrow range.
- There are few gaps in the sequence of keys currently in use.

The first assumption assures that minimal processing overhead will be consumed by access operations. In general, the more complex the conversion procedure, the more processing overhead will be added to each record access. The last two assumptions are closely related in that both help to ensure little or no wasted storage space. The first assumption guarantees that the fewest number of storage locations must be allocated to the file. The second guarantees that few allocated storage locations will be empty. For example, if the first currently used key value is 100000 and the next currently used key value is 100423, 422 storage locations in the file are currently empty. Empty storage locations are minimized when keys are assigned in sequence and when old keys are reused as soon as possible (e.g., the account number of a deleted customer is assigned to the next new customer).

Indexed file organization is frequently used when all the assumptions needed to assure successful use of direct organization cannot be met. It can also be used to support both sequential and random access to the same file. Indexed organization uses a table of key values and storage locations that is stored separately from the records of the file. The index contains an entry for each currently used key value and correlates it with a logical record number or physical storage location. The entries are generally sorted by ascending key value, to allow efficient searching of the index. In some file management systems, the index is stored separately from the data records in an entirely separate file. In others, storage blocks for the index are allocated at the beginning or end of the file.

Indexed file organization generally requires more storage space than sequential or direct organization. This is because both the index and the data records must be stored. In addition, accesses using the index require processing overhead to search for key values in the index. However, these disadvantages are often offset by the increased flexibility of file access, made possible by indexed organization.

Note that each type of file organization makes certain assumptions about file contents and internal organization. Because the assumptions and structures differ, the method of file organization must be declared when a file is created. This allows the file management system to initialize the appropriate storage structure for that type of organization. A file's organization method is typically stored within the directory for use by the operating system when performing file I/O operations.

File Access Methods

Access to file contents can vary by the units of file I/O and the order (or lack thereof) in which those units are read or written. Virtually all file management systems support byte-oriented file I/O. In this form of I/O, a file is assumed to consist of a sequence of byte values. No assumptions are made by the file management system as to the interpretation of those bytes or their organization into higher level data structures (e.g., arrays and records). Some file management systems also support record-oriented file I/O. In this form of I/O, entire records are read or written as integrated units. Some method must be provided to mark the start (and end) of records within the file. This is typically implemented using a control character such as ASCII 30.

A file management system can support a variety of methods for accessing file contents. Some of the simpler methods that are typically supported are

- Sequential
- Direct (random)
- Indexed

Depending on the file management system, a variety of file organization methods can be used to support file access methods. Note that an exact equivalence between file organization and access method is not always required. For example, direct access can generally be made to a sequential file if the range of keys and storage location addresses is known in advance. Similarly, sequential access to an indexed file can be implemented if the index is stored after the data records or in a separate file. In spite of these possibilities, access to file contents is generally most efficient when the organization and access methods are the same.

Sequential file access is by far the simplest type of access to implement. Sequential access refers to reading and writing file contents in the order in which they are stored. With sequential file access, the file management system keeps track of the current position within the file, and maintains an internal pointer for this purpose. When a file is opened, the pointer is set to the first byte or record in the file. Subsequent read or write operations advance this pointer to subsequent bytes or records of the file. Thus the next byte or record to be read/written is always represented by the value of the pointer. In general, the pointer is always incremented in a positive direction, and it is not possible to advance it more than one position at a time. Thus, to read record 37 in a file, the previous 36 records must be read first in order to advance the pointer to the proper position.

Direct file access allows the pointer position to be specified directly by the application program. Thus, to read record 37 in a file, the application program issues a service call to the file management system that specifies the position of the desired record. It is not necessary for the application program to directly manipulate any of first 36 records in the physical sequence. If record-oriented file I/O is not supported, direct access can be implemented on byte positions within the file. In this case, the pointer value represents the current byte instead of the current record within the file.

Direct file access allows certain types of file I/O operations to be implemented much more efficiently than with sequential access. For example, consider a file of 10,000 employee records sorted by ascending Social Security number. Now consider an application program that accepts a Social Security number from a user, searches the file for the record with that Social Security number, and displays it to the user. If sequential file access is used to search the file, each record must be sequentially read and its Social Security number compared to the one input by the user. The search stops when a match is found. This method of search is called a *linear search*. On average, half the number of records in the file (5,000 in this example) will have to be searched for each Social Security number input by the user. For large files, the searching overhead can seriously degrade performance due to the large number of extraneous records that must be read.

With direct file access, this same search could be implemented using a *binary search*. In a binary search, the number of records in the file must be known in advance. This information is typically stored in the directory by file management systems that support record-oriented I/O. It can also be computed based on the length of the file (in bytes) and the length of a record (in bytes), if record length is fixed. The search begins by directly accessing a record in the middle of the file (record number 5,000 in this example). The Social Security number of that record is then compared to the Social Security number input by the user. If they match, the search is over. If the input Social Security number is greater than the number in the record just read, the upper half of the file is then searched (i.e., the next record read will be number 7,500). If the input Social Security number is less than the number in the record just read, the lower half of the file is then searched (i.e., the next record read will be number 2,500). The search proceeds iteratively in this fashion with the number of records currently under consideration cut in half on each iteration. For the file of 10,000 records in this example, a match should be found by executing no more than 14 read operations (n such that $2^n \geq$ 10,000).

Indexed file access is a variation of record-oriented direct access in which a requested record is specified by the value of a key field, rather than by position. Note that in the previous example, the file management system was unaware of the key value corresponding to each logical record. The key value was compared to the search value by the application program, not by the file management system. With indexed access, the job of searching and comparing key values is moved from the application program to systems software. The application program requests a record by key value, and the file management system is responsible for whatever processing is necessary to locate the corresponding record.

An Example

The following example is typical of storage allocation and file I/O procedures in simple file management systems. In fact, the procedures and data structures described are almost identical to those of MS-DOS. Assume that the directory entries in Table 12.2 represent actual files stored on a disk drive. The storage allocation (block) size used by both the disk drive and the file management system is 512 bytes. Blocks are allocated to files nonsequentially, as shown in Figure 12.6 and the file allocation table shown in Table 12.3. Assume that records are the unit of file input and output, and that each record is 55 bytes in length. Assume further that records are sequentially stored within the blocks allocated to the file.

Sequential access. In response to any read operation performed by the application program, the file management system must perform the following tasks:

- Determine the physical disk block containing the desired record.
- Load that block into the buffer (if it is not already there).
- Copy the portion of the block containing the desired record to the application program's data area in memory.
- Increment the pointer to the current record in the file.

For the first read operation, the first disk block allocated to the file will always contain the first record. The first block allocated to FILE3 can be determined from its directory entry. Thus the first byte of the first record in FILE3 is contained in physical block 5. The file management system will issue a read request to the disk controller for physical block 5 and load that block into the buffer when it is ready. Because the application program is reading the first record in the file, the record will begin in the first byte (byte 0) of the block and buffer. Thus the file management system will copy 55 bytes from the buffer to the application program's memory region starting at byte 0. At the conclusion of the operation, the file pointer will be incremented (from 1 to 2).

Subsequent read operations require calculations to determine the disk block containing the record to be read and the byte offset within that block. These calculations are performed using the record size, the pointer value, and the contents of the file allocation table. For the second record, the calculation is as follows:

$$\frac{(\text{Pointer} - 1) \times \text{Record.Size}}{\text{Block.Size}} = \frac{(2-1) \times 55}{512} = 0, \text{ remainder } 55$$

Thus, the next (second) record begins in logical block 0, byte offset 55. From the directory and file allocation table, it can be determined that logical block 0 (the first block allocated to the file) corresponds to physical block 5. That block is already in the buffer, so the file management system copies 55 bytes starting at offset 55 and increments the pointer value.

Direct access. Now assume that direct access instead of sequential access will be performed. As before, the file management system will allocate a memory buffer when the FILE3 is opened. However, no pointer need be initialized for direct access. Note that although the physical blocks of the file can be scattered around the disk in any order, the nature of the access to those blocks is sequential. This is because the links for each file in the file allocation table are sequentially organized. Also, the bytes and records within each block are sequentially stored. Thus, the example below represents a simulation of direct access on a sequentially organized file.

Assume that the first read operation requests record 37 of the file. As before, the calculation is

$$\frac{(Pointer - 1) \times Record.Size}{Block.Size} = \frac{(37-1) \times 55}{512} = 3, \textit{ remainder } 499$$

Thus, record 37 begins in logical block 3, offset 499. To determine the physical block corresponding to logical block 3, the file management system must follow the chain of entries in the file allocation table for FILE3. Logical block 0 corresponds to physical block 5, as recorded in the directory. Logical block 1 corresponds to the pointer entry in entry 5 of the file allocation table, or physical block 14. Logical block 2 corresponds to the pointer entry in entry 14 of the file allocation table, or physical block 16. Logical block 3 corresponds to the pointer entry in entry 16 of the file allocation table, or physical block 19.

Thus, record 37 starts in physical block 19, byte offset 499. The file management system will load this block into the buffer and begin the transfer of 55 bytes to the application program memory region. However, all of the record contents are not contained within the buffer. While copying, the file management system will come to the end of the buffer before it has finished copying 55 bytes. The system must detect this condition, issue a read request for the next physical block, and resume the copy operation at the beginning of that block. Thus, the system must follow the chain of pointers in the file allocation table from physical block 19 to the next physical block (20). It must issue a read request for this block and load it into the buffer. It then copies the first 42 bytes of the buffer to complete the read operation.

DIRECTORY CONTENT AND STRUCTURE

A directory contains descriptive data about files contained in a storage device or subset thereof. This data is usually accessible by both the operating system and the user. Directory content can include the following:

- Filename
- Location
- Ownership
- File Type
- Size
- Protection

A directory is normally implemented as a table that holds pointer values indicating the device and physical address at which each file's allocated storage space begins. The directory includes the size of the file expressed in bytes, words, blocks, or tracks. The maximum size permitted for the file can also be stored.

Protection data includes file ownership and any access controls that apply. For example, if a file is write protected, a write operation cannot be performed except under predetermined conditions (e.g., alterations authorized by the file owner or the system administrator). A usage count indicates the number of processes that are currently accessing the file. The file management system can also keep track of the date, time, and process identification of each file access operation. Separate sets of data are stored in the directory for dates of file creation, most recent modification, and most recent access.

Directory Structure

Early microcomputers used relatively simple directory structures. This simplicity reflected the single-user nature of the systems and their relatively small secondary storage capacity. Within such a system, one or two disk drives might serve a single user. The directory structure is relatively straightforward, because one directory is used for each storage device. To access a desired file, the user must keep track of the device on which a desired file is stored and must then initiate a search within the corresponding directory. Such a directory structure is called a *single-level directory*.

Within larger systems, the files of many users can be allocated among many storage devices. For example, the files stored in a large computer system might encompass hundreds of disk platters and thousands of tapes. To facilitate search and retrieval of files within such a large system, it becomes necessary to build directories that keep track of other directories. This is generally implemented through *hierarchic directory structures*, or multilevel systems of directories. Types of hierarchic directory structures include

- Two-level directories
- Tree structures
- Acyclic graphs
- General graphs

Two-Level Directories

The diagram in Figure 12.9 shows a *two-level directory* structure. The file management system supports multiple users who share access to a common set of files and devices. One directory, called a *user directory*, is maintained for each user. At the next level, a *root directory* contains the device locations of each user directory.

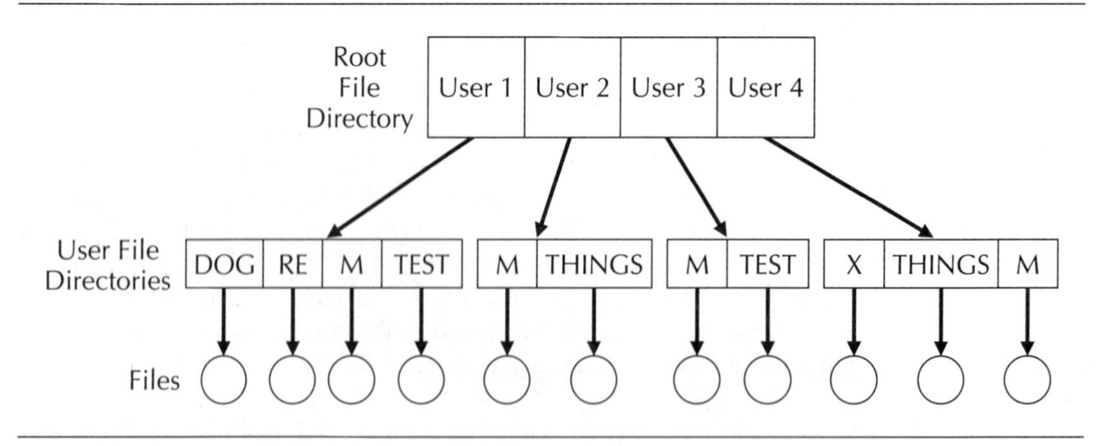

FIGURE 12.9 A two-level directory structure.

Files listed in a user directory have been created by (and are owned by) the associated user. In addition, the directory can also hold pointers (also called links) to files owned by other users, to which the user of the directory has access. The pointer to a file that resides under another user's directory is the user name combined with the filename. This combination is called the *path name* to the file. Thus, an access request to another user's file first searches the device location under the user name in the root directory, then branches to that user's directory and searches for the file location.

Within a multi-user file management system, users might not have knowledge of filenames in other directories to which they do not have access. So, it is possible that two users will coin identical filenames independently. This situation can be permitted and creates no conflicts, as long as the following restrictions apply:

- Duplicate filenames are not permitted within the same user directory.
- The path name (user name and filename) of each file in the system must be unique.

Tree Structure Directories

A more complex system is a *tree structure directory,* as shown in Figure 12.10, which encompasses multiple levels of directories. Access paths, or branches, proceed downward from a single (root) directory at the highest level. At lower levels, a separate subdirectory is maintained for each device, or node. Each node can be an individual file or another directory. Directories for individual users or groups are also located at lower levels of the hierarchy. These directories can contain other directories, thus allowing users to hierarchically organize their own files.

For each process or user, systems software maintains a pointer to the directory that is currently being accessed. That directory is called the *current directory* or *working directory*. Each user in the system normally has a default working directory, or *home directory*. When a user logs in interactively or runs a batch process, this home directory is normally made the default current directory. The user or process can subsequently change the current directory by issuing an explicit command or the appropriate service call.

Tree structure directories serve to implement several desirable functions of a file management system, including

- Directory search
- Partitioning of groups of related files
- Access controls

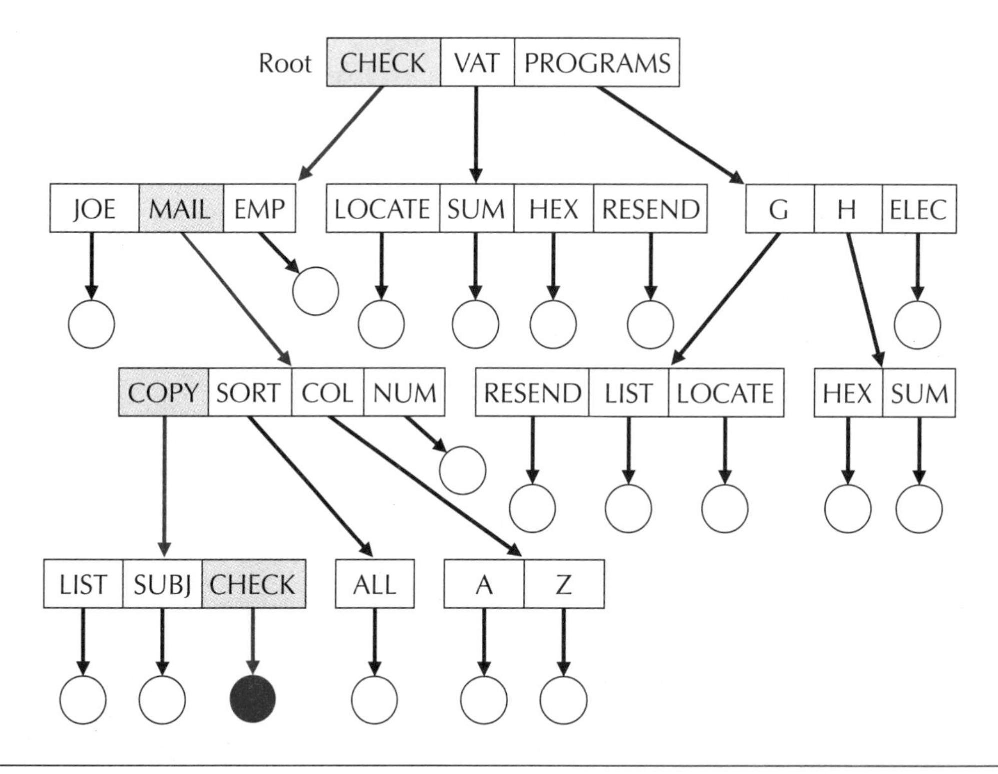

FIGURE 12.10 A tree-structured directory.

Searching for a desired file or accessing a particular node is facilitated by the tree structure. An access path represents the shortest route that traverses the network of directories to reach the desired resource. If the system were organized under a single directory in a linear structure, all preceding entries would have to be searched to find the desired resource. Within a tree structure, each branch selection eliminates a significant portion of the available routes at that level and greatly increases the efficiency of the search.

In a tree structure, the owner of a directory is deemed the owner of the directories and files beneath and connected to it. Normally, a directory would not include pointers to other directories or files at the same level or a higher level in the hierarchy. Thus, the system is partitioned according to the access paths defined by the directories. Because a user normally owns his own home directory, all files and subdirectories contained therein are also owned by him. For example, if the user SMITH owns the directory CHECK in Figure 12.10, all files in that directory and any of its subdirectories (i.e., MAIL, COPY, SORT, and COL) are also owned by SMITH.

The partitioning of the structure according to its branches can be used to implement access controls. For a path to be valid, it must proceed from the current directory downward through its own subdirectories. Therefore, access will be restricted to the resources owned by each node, and sharing will be prohibited. The level at which a user's current directory is created within the system can be used to assign the user a level of access authorization.

Within the hierarchy of directories, names of access paths can be specified in one of two ways:

- Complete
- Relative

A *complete path* name begins at the root directory and proceeds through all nodes along a path to the desired file. Such a reference is also referred to as a *fully qualified file reference*, because the entire access path is fully stated (qualified). For example, the name

ROOT/CHECK/MAIL/COPY/CHECK

is a fully qualified file reference for the file CHECK in Figure 12.10, where the symbol / separates directory names.[2] A *relative path* name begins at the level of the current directory. In a conventional tree structure, the assumption is that the path proceeds downward through subdirectories contained within the current directory.

Graph Directories

A less restrictive form of a hierarchic directory is an *acyclic graph directory,* an example of which is shown in Figure 12.11. In effect, this structure is a tree in which resource sharing is permitted. Access paths still must proceed downward, and sharing is restricted to nodes at the level of the current directory or below. Sharing can be implemented by placing the symbolic name and location pointer of a resource in the directory of each user for which access has been authorized. For example, if one user owns the directory THES and another user owns the directory CHECK, both will have access to the common resource SUM. Thus, multiple access paths can lead to a given node.

This structure is said to be acyclic because each access path is permitted to transverse a node only once. Otherwise, a cycle would result. Cycles are inefficient because unnecessary processing overhead might be incurred in searching the same directory more than once. Cycles also can induce processing errors because routines that set counters or flags might update the same location repeatedly.

[2] In many operating systems, a leading / or \ is a shorthand notation for the root directory. For example, the string /CHECK/MAIL/COPY/CHECK is the Unix equivalent of the complete path shown above.

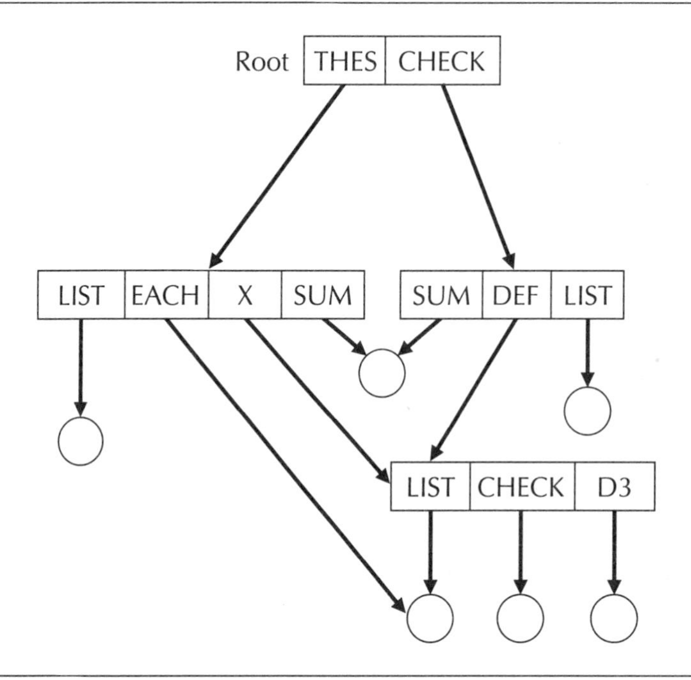

FIGURE 12.11 **An acyclic graph directory structure.**

A less restrictive variation of a hierarchical directory structure is the *general graph directory*, as shown in Figure 12.12. Within such a structure, access paths are permitted in any direction. As a result, cycles can be set up within the system. Although a benefit of this approach is its flexibility, the key tradeoff over acyclic structure is added complexity. Extensive software controls must be implemented within the file management system to assure that cycling does not cause inefficiency or errors.

Directory Implementation

When a storage device is initialized, the file management system must build a reserved storage area for the device directory. In some file management systems such as Unix, a directory is simply a special type of data file. When a file is created or modified, the file management system will make a corresponding entry in this directory, including the symbolic name of the file and a pointer to its starting address. In implementing a directory, its logical structure must be represented in a physical list or table. Physical methods of implementation include

- Linear lists
- Sorted lists
- Linked lists

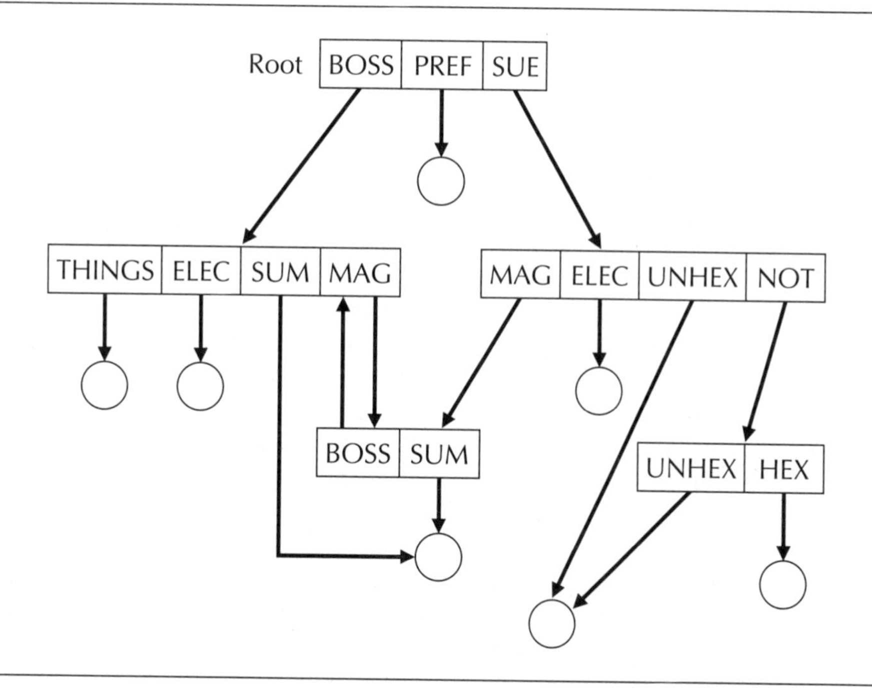

FIGURE 12.12 **A general graph directory structure.**

A directory for a relatively small storage device, such as the volume directory for a diskette, might be maintained as a *linear list*. To find a desired symbolic filename in such a linear list, all items must be searched sequentially. The processing overhead required for this operation might not be prohibitive, as long as the list is short. A directory built as a *sorted list* might be maintained in alphabetic sequence, by symbolic filename. Because the order of the list is predefined, a search for a symbolic name might be implemented as a binary search. This type of search is extremely efficient and fast. However, the tradeoff of maintaining a sorted list is the processing overhead required to maintain it in order. Each time a file is added, deleted, or renamed, the entire list must be sorted.

Maintaining a directory as a *linked list* can be highly efficient, because changes to the list involve updating a symbolic filename and a set of links. However, searching for an individual file within the directory is less efficient. To search for a desired file, all links in the list must be tested sequentially. This approach becomes cumbersome as the number of file entries grows.

The creation and manipulation of files is implemented within the command and service layers of the operating system. File creation by end users is typically accomplished using the facilities of a utility called an *editor*. Editors are typically used to create files containing source code and job control language commands, and sometimes to create data files for application programs. Editors for such files are normally called *text editors*. Other editors (called *binary editors*) allow a user to make changes in other types of files, such as executable files.

Once created, a file can be manipulated by a number of command-layer functions. Examples of these functions include copying, renaming, and deleting files. The command-layer facilities provided to end users are built on a basic set of functions within the service layer. Functions such as file creation, deletion, and renaming are typically implemented as a single service call. A text editor will use these service calls when performing actions such as creating a new file or renaming an old (backup) version of an existing file. These service-layer functions are also available to programmers in higher level languages. In some languages such as C, direct access to operating system file manipulation service calls is provided. In others, such as COBOL and FORTRAN, file manipulation commands in the symbolic programming language are translated into corresponding operating system service calls by a compiler or an interpreter, or via the link editing process.

File Operations

A file is created when it is defined by a user (its owner) and when an area of physical storage is allocated to it. This function can be performed directly by the user through the command layer (e.g., with an editor or as the result of a copy command). It can also be performed indirectly via a call to one or more service-layer routines by an application program. In some cases, a user can be deemed the owner of a file but might be unaware of its existence. For example, a process within an application program could create a file as a temporary storage area, or work file. In a practical sense, the user of the program has indirect control of the data content and can be considered its owner.

When a file is created, its creator must give it a symbolic (or file) name. The filename identifies the file at the operating system level and can be referenced within the source code of an application program. Such a reference generally becomes part of a call to service-layer file-handling routines. The filename can also be incorporated in explicit commands issued by the user directly through the command layer (e.g., a copy or delete command). Various program statements and user commands permit operations on the file as a whole or on portions of its data content.

Under most file management systems, a file is deleted by removing its name or identifier from the directory of the corresponding storage device or volume. Thus, in a technical sense, file content is not erased. The delete operation merely makes the file's storage area available for reallocation. The space is reallocated as other files are created, and the data content of the previous file is actually destroyed only when the reallocated space is overwritten.

Because the content of a file is not destroyed immediately by a delete operation, it is possible to restore the file as long as the space has not been reallocated. Data content can be recovered by recreating the filename in the directory and reestablishing the pointer values that associate it with the storage area. This process is sometimes referred to as an undelete operation.

Deletion is implemented this way primarily for the sake of speed. However, security could be sacrificed, because unauthorized persons who have sufficient technical knowledge might be able to gain access to a deleted file by rebuilding its entry in the directory.[3] To prevent such access, some file management systems include an optional *file destruct* capability. This operation not only removes the filename from the directory, it also overwrites the storage space with null values. The tradeoff for greater security is the increased time required to perform the erasure.

In renaming a file, its symbolic name is replaced with a new name in the volume or device directory. Renaming can occur at the discretion of the user, perhaps for convenience of reference. Or the file system itself can cause a file to be renamed. For example, it might be necessary to rename the prior version of a file that has been updated to prevent its use by current programs.

A file can be moved by two means. The first is actually a combination of two operations: copying and deleting. When a file is copied, its records are brought into main memory, one after another, and then written in the same sequence to another storage area or device. In a file move, the original file is deleted immediately after the copy operation. Effectively, these two operations move the file from one storage area or device to another.

The implementation of a file move in this manner can consume a considerable amount of system resources if the file is large. Every record of the original file must be both read and written. An alternative implementation is to leave the contents of the file in their original location and simply move the corresponding directory information. However, this method is possible only if the original and new locations of the file are within the same storage device or volume.

[3] An interesting example of this security problem occurred when a large number of computers owned by the U.S. Department of Justice were sold to a computer dealer. The computers were confiscated a short time later when it was discovered that the contents of the disks had not been wiped clean. The contents included a considerable amount of data (including names of suspects and informants) about ongoing criminal investigations.

Directory Operations

Creating and deleting directories can be accomplished by several means. For files, a service call generally exists for directory creation and another for directory deletion. These service calls can be used directly by systems or application programs or indirectly (by a user) through command-layer facilities. One example of directory creation by a system program is the creation of a home directory for a newly created user account. Typically, a program is provided for use by system administrators to add, modify, or delete user accounts. Among other functions, this program will generally add, modify, or delete the home directories associated with user accounts. Directory creation/deletion can also be included in a system program that installs (or removes) large software packages. Complex software typically consists of many files, organized in a multilevel directory structure. Thus, installation programs must be capable of creating and manipulating directory structures in addition to files.

In systems that support tree-structured or graph-structured directories, directory creation, modification, and deletion abilities are typically provided to end users through command-layer facilities. At a minimum, the functions of directory creation and deletion are provided. In addition, functions such as directory renaming, copying, and movement can also be provided. These functions might be quite complex to implement, because operations must be performed not only on a single directory, but on subdirectories as well. For renaming and movement operations, the uniqueness of path names must be guaranteed.

The most common directory operation is the listing operation. At a minimum, a directory listing includes the symbolic names of files and subdirectories. In most systems, all other directory information (e.g., size, location, usage data, ownership, and type) might also be listed. Listing facilities typically provide a number of alternate presentation formats and options (e.g., column-oriented lists and sorted output). As with other directory operations, these operations are implemented within the service layer and can be used directly by application programs or by end users through command-layer facilities.

FILE INPUT AND OUTPUT OPERATIONS

The operation of opening a file can be performed only on a file that has been created in a previous operation. Opening a file makes its content available for access by the user or requesting process. Within the operating system, opening a file means that it is allocated logically to a specific process. Typically, the operating system performs the allocation by entering the symbolic name of the file in a table of open files that is maintained for each process. In addition, other housekeeping chores such as allocating buffer space in primary storage are also performed at this time. The file management system will normally create and maintain a data structure called a *file control block* for each open file. This data structure stores information about the file, including current position, the number and location of I/O buffers, and the current state of each buffer (e.g., empty, full, modified, or not modified). This data structure is updated (as necessary) as file I/O operations are performed.

The operation of closing a file severs its logical relationship with the current process. Within the file management system, file closure is accomplished by removing the symbolic name of the file from the allocation table for the current process. Other resources associated with the file/process connection (e.g., buffers) are released. The file control block determines these resources. It also determines if any I/O operations must be completed (e.g., copying modified buffers to disk). By closing a file, the file management system releases it and its related resources for reallocation to other processes.

In a *read operation*, the read/write head of the storage device is directed to seek a specific record location. Then the data stored there is moved to main memory for access by the CPU. If a process is permitted to *write* to a file, it could cause data to be recorded at addresses on the storage device that are allocated to the file. Both read and write operations are normally implemented via service calls. The service routines initiate a data transfer through the corresponding I/O port.

Several kinds of *edit operations* are generally available for modifying data within files. Typically, a record to be modified is brought into main memory, and its data is manipulated to produce the desired change. Then the record is rewritten to a storage device. Depending on the physical organization of the file or device, the record could overwrite the prior version, or the edited version could be written to a new location and the prior version retained as part of a backup system. Specific edit operations include

- Update
- Append
- Insert
- Delete

In an *update operation*, data within a field or record is replaced. Updating can be performed in place or by rewriting the record or file to another device. In an *insert operation*, a new data field is added within the format of a record; or a new record can be added within the sequence of a file. In an *append operation*, data is added at the end of the existing record or file. This operation moves the physical end-of-file marker, or indicator, to the first available record position after the appended portion. In a *delete operation*, a record is logically or physically removed from a file.

Read and Write Operations

The set of available read and write operations and their exact implementation depend on the type of file being read and its method of file organization. In general, sequential access to a sequentially organized file is the simplest and most efficient to implement. Direct access to a sequentially organized file is a bit more complex but still relatively straightforward. Access to indexed storage structures generally require the most complex I/O operations.

File I/O operations can be defined as either record-oriented or byte-oriented. In record-oriented I/O, an entire logical record is the unit of data transfer between the file management system and an application program. In byte-oriented I/O, data is transferred between the file management system and an application program as a sequence of bytes. In this type of file I/O, the file management system is unaware of the logical structure of the data. Thus, all conversion between logical record structures and streams of bytes must be performed within the application program.

Sequential access. A set of file access service calls to support byte-oriented sequential file access might be

read_byte(FILE,DATA)

write_byte(FILE,DATA)

read_n_bytes(FILE,N,DATA)

write_n_bytes(FILE,N,DATA)

where FILE is a file identifier (typically an integer returned by a file open service call), *N* is an integer containing the number of bytes to be read or written, and DATA is the memory address of a data element or structure containing the data after reading or before writing. Data is read or written entirely as sequences of byte values, with no regard to their interpretation or logical structure. For example, the service call

read_n_bytes(FILE,4,CUSTOMER_BALANCE)

simply reads the next 4 bytes from the file and copies them into memory starting at the address of the data item CUSTOMER_BALANCE. A field such as CUSTOMER_BALANCE is probably a real number (4 bytes in length, in this case), but that fact is unknown by (and irrelevant to) the file management system.

With all methods of sequential file access, the file management system must keep track of the current position within the file. For byte-oriented data transfer, this position is the sequential number of the most recently read (or written) byte in the file. Although tracking position in this manner is not strictly necessary for simple sequential access, it is needed if input is blocked or buffered, as described in a later section of this chapter.

A simple set of service calls for record-oriented file I/O is

read_record(FILE,DATA)

write_record(FILE,DATA)

where DATA is the starting address in memory of a data structure comprising a record. In file management systems that support record-oriented I/O, some method must be provided for marking the end of one record and the beginning of another within a file. Typically, a control character (e.g., ASCII 30_{10}) is placed after each group of bytes that comprises a record. A read_record service call causes the file management system to sequentially read all bytes up to and including the end-of-record control character. These bytes are sequentially loaded into memory. The write_record service call copies bytes from memory to a file up to and including the end-of-record control character.

Sequential file I/O is the most basic method of file access and it is useful in a wide variety of situations. Many batch application programs process input records sequentially and generate sequential output files. The operating system itself uses sequential I/O to perform routine functions such as loading the executable image of a program from a file into memory. However, many types of information processing require nonsequential file access methods.

Direct access. Direct access in byte-oriented file I/O refers to the ability to read or write any byte in the file while ignoring all other bytes. Direct access in record-oriented file I/O refers to the ability to read or write any record in the file while ignoring all other records. With such access methods, the concept of current position (or a file pointer) is irrelevant. Read and write operations are not restricted to the record or byte at the position following the current pointer. They can operate on any portion of the file's contents.

A typical set of service calls to support direct file I/O are

read_n_bytes(FILE,POSITION,N,DATA)

read_record(FILE,POSITION,DATA)

write_n_bytes(FILE,POSITION,N,DATA)

write_record(FILE,POSITION,DATA)

where the parameters FILE, *N*, and DATA are the same as previously defined for sequential I/O. The parameter POSITION is an integer representing the position within the file where reading or writing is to occur. This integer might represent a sequential byte or record position, depending on the type (byte- or record-oriented) of file I/O operation. For example, the service call

read_record(FILE3,37,RECORD)

will cause record 37 of FILE3 to be read and copied to the data structure RECORD within the application program's data area.

Indexed access. Indexed access is similar to direct access except that records are specified by key value instead of position. A typical set of service calls to support indexed file I/O might be

read_n_bytes(FILE,KEY,N,DATA)
read_record(FILE,KEY,DATA)
write_n_bytes(FILE,KEY,N,DATA)
write_record(FILE,KEY,DATA)

where the parameters FILE, *N*, and DATA are the same as previously defined for sequential I/O. The parameter KEY is a data item that matches the index key used to access file contents. Typically a key field is an integer or character array. For a read operation, the system will search the index to find the corresponding record position, then issue another read command using direct access. Write operations require that the index be updated, as discussed below.

Edit Operations

As with access operations, the complexity of update operations depends heavily on the underlying organization of the file. These operations are typically simplest for sequentially organized files and more complex for direct and indexed files.

Edits of sequential files. For sequential files, updating an existing record is relatively straightforward. The record to be modified is first located and read into memory. The contents of the record (in memory) are altered, and then written back to the same storage location. This operation is commonly called an *update in place*.

Append and insert operations on sequential files are similar to those same operations on arrays (see Chapter 4). For an append operation, a new storage location that is large enough to hold the appended data must be allocated to the end of the file. In many file management systems, a special type of file access called *append mode* is declared when a file is opened. On receiving the appropriate open service call, the system opens the file and positions the pointer at the very end. Subsequent write commands automatically cause storage to be allocated at the end of the file and new data to be written there.

Insertion is substantially more difficult to implement. For example, consider the insertion of a new record after record 500 in a file of 1000 records. The following procedure is required:

- Allocate additional storage space at the end of the file to hold a new record.
- Starting at record 1000 and working backward to record 501, copy each record to the next logical position (e.g., record 1000 is copied to the 1001th position).
- Write the new record at logical position 501.

This procedure is extremely inefficient, especially with large files. For this reason, few file management systems directly support the insertion of new data into sequentially organized files. For similar reasons, deletion of records is also generally unsupported for sequentially organized files. Implementation of these operations can be performed in application software by copying a file into memory as a linked or indexed data structure, performing the insert and delete operations, and then writing the contents of memory to a newly created file.

Edits of nonsequential files. Update operations on direct files are implemented in essentially the same manner as for sequential files. That is, the record to be updated is located, read into memory, altered (in memory), and written back to the same location in the file. Append operations are also similar to those for sequential files. For both direct and indexed files, new space is allocated at the end-of-record storage area, and the new record is written to that location. For indexed files, an entry for the new record must be added to the index, possibly necessitating the allocation of additional space to the index itself.

Insert operations are generally not supported for direct files for reasons similar to those of sequential files. If it were supported, the correspondence between key values and record positions would be altered, necessitating a redefinition of the entire access scheme. This is not a problem for indexed files, because any newly created record can be placed at the end of the record storage area of the file. Thus, in an indexed file, there is little logical difference between an append operation and an insert operation. The differences that do exist are confined entirely to the update of the index (as opposed to the records of the file). For an append operation, a new index entry is created at the end of the index. For an insert operation, the index entry must generally be created in the interior of the index. If the index is organized as an array (or table), the index must be updated in essentially the same fashion as described above for record insertion in sequential files.

Delete Operations

Within a file, selected records or fields within records can be deleted in edit operations. The exact implementation of a delete operation varies according to both the physical and logical organization of the file. Options include

- Logical delete
- Physical delete

A *logical delete* is performed by marking a record as deleted. This can be accomplished by a Boolean field attached to each record. The value of the field is normally false, but is set to true if the associated record is deleted. A logical delete can also be implemented by overwriting the record (or byte) sequence with a predefined control character sequence. In either case, the logical delete should be detected in any subsequent read operations. If the record is to be retrieved based on an index or key, the file management system should return an error message stating that "no such record exists." If the records of the file are read sequentially, the file management system should skip over the logically deleted record as though it didn't exist. Note that the physical record still exists on secondary storage.

Over time, many deleted records can accumulate, resulting in a substantial amount of wasted storage. To counteract this problem, logically deleted records must be physically deleted on a periodic basis. This can be accomplished by a special utility program or by performing a file-by-file dump, as described later. The process of physically deleting logically deleted records is referred to as file compaction or file packing. The file is said to be compacted, because storage allocated to logically deleted records is freed, resulting in less storage allocated to the file.

Periodic file compaction (as opposed to immediate file compaction) improves the efficiency of edit operations. Because compaction requires manipulating every record in a file, it consumes a substantial amount of computational resources (processor and I/O operations). It is normally impractical to perform compaction every time a record is deleted, particularly if the file is large or if delete operations are numerous and frequent.

FILE SECURITY AND INTEGRITY

Computerized information systems can be viewed as tools for creating and maintaining large collections of data. Because data is an important and valuable organizational resource, a file management system must provide facilities to prevent the loss, corruption, and unauthorized access to data stored in files. Note that these protections must also extend to files containing programs, because they are both a resource and the means by which data is stored, modified, and viewed. Access to files by multiple users also raises issues of security and accidental corruption of data. Thus, complete file protection in a multi-user system must provide facilities for file sharing while preventing data corruption. These topics are further discussed in the following sections.

Ownership and Access Controls

In any multi-user system, ownership of system resources is an important feature that must be supported by systems software. Within the file management system, the concept of ownership extends to files and directories, and possibly to entire secondary storage devices. Access controls for files (and other resources) must be based on a separate system of user identification and authorization. Typically, this is provided through the operating system by a system of user accounts, login, and logout processing. Each user must have an account with a unique identifier (e.g., a user name or account number). Initial access to the system is controlled by login processing. A user must enter an account identifier and a password to establish and authenticate his or her identity. Once established, user identity is made available to the file management system for access control purposes.

Many schemes of access control and user privilege can be implemented. A relatively simple system, such as that used by the Unix operating system, separates access controls into four levels. From highest to lowest these levels are system (superuser), user, group, and public. System-level access is restricted to a specific user account assigned to the system administrator. This user has unrestricted access to system resources, including all files, directories, and secondary storage devices.

Individual users are deemed to be the owners of files that they create. They have the ability to grant or deny file access privileges to other users. Groups are predefined sets of user accounts. They can represent project workgroups within an organization, all of the members of an organizational unit, or any other grouping criterion implemented by the system administrator. In Unix, each user must belong to one and only one group. In other systems that implement group identification, users can belong to many different groups. The public is simply a term referring to all known users or accounts.

An individual user can control the level of access provided to files he or she owns. In Unix, three types of file access privileges are defined. These include read, write, and execute access. Read access allows a user or a process to view the contents of a file. Write access allows a user or a process to alter the contents of a file or to delete it altogether. Execute access allows a user or a process to execute a file, assuming that the file contains an executable program or job control stream. A file owner can reserve any of these access privileges for himself, thereby denying those access privileges to all other users except the system administrator. If the user desires, any access privilege (or set of privileges) can be extended to other members of the group or to the public. Thus, for example, a user might grant read and write access to the group but only read access to the public. The user could even deny certain access privileges to himself. For example, a user might deny himself write access to prevent accidental deletion of an important file. Of course, the user could alter those access controls at any time.

In addition to providing a system of access controls for files, most file management systems also provide similar access controls for directories. Users are deemed the owners of their home directories, and any directories that exist below them in a hierarchical directory structure. Typically, access privileges for reading (listing directory contents) and writing (altering directory contents) are defined. As with file access privileges, these can be individually granted or denied to a group or to the public. Directory privileges provide additional security for individual files. Read permission to a directory is denied to all other users, thus making it difficult or impossible for others to even know of the existence of a file.

Access controls are normally implemented within the service routines that access and manipulate files and directories. For example, a request to open a file will first check the access controls that exist for the file, then determine if the requesting user has the access privilege to read the file. Service routines for record editing, record deleting, and file deleting will also implement checks of access controls. Note that for purposes of verifying access privilege, a process executed by a user generally inherits that user's access privileges (or lack thereof). Although access controls are a necessary part of file manipulation and I/O in most systems, they impose additional overhead on many file manipulation operations. In some file management systems, the system administrator can choose among several different levels (degrees of enforcement) for file access controls. These allow an administrator to balance overall file management system performance against the relative need for file security.

A file management system must also provide additional controls to ensure that normal file management system controls are not circumvented. Typically, these controls are placed on secondary storage devices and on the data structures directly related to them (e.g., file allocation tables and master directories). Such controls prevent unauthorized access to data and programs via direct access to the secondary storage devices on which they are stored. A file management system can also implement a system of logging access to files. Individual access to files (e.g., file opens, writes, creations, and deletions) can be automatically recorded in a log file that can be viewed later by the system administrator. Such a system provides an audit trail to identify unauthorized access to files and directories. As with other access controls, logging imposes a performance penalty on all file accesses.

Multi-User File I/O

A file management system that allows multiple users or processes to simultaneously access a single file must provide additional access controls to prevent incorrect (and unintentional) corruption of file contents. This need arises from timing problems inherent in multi-user file access. As an example, consider the following sequence of operations:

```
file_open(file)
read_record(file,record)
/* change record contents */
write(file,record)
file_close(file)
```

When the file_open call is executed, a buffer will be allocated for file I/O. The read_record call then causes a record to be copied from secondary storage to the buffer and from there to the program's data area. At this point, the program alters the contents of the record (i.e., changes the value of one or more fields). The program then executes a write_record call, which causes the modified record contents to be copied from the program's data area to the file buffer. The buffer will be flushed (copied to disk) when the file_close call is executed.

By itself, this file manipulation sequence causes no inherent processing problems. However, when multiple programs execute this same sequence within the same time frame, unintended results could be obtained. For example, consider the same programming sequence executed by two different programs, as depicted in Figure 12.13. The execution of these commands results in a problem commonly called the *lost update*. The problem arises from the fact that each process modifies a copy of the record within its own data area. At time 2, process 1 reads a record that is ultimately copied to its own data area. At time 3, process 2 reads that same record, which is ultimately copied to its own data area. Both processes modify the record copy in their own data area and then use a WRITE_RECORD call to copy the record back to the file. However, neither process coordinates its activity with the other. As a result, the modifications made by process 1 are lost because the WRITE_RECORD and FILE_CLOSE calls issued by process 2 (at times 5 and 6) causes the modifications made by process 1 to be overwritten.

Time	Process 1	Process 2
1	`file_open(file)`	
2	`read_record(file,record)`	`file_open(file)`
3	`/* alter record */`	`read_record(file,record)`
4	`write_record(file,record)`	`/* alter record */`
5	`file_close(file)`	`write_record(file,record)`
6		`file_close(file)`

FIGURE 12.13 A simple file update procedure executed by two different processes with a time lag.

To prevent lost updates, a file management system must provide a mechanism to prevent multiple processes from attempting to update the same record at the same time. The general means of implementing such a control is through use of locks. In essence, a lock prevents access to a resource until the lock is removed. Locks can be applied to an entire file or to individual records within a file. If *file locking* is used, the file management system locks a file when a process issues a file_open call. The process that issued the call is allowed access to the file, but attempts by other processes to open that same file are denied. The lock is released when the process that opened the file issues a file_close call.

File locking is relatively simple to implement in a file management system. However, its use can impose severe performance penalties, particularly when many processes require access to the same file. File locking allows only one process at a time to access a file, thus drastically reducing the amount of I/O that can be performed on any one file. Another form of locking that addresses this shortcoming is *record locking*. In record locking, the file management system allows multiple processes to access the same file. However, it prevents multiple processes from simultaneously modifying the same record by applying locks to individual records. Note that due to the use of memory buffers for file I/O, the unit of locking is generally a physical record (block) rather than a logical record. Record locking allows many processes to access the same file, as long as they access differing records. However, record locking is substantially more complex to implement as the file system must track every file access (not just opens and closes) and maintain a potentially large set of locks.

As discussed thus far, locking is a process implemented without the direct knowledge of application programs. That is, application programs are unaware of the locking process and contain little or no programming dependent on them. To support this independence, the file management system must provide a mechanism to delay the execution of an access to a locked resource. This can be implemented by queuing access requests to locked processes and blocking the requesting processes until the lock is released. In this way, application programs are unaware of the locking mechanism (although the user could be aware of the delay).

A simpler option is to simply fail the access request—that is, return an error code to an application program when it requests access to a locked resource. This method transfers much of the processing overhead of locks from the file management system to application programs. Typically, an application program will test each access for failure due to a lock and implement procedures to retry the access at a later time. Because of the complexity required of application programs, it is generally preferable to implement access spooling and/or retries within the file management system.

SYSTEM ADMINISTRATION

The management of a collection of data files is a major function of *system administration*. In a broad sense, system administration encompasses all aspects of keeping a system and its files operational, secure, and reliable. If the collection of files is integrated under a database management system (DBMS), a separate *database administration* function is usually needed to deal with the technical details of maintaining the database. Specific administration functions supported by most large-scale file management systems include the following:

- File migration (version control)
- Automatic and manual file backup
- File recovery

These functions are fully discussed in the following sections.

File Migration

When a user alters the contents of a file, the unaltered (original) version is generally overwritten by the new version. However, it is often desirable to maintain the original version as a protection copy. Many popular application tools such as word processors and text editors automatically generate such copies. For example, the standard MS-DOS text editor will automatically rename the original version of a file with a new name that has the same prefix, followed by the filename extension .BAK. Many application programs are also written to preserve original versions of files. For example, the original version of a bank account master file can be copied prior to processing daily transactions such as deposits and the like. The application program then uses a separate file of transactions to update the copy of the account master file. This process is depicted in Figure 12.14. The original account master is commonly called the *father;* the copy that has been updated to reflect new transactions is called the *son*. After another set of transactions is processed, the father becomes the *grandfather*, the son becomes the father, and the new copy of the account master file becomes the new son.

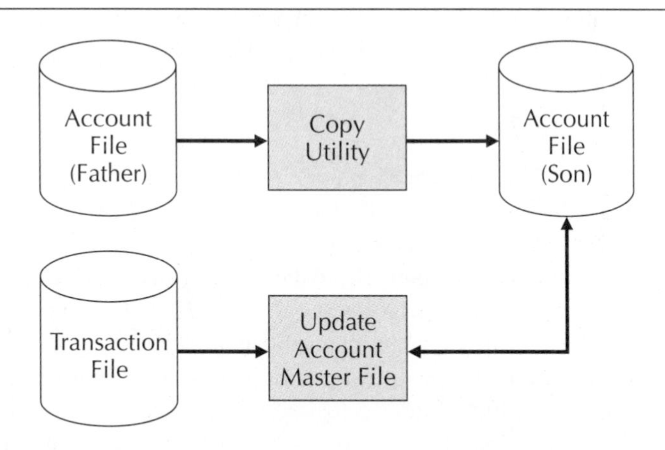

FIGURE 12.14 The creation and use of father and son file versions in a master file update.

In a large-scale file management system, the process of naming and storing original versions of altered files might be automated. That is, older versions of files can be created and stored under control of the file management system instead of application software. In such systems, it is common to attach a version number to each filename. When a file is created, it is assigned a version number of 1. The first time it is altered, the altered copy is assigned a version number of 2, and so forth. The file management system automatically generates and stores copies and tracks version numbers by storing them within a file's directory. When desired, a user or application program can access an older version of a file by explicitly referring to its version number.

As files are altered, their older versions will accumulate on secondary storage. This can quickly swamp the secondary storage capacity of a computer system, especially when files are large and/or frequently altered. To compensate for this, a file management system will typically implement a system of *file migration*, a systems management technique by which the storage cost and performance of each file version is balanced with user demand. As a file version becomes increasingly outdated, demand will decrease. As demand decreases, the file version should migrate from online to offline storage, and eventually to an archive. Ideally, a file copy should reside at a point in the hierarchy at which an expected frequency of access can be met at the lowest possible storage cost.

The migration of files is logged so that the location of a version can be determined to facilitate recovery, if necessary. File migration is typically a semi-automated process. The generation and tracking of file versions is performed automatically. The timing of the movement of older versions to offline storage can also be determined automatically, subject to some control by the system administrator. However, the physical movement of old versions to offline storage is typically integrated with normal manual backup procedures.

File Backup

Besides tracking prior versions of files, the file management system should provide facilities to generate backup copies of current versions of files. Typically, these facilities consist of a set of utility programs that can be run manually or scheduled to automatically execute on a periodic basis. Backups are needed in case of loss or damage to the original files. Loss or damage can result from a number of causes, including accidental alteration or erasure, intentional alteration or erasure (perhaps due to a virus), and the partial or total physical failure of a secondary storage device (called a "head crash"). Note that backups can protect not only against loss or damage to files themselves, but also to the directory and storage allocation data structures needed to access them.

Automated backups performed by a file management system and/or system administrator usually include the following options:

- Full backup
- Incremental backups

When *full backup* is performed, the file management system copies (or dumps) all current file contents to auxiliary storage devices. This strategy assures that protection copies exist for all active files. However, in any large computer installation, this process will consume a significant amount of system time. For this reason, full backups are usually performed at relatively long intervals (e.g., weekly) and during off-peak hours.

An alternative is to perform *incremental backups*. In such operations, the file management system tracks files that have been active during the current processing cycle. Typically, the system will maintain a record of the date and time of the most recent backup. Directory entries typically store the date and time of the most recent modification to each file. An incremental backup facility checks the date and time of the most recent backup and compares that time to the directory entry of each file. Only those files that have been modified since the most recent backup are copied. Most large-scale file management systems use both strategies. For example, incremental backups might be done at the end of each business day, with full backups run each weekend.

Backup methods. In performing a full dump, the file management system causes an entire secondary storage device, volume, or portion thereof to be copied to an auxiliary storage device. The operation can be performed in physical or logical order. If copying is in physical order, the entire content is copied block by block (or track by track). This includes not only the files, but storage allocation and directory data structures as well. Thus, the physical organization of the backup will be identical to the physical organization of the source. Such copies are called *mirror-image backups*.

If the order of the dump proceeds logically, symbolic filenames are passed to the copying routines in the order in which they appear in the directory. Thus, copying is said to proceed file by file. In performing a file-by-file dump, the copying routine must create each file on the destination device before the file can be written. These actions have the effect of reallocating the physical organization of the files on the destination device and of creating new storage allocation and directory data structures. The resulting copy can be more efficiently organized than the original. For example, blocks of a single file that were randomly scattered over a storage medium can be written to the copy in sequential order. In addition, logically deleted records can be physically deleted during the backup process.

Backup copies should generally be made to a storage device or volume different from the original. For example, it makes little sense to back up files on one partition of a hard disk to another partition of the same disk drive. Both the original and the backup would be lost in the event of total failure of the drive. For backups made to removable media (e.g., tape or removable disk), the copies should be stored in a distinct physical location. Large computer centers typically store backup copies in a separate building or site to minimize the probability of a disaster such as a fire destroying both copies. With the advent of large-scale computer networking, backup copies are frequently transmitted directly to remote storage facilities via data communication lines. This eliminates the cost, delay, and danger inherent in the physical transportation of storage media.

Transaction Logging

Another form of automated backup is *transaction logging,* also called *journaling.* The use of the term *transaction* in this context should not be confused with the more generic meaning of the term (e.g., a business transaction). In this context, a transaction is any single change to the content of a file—a newly added record, a modified field or the like. In a file management system that supports transaction logging, all changes to file content are automatically recorded in a separate storage area. Thus, for example, the execution of a write_record service call causes the file buffer to be modified, and that modification is also written to a log file. To be most effective, the log entries should be immediately or frequently written to a physical storage device (as opposed to a buffer).

Transaction logging provides a high degree of protection against loss of data due to program or hardware failure. When an entire system fails due to a power failure or fatal systems software error, the contents of buffers for active files are lost. Because these buffers might not have been written to physical storage prior to the failure, the contents of those files become corrupted. In addition, any content changes within the buffers are also lost. Transaction logging provides the ability to recover most or all of the lost changes and to repair corrupted files. When the system is restarted, the contents of the transaction log are reviewed and unflushed buffers are written to their respective files. Transaction logging is commonly used in large-scale online transaction processing systems. It is generally not used in other situations due to the substantial processing and storage overhead it requires.

File Recovery

The provision of adequate facilities for backup is not sufficient protection for files. They must be supplemented by a reliable set of recovery procedures. Typically, these procedures consist of both automated and manual components.

For example, the replay of a transaction log and subsequent repair of a damaged file is an example of an entirely automated recovery procedure. Recovery procedures utilizing full or incremental backups stored on removable media generally rely on manual procedures, to at least some degree. An exception to this rule is the use of fully automated "jukebox" systems for tape storage and mounting.

Typically, the file management system maintains logs of backups to aid in locating backup copies of lost or damaged files. Utility programs can be used to search these logs for particular files or groups of files. These logs identify the volume on which the backup copies are located. At the time of the backup, some form of volume identification should have been written to the volume itself and on a manually prepared volume label. The system administrator is responsible for locating the appropriate volume and mounting it in the appropriate device. The system will read the automated label to verify that the correct volume has been mounted before beginning recovery operations.

Recovery procedures for a crashed system or physically damaged storage device are generally more sophisticated and highly automated. In these cases, damage might have occurred not only to files, but also to directories and storage allocation structures. An automated facility is provided to reconstruct as much of the directory and storage allocation data structures as possible. A consistency check must be made to ensure that

- All storage locations appear within the storage allocation data structure.
- All files have correct directory entries.
- All storage locations of a file can be accessed through the storage allocation data structure.
- All storage blocks can be read and/or written.

Performing the consistency checks and repair procedures generally consumes a great deal of time (anywhere from a few minutes to several hours), but it mitigates the need to do large amounts of data recovery from backup copies, and thus minimizes the amount of current data that is lost. It could also prevent the need to reinstall systems and application software.

DATABASE MANAGEMENT SYSTEMS

In overall function, a database management system, or DBMS, does not differ from the file management system covered in this chapter. A DBMS, like a file management system, is a set of programs that manages a collection of automated data stores. The primary differences between them, as discussed below, include the provision of multiple logical views of data and extension of a logical structure to incorporate data from multiple files.

In the case of a file management system, data is structured within files, each of which is composed of a sequence of records. Record format and sequence have both logical and physical structures. In an unblocked file management system, the logical structure of files and records coincides with their physical structure. If the system is blocked, multiple logical records will usually be contained within a single physical record. Data transfers and storage within the computer system are managed at the level of physical records, from which the file management system must extract logical records for presentation to the requesting user or program.

Under a file management system, each application generally requires a different logical data structure. In turn, each logical data structure requires a different file design and physical implementation. For example, consider two common applications: accounts receivable (A/R) and sales prospecting. For the A/R application, an essential record structure will be the customer account record. The A/R application program is used to maintain files of these records, possibly arranged in descending order of outstanding balances. The other application, sales prospecting, might use a file containing a mailing list for all customers. Each record within the file would contain a customer name and address.

Because the A/R records already contain the data needed for the mailing list, it might seem desirable and efficient to share customer names and addresses with the sales application. However, considerable processing overhead will be required to extract the names and addresses from the A/R files and then to sort the records into the order required by the sales application. For example, it might be necessary to sort the list alphabetically by customer name or numerically by ZIP code. Although such an approach might be used to build the mailing list initially, it would be impractical to incur this much processing overhead each time the sales application was run. So, under most file management systems, separate file and record structures would be designed and maintained for each application.

From this example, two key disadvantages of traditional file management systems become apparent. First, redundant copies of data items will exist among the system's collection of files. This data redundancy is potentially wasteful of valuable storage space. Second, file design remains a major task each time a new application is developed. This amounts to redundant development effort, because the required data already could reside within the existing collection of files.

DBMS techniques largely eliminate these barriers and facilitate the sharing of an organization's data resources among multiple applications. These techniques permit a database to be managed as a pool of seemingly discrete data elements, any combination of which can be specified for use by an application program.

From a technical viewpoint, a DBMS can present such a simplified view to the user only by integrating many underlying complexities. From this perspective, a DBMS does not supplant traditional file management techniques, but extends them to a higher logical level. That is, a DBMS integrates the following levels of hardware and software:

- Data definitions
- Data model
- Physical access and storage implementations

Data Definitions

In the application examples cited above, explicit *data definitions* are contained within each application program. By contrast, under a DBMS, data elements are defined within the DBMS itself. Under a DBMS, logical representations of data are highly flexible. The tradeoff for this flexibility is the considerable amount of effort required to define and install a database. This data definition step involves establishing an organization-wide schema for the database. The schema specifies the types and formats of data elements, as well as processing operations that could be performed on each type. For each new application, then, it will be necessary to specify only a subschema, or subset of these data elements, as well as the logical formats in which the data will be presented to users. As maintained under a DBMS, these logical representations might bear little resemblance to the ways in which data is actually manipulated and stored at lower levels of the system.

Data Models

The internals of a DBMS are organized according to a *data model.* A data model is a generalized set of logical structures and operations. That is, a data model describes how the database will be managed without specifying the details of physical implementation, which are dealt with at lower levels of the system.

In today's computer marketplace, DBMS software is usually offered as an integrated, packaged product. In general, each DBMS package implements one type of data model. For the most part, databases organized around differing data models are fundamentally incompatible. Common data models include

- Relational data models
- Network data models

The logical structure of a *relational data model* is that of a table, as shown in Table 12.4. The table is composed of rows, or *tuples,* and columns, or *attributes.* The table itself is called a *relation.* A relational database can be built from one or more such tables. Typically, each table represents a class of objects such as customers or inventory items.

Specific processing operations are predefined for data held in such tables. In a *select operation,* a new relation is built from designated rows. In a *project operation*, a new relation is formed from designated columns. In a *join operation*, two relations are combined to form a new relation. These basic, powerful operations can be used to extract information from data held in such tables.

TABLE 12.4 The logical structure of a relational data model.

Customer_No.	Customer_Name	Initials	Address	City	State	Zip	SSN
P105432	Abhom	SL	23 Broad	Salem	OR	03955	105-35-4911
P907216	Abhom	MK	23 Broad	Salem	OR	03955	362-10-1455
P018715	Jones	PK	101 Willow	Candoo	VA	21047	244-17-7865
P403228	Wabash	BR	1407 Tree	Polk	MD	20077	504-87-3241
B019754	Billings Corp.		3 Industry Ct.	Vale	NY	01829	394-62-4197
P248901	Mark	JL	49 Cram	Dundee	IL	43605	211-98-5651

As shown in Figure 12.15, a *network data model* resembles a tree structure. Each node within a network data model is a logical record type. Under network organization, any record type can have zero (or more) data elements. Therefore, it becomes possible to have an empty node that serves as a link among other nodes. Furthermore, record types can be related to, but are not necessarily derived from, one another. Related records are grouped into *sets*. One, the controlling record, is deemed the owner of the set, to which the other records are members. This type of organization is potentially more complex than relational database structures, with the tradeoff that some types of operations might be more difficult to implement.

Physical Access and Storage Implementation

At the level of physical access and storage, a DBMS is usually implemented through the computer's operating system. That is, at the physical level, a DBMS can be built on a conventional file management system. Tables or record structures can be implemented as physical records and files. Allocation is handled by the file management system, and file descriptions are generated by the DBMS. Thus, symbolic filenames and descriptions at the operating system level might not correspond with logical structures that would be meaningful to the user.

A DBMS can be used to integrate multiple file access methods. This can be possible if the operating system provides for different file types and access methods. For example, an operating system might support both indexed sequential and direct access methods, either of which might be used by the DBMS.

Ultimately, the access method used will be transparent to the DBMS user. The physical implementation is up to the DBMS designer. The basic design challenge

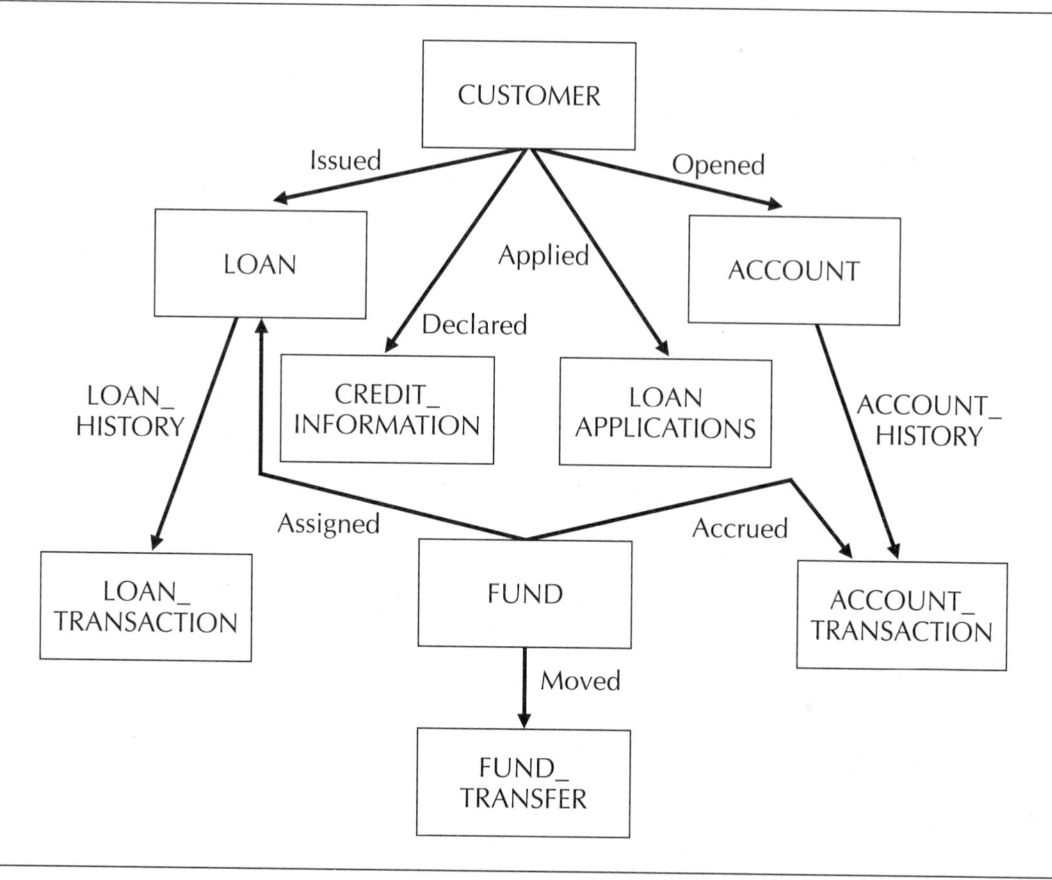

FIGURE 12.15 The logical structure of a network data model.

is to use the I/O and file-handling options of the file management system to maximize the performance and efficiency of DBMS processing. In some cases, it might be necessary to circumvent functions of the operating system and to build some file-handling functions directly into DBMS software. For example, it might be necessary to circumvent dynamic reallocation of file storage by the operating system. Otherwise, DBMS performance might be affected adversely as files become highly fragmented and the number of required accesses increases for each search.

An in-depth treatment of DBMS technology is beyond the scope of this text. More material on this subject can be found in other texts devoted solely to that topic. The objective of the discussion in this chapter is to show that, at a technical level, a DBMS does not replace the operating or file management systems of the computer. Rather, data access functions at the physical level are implemented through the file system. DBMS software is added "on top of" these capabilities and extends the file management system to higher, more abstract, logical levels.

SUMMARY

A file management system is responsible for controlling access to data and programs on secondary storage devices. The file management system is normally part of the operating system and can be supplemented by external utility programs and/or a database management system. It is organized and implemented according to the principle of software layers. High-level requests are received by the file management system through the service layer. These requests are translated into low-level requests to secondary storage device controllers. These low-level requests are implemented by the kernel and the hardware.

The organization of the file management system via software layers allows the file structure presented to users and applications to differ from the physical structures used to store data. The user (or application level) view of the data in a file is called the logical file structure. Logical file structure differs from physical structure in organization and access method. The file management system is responsible for translating operations on the logical structure into operations on the corresponding physical structure. The separation between logical and physical structures allows users and applications to manipulate files without knowledge of their underlying physical organization on mass-storage devices.

Allocation of physical storage locations and the management of physical I/O operations are key tasks of the file management system. Physical I/O operations are normally implemented using some combination of blocking and buffering. Blocking refers to the grouping of multiple logical records into a single physical record for device-level storage and I/O. Buffering refers to the use of memory for the temporary storage physical records. The combination of these I/O methods reduces the number of physical device accesses required to support ongoing processing. This generally results in improved program efficiency.

Storage locations on physical devices must be allocated to individual files and directories as needed. The principles of secondary storage allocation are similar to those of memory allocation. Possible allocation methods include contiguous (sequential), direct (random), and indexed. Each method represents a different tradeoff among the efficiency of read, update, and search operations. The allocation of storage locations to files is tracked in a data structure normally called a file allocation table. Various allocation methods imply differing structures for this table and differing methods of table search and update.

Directories are maintained by the file management system to locate individual files and store information about them. This information includes the filename, type, and storage location(s), and can also include size, usage, and security information. Because the number of files contained within a typical computer system is very large, complex directory structures such as trees and graphs are commonly used. Tree structure directories, the most common, allow directories to be nested within other directories to an arbitrary depth. This method of directory organization makes search operations efficient and can be used to implement file access restrictions.

A file management system provides many programs for file manipulation. These programs can be executed directly from the command layer or indirectly via application program requests to the service layer. File manipulation operations can be classified as full file manipulation (e.g., create, copy, and delete), file I/O (e.g., read, modify, or write records or fields), and directory operations (e.g., list and create). File I/O operations vary, depending on the method of file organization and the method of access.

In a multi-user information system, the file management system must normally support file sharing and file protection. This requires the system to implement some form of file ownership and a strategy for setting and enforcing access privileges. The right to read, update, execute, or delete a file can be granted by a file owner to other system users. File manipulation and I/O routines must verify access permissions as they execute. These verifications increase file security at the expense of additional processing overhead. The file management system must also protect against unauthorized access directly to secondary storage devices. Such access could allow a user to bypass established security provisions.

Reading and writing of a file by multiple users is coordinated through the file management system. The system must ensure that multiple updates do not interfere with one another and that each file user always receives current and consistent data. This is generally implemented through a file- or record-locking procedure. A user that needs to modify the contents of a file places a lock on the entire file or on the record to be modified. While locked, access to the file or record is denied to other users. Once an update is complete, the lock is removed, thus making the data available to other file users.

System administration refers to the protection of system data and to ensuring maximal availability of system resources to users. Many system administration functions are implemented through the file management system. These functions include monitoring access controls and user accounts, file migration, and backup/recovery operations. The system administrator is responsible for the migration of old versions of files to backup and offline storage. He or she is also responsible for creating periodic backups of files for use in the event of accidental damage. Much of the process of file migration and backup is automated within the file management system, but the complete implementation of these tasks, especially of data recovery operations, requires active human intervention.

A database management system (DBMS) presents logical data structure different from that of a file management system. Multiple logical structures can correspond to a single physical structure, and logical structures can encompass multiple physical files. A DBMS is typically implemented as a systems software extension to an existing file management system.

Key Terms

acyclic graph directory
append mode
append operation
attribute
binary editor
binary search
block
blocking
buffer
compaction
complete path
current (working) directory
data definition
data model
database administration
delete operation
direct file access
direct file organization
edit operation
editor
father
field
file
file allocation table
file control block
file control layer
file close operation
file destruct
file compaction (packing)
file locking
file open operation
file migration
file management system
file type
full backup
fully qualified file reference
general graph directory
grandfather
hierarchic directory structure
home directory
incremental backup
indexed file access
indexed file organization
insert operation
join operation
journaling
linear list
linear search
linked list
logical delete
logical file view (structure)
logical record
lost update
mirror-image backup
network data model
path name
physical file structure
physical record
project operation
read operation
record
record locking
relation
relational data model
relative path
root directory
select operation
sequential file access
sequential file organization
set
single-level directory
son
sorted list
storage allocation
storage I/O control layer
system administration
text editor
transaction logging
tree structure directory
tuple
two-level directory
update in place
update operation
user directory
write operation

Vocabulary Exercises

1. Data is generally moved to/from secondary storage devices in _____ or _____.
2. A _____ is generally the unit of file I/O to and from an application program, whereas a _____ is generally the unit of file I/O to and from a secondary storage device.
3. _____ describes the number of _____ contained within a single physical record.
4. In a _____ directory, a file can be located within no more than one directory. This restriction does not apply in a _____ directory.
5. _____ is the simplest but least flexible method of file access.
6. A _____ consists of a master (root directory), one or more subdirectories, and a filename.
7. The logical structure of data within a database management system is represented by a _____.
8. _____ allows records to be directly retrieved by specifying the value of a key field.
9. A _____ marks a record as removed without physically removing it from a file.
10. _____ directory structures allow directories to be located within other directories.
11. A _____ deletes directory and other references to a file and releases allocated storage locations. A _____ also does this and, in addition, destroys the data content of all storage locations that were allocated to the file.
12. In a _____, all files within a directory or storage device are copied to backup storage. In a _____, only those files altered since the last backup are copied to backup storage.
13. A _____ writes a new record at the end of an existing file.
14. When a file is _____, logically deleted records are physically removed from storage.
15. In _____, a record can be quickly accessed by specifying its location within the file, ignoring all other records in the file.

16. _____ or _____ is frequently used to reduce potential data loss in large online transaction processing systems.

17. A _____ records the allocation of storage locations to specific files.

18. Lost updates can be avoided by the use of _____ or _____. The latter method requires more processing overhead than the former, but generally allows a greater number of processes to access the file simultaneously.

Review Questions

1. What is the difference between the logical and physical structure of a file?
2. What is file migration? What is its relationship to the storage hierarchy?
3. How is file deletion normally accomplished? What security problems might arise from this method?
4. By what methods can a file be moved? What are the comparative advantages and disadvantages of these methods?
5. What is the difference between a full dump and a selective (or incremental) dump?
6. What is the difference between a logical delete operation and a physical delete operation?
7. What are the advantages and disadvantages of maintaining the logical order of records using a linked list?
8. How is the concept of software layers normally implemented within a file management system?
9. What are the advantages and disadvantages of an operating system that implements file typing?
10. Explain the structure of a tree structure directory. What are its disadvantages, compared to graph structure directories?
11. What is blocking?
12. By what methods can storage device blocks be allocated to files? What are the comparative advantages and disadvantages of each?
13. What levels of access privilege can exist for a file?
14. For what reason(s) might file locking be implemented?
15. What is a database management system? How does it differ from a file (management) system?

Research Problems

1. To increase efficiency of application program execution, some file management systems utilize a large number of file organization and access methods, each optimized to a very specific type of processing. The MVS (IBM mainframe) operating system contains such a file management system. Investigate this file management system to determine the various methods of file organization and access that are supported. For what type(s) of application program file manipulation is each method intended?
2. The Unix operating system and its embedded file management system were not originally designed to support large-scale transaction processing systems. For this and other reasons, Unix lacks many aspects of file management that would support such an environment (e.g., transaction logging, indexed files, file migration, and so on). Identify the specific deficiencies of the Unix operating system with respect to large-scale online transaction processing. How do/might application developers overcome these deficiencies?

13 Application Support and Control

Chapter Goals

- *Describe service-layer functions for process control and memory allocation.*
- *Describe service-layer functions for multiple-process synchronization and communication.*
- *Describe service-layer functions that support character, full-screen, graphic, and window-based I/O.*
- *Describe interactive and batch control of software with command languages.*

SERVICE FUNCTION OVERVIEW

Recall that much of the operating system can be thought of as a collection of general-purpose utility programs. These programs are designed to provide an interface between users (or their application programs) and low-level hardware functions. Application programs use these services directly through procedure calls to the operating system service layer. Users access these same services, either indirectly through application programs or directly through the command layer.

Service functions provided to application programs can be roughly classified into the following categories:

- File manipulation
- Process and memory functions
- I/O functions

Brief descriptions of each of these categories are given in Chapter 11. Details of the file manipulation operations provided by the operating system are provided in Chapter 12. One or more service calls exist to implement each of the file manipulation and file I/O operations described in that chapter. These calls provide application programs with all of the file manipulation operations generally available to users through the command layer.

PROCESS AND MEMORY SERVICE FUNCTIONS

Services for process control, communication, and memory allocation provide flexibility in the design and execution of application programs. In earlier generations of operating systems, it was assumed that each application program was a stand-alone entity. Communication between programs was assumed to take place through I/O operations to temporary or permanent files. Memory requirements were assumed to be static. That is, a program was allocated a region of memory when loaded, and that region did not change in size during the execution of the program.

As both application and systems software became more complex, these assumptions became increasingly more restrictive. Programs might need to have additional memory allocated to them during execution or might want to release memory. Large application systems might be composed of many programs, each of which might communicate directly with other programs (i.e., without the use of temporary files). Rather than load all application processes at once, it might be desirable to allow executing processes to request the creation of new processes or the termination of processes no longer needed.

Such capabilities are commonly incorporated into modern operating systems. They provide substantial flexibility for the organization of large application systems into groups of cooperating programs or processes. They also utilize machine resources more efficiently by allowing processes to be created and terminated, and memory to be allocated and freed, on an "as needed" basis.

Memory Allocation

Memory allocation services are provided to allow memory to be allocated to (or released from) a process during execution. These services are extremely useful when the storage requirements of a process cannot be determined in advance. Consider, for example, a text editor or a word processor. As the user adds text to the document being edited, the program needs additional storage for that text. Similarly, the deletion of text reduces the program's storage needs.

One approach to addressing this variability is to provide an initial allocation of memory to hold a small- or medium-size document. If the document actually being edited doesn't fill this space, the excess memory is unused (wasted). If additions to the document cause it to overflow the storage initially allocated, the excess can be written to secondary storage through file-service operations.

While this approach does work, it can result in wasted memory or in slow program operation due to secondary storage accesses. An alternative is to allow the program to request additional memory as text is added and to release that memory as text is deleted. The program could, for example, request an additional block of memory (e.g., 4 Kbytes) each time added text exceeds the current memory allocation. Those same memory blocks could be released when no longer needed. Allocating and releasing memory in blocks, as opposed to smaller units of storage, reduces the overhead associated with memory allocation and reallocation service calls.

Implementing this capability requires the provision of at least two service calls:

allocate_memory(AMOUNT_REQUESTED,STARTING_ADDRESS)

free_memory(STARTING_ADDRESS, SIZE)

The allocate_memory call is a request to the operating system to allocate a memory block of the requested size to the calling process. If the memory is available, the operating system will update the appropriate memory and process control data structures and return the starting address of the block to the process. A facility must also be provided for returning error conditions such as "additional memory not available."

The free_memory service call informs the operating system that the block of memory identified by the STARTING_ADDRESS and SIZE parameters is no longer needed by the process. The operating system will update the appropriate process and memory control data structures to indicate that the block of memory is now free for use by other processes.

Process Creation

It is often desirable to allow a process to request that another process be loaded and executed. Consider, for example, a general-purpose utility for sorting data stored in variable-size arrays. There are many applications that might use such a utility during their execution (e.g., text editors, programs that maintain indices in memory, and so on).

One method of providing these sort facilities is to implement sorting algorithms within every process that might need to sort data. This method requires no process creation services but requires that redundant code (the sort instructions) be included in many different programs.

An alternative method is to provide a mechanism by which an executing program can request that a sort utility be loaded and executed. Note that a mechanism is also required to send data to be sorted from the original process to the sort process and to send the sorted data back to the original process. (The latter capability is discussed in the next section.)

Process creation is normally implemented by a service call like this one:

create_process(NAME, LOCATION, ARGUMENTS, PROCESS_ID)

This call directs the operating system to load the executable program indicated by the NAME and LOCATION arguments. The process executing the service call is called the *parent process*; the process that is loaded is called the *child process*. Additional ARGUMENTS (e.g., the order of a sort—ascending or descending) can be passed to that process as they would if the process were invoked from the command layer. Once loaded, the new process is managed by the operating system like any other process. It changes states, executes, and terminates via the normal process management facilities of the operating system.

Interprocess Communication

Cooperation between executing processes requires some mechanism for synchronizing their execution. The nature of this synchronization depends on the nature of the data passed between the two processes and the relationship between their execution cycles. A number of interprocess relationships are possible, including the following:

- One process executes another as a subroutine, with no data passed in either direction
- One process executes another as a subroutine, with data passed in one or both directions
- Each process executes one process cycle in synchronization with the other process

Each of these scenarios requires a different approach to interprocess synchronization and data communication.

Interprocess signals. In the first scenario, support for *interprocess synchronization* is relatively simple. The parent process executes a service call to request that the child process be loaded and executed. Synchronization is implemented by a *signal* sent from child to parent, indicating the termination of the child process. The signal can be implemented within the child process or through the process management facilities of the operating system. The use of such a signal requires a mechanism by which the parent process waits for the completion (termination) of the child process.

The following program statement is an example of the creation of a child process from within a parent process:

create_process("CHILD","/USR/BIN",PROCESS_ID)

In this statement, the parent process requests the operating system to create a new process. The name of the process (as it is known to the file management system) is "CHILD", and it is located in the directory "/USR/BIN". The operating system will search the directory for the file, load its contents into memory, update the appropriate process control structures, and place the new process in the ready state (thus enabling execution). An identification code for the new process is returned to the parent through the PROCESS_ID parameter.

If the parent wants to wait for the completion of the child process before continuing its own execution, it must execute an operating system call such as

wait(PROCESS_ID)

immediately after executing the create_process call. The wait function causes the operating system to place the parent in a blocked state, pending the termination (through an error or normal exit) of the process identified by the PROCESS_ID parameter of the wait call. Thus, synchronization is implemented as a minor extension of the normal process management facilities of the operating system.

Many operating systems provide facilities to return an error or exit status when a process terminates. These are generally implemented as numeric codes that are communicated to the command layer or to other processes. If such codes are implemented within the operating system, a condition (code) is returned to a parent process when a child process terminates. For example, the service call

wait(PROCESS_ID,EXIT_CODE)

will cause the parent process to be blocked pending the completion of the child process. The exit code of the child process will be communicated to the parent through the EXIT_CODE parameter. The parent can test that code to determine whether the child process executed successfully.

More complex signals are required when processes must coordinate the execution of individual process cycles. Consider, for example, two cooperating processes. Process 1 is responsible for retrieving input data from an I/O or secondary storage device and storing that data in a shared region of memory. Process 2 is responsible for performing some transformation on the data and sending it to an output device. Each process has a repetitive cycle. Process 1 repetitively reads input records and stores them in memory; process 2 repetitively transforms and outputs those records.

Signals are required to coordinate the operations of each of these processes. Process 2 cannot process a record until process 1 has read and stored it. Similarly, process 1 cannot store the next record until process 2 has retrieved the previous record. Coordination between these two processes can be implemented by a pair of signals. Process 1 will send a signal to process 2 after each record is stored in shared memory. Process 2 will send a signal to process 1 each time a record is retrieved from shared memory. These signals will prevent process 1 from overwriting shared memory with a new record before process 2 has retrieved the old record. The cooperation is illustrated in Figure 13.1.

Process 1

```
read(record)
while not end-of-file
      wait(process 2)
      store(record)
      signal(process 2)
      read(record)
end while
store('eof')
```

Process 2

```
signal(process 1)
wait(process 1)
retrieve(record)
while (record ≠ 'eof')
      signal(process 1)
      transform(record)
      write(record)
      wait(process 1)
      retrieve(record)
end while
```

FIGURE 13.1 **An example of two cooperating processes, with cycles coordinated through the use of interprocess signals.**

As in the prior example, the operating system will place each process in a blocked state during a wait instruction, pending the arrival of a signal from the other process. Note that coordination of the processes in the event of an end-of-file condition could also have been implemented through signals. A separate signal could have been used for this condition, or a single signal with multiple values could have been used for all synchronization activities. Figure 13.2 illustrates the latter alternative.

Process 1

```
read(record)
while not end-of-file
      wait(process 2)
      store(record)
      signal(process 2,'ready')
      read(record)
end while
signal(process 2,'eof')
```

Process 2

```
signal(process 1,'ready')
wait(process 1,condition)
while (condition = 'ready')
      retrieve(record)
      signal(process 1,'ready')
      transform(record)
      write(record)
      wait(process 1,condition)
end while
```

FIGURE 13.2 **An example of two cooperating processes coordinated through the use of a multivalued signal.**

The implementation of interprocess signals is an important and basic function of most operating systems. This is because signalling is a general method for controlling access to shared resources. Note that in the previous two examples, signals are used not only to synchronize the processes, but also to coordinate their use of a shared resource (the shared region of memory used for data communication).

The operating system must implement similar controls over access to many shared resources. Access to the CPU, memory, secondary storage, and I/O devices by multiple processes requires mechanisms for preventing conflict. An operating system implements these mechanisms by the same basic signalling methods used in the previous examples. However, the use of signals for these resource management functions is generally hidden from executing processes other than the operating system itself.

Interprocess data communication. It is unusual for data communication to be implemented entirely through signals and shared memory. In general, the implementation of shared memory shown in the previous example requires both processes to be compiled and/or linked together. When linked together, the two processes are in fact a single process for purposes of process management within the operating system.

Interprocess data communication is normally implemented by regions of shared memory or secondary storage that are created and controlled by the operating system in response to service requests by processes or instructions received through the command layer. The common term for such a region of shared storage is a *pipe* (i.e., a data pipeline between processes), although some operating systems use different terminology. A conceptual representation of a pipe is shown in Figure 13.3.

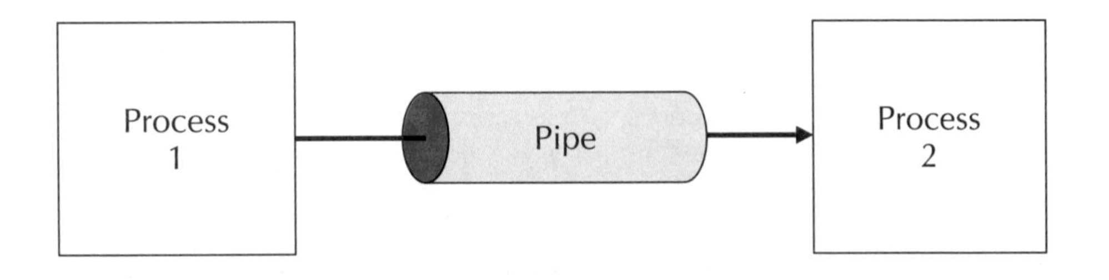

FIGURE 13.3 A conceptual diagram of a pipe between two processes, with one-way data flow.

The implementation of a pipe requires mechanisms whereby a process can request the creation of pipe between itself and another process. Mechanisms are also required to allow data to be written into the pipe or read from the pipe. Thus, a set of service calls such as the following could be implemented:

open_pipe(PROCESS_ID,DIRECTION,PIPE_ID)

close_pipe(PIPE_ID)

read_pipe(PIPE_ID,DATA)

write_pipe(PIPE_ID,DATA)

The open_pipe call directs the operating system to establish a region of shared storage for communication between the calling process and the process identified by PROCESS_ID. Depending on the particular operating system, this shared storage can be implemented in memory or as a temporary file. To allow a process to use multiple pipes simultaneously, the operating system returns PIPE_ID as an identifier for use in subsequent pipe operations. The close_pipe call directs the operating system to release the area of shared storage.

The argument DIRECTION in the open_pipe call tells the operating system whether data will flow to or from the calling process. It can also be possible to use the pipe for bidirectional data communication, in which case possible values of DIRECTION are TO, FROM, or BOTH.

The write_pipe call directs the operating system to move the information stored in the data structure DATA to the storage region allocated to the pipe. The read_pipe call directs the operating system to move data from the storage region to the data structure referenced in the DATA argument. Note that the location of the storage region, the movement of data into and out of that region, and the data structure(s) used to implement the pipe are entirely under operating system control.

In some operating systems, input to and output from pipes is implemented in a manner similar to input to and output from files. That is, once created, a process interacts with a pipe in essentially the same manner as it interacts with a file. Thus normal functions such as reading, writing, buffering, and end-of-file conditions are the same for pipes as for files. The advantage to such an approach is that it can be used both for data communication and for synchronization.

Figure 13.4 shows the example from Figure 13.2, modified to use pipes for both data communication and process synchronization. Note that most of the explicit signal and wait commands have been eliminated. The signal to coordinate reading from and writing to shared memory is completely replaced by I/O to and from the pipe. Attempts by process 1 to write to a full pipe will cause the operating system to block the process until process 2 has consumed some or all of the data in the pipe. Similarly, an attempt by process 2 to read from an empty pipe will cause the operating system to block the process until data is written into the pipe by process 1. The end-of-file condition is handled by explicitly reading and writing the end-of-file marker to and from the pipe.

Process 1

```
open_pipe(process 2,"to",pipe_id)
signal(process 2,pipe_id)
read(record)
while not end-of-file
      write_pipe(pipe_id,record)
      read(record)
end while
```

Process 2

```
wait(process 1,pipe_id)
read_pipe(pipe_id,record)
while (not eof(pipe_id))
      transform(record)
      write(record)
      read_pipe(pipe_id,record)
end while
```

FIGURE 13.4 **An example of two cooperating processes using a pipe for synchronization and data communication.**

INPUT/OUTPUT SERVICE FUNCTIONS

The service layer will normally provide a set of service calls to allow processes to interact with input/output devices and, therefore, directly with users. The set of services provided has changed dramatically over the last two decades as the nature of human and computer interaction has changed.

Character-Oriented I/O

Early operating systems managed input and output in terms of streams of individual characters. This view of I/O mirrored the capabilities of early I/O devices such as printers, card readers, and teletypes. Such devices were limited to the movement of printable characters to and from the computer. Very few specialized control functions were required for this type of I/O. Those control functions that were provided were encoded as nonprinting characters (e.g., form feed, carriage return, line feed, and so on).

The service calls required to support this type of I/O were relatively simple and few in number. Examples of these calls include

read_character(DEVICE_ID,CHARACTER)

write_character(DEVICE_ID,CHARACTER)

Both of these calls interact with an I/O device (or device controller) attached to a specific I/O port. The device is identified by the DEVICE_ID argument. In simple operating systems, this identifier can correspond to the bus port to which the device is attached. The argument CHARACTER is usually a memory address where the character is stored (or should be stored once input).

Because characters are generally read and written in groups (e.g., words, fields, lines of a report) service calls are normally provided to read and write entire strings as well as individual characters. The following service calls provide these additional capabilities:

read_string(DEVICE_ID,STRING)

write_string(DEVICE_ID,STRING)

As in single character I/O calls, DEVICE_ID identifies a specific I/O device. The argument STRING is generally the starting address of a character array in memory.

Note that string input requires some mechanism for identifying the end of one string and the start of another. In many operating systems, a special character is placed at the end of each string array to signify its end. Unix, for example, terminates character arrays with an ASCII 0 (null) character. Such a convention allows string I/O to be implemented as a loop containing character I/O commands. For example, the call write_string might be implemented as shown in Figure 13.5.

A similar strategy can be used for the input and output of printed text lines. In most operating systems, text lines are character strings terminated by a special "end-of-line" character such as a line feed or carriage return. Thus, a service call to output a text line could be implemented as above by changing the test for a null character into a test for the end-of-line character.

```
procedure write_string(device_id,string_start_address)

begin
    i=0
    character=string_start_address
    while (character ≠ 0)
        write_character(device_id,character)
        i=i+1
        character=string_start_address+i
    endwhile
    write_character(device_id,character)
end
```

FIGURE 13.5 A procedure to output a null terminated string.

Full-Screen I/O

With the introduction of interactive full-screen terminals, I/O operations expressed entirely in terms of characters became increasingly cumbersome to use. This is because I/O to or from a full-screen terminal might not flow in a linear sequence from left to right and top to bottom. Full-screen I/O could require data to be displayed at many different screen locations simultaneously. Input from such terminals often requires printing a blank form on the screen and then allowing the user to move (tab) from field to field to input individual data items. Alterations to individual fields require the ability to move the input cursor directly to a particular position on the screen, without altering other characters already printed there.

Other characteristics of full-screen I/O further complicate matters. Most full-screen I/O devices provide special display features such as underlining, inverse video, bold intensity, half intensity, and invisible characters. All of these require complex control capabilities. There are also differences in the ways such features are implemented by the designers and manufacturers of various I/O devices. Most I/O devices implement control of these capabilities through special sequences of (control) characters. An American National Standards Institute (ANSI) standard exists for the character-control sequences to utilize these features. A partial listing of the ANSI control code sequences is given in Table 13.1. However, not all manufacturers adhere to the ANSI standard, and many devices implement commands and display features that are not included in the standard.

Many operating systems implement a set of service calls to access full-screen display features. For example, one or more service calls might be provided to implement each of the functions in Table 13.1. However, an additional mechanism is required to allow processes to perform full-screen I/O to a variety of I/O devices.

TABLE 13.1 A portion of the ANSI standard control-character sequences for full-screen, charater-based displays. The string <escape> stands for ASCII control character #27

Feature or Command	Control-Character Sequence
Position Cursor	<escape>[#;#H
Underline On	<escape>[4m
Inverse Video On	<escape>[7m
Bold On	<escape>[1m
All Attributes Off	<escape>[0m
Erase to End of Line	<escape>[K
Erase Entire Screen	<escape>[2J
Cursor Up	<escape>[#A
Cursor Down	<escape>[#B
Cursor Left	<escape>[#D
Cursor Right	<escape>[#C

In most operating systems that support interactive full-screen I/O, a mechanism is provided to inform the operating system as to the type of full-screen I/O device being used. This can be implemented in a number of ways, including

- Asking the user to provide a device type or identifier during login processing
- Querying the I/O device directly (most full-screen terminals will transmit an identification code in response to the proper control sequence)
- Specifying a default device in a file of initialization information and allowing the user to alter this information, if needed

Once the I/O device type is determined, that information is stored for later use by full-screen I/O service routines. Whenever a process requests a full-screen I/O function, the service routine will test the I/O device type to determine the proper control sequence to implement that function on that particular device. To improve efficiency, a table of control sequences is sometimes created and stored in memory when the I/O device is first identified. All full-screen service routines then utilize entries in this table instead of testing the device identifier each time they are called.[1]

[1] For example, Unix uses a variable called TERMCAP (terminal capability) to store a list of full-screen capabilities and control sequences in memory. These are utilized by I/O service routines when performing full-screen and editing functions.

Graphic I/O

The use of graphic I/O devices substantially increases the complexity of I/O processing and, thus, the number and complexity of I/O service calls. Note that although full-screen I/O requires special features and formatting controls, all communication with the I/O device consists of characters (i.e., both the text to be printed and the control command sequences consist of characters). Graphic I/O devices present control complexities that are difficult to address entirely through the use of characters.

In general, graphic I/O devices generate images in terms of pixels rather than characters. Individual pixels of the display can be manipulated in terms of binary values such as black or white, gray scales, or color codes such as RGB intensities. As discussed in Chapter 8, communication with graphic devices can use bit maps, vector commands, a display language, or some combination of these. As with full-screen I/O devices, there is considerable variation among devices and manufacturers in display capabilities and the methods of controlling those capabilities.

Unlike full-screen I/O, there is no standard method of controlling graphic displays. Particularly for devices that use bit-mapped and vector command communication, the differences among devices are virtually irresolvable. There is some trend toward standardization in display language-driven graphics. The Postscript display language has arisen as a clear leader in the competition for a standard graphics display language. However, there has been no acceptance of Postscript by any standards organization and there are significant competitors, such as Hewlett-Packard Graphics Language.

Because of the variation and lack of standardization in graphic I/O, few operating systems provide a set of graphic I/O service calls. However, many operating systems provide special utilities to utilize specific graphic I/O devices for specific operations such as character-based printing and plotting. Capabilities are generally limited, and input tends to be program-specific.

Application programs must generally interact with graphic I/O devices without direct assistance from the operating system. In most cases, the application program must understand the control language and capabilities of the device and must communicate with it using the standard character-based I/O services provided by the operating system. Differences among devices must be programmed into the application.

A few operating systems (those that implement window-based I/O, in particular) do implement standard graphic I/O routines within the service layer. These are generally standardized around a display language such as Postscript. Service calls are provided for operations such as font selection, line drawing, and bit-mapped display. The application program utilizes the standard display language when using these service calls. If the actual device does not support the display language, the operating system must provide a translation from the display language to control sequences that the device understands. Such a translation is seldom exact, resulting in displays that are not exactly as the application program intended.

For example, an operating system can standardize graphic display service calls on Postscript. An application program that wants to display text on such a device must specify a Postscript font and utilize Postscript formatting commands when interacting with graphic I/O service routines. If the actual device is not Postscript-compatible, such as a Hewlett-Packard Laserjet printer, the operating system must translate Postscript commands into Laserjet commands. However, because fonts and formatting capabilities differ, the location and appearance of the text on the printed page will not be exactly the same as if the output device were Postscript-compatible. This difference could be unacceptable in some situations.

Window-Based I/O

Many operating systems now support the implementation of interactive interfaces using windows. Figure 13.6 shows an example of a window-based display. Window-based I/O differs from graphic and full-screen I/O in a number of ways:

- Division of a display into multiple output regions (windows)
- Use of colors and background patterns
- Ability to use various font styles and sizes
- Display of purely graphic images
- Provision of standard facilities for menu display and control (e.g., popup menus, push buttons, scroll bars, etc.)

There is no standardized set of windowing capabilities, but those listed above are available in most windowing software.

A number of window interfaces have been standardized and included in operating systems. These include Microsoft Presentation Manager (embedded in OS/2), X Windows (supported in many Unix-based operating systems), and Microsoft Windows (supported as an extension to MS-DOS).

The control of window-based I/O combines many of the features (and complexities) of full-screen and graphic I/O and adds some of its own as well. As with full-screen I/O, a set of service routines is provided by the operating system to allow application programs to utilize these services:

- Definition of window size and display characteristics (e.g., background color, scrolling capabilities, font style and size, available control functions, etc.)

- Definition and display of menus and the input of user menu selections
- Display of graphic images based on bit maps, vector commands, or a display language
- Standardized facilities for window control, including resizing, movement, overlap with other windows
- Copying of text and/or images between windows (i.e., "cut-and-paste" capabilities)
- User input via a keyboard or a pointing device (e.g., a mouse, a trackball, or a digitizing tablet)

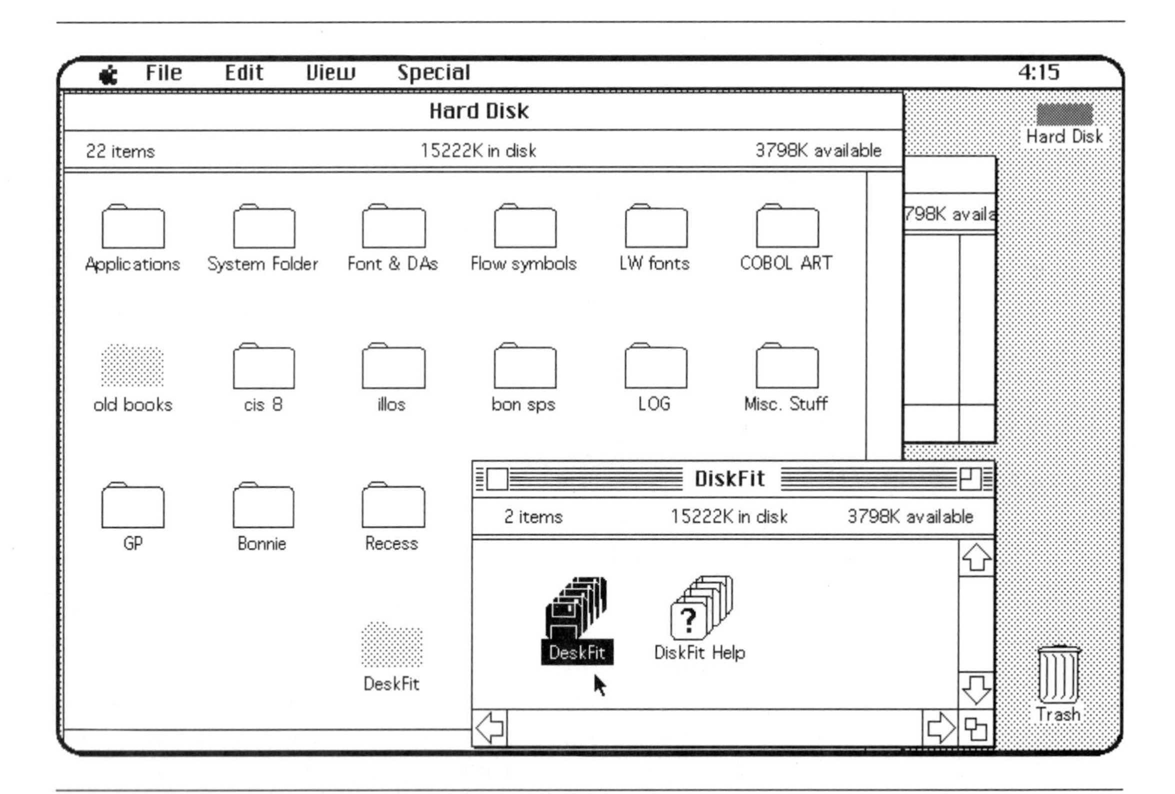

FIGURE 13.6 **A sample graphic display using windows.**

As might be surmised from the variety of capabilities listed, the number and complexity of service calls that implement window-based I/O is extremely large. The number of service calls is typically more than 100, and a variety of parameters must be specified for each call. In addition, the operating system must provide translation facilities between these standardized service calls and the specific commands necessary to implement them on various graphic display devices. Because of these complexities, window-based operating systems generally consume a substantial amount of hardware resources (e.g., CPU cycles for translation, memory for display images, disk space for service routines and fonts, etc.).

The organization of software components within a typical window-based interface is shown in Figure 13.7. The display manager is responsible for interactions with a particular graphic I/O device. Note that this is the only component that depends on the implementation specifics of the I/O device. The window manager is responsible for basic windowing functions. These include window display, sizing, menu display, mouse movement, and other related functions. The window manager provides display services for all system or application programs. These programs interact with the window manager via system service calls. With this organization, the command processor is simply another process using the services of the window manager for I/O functions.

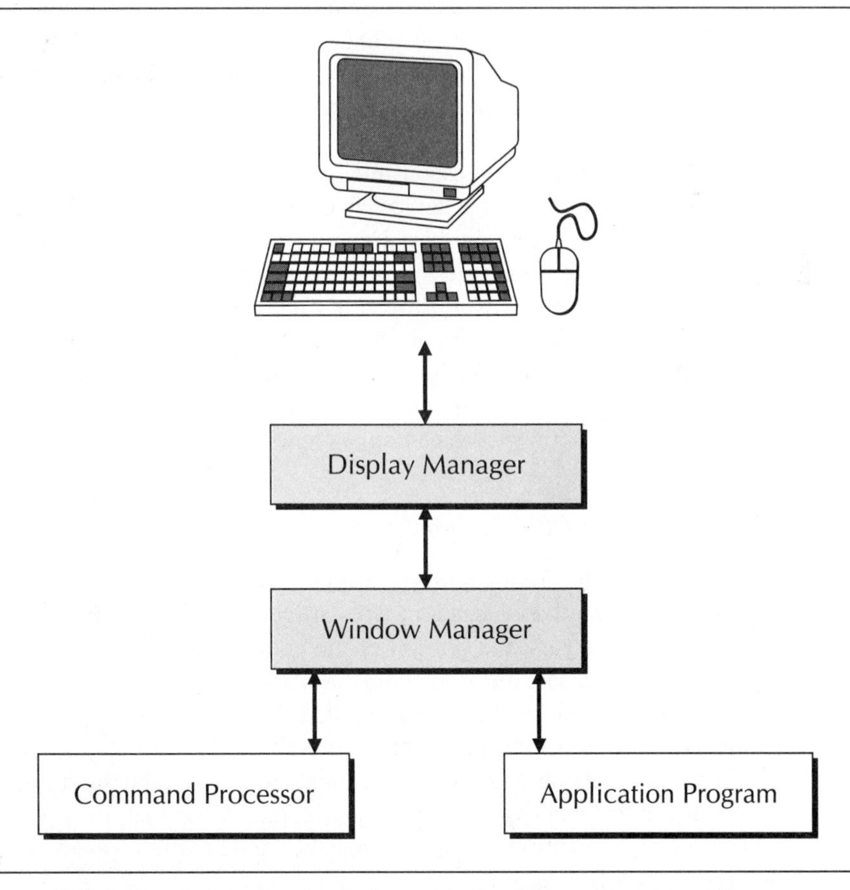

FIGURE 13.7 **The software components of a typical window-based interface.**

THE COMMAND LAYER

The command layer provides a user interface for controlling hardware and software resources. The primary mechanism for this control lies in the user's ability to execute system utilities and application programs. Other functions of the command layer include these:

- Provide facilities for both batch and interactive program execution
- Provide facilities for complex control of multiprocess applications
- Provide interactive capabilities for file manipulation and management

As with other aspects of operating system architecture, there is considerable variation as to the exact features provided and the details of their implementation.

The organization of the command layer and its relation to system and application programs is shown in Figure 13.8. The exact organization is somewhat dependent on the type of user interface. As with application programs, the command layer user interface can be character, full-screen, or window-based. Character-based interfaces are also called *command line interfaces*; full-screen interfaces are sometimes called *forms-based interfaces*. The organization of command layer components shown in Figure 13.8 is typical of command line and forms-based interfaces.

The display manager is responsible for the actual display of characters and/or images on the video display device. It is also responsible for keyboard and/or mouse inputs. For a window-based command interface, an additional layer of software (the window manager) is interposed between the display manager and the command interpreter, as illustrated in Figure 13.7. Input is given to the command interpreter, which is responsible for recognizing commands and program names. Depending on the input, the command interpreter could call an internal command processor or a program loader. The program loader loads and executes system and application programs not contained within the command layer itself.

Batch and Interactive Services

The facilities of the command layer can be used to execute programs and utility functions in interactive or batch mode. Particularly in command line and forms-based command layers, the portion of the command layer dedicated to interactive service is called the *command interpreter*. The characters and words interactively entered by the user are called the *command language*. The set of commands used in batch mode are commonly referred to as a *job control language*. In many operating systems, the interactive command language is a subset of the job control language. In others, particularly those that use a window-based interface, they are entirely separate languages and are processed by entirely separate systems software components.

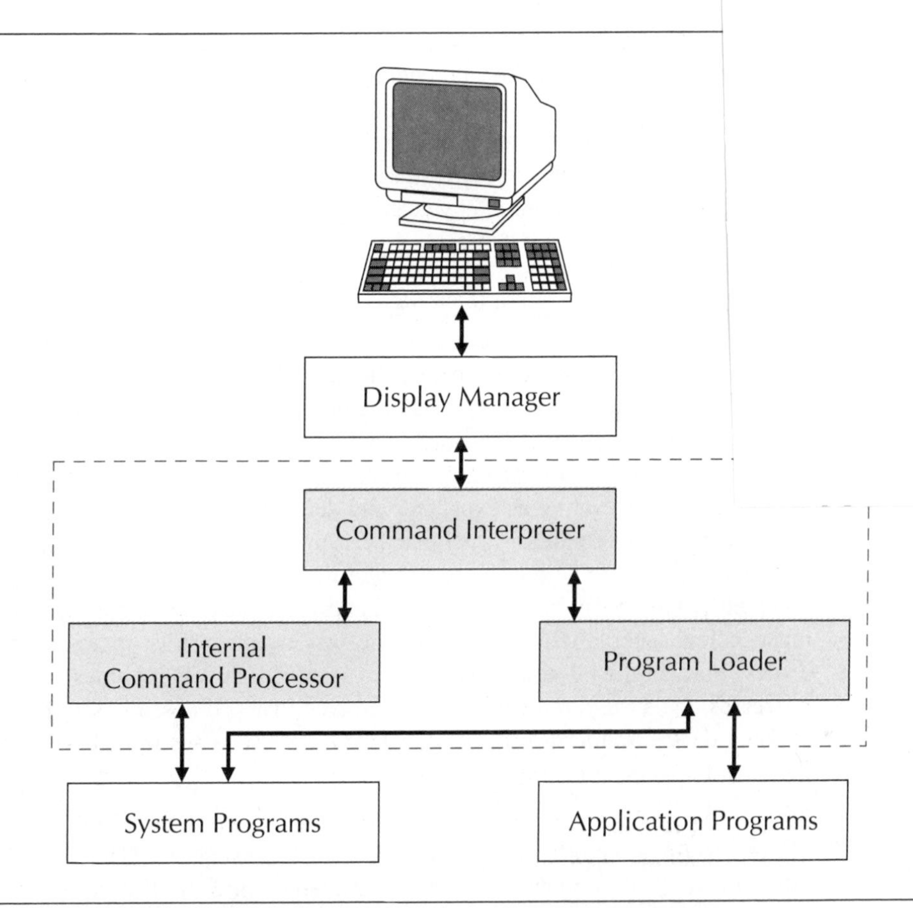

FIGURE 13.8 The components of the command layer (shown within the colored dotted line) and their relationship to other systems software components.

Batch service. Many application programs and system utilities are designed to run without user interaction. Examples of application programs that normally run in batch mode include periodic reporting, file maintenance activities, and transaction processing activities that do not utilize online input—for instance, payroll and accounts payable systems. Examples of system utilities that are typically run in batch mode are periodic reporting programs, such as user accounting, and file backup procedures.

For various reasons, many large application systems are partitioned into multiple programs or processes. An entire collection of related programs that are run within the same time frame (e.g., all of the programs executed for monthly payroll processing) is called a *job.* Each individual step (or program) in a job is called a *job step.* The set of commands necessary to control and execute a series of job steps is called a *job stream.*

Dividing a job into job steps can be advantageous for many reasons, but it creates problems in the control of the entire job stream. Control mechanisms must be provided for building job streams, requesting job execution, controlling job sequences and relative priorities, and specifying alternative machine actions in case of errors or contingencies. Providing these control mechanisms adds substantially to the complexity of a job control language. In many operating systems, the complexity of the job control language that controls batch program execution is as great as the complexity of the programming language statements that define individual programs.

Within a batch system, a user exerts no direct control over job execution or sequencing after jobs have been submitted to the command layer or computer operator. Similarly, a system operator in the computer room does not become involved in these details once a batch run has been set up and initiated. The operator will be concerned primarily with seeing that the required resources (tape drives, I/O devices) actually are made available, monitoring the system to assure that it is operating as expected, and dealing with any errors as they occur.

Under normal conditions, neither the user nor the system operator would be expected to interfere with a job run. Accordingly, all machine actions required to complete each job run must be specified entirely in advance. These specifications must include alternatives for all possible processing results, including probable errors, that might be generated. For example, if a job includes cooperating processes, the system must be given directives for each possible set of outcomes. To provide for uninterrupted job flow, recovery processes must be specified for the most likely error conditions.

Interactive service. Within an interactive multi-user system, each user is capable of reacting to conditions while processes are executing. Thus, the user can correct errors the moment they are encountered. In a multi-user system that is supported by a timesharing host, there remains a need for a system operator. The primary responsibility of this operator will be to assure that system resources are being made available and are sufficient to meet demand.

Because the user of an interactive system can be more involved during execution, processing requests could contain less detailed specifications than would be presented to a batch system. Accordingly, command languages for interactive systems tend to be less formal and less complex than job control languages for batch systems.

Command Structure and Interface

As stated earlier, interactive command interfaces can be command line, forms-based or window-based. Regardless of the type of interface, the data the user must provide to the command layer is essentially the same. The primary differences lie in how the user is prompted for input and the manner in which input is provided.

Command structure. The format of a typical command in a command language is a line of text (one or more characters terminated by a carriage return) of the form:

Command Arguments Options

The command field references the operation to be performed. Argument(s) refer to objects to be manipulated by the command, such as files. Options provide a mechanism for altering the normal (default) execution of the command. Some command languages require that all options be specified. In others, default values are assumed if their corresponding options are not present on the command line. In some operating systems, options can precede arguments, or they can be interspersed with arguments in the command line.

The command is recognized as an explicit instruction by the command interpreter. The command interpreter will normally test the first word of each input line to determine if it matches one of its internally recognized commands. If it does, the corresponding instructions within the command processor will be executed, or a corresponding program stored on disk will be loaded and executed.

If the command is not recognized by the command interpreter, it is assumed to be the name of an executable program (or file of commands) held in secondary storage. The command interpreter will initiate a search to find a file that matches the command name. If found, the file (program) will be loaded and executed. This is the normal mechanism for executing user application programs.

The search process for executable files is generally subject to some user control. The command interpreter will normally have a predetermined list of devices and/or directories to be searched for commands. This list is often specified in an initialization file that is read by the command interpreter during user login. By modifying the contents of this file, the user can alter the default search process to include additional devices or directories or to exclude one or more default devices or directories. This list of devices and directories is called a *path* or *search path.*

In some cases the user can specify the device and/or directory containing an executable program as part of the command syntax. Typically, the device and/or directory are placed prior to the command name on the command line. The command interpreter will detect this information and will search the corresponding device/directory for a filename corresponding to the command name.

Arguments to commands are normally references to data to be manipulated by those commands. For example, the argument C:\ in the MS-DOS command

DIR C:\

instructs the operating system to display a directory listing of the root directory (\) of device C:.

Some commands use multiple arguments. For example, the MS-DOS command

COPY C:\FILE1.TXT D:\

instructs the operating system to locate FILE1.TXT in device/directory C:\, make a copy of the file, and store the copy in device/directory D:\ (with the same name as the original).

Many commands recognize options on the command line to control various processing features. For example, the MS-DOS command

DIR C:\ /P

will produce a directory listing and will pause after each screen of text is displayed. This allows the user to see the beginning of a long listing before it scrolls off the top of the screen. The option /P modifies the default behavior of the DIR command.

Other operating systems use different syntactic conventions for options. Unix, for example, uses options of the form

–option value

where –indicates that an option identifier follows, option is a character or string representing the option, and value is a field identifying the value of the option, if needed. For example, the Unix command

lpr –d printer1 file1.txt

instructs the operating system to print the file file1.txt. The option –D printer1 alters the default destination for the file to printer1.

In MVS, options are normally specified in the form

option1=value, option2=value, option3=value

where option is a keyword indicating the option being set, and value is the value of the option. The sign = separates the keyword from the value, and the , symbol separates multiple options on the same command line.

Forms-based command interfaces. The complexity of specifying options and arguments in a command line interface often leads to errors in user input. The user must not only memorize the names of valid commands, but also the type and syntax of arguments and options. In essence, the user is speaking to the computer in an artificial control language. Forms-based command interfaces address some of the problems and complexities associated with command line interfaces. At its core, a forms-based interface is based on a command language. That is, a user must still memorize a basic set of command names. However, the forms-based interface provides a much easier mechanism for the user to input command arguments and options.

Once a command name has been entered at the command prompt, the command interpreter places a form or menu on the screen. In the case of a form, the user is presented with a set of blank fields into which he or she can place argument names and option values. These blank fields are each preceded by the name of the argument or option. Thus, the user need not be concerned with the placement (syntax) of these fields within a command line. In the event of errors, an appropriate error message can be displayed (often at the bottom of the screen) and the cursor moved automatically to the field(s) that need to be changed. An example of a forms-based interface is provided in Figure 13.9.

```
------------------------- EDIT - ENTRY PANEL -----------------------
COMMAND      ===>

ISPF LIBRARY:
   PROJECT ===> ABC.USER12
   GROUP   ===> XYZ        ===>            ===>          ===>
   TYPE    ===> SOURCE
   MEMBER  ===>                (Blank or pattern for member selection list)

OTHER PARTITIONED OR SEQUENTIAL DATA SET:
   DATA SET NAME    ===>

   VOLUME SERIAL    ===>            (If not cataloged)

DATA SET PASSWORD   ===>            (If password protected)

PROFILE NAME        ===>            (Blank defaults to data set type)

INITIAL MACRO       ===>            LOCK        ===> YES    (YES, NO or NEVER)

FORMAT NAME         ===>            MIXED MODE ===> NO     (YES or NO)
```

FIGURE 13.9 A forms-based interface for a file management utility in IBM Structured Programming Facility (SPF).

A forms-based interface is sometimes supplemented with a menu (list) of argument and option values. For example, the user's command to list a directory might be provided with a list of directories from which to choose. The user might also be provided with a list of options or option values such as sorted listing, or short or long display.

Window-based command interfaces. A window-based interface is generally easier to use than either a command line or forms-based interface. This is due to the extensive use of menus and graphic images. In general, all of the commands available to the user are provided in menus. The large number of commands could necessitate a hierarchical layering of menus. For example, an initial menu can list choices such as file operations, word processing applications, and accounting applications. Each of these options is in turn another menu listing individual commands (copy, move, rename, delete) that can be executed. An example of layered menus appears in Figure 13.10.

A window-based interface thus extends the use of menus and prompts beyond the realm of arguments and options to the commands themselves. Graphic images are often used to supplement these menus for further ease of use. Visual metaphors (called *icons*) represent certain commands or objects to be manipulated. For example, the files available for editing might be shown as pictures of file folders, with a filename displayed on the tab of each folder. The use of visual metaphors increases the interface's ease of use.

Finally, window-based interfaces seldom require the user to enter character or line-oriented input. The user selects icons and images using a pointing device such as a mouse or a trackball. Arguments and options are then displayed on the screen, and the user can select or manipulate these with the pointing device. Thus, the user need not remember the name of commands or options. Issues of language syntax are also largely irrelevant.

Extended Job Control Command Facilities

The command facilities discussed thus far are generally adequate for interactive control of programs, but additional facilities are required for batch job control. These include the ability to specify the execution of multiple programs, repetitive execution, conditional execution, and other complex control features.

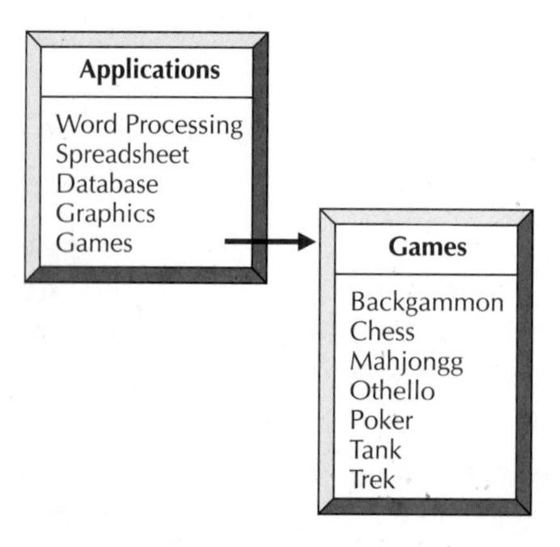

FIGURE 13.10 **An example of a layered menu. Each choice on the Applications menu represents another menu offering another set of choices.**

Control structures. The command line examples in previous sections all consisted of single commands. It is usually necessary to execute multiple commands to complete a set of interrelated tasks. In addition, the execution of some commands can be optional or repetitive. A command processor must provide some sort of control structure(s) to allow for possibilities such as these.

The basic control structures common to most programming and command languages are

- Sequence
- Selection
- Repetition

Full control capabilities with respect to program invocation and control require a command language with these control structures.

Sequence control simply refers to the sequential execution of multiple programs. For various reasons, it is often desirable to construct an application system as a linear series of separate processes. Figure 13.11 shows an example of a batch transaction processing system organized in such a fashion. Sequential execution can be implemented simply by invoking each program, one at a time, at the command processor prompt. However, this solution requires human intervention (typing the next command) at each step in processing.

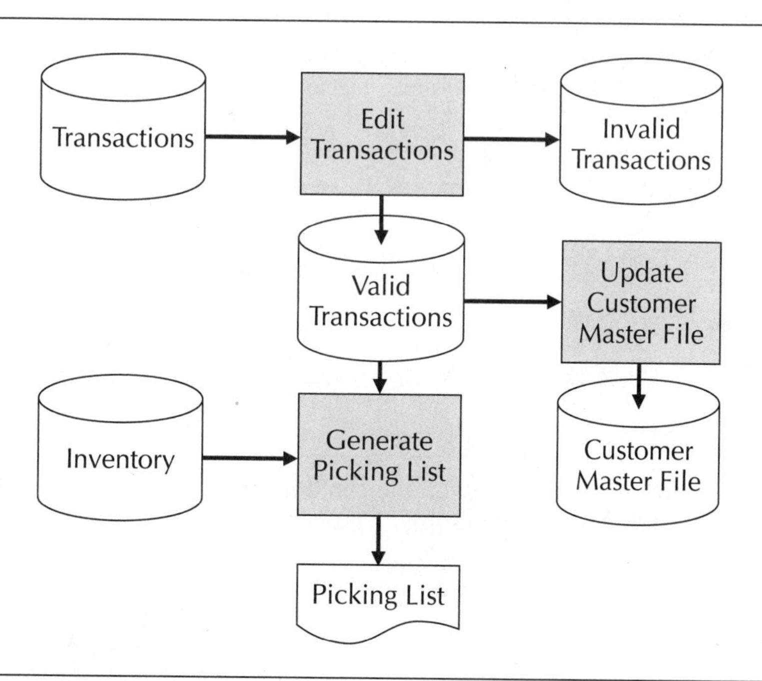

FIGURE 13.11 A batch transaction processing system consisting of three sequential processes.

Most command interpreters accept input either interactively (from the command prompt) or in batch mode from a file of commands. The commands necessary to execute each processing step are stored in a file and given to the command interpreter as input. This eliminates the need for a user or operator to type each individual command after the previous processing step terminates. An example of such a command file is shown in Figure 13.12.

```
/*   Batch Command File - Daily Transaction Processing   */

RUN edit_transactions
RUN update_customer_master_file
RUN generate_picking_list
```

FIGURE 13.12 A batch command file. Each program is executed, one after the other, without user intervention.

Selection refers to the ability to select alternate processing paths, depending on stated conditions and/or the results of previous processing steps. This is usually implemented as some form of *if statement* or *case statement*, as in many programming languages. An example of conditional processing is shown in Figure 13.13. In this example, the system date is tested to determine if the current date is the last day of the year. If the test is successful, a year-end processing step is executed in addition to the normally executed processing step.

```
/*   Batch Command File - Daily Transaction Processing   */

RUN edit_transactions
RUN update_customer_master_file
RUN generate_picking_list
if (sys_date = 'December 31') then
   RUN end_of_year_program
endif
```

FIGURE 13.13 A batch command file with the conditional execution of END_OF_YEAR_PROGRAM.

Selection is often used to detect errors during processing. As discussed earlier in the chapter, most operating systems generate an exit or status code for each terminated process. An example of a command file that tests exit codes is shown in Figure 13.14. The exit status for the most recently terminated process is stored in the variable $exit_code by the operating system. It is tested after

```
/*   Batch Command File - Daily Transaction Processing   */

RUN edit_transactions
if ($exit_code = 0) then
   RUN update_customer_master_file
   if ($exit_code ≠ 0) then
      echo "failure in process update_customer_master_file
      echo "exit_code=" $exit_code
   endif
   RUN generate_picking_list
   if ($exit_code ≠ 0) then
      echo "failure in process generate_picking_list"
      echo "exit_code=" $exit_code
   endif
else
   echo "failure in process edit_transactions, exit code=" $exit_code
   echo "remaining processing steps aborted"
endif
```

FIGURE 13.14 **A batch command file with nested conditional executions. The exit status of each program is tested to determine whether to execute the next step or print an error message.**

the first processing step to determine whether remaining processing steps should be executed (the file valid_trans. might not be correct if the process edit_transactions fails). In the event of failure in any process, an error message is generated to the user or operator, identifying the failed process and its exit code. This information can be used to initiate corrective action.

Repetition control structures allow one or more processing steps to be executed multiple times. They come in many varieties, including

```
For-each (list of items) do (processing)
While (condition) do (processing)
Repeat (processing) until (condition)
```

As with other control structures, the exact syntax and semantics varies widely between operating systems. Figure 13.15 shows the example from Figure 13.14, recoded (and simplified) using a FOR-EACH control structure.

```
/*   Batch Command File - Daily Transaction Processing    */

$LIST=(edit_transactions,update_customer_master_file,generate_picking_list)
for each $PROGRAM in $LIST do
    RUN $PROGRAM
    if ($exit_code ≠ 0) then
         echo "failure in process" $PROGRAM
         echo "exit_code=" $exit_code
         halt
    endif
end for
```

FIGURE 13.15 **A batch command file with nested conditional executions. Sequential execution and testing of exit code is implemented with a FOR-EACH control structure.**

Input/output redirection. It is often desirable to alter the source and/or destination of a program's input or output data. The command interpreter itself is an example. When executing interactively, input is received from a keyboard or other interactive I/O device. When executing in batch mode, input is normally received from a file of previously stored commands.

The ability to change the source of input data or the destination of output data is referred to as *input/output redirection.* It requires a mechanism for stating alternate data sources and destinations on the command line. It also requires that processes be written in such a way that they can accept input from (or generate output to) a variety of sources and destinations.

To implement this I/O flexibility, programs will often utilize standard input and standard output devices. That is, they will read input from or write output to a standard or generic device. The operating system is then responsible for routing accesses to these standard devices to actual physical devices while the program is executing. By default, the user's terminal or console is usually used for both the standard input and standard output. Thus, executing the command

RUN application_program

would cause the program to be loaded and executed, with all I/O routed to the user's terminal.

If the user desired to provide input from a file, he or she might enter:

RUN application_program < input_file

In this example, the symbol < directs the operating system to perform input redirection. The operating system will load and execute the program and will assign the file INPUT_FILE as the standard input. Similarly, the command

RUN application_program > printer1

would load and execute the program and instruct the operating system to direct output to the standard output device to PRINTER1 (a symbolic name for a printing device).

These facilities provide flexibility in the I/O sources and targets of application programs. Note, however, that in each of the above examples, the redirected files and devices are character devices. Redirection to graphic devices is substantially more complex.

I/O redirection is often combined with the piping facility of the operating system. For example, the command

RUN program1 < file1 ¦ program2 > printer2

instructs the operating system to load and execute PROGRAM1 using the contents of FILE1 as the standard input and routing standard output into a pipe. In this example, the pipe is represented by the ¦ symbol, as it is in Unix and MS-DOS. The pipe is the standard input for PROGRAM2 (executing concurrently), which in turn sends its output to PRINTER2.

Note that in contrast to earlier examples of pipes, neither program is aware that is using a pipe. Thus, the programs contain no explicit calls to create or use pipes. In essence, the operating system is "fooling" the programs into thinking that they are communicating with real I/O devices.

Resource declarations. In many job control languages, it is necessary to specify the resources that will be required to successfully execute a job stream. Examples of these resources include memory, temporary disk space, specific files, and specific I/O devices. Any such statement of resource needs in a job stream is termed a *resource declaration*.

The use of resource declarations has two important benefits. First, it allows the operating system to determine whether sufficient resources exist to successfully complete a job before execution has begun. Normally, the operating system will examine the resource declarations and attempt to obtain and "lock" those resources before the first job step is executed. This saves resources that might be wasted if a later step were to fail due to the lack of a required resource.

The other benefit of explicit resource declaration lies in the provision of additional information to the operating system for resource management. Resource declarations allow the operating system to "plan ahead" for future resource needs. By explicitly stating resource needs at the beginning of a job, the operating system is able to determine whether sufficient resources exist to execute all pending jobs. If resources are insufficient, the operating system can attempt corrective action(s). These might include delaying the execution of one or more jobs or shifting resources from one use to another.

The alternative to explicit resource declaration is *dynamic resource allocation*. In this type of resource allocation, the operating system responds to resource demands as they occur. For example, an operating system using dynamic resource allocation doesn't attempt to determine the maximum memory that will be used during a job. It simply waits for programs to request memory, then attempts to fill those requests.

In general, dynamic resource allocation requires less sophistication on the part of the operating system. However, it allows the possibility that jobs could fail due to unavailability of required resources. The desirability of explicit resource declaration as compared to dynamic resource allocation depends on the severity of resource constraints and the sophistication of the operating system's error correction facilities. If resources are not severely constrained, it is unlikely that a job will fail due to lack of resources. In this case, explicit resource declarations cause needless processing overhead. If resources are constrained, resource declarations are more valuable.

However, that value is dependent on the mechanisms by which the operating system handles failures due to resource limitations. If the operating system delays scheduling the job, no user or operator intervention is required. But if the operating system simply fails (terminates) the job, user or operator action is required. In the latter case, the only advantage to explicit resource declarations is that the failure occurred earlier than it would have otherwise and that no resources were "wasted" on a partially completed job.

SUMMARY

System service functions for process control, memory allocation, and interprocess communication provide flexibility to application developers. Memory allocation and deallocation service calls allow the data areas of executing processes to grow "on demand." This makes such processes more flexible to changing storage needs and could result in more efficient process execution and overall memory allocation.

Application systems may be composed of multiple processes that can be combined on an "as needed" basis. Service calls for process loading and execution support this by allowing one process (the parent) to load and execute another process (the child) as a subroutine or cooperating peer. This facility allows the reuse of utility programs and allows complex software to be built from small, special-purpose modules.

Multiple (cooperating) process execution requires mechanisms for interprocess synchronization and data sharing. Interprocess synchronization is addressed through the provision of a signalling mechanism. Processes can send signals to one another, and process execution can be synchronized by waiting for such signals. Data can be sent between processes via signals or pipes. Pipes are shared regions of memory under the control of the operating system. Data is written into a pipe by one process and read from it by another process. A process interacts with a pipe in the same manner as an I/O device or file.

I/O service functions support application software by hiding the complexity of interaction with I/O devices. The exact services provided depend upon the type of I/O devices supported and the complexity of control of and communication with those devices. I/O devices can be classified as character, full-screen, graphic, and window-oriented.

Character-oriented devices communicate solely in terms of characters and have minimal control requirements. Service routines that support character-based I/O are relatively simple and few in number. Full-screen I/O extends character-based I/O with the inclusion of specialized commands for character placement and special display features. Functions such as cursor control, display intensity, and others require additional service functions and operating system complexity.

Graphic I/O devices utilize pixel-oriented (as opposed to character-oriented) I/O. Due to a lack of standardization among graphic I/O devices, service functions to support them are minimal in many operating systems. Those systems that do support them tend to be standardized on a graphic display language such as

Postscript. Window-based I/O is an extension of graphic I/O that includes predefined functions for multiple output regions (windows), font control, image display, and standardized facilities for menu display and user interaction. These facilities require a large number of different service routines and consume a great deal of computer resources due to their complexity.

The command layer provides a user interface for executing application and systems software. The command layer supports batch processing, interactive processing, or both. An interactive command layer is designed to support one of the I/O modes previously described. A software component that supports a command language (usually through a character or full-screen interface) is called a command interpreter.

A batch command language is often more complex than an interactive command language. This results from the complexity of executing multiple programs and from the "unattended" nature of batch execution. Batch command languages typically include control structures similar to those provided in programming languages. These structures allow the expression of complex control capabilities such as multiple (sequential) program execution, repetitive processing, and conditional processing. Command languages also support the alteration of program I/O through the use of standard I/O redirection and pipes.

Some batch command languages require the explicit declaration of required hardware resources. This allows the operating system to plan resource allocation more efficiently. It also guarantees that a job will not fail during execution due to the unavailability of required resources. An alternative method is not to require explicit resource declarations and to allocate resources to processes as they are requested. This strategy is called dynamic resource allocation.

Key Terms

case statement
child process
command argument
command interpreter
command language
command line interface
command option
dynamic resource allocation
forms-based interface
icon
if statement
interprocess synchronization
I/O redirection
job
job control language
job step
job stream
parent process
pipe
repetition control
resource declaration
search path
selection control
sequence control
signal

Vocabulary Exercises

1. _____ describes the use of operating system facilities to alter the source or destination of data to or from a process.

2. A _____ command interface prompts the user for command parameters and options by displaying a full-screen command template.

3. A window-based command interface can use _____ to represent files, programs, or other objects that can be manipulated by the user.

4. A _____ is a single processing step (usually a complete program) within a _____.

5. Some job control languages require explicit _____, thus allowing the operating system to allocate needed resources before job execution begins.

6. A _____ allows the movement of data between two cooperating processes.

7. The control structures supported by most job control languages include _____, _____, and _____.

8. When an executing process requests the creation of another process, the requesting process is called the _____ and the newly created process is called the _____.

9. _____ can be implemented user signals and/or pipes.

10. A _____ interface requires the user to type the command, parameters, and options in response to a prompt.

11. An operating system using _____ does not require resource declarations in its job control language.

Review Questions

1. How does the provision of memory allocation services improve the efficiency of memory utilization?
2. What is a signal? For what is it used?
3. What signals are required to synchronize the program cycles of two cooperating processes?
4. What is a pipe? In what ways is it similar to a file?
5. In what ways is full-screen I/O more complex than character I/O?
6. In what ways is graphic I/O more complex than full-screen I/O?
7. Of what value are standard methods of I/O device control (e.g., ANSI full-screen command sequences)?
8. Why are window-based command interfaces easier to use than forms-based or command line interfaces?
9. What is the difference between batch and interactive job execution? What command language features and capabilities are required to support batch job execution?
10. What is a search path?
11. What control structures are normally present in a job control language?
12. What are standard input and output devices? What is I/O redirection?
13. What are job control language resource declarations? For what are they used?

Research Problems

1. Input/output redirection and pipes are provided by many operating systems. For example, both Unix and MS-DOS provide these capabilities. However, the methods by which these operating systems implement these capabilities differ substantially. Investigate the implementation of pipes and standard I/O device redirection in both Unix and MS-DOS. Consider a sequence of processes, I/O redirections, and pipes, such as

   input_file < program_1 ¦ program_2 ¦ program_3 > output file

   Compare and contrast the methods by which each operating system implements this sequence. Which method of implementation is more efficient? Why?

2. As stated in the chapter, many batch job control languages require the use of explicit resource declarations. The job control language of MVS (an IBM mainframe batch operating system) is one such language. Investigate the use of explicit resource declarations in this JCL. What resource requirements must be declared? How (by what commands or statements) are they declared? How is this information used by the operating system? Are facilities provided to recover from errors generated as a result of resource declaration statements?

PART

4

Advanced Topics

14 *Computer Networks and Distributed Systems*

Chapter Goals

- *Describe the costs and benefits of distributing computer resources.*
- *Describe specific methods of resource sharing for I/O devices, mass storage (files), and processing (CPUs).*
- *Describe the structure and organization of network software and hardware.*
- *Describe common physical attributes of networks, including topology, access protocol, and communication hardware.*

WHY DISTRIBUTE COMPUTER RESOURCES?

Consider networking circa 1974. A corporation with several regional offices operates a large mainframe computer at its home office. The majority of its application systems run in batch mode, but many of the newer ones use online data entry. To provide better service to the branch offices, a number of measures are taken:

- *The provision of one remote job entry (RJE) computer at each branch office.* Each RJE computer has a card reader, card punch, and line printer. Each is connected to the home office mainframe by a 9600-baud modem over a leased telephone line.
- *The provision of staff (keypunch operator, computer operator, programmer/analyst) and supporting equipment (keypunches, card sorting and collating machines) at each branch office.*
- *Placement of three data entry terminals in each branch office.* These are connected to the home office via 1200-baud modems over normal telephone lines and are used for online data entry.

Although costs are high, turnaround time for many applications is vastly improved. Documents no longer need to be transmitted by overnight mail. Data capture occurs locally and is transmitted to the mainframe via online terminals or through the RJE facilities. Output is sent directly to the RJE facilities. Local support personnel build and maintain a few custom applications at the branch offices using RJE facilities.

Ten years later, the home office operates four of the largest available mainframes, each of which has the maximum possible amount of disk storage, memory, and terminal connections. A minicomputer is in use at each branch office for both local processing and for RJE. These machines support many local applications, as well as data capture for applications at the home office. Some online terminals at the branches are connected directly to the local minicomputer and some are connected via modem to one of the home office mainframes.

The following problems exist with this configuration:

- *Data communication capacity is near saturation.* A great deal of data transmission occurs between the branch and home offices. Home office applications require access to local data stores. Large batch jobs (through the RJE facilities) copy files from the branch minicomputers to the home office.
- *Maintenance of distributed applications is problematic.* Changes to distributed programs must be made separately on each minicomputer. Installation of new or revised application systems often takes weeks or months.

- *Online terminal communication is insufficient to meet peak demand.* At the branch offices, management wants to be able to shift terminals between computers as needed, but this is not feasible with the current configuration.
- *Desktop computers are starting to be used in some branch offices.* Mechanisms are needed to move data between these machines and the minicomputers and/or mainframes.
- *The I/O and processing requirements of the largest application systems are nearing the capacity of the largest available mainframes.* Hardware technology is advancing rapidly, and the company upgrades to the latest models quickly. In spite of this, the corporation anticipates that its largest online applications will exceed the capacity of the largest available mainframes within a year or two.

These problems were typical of organizations in the late 1970s and early 1980s. The answer was always the same: buy a bigger machine. When that didn't work, buy another big machine. Eventually, the problems overwhelmed even the largest computer installations. To address these problems, improvements in data communication technology and the introduction of networked computing were required.

Distribution of Computing Resources

It is very difficult to discuss the topic of distributed systems or networked computing without identifying the goals of networking and exactly what portions of a system are to be distributed. Unfortunately, the terms *distributed system* and *network* are too often used without context or specificity. Because there are many types of networks and many ways to distribute computing resources, these terms are generally too vague to be useful.

The goal of networking computer hardware and systems can be stated quite simply: to allow communication between computing devices. But this definition fails to address higher level issues of networking, including

- What is to be communicated?
- What additional capabilities does networking make possible?
- What are the benefits of these capabilities?
- What are the costs of these capabilities?

Networking provides a basic communication capability between computer systems. This communication can be an end in and of itself. For example, electronic mail, news distribution, and bulletin boards are services based solely on communication capabilities. The communication capability can also be a basis on which to provide more complex services. These are described in detail below.

Benefits of Distribution

Most of the benefits of networking are derived from the additional capabilities made possible by intercomputer communication. These benefits arise from the ability to share resources among many networked computers. In particular, the following resources can be shared:

- Data
- Programs
- I/O devices
- Mass-storage devices
- CPU services

The ability to share programs and data across a network is normally implemented via a *distributed file system.* Access to files containing data or programs is provided via network connections. In general, distributed file systems provide access to a common set of files. This capability can reduce redundant storage of data and programs. This in turn reduces the problems inherent in maintaining file consistency (e.g., redundant updates) and reduces the total amount of mass-storage space required within the network.

In general, the ability to share common hardware devices allows a reduction in total hardware investment. Unnecessary redundancy in computer hardware can be avoided, often saving a substantial amount of money. This is true of all types of computer hardware, but is often most pervasive with I/O devices. It is often difficult for individual users to justify the cost of expensive I/O devices such as high-capacity laser printers, color plotters, scanners, and high-speed communication devices. However, these devices become more affordable on a per-user basis when shared by many users through a network.

In addition to its use for distributed file systems, shared mass storage has other advantages. Many programs require large amounts of temporary storage during execution. Replicating this storage across multiple computer systems can be very expensive. Providing a common pool of such temporary storage can substantially reduce total mass-storage costs.

Access to shared CPUs can be a benefit in several ways. Recall from Chapter 2 that cost-effective computing solutions tend to use the smallest possible class of hardware. For example, providing three microcomputers is often a more cost-effective solution than providing a single minicomputer. Although such tradeoffs are valid in an aggregate sense, they do not address issues such as peak demand and irregular need for access to specialized computing resources. There could be times when a user needs to execute an application that exceeds the capacity of his or her own microcomputer, or there might be times when a user needs access to a substantially more powerful machine (e.g., to execute a simulation program on a supercomputer).

Networking provides a basic capability to address these types of problems. A user that needs access to a larger CPU might be able to access such a CPU over the network. Input data can be routed over the network to a process executing on another CPU and the processing results routed back to the user's computer. A user can access local machines that have excess capacity or a distant computer center that provides access to highly specialized and/or expensive hardware. In either case, the user can address most computing needs with local (relatively inexpensive) hardware and peak and/or specialized needs through access to shared resources.

Costs of Distribution

Although the benefits of networking and distributed computing are numerous, there are many costs to be considered. First and foremost is the cost of the network itself. As with most other types of computer hardware, the cost of a network tends to rise nonlinearly in relation to its capacity. Thus, the cost of a network for 50 computers could be substantially more than five times the cost of a network for 10 computers.

Networks consist of a variety of hardware and software elements, each of which must be purchased and maintained. Maintenance costs apply not only in the case of equipment failure, but also in response to changes in the composition of the network. Maintenance costs associated with such changes tend to be high because networks tend to be dynamic (resources are frequently added, changed, and deleted).

A less tangible cost of networked computing lies in the loss of local control over hardware and software. One of the reasons for the tremendous growth in the use of smaller machines has been users' desire to exercise more control over their own computing, yet networking often dilutes that control. Users require access to data, programs, and hardware devices that they themselves do not own or control. Access must be negotiated between resource owners, and changes in those resources require changes in user computing procedures.

Another cost of networked computing lies in the inefficiency inherent in intercomputer communications. In spite of rapidly advancing communication technology, communication among hardware devices within a single computer system is still much faster than communication across a network. As resources are further and further distributed, communication inefficiencies become a drag on the performance of each individual computer on the network. This can be addressed by adding additional network capacity, but the user then faces costs that rise nonlinearly, as described earlier.

Although the aforementioned costs are real and significant, modern experience has shown that they are often overshadowed by the benefits of networking.

In addition, the performance gap between intracomputer and intercomputer (i.e., network) communication continues to decline. The cost of network capacity is also falling rapidly. It is up to each network designer and administrator to determine the tradeoff that is best for their particular computing needs.

NETWORK SERVICES

Although all network services rely on basic intercomputer communications, the type and complexity of these services varies substantially. Network services can be roughly classified into the following categories:

- Access to shared I/O devices
- Terminal-to-host communications
- User-to-user communications
- File transfer
- File sharing
- Distributed processing

These services are listed in order of increasing complexity. In this case, complexity refers to both the complexity of the software that provides the services and the hardware resources consumed by that software.

Extended Communication Services

Many common networking services are relatively simple extensions to the services available to the users of a large multi-user computer. Access to shared I/O devices is the best example of these. In a multi-user computer system, users compete for access to shared peripheral devices. For example, several users can generate output destined for a laser printer, but only one such user can access that printer at a time.

A multi-user operating system must provide software to control access to shared I/O devices. In the case of a printer, the operating system provides a mechanism for spooling users' printed output to temporary storage. Another mechanism continually monitors that storage for additions and sends them, one at a time, to the printer. The former mechanism is generally called a *spooler;* the latter is called a *scheduler*. The organization of these components is depicted in Figure 14.1.

These components can also be referred to by a more general set of terms. The spooler and scheduler are designed to provide shared access services to a common resource (the printer). These types of processes are commonly called *servers* because they provide services to user (or other) processes on request. The user processes are referred to as *clients*. They are users of the service provided by the server process(es).

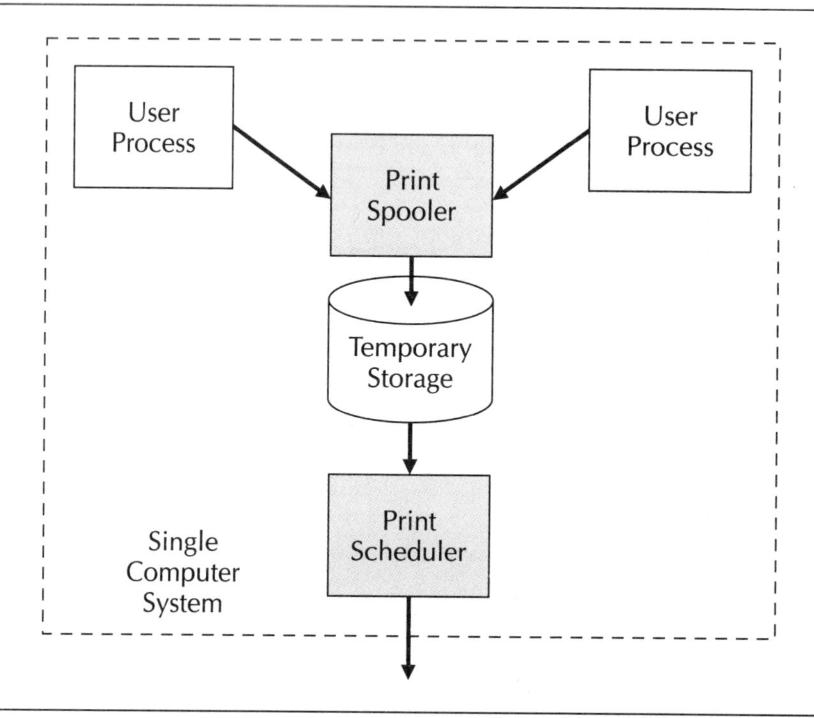

FIGURE 14.1 **Print spooling and scheduling within a single computer system.**

In a large multi-user computer, both server and client processes are executed within the same machine. In a networked computer system, they can execute on different machines. For example, various user processes can be executing concurrently on individual users' microcomputers. The print spooler and scheduler may be executing on a print or network server. This organization, depicted in Figure 14.2, is an example of a *client-server architecture.*

When executing on a single computer, client and server processes communicate through the interprocess communication facilities of the operating system (e.g., signals, pipes, and shared memory or files). When executing on different computers, they communicate through the software and hardware facilities of the network. Ideally, the use of a network is transparent to client processes. For example, a user process that generates output destined for a printer should be unaware that the printer is not directly connected to the local computer.

Network transparency requires modifications to the operating system of each computer on the network. In the example of remote printer access, the operating

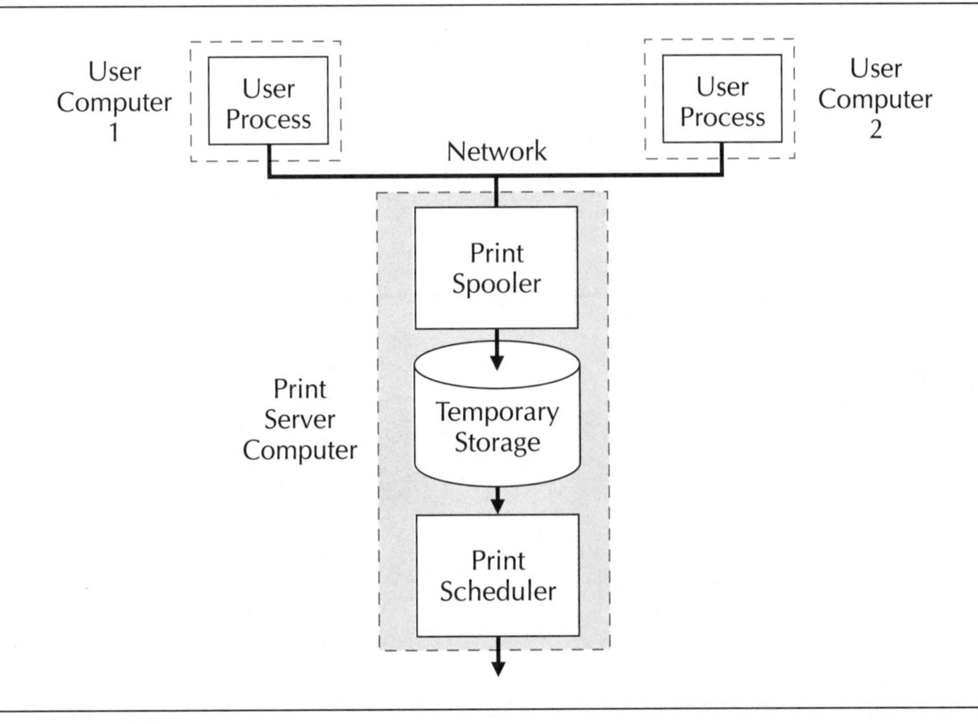

FIGURE 14.2 **Print services implemented on a network server using a client-server architecture.**

system must capture requests for printer access and reroute them to the appropriate node of the network. This requires a modification to the normal printer service routines provided by the operating system and the addition of network interface services. These software components are depicted in Figure 14.3.

Mail and bulletin board services present problems similar to those of I/O device access. Each of these services requires access to an area of storage where messages are stored. Users gain access to these storage areas by invoking a mail or message utility. When implemented within a single machine, these utilities combine aspects of both server and client processes. They are client processes to the extent that they provide user-oriented facilities (e.g., creation and editing of messages). They are server processes to the extent they provide access to message storage areas.

The distinction between server and client becomes much clearer when message passing takes place over a network. In a networked system, message storage can be provided on each machine (for that machine's users) or it can be provided

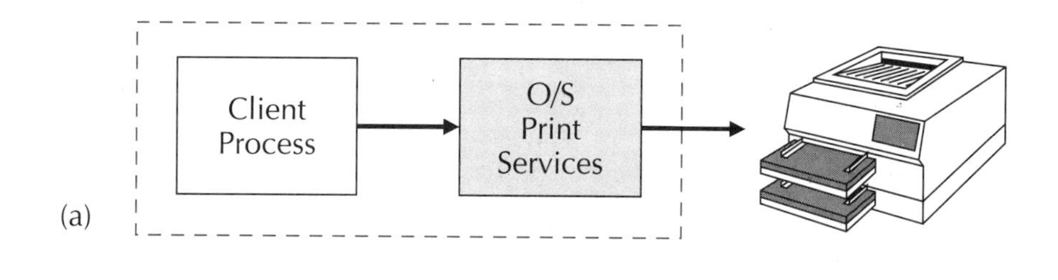

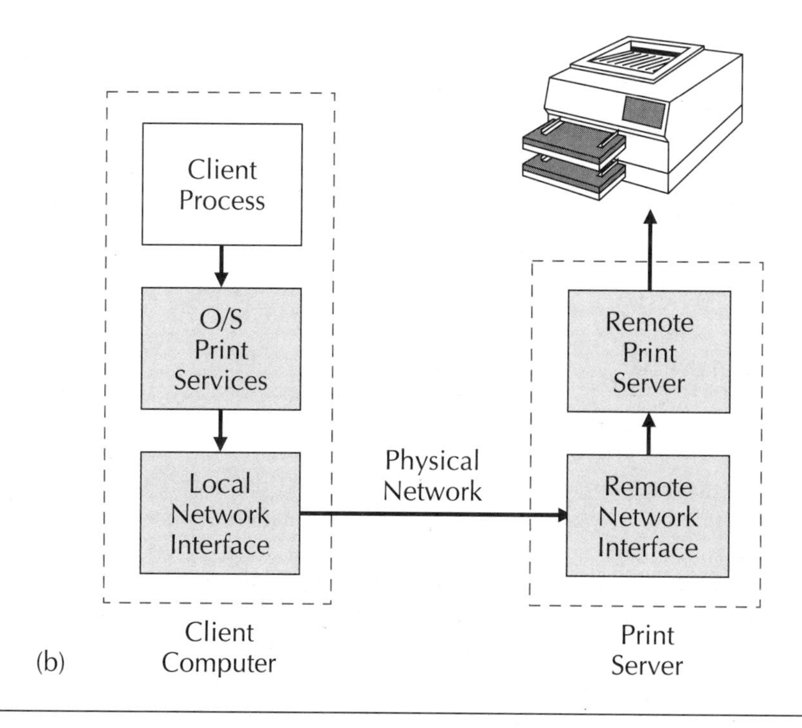

FIGURE 14.3 Comparison of the software components needed to implement print services in (a) a single machine architecture and (b) a client-server architecture.

on a single machine in the network. In the latter case, the machine that stores messages and provides access to them executes the server process(es). The machines that users use to send or receive messages executes the client processes.

As with remote printing, extensions to normal mail facilities are required to interact with the network. Outgoing messages must be redirected from their normal destination to the server. This requires mail processes to interact with network interface services in a manner similar to that depicted in Figure 14.3.

Distributed File Systems

Sharing of programs and data in a network requires the provision of a mechanism for sharing files. File sharing can be implemented in several ways. These include

- Explicit file transfer
- Transparent file transfer
- Transparent file access

The term distributed file system can be applied to any of these methods, although the latter is the most common.

In *explicit file transfer*, the user must explicitly request the transfer of a file from a remote machine to a local machine. The user must know on which machine the file is stored, and must use a specific software utility to request the transfer. The file is, in essence, copied from one machine to another over the network. If the user desires to update the shared copy of the file, the local version must be copied back to the remote machine after it is modified.

Transparent file transfer eliminates the need for the user to request a transfer or even to know that the file does not reside locally. When a user (or user process) requests access to a remote file, the transfer is initiated automatically by the operating system. The implementation of the transfer mechanism has some similarity to the implementation of the remote print server described earlier. File access requests to operating system services are redirected to a remote file server.

In particular, a modification to the operating system service for opening and closing files is required. When an open operation is requested, the operating system must determine if the file exists locally. If not, a transfer request must be initiated to the remote machine that contains the file. This implies that the local file system contains information about files stored on remote machines or that it can request such information over the network.

When a file is closed by a user process, a transfer in the opposite direction must be initiated. Thus, a user's access to a remote file requires copying the file twice (once when the file is opened and again when it is closed). Transparent file transfer can be extremely inefficient, especially if the user needs only a portion of the file. For example, assume that a user needs to update a single record in a large file. Assume further that the file is organized for direct access (e.g., using an index). The mechanism described above requires that every record of the file be copied from the remote machine to the local machine and back again. Yet the user needs access to only one record. Because of this inefficiency, transparent file transfer is seldom used in modern networks.

Transparent file access solves the efficiency problems associated with transparent file transfer. Instead of copying an entire file and then operating upon it locally, only those records that are needed are transferred. In addition, modified records are immediately transferred back to the remote location. Thus, little or no local storage is allocated to the file, and a minimal amount of network communication is required.

The advantages of transparent file access come at the expense of additional complexity in all file manipulation operations. Not only must the open and close functions for files be modified, every file access service call requires network access. Each read or write operation requires network access to the remote file. Thus, transparent file access reduces network data communication at the expense of increased complexity of O/S file service routines.

Distributed Processing

Distributed processing refers to access by a client process to the CPU resources of a server machine. It can be implemented in a number of ways and with various degrees of transparency. Some possible implementations include

- Remote login
- Remote process execution (explicit)
- Remote process execution (transparent)
- Load sharing

Remote login facilities are normally provided in networks to allow users full access to the facilities of a remote machine. In essence, the user interacts with the remote machine as though he or she were using a terminal attached directly to that machine. The network serves as the data communication facility and a process on the local machine provides an interface to the user's I/O device.

The user must explicitly request a connection to the remote machine via a software utility specifically designed to support remote login. Once a connection has been established, all input from the user's input device (e.g., a terminal) is redirected to the remote machine over the network. Any processes initiated by the user are executed on the remote machine. Output from those processes is routed back through the network to the user's local machine and from there to his or her I/O device.

A comparison between this type of interaction and a direct connection is shown in Figure 14.4. Note that the local command processor is not used when the remote login is in progress. It is replaced by a local version of a remote login process. The function of this process is to translate terminal I/O between the normal local terminal interface and the network. Because of incompatibilities between terminal I/O devices and the I/O services available on various machines, some translation is normally necessary. Typically, the remote login process on the local machine and the remote terminal interface on the remote machine will negotiate an appropriate I/O protocol when a connection is first established. Each process will then perform any necessary translation between this protocol and local protocols during all I/O operations.

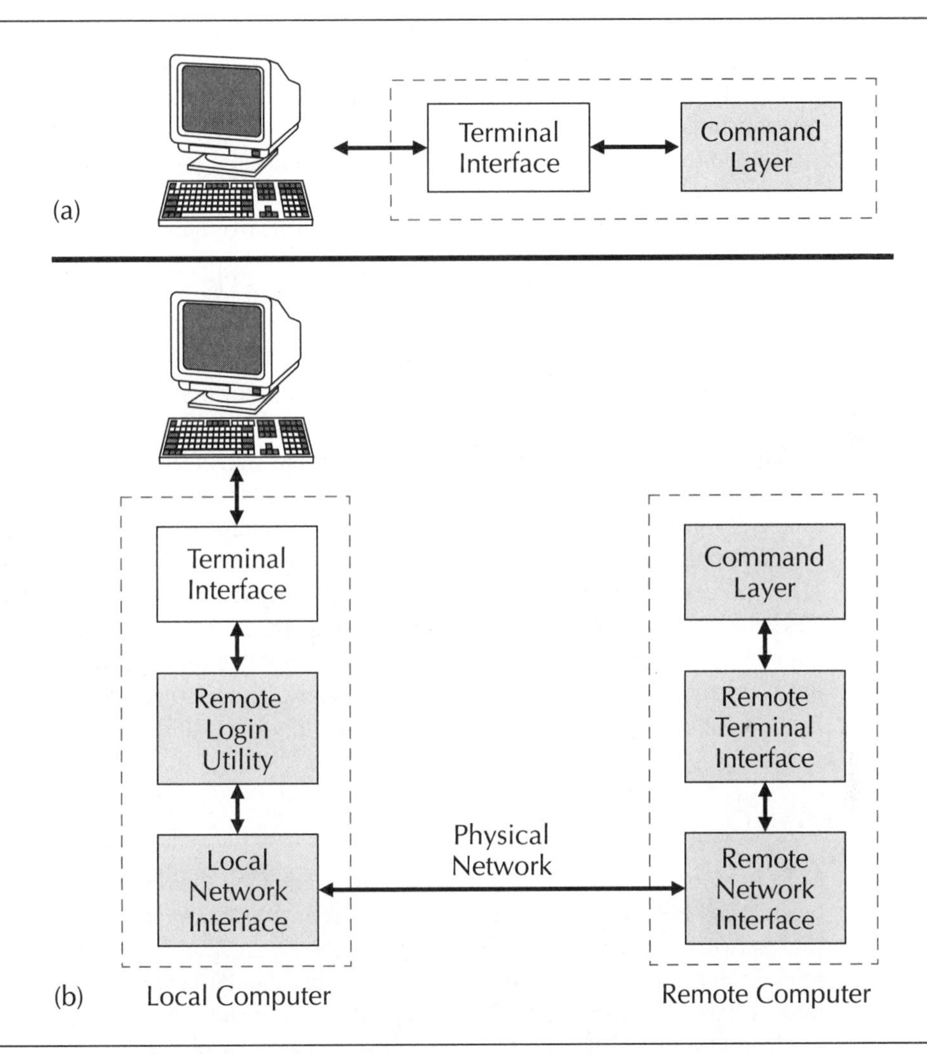

FIGURE 14.4 The software components necessary to implement a terminal interface (a) on a single machine and (b) on a client-server architecture using a remote login utility.

A user can explicitly request the execution of a process on a remote machine. If such a facility is provided, the formalities of user login on the remote machine are normally bypassed. Thus, the interface to the command layer on the remote machine is also bypassed, and a single process is loaded and executed. As before, I/O protocols must be negotiated and appropriate translation must take place. But the I/O connection exists only while the remote process is executing. The connection is terminated as soon as the remote process terminates and final output is received on the local machine.

Remote process execution can also take place transparently (i.e., without the user's knowledge). It is generally programmed directly into an application or system utility. Note that this is a different arrangement than the normal client-server relationship. In a client-server relationship, the server processor is continually active on the remote machine. It resides on the machine in a blocked state, pending the arrival of a service request on the network. In contrast, a remotely executed process is not loaded and executed until a request is received and is terminated once the request is satisfied.

Remote process execution facilities can be used to implement *load sharing,* or load balancing, across a network. Load sharing occurs when a busy processor asks another processor to execute a process for it. Thus, processing load is shared (or balanced) between the two machines. Note that in the previous examples of remote process execution, the process to be executed resided on the remote machine. Only processing requests and I/O were transferred over the network. However, when remote process execution is used for load sharing, an executable image of the process is normally transferred between machines over the network.

In sophisticated load-sharing systems, it is possible to move a partially completed process between processors. Such a capability requires that the participating CPUs have compatible architectures (instruction sets, registers, and the like). It also requires operating system compatibility, to the extent that the executing process utilizes operating system services. For these reasons, load sharing is normally implemented only in tightly integrated networks of fully compatible computer systems.

NETWORK LAYERS

Most modern networks are implemented in layers. One such layering method has been issued as a standard by the International Standards Organization (ISO). The ISO network layer model is referred to as the *Open Systems Integration (OSI)* model. Not all networks adhere strictly to the standard, but most are at least loosely based on its concepts. The OSI model is depicted in Figure 14.5.

The OSI model was conceived to provide a degree of standardization among networks and to provide a maximum amount of network transparency in each layer. The concept is similar to the software-layer model described in several chapters of this book, although the OSI model is specified in substantially greater detail.

Each layer uses the services of the layer immediately below it for all network accesses. Thus, layer n uses the services of layer $n - 1$ and only those services. The services and implementation of layers $n - 2$ through layer 1 are unknown to layer n. In addition, the implementation of layer $n - 1$ is unknown. Only the services and the mechanism(s) by which services are requested are known. The function of each layer is discussed in detail in the following sections.

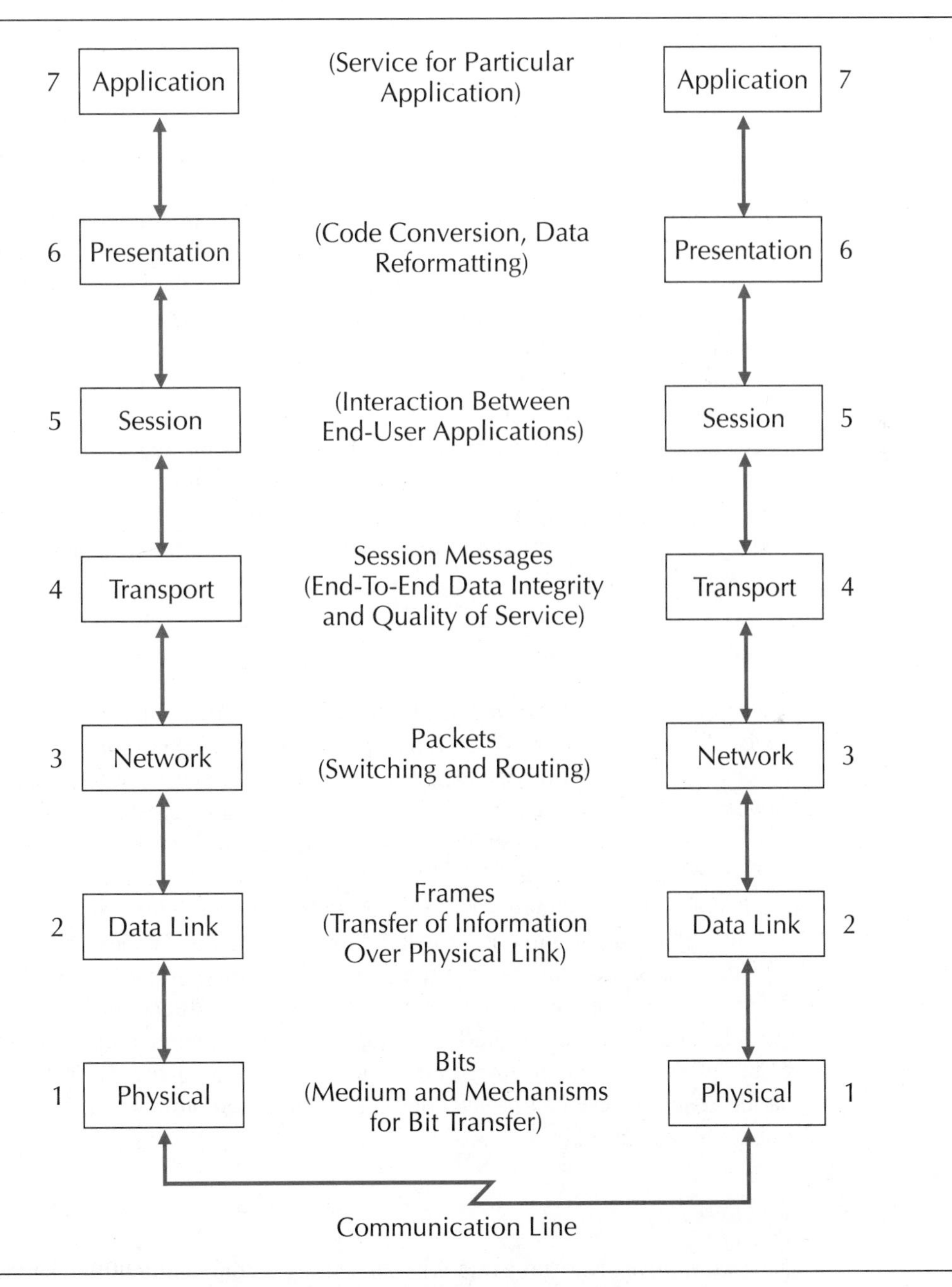

FIGURE 14.5 **The seven layers of the OSI network communication model.**

Application Layer

The *application layer* refers to any software that generates high-level requests for network services. The application layer consists of network utilities used directly by end users (remote login and mail programs) as well as network services embedded in the operating system service layer (access to remote files). The OSI application layer accepts requests for service and forwards them to the appropriate network node for service. Each command accepted from a user or program is translated into an equivalent set of *network messages*. These messages are transmitted to the application layer of a server process over the network. The result of processing the message is sent back over the network.

Note that the application layer of the client communicates with the application layer of the server and vice versa. This is true for each layer of the OSI model (not just the application layer). Thus, equivalent layers on the client and server are, in fact, cooperating processes. As discussed in Chapter 13, cooperating processes must be synchronized. Message passing through the network is the mechanism by which this synchronization, as well as required data communication, is implemented.

Presentation Layer

The *presentation layer* is responsible for all communication with the user's I/O device. Some translation is normally required to display output generated by a server on a user's I/O device. Translation is required because of incompatibility between local I/O devices and remote servers. Typically, a network protocol will define a generic display device. All communication destined for display on a user's terminal is encoded for this generic device. The presentation layer is responsible for translating user input into generic device input and for translating generic device output into output specific to the user's display device.

The presentation layer is not required for some network services. For example, reading from or writing to a remote file requires no direct interaction with the user's local I/O device. Similarly, a request by a user's application program for a remote process execution might not require user-oriented I/O. Network services that do require presentation-layer services include remote logins and interactive user application programs.

Session Layer

The *session layer* is responsible for establishing a connection with a server. This connection is initiated by sending the server a message requesting a connection. As part of this process, the client and server session layers will normally establish one or more parameters of the communication protocol. These can include character coding, asynchronous or synchronous message passing, and other communication parameters. Once a connection is established and the protocol negotiated, the session layer simply passes messages received from layers above to layers below and vice versa.

Transport Layer

The *transport layer* is responsible for translating messages according to the transport protocol of the network. In most networks, the major portion of this translation involves the division of messages into packets. A *packet* is a unit of data transmission on a network. The format of a packet can vary widely, depending on the transport protocol in use. However, the logical content of packets is usually quite similar. The contents of a packet include

- The message (or portion thereof)
- Identification of the sender
- Identification of the receiver
- Error-detection data
- Packet-sequencing information

Identification of the sender and receiver is normally included in coded form. In general, senders and receivers are identified in terms of their physical network location and a process identifier. Each is identified by a symbolic or numeric field within the packet. This information can be used by intermediate nodes of the network to route messages to their destination and to route error notification back to the sender. Process identifiers allow low-level network operating system processes to route packets to the transport layer of the appropriate process.

To protect against errors in network transmission, packets usually have some form of error-checking data encoded within them. This can take numerous forms, including parity checking, cyclical redundancy checks, and others. Several of these forms were discussed in Chapter 7. Error-checking information is usually encoded in the header or trailer field of the packet by the sending transport layer. The receiving transport layer recomputes the error-checking data and compares it against the version stored in the packet to determine if a transmission error has occurred. If an error is detected, the receiver will transmit a message to the sender requesting retransmission of the packet.

Because messages can be decomposed into many packets, some method must be provided to correctly reassemble them at their destination. A number of schemes can be used for this, of which the following is typical:

- Packets are sequentially numbered and the number is stored in the packet header field.
- A coded field in the first packet identifies it as the first packet.
- A coded field in the last packet identifies it as the last packet.

This information is sufficient to allow the sender to determine if all packets in a message have been received and to reassemble them in the proper order.

In many networks there is no guarantee that packets will arrive in the same order that they were sent. Thus, sequencing information is a necessity. Packets are simply decoded in sequential order, according to their identification numbers. Specially coded first and last packets indicate the start and end of a message. By examining the sequence numbers of packets received, the receiver can determine the identity of missing packets and request their retransmission.

Network Layer

The *network layer* is responsible for routing individual packets to their proper destination. Remote locations are normally known to users or applications by a symbolic name. The network layer must convert this name into a physical address before a packet can be sent. In some network topologies, this is a trivial exercise. The sender simply looks up the address of the receiver in a table or file, places that address in the packet, and transfers the packet to the data link layer. This scenario is typical of routing within *local-area networks*, which are networks of fewer than 100 nodes, usually located within a single floor or building.

Routing can be much more complicated in *wide-area networks*. Because of the large number of nodes in such networks, it is generally impractical for every machine to maintain a local data store containing the names and addresses of every machine connected to the network. There are several strategies to resolving this dilemma:

- *Certain machines on the network can be designated name servers*. Every network node knows the physical address of one or more name servers and can request an address lookup from any of them.
- *Certain machines on the network can be designated message-forwarding machines for a range of symbolic names*. Packets can be sent to these machines that will, in turn, send them to their final destination or to another forwarding machine.

In either case, knowledge of symbolic names and addresses is limited to a relatively small number of machines on the network.

The network layer can also be responsible for forwarding messages to other machines. In the second example cited above, the network layer of the intermediate machine would be responsible for forwarding the message to its final destination. In certain types of networks, all machines must be capable of message forwarding. Such a network usually implies the use of name servers for address-lookup functions.

Data Link Layer

The *data link layer* is responsible for transmitting a packet to the network interface device of the computer (i.e., the physical layer). Any translation between the format of a packet (as used in higher layers) and the specific requirements of the hardware device must be performed here. If, for example, the network interface device communicates with the network in blocks of 256 bytes, the data link layer is responsible for breaking up packets into groups of 256 bytes. Issues such as data blocking, direct-memory access, and interrupt handling are dealt with in this layer.

Physical Layer

The *physical layer* is the layer at which communication among devices actually takes place. The network interface device is responsible for interacting with the data link layer and for placing packets (or pieces thereof) on the network. This can involve a number of physical transmission functions, including blocking, character or block framing, digital to analog conversion, bus or direct-memory interfaces, and physical network access protocols. Most of these issues were discussed in Chapter 7. The issue of physical network access protocols is discussed in the next section.

NETWORKING ARCHITECTURE

There are numerous issues to be resolved when designing a network architecture. Some of these are

- Topology
- Network access protocol
- Communication channel hardware
- Data communication capacity

There are several alternatives within each of these categories. Thus, it appears that a network designer is faced with a large number of possible combinations.

Fortunately, there has been some standardization in the area of network architecture. The *Institute of Electrical and Electronics Engineers (IEEE)* has taken a leading role in establishing such standards, and they have been widely adopted by manufacturers of networking hardware and software. These standards are discussed more fully in a later section.

Network Topology

There are three primary options for network topology:

- Star (or hub)
- Ring
- Bus

The *star topology* uses a central node to which all other nodes are connected. This topology requires all communication to pass through the central node. Because this can overload the processing and data communication capabilities of the central node, the star topology is seldom used in modern networks. It has been largely supplanted by the ring and bus topologies, which are illustrated in Figures 14.6 and 14.7, respectively.

The *ring topology* (Figure 14.6) requires that each network node be connected to two other nodes. The entire network must form a closed loop (or ring).

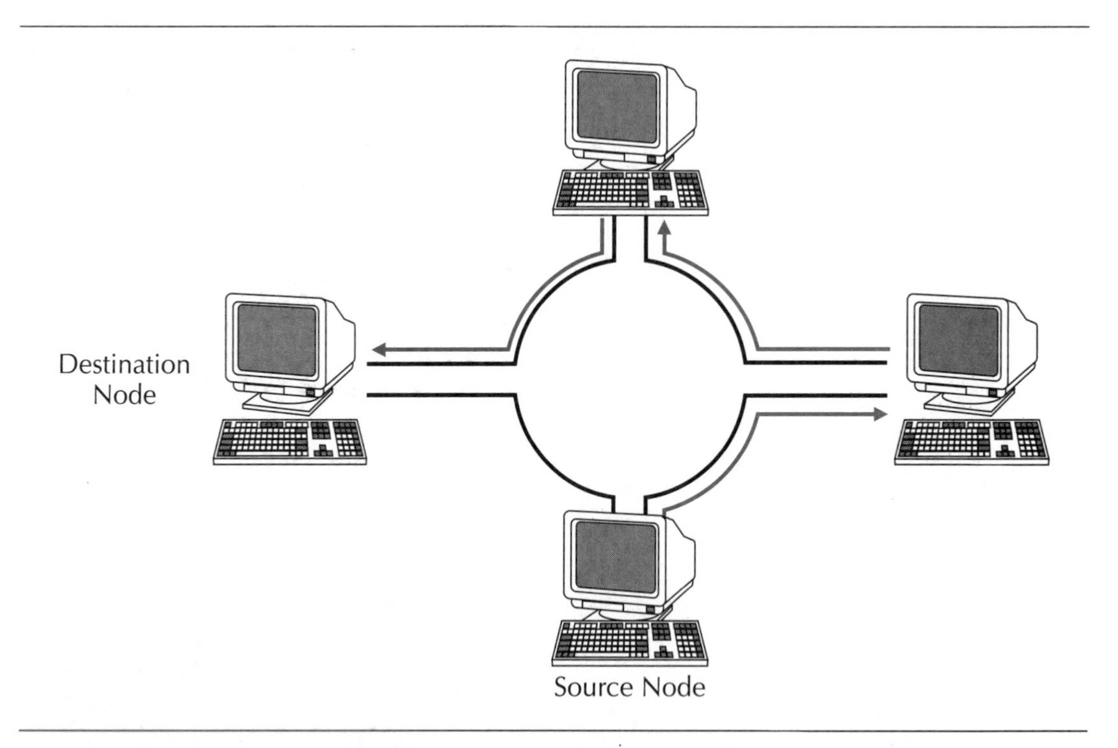

FIGURE 14.6 A ring network topology. The arrows show the flow of a message from sender to receiver. Note that the message travels in one direction around the ring.

Messages are passed through the network in one direction. Each node on the ring acts as a *repeater*. Messages received by a node are retransmitted to the next node. A message received by a node that is addressed to that node is not retransmitted. Thus, a message travels around the ring, retransmitted by intervening nodes, until it reaches its destination.

The *bus topology* (Figure 14.7) requires that each node be connected to a common transmission line (i.e., the bus). A node's connection to the bus is referred to as a *tap*. Each node detects messages passing on the bus via the tap. If the message is not intended for a node, it is simply ignored. A terminating resistor is located at each end of the bus. This device absorbs any signals it receives. Data movement is bidirectional. A node sends a message through its tap onto the bus. The message then travels in both directions until it reaches the terminating resistors at each end of the bus.

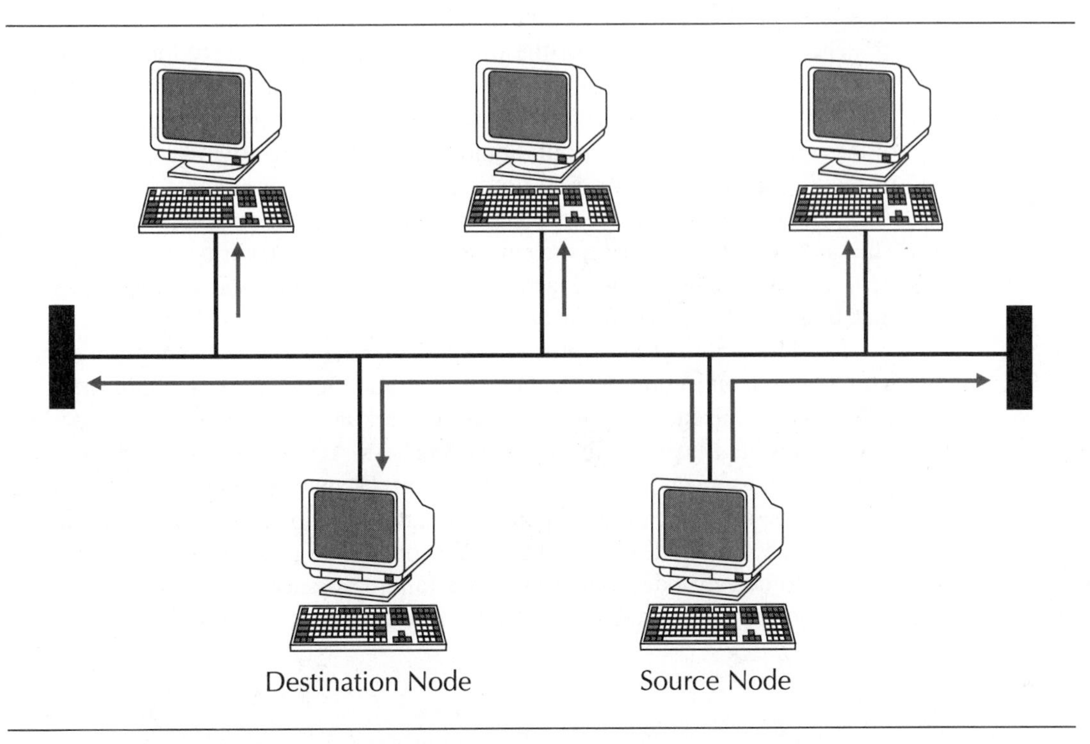

FIGURE 14.7 A bus network topology. The arrows show the movement of a message from sender to receiver. Note that the message moves down the bus in both directions from the sender's bus connection.

Each topology has distinct advantages and disadvantages. The advantages of the bus topology are simplified wiring and network interface hardware. The location of the bus is not restricted in any way by the location of nodes attached to the bus. Each node is simply attached to a tap on the bus by a direct connection. The network interface used by each node is primarily a passive device. It simply detects messages on its bus tap.

The primary disadvantage of the bus topology arises from the use of taps and passive node interfaces. Each tap removes some signal energy from the bus, thus decreasing the signal strength beyond that tap. The more taps, the more loss of signal strength. At some point, signal strength degrades to the point where reliable communication is impossible. Thus, the number of nodes on the network and the total length of the bus and taps must be limited. To counteract this problem, a bus network can be broken into segments, and repeaters can be used to connect the segments. The repeater simply retransmits messages from one bus segment onto another bus segment. Thus, by using a sufficient number of properly placed repeaters, the effective length of a bus network is virtually unlimited.

The primary advantage of the ring topology is a relatively long maximum network length. Because each node retransmits a message, there is little or no degradation of signal strength as a message traverses the ring. The disadvantages of the ring topology are susceptibility to failure and difficulty in adding, deleting, or moving nodes. Because each node is connected to two others, the addition, deletion, or movement of a network node requires two connections to be changed. This fact, and the basic ring organization, make network wiring a difficult task. Also, because each node is an active repeater, the failure of any node breaks the ring, thus disabling the entire network.

To address both of these problems, most ring networks use some form of network control/connection device, commonly called a *media-access unit (MAU).*[1] This is a centralized device to which all network nodes are attached by a pair of connections, as illustrated in Figure 14.8. The MAU serves two functions. The first is to act as a centralized connection point. Changes to nodes on the network can be made at this centralized point. The latter function arises from error-detection circuitry within the device. The MAU continually monitors network communication to detect node failures. When a failure is detected, the MAU reroutes communication around the failed node so that the ring is not broken.

[1] The term *media-access unit* is derived from IBM token ring networks and is referred to by the acronym *MAU*. However, that acronym is also used by other vendors and sometimes with other network topologies. For example, Novell networks use the term mainframe-addressable unit. The discussion of MAUs in this chapter refers solely to media-access units as used within ring networks.

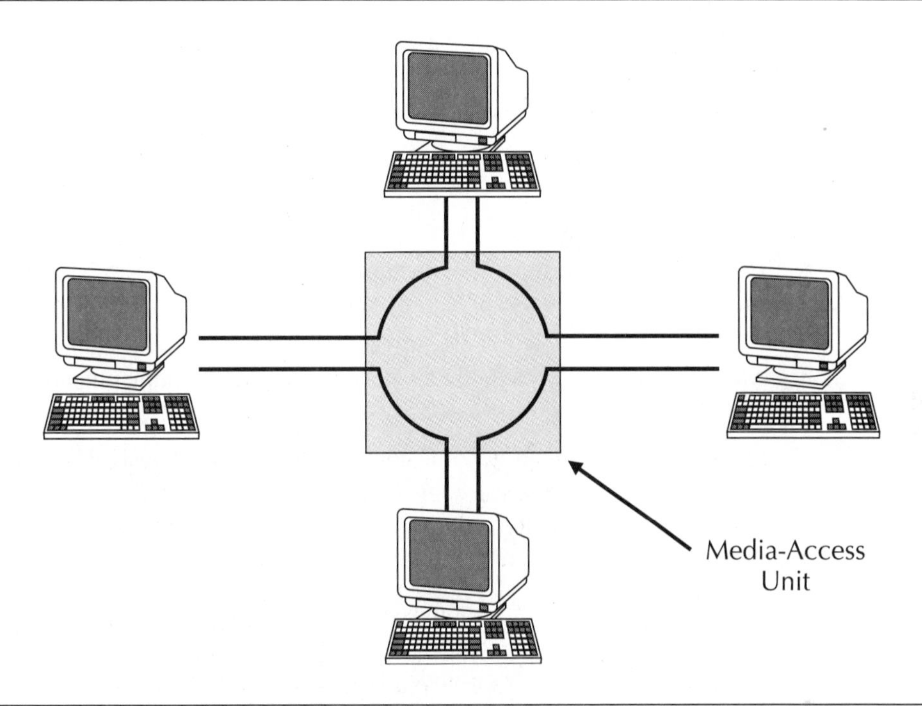

FIGURE 14.8 **A ring network using a media-access unit.**

The network interface for a node on a ring is an active device (messages must be retransmitted). In contrast, the network interface for a node on a bus is a passive device (messages are simply received). The difference between the two is one of complexity and cost. Active devices are more complex and, therefore, more expensive. In addition, media-access units used in ring networks add complexity and expense. Thus, the hardware needed to implement a ring network is generally more expensive than for a comparable bus network.

Media-Access Protocol

A *media-access protocol* controls transmission by nodes on the network. In general, only one node at a time can actively transmit. If more than one node attempts to transmit simultaneously (or nearly so), the messages will mix, producing noise or garbage. This is referred to as a *collision*. The two most common methods (access protocols) for avoiding collisions are

- Carrier Sense Multiple Access/Collision Detection (CSMA/CD)
- Token passing

CSMA/CD is an access protocol commonly (although not exclusively) used in bus network topologies. The basic strategy is not to avoid collisions, but to detect their occurrence. If a collision is detected, the transmitting nodes must retransmit their messages. The basic protocol is as follows:

- A node that wants to transmit listens (carrier sense) until no traffic is detected.
- The node then transmits its message.
- The node listens during and immediately after its transmission. If abnormally high signal levels are heard (collision detection), the node ceases transmission.
- If a collision was detected, the node waits for a random time interval and then attempts retransmission.

The *token passing* access protocol is commonly (but not exclusively) used in ring network topologies. It uses a control message called a *token*. The node that "possesses" the token is allowed to originate messages on the network. All other nodes can only receive or retransmit. A node gains possession of the token by receiving it from another node. It must relinquish the token (transmit it to another node) after a specified time interval. The order in which the token is passed between nodes might differ from the physical order of nodes on the network.

The advantage of CSMA/CD lies primarily in its simplicity. There are no control messages (tokens) to be passed between nodes and, thus, no inherent ordering of nodes in the network. There is also no tuning to be performed, because all nodes compete for network access on equal terms.

The disadvantage of CSMA/CD lies primarily in its efficiency, relative to other access protocols. Because this method does not avoid collisions, a portion of the network transmission capacity is wasted. Each collision is, in essence, a missed opportunity to transmit a message. As network traffic increases, the number of collisions also increases. At sufficiently high levels of communication traffic, network performance starts to decrease due to excessive numbers of collisions and retransmissions. The effect of this phenomenon on effective communication capacity is depicted graphically in Figure 14.9.

Token passing avoids the potential inefficiencies of CSMA/CD. Because collisions are prevented entirely, no transmission capacity is wasted due to collisions and retransmissions. A small portion of network capacity is needed to transmit the token between nodes, but it is generally insignificant compared to total data transmission capacity.

Other advantages of token passing include the ability to tune network performance and a superior ability to support real-time data transmission requirements. Network performance can be tuned by varying the time interval after which a node must relinquish the token to increase the network capacity allocated to

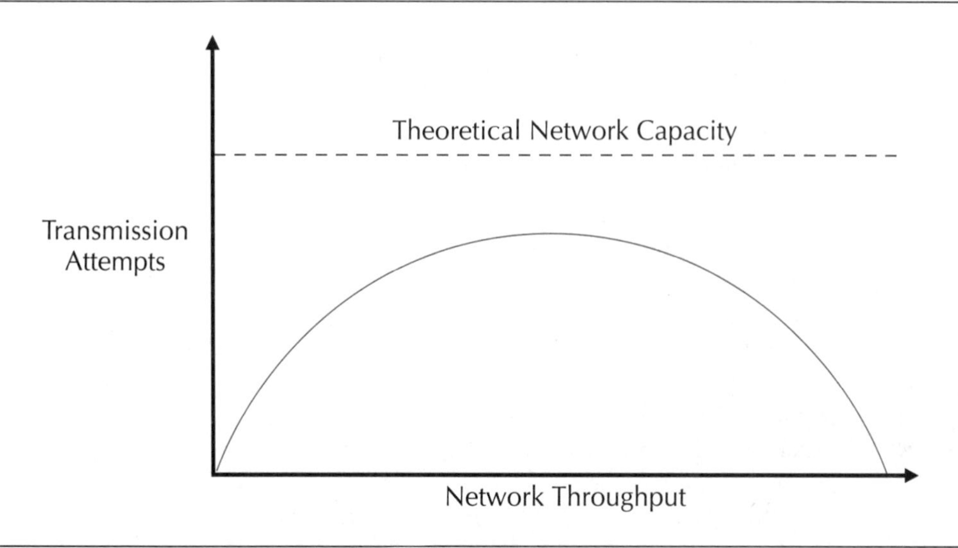

FIGURE 14.9 **The relationship between network throughput and attempted transmissions using the CSMA/CD media-access protocol.**

nodes such as file servers that need to transmit large amounts of data. Because each node has a limited time during which it may possess the token, the maximum time any node must wait to receive the token can be calculated. Thus, the maximum time interval a node must wait to transmit messages is simply the sum of the maximum times that all other nodes can hold the token. This predictability is essential in many real-time applications.

The disadvantages of token passing are due to its complexity. It is somewhat more susceptible to failure and requires special procedures when the network is started. Some device must be responsible for creating and transmitting the token when the network is started. Furthermore, each node must know the next node in the token-passing sequence. The failure of a node requires the previous node to bypass the failed node. Startup and failure modes require additional control strategies or the intervention of a network control device such as a media-access unit.

Communication Channel Hardware

The hardware required to implement network connections consists of devices to conduct messages and devices to interface to computer hardware. The exact hardware required to implement a network depends on the type of network (e.g., local- or wide-area), the network topology, the number and type of attached computer equipment, the method of data transmission, and other factors. Each class of hardware is discussed in detail below.

Signal propagation hardware. Various possibilities exist for interconnecting nodes on a network. Some of the more common include

- Twisted-pair wire
- Coaxial cable
- Fiber optic cable

Each of these has advantages and disadvantages relative to the others. Also note that each transmission conductor is generally capable of either analog or digital communication.

Twisted-pair wire is a very common transmission medium for local-area network connections. It consists of two conductive wires that are twisted around one another. The wires might or might not be enclosed (shielded) by nonconductive material. The primary advantages to twisted-pair are its low cost and ease of installation. Its primary disadvantages are a high susceptibility to noise and limited transmission capacity.

Coaxial cable consists of a single conductor surrounded by a metallic shield. The entire cable is wrapped in a nonconductive material. Due to its construction, it generates very little external electromagnetic interference. It is also relatively immune to external interference. Thus, several cables can be run together without interference. It also has a relatively high data transmission capacity. This has made it a transmission medium of choice where high bandwidth and low noise are essential, such as in cable television.

Its primary disadvantage is its cost and ease of use. Due to its construction, special adaptors and connectors must be used when connecting cables or attaching cables to devices. It is comparatively costlier than twisted-pair wire, but not nearly as expensive as fiber optic cable. It is somewhat less flexible than twisted-pair and could be more difficult to install.

Fiber optic cable consists of one or more strands of light-conducting filaments such as fiberglass. Because light waves (rather than electricity) are used as the basic transmission medium, electromagnetic interference is not a problem. Thus, fiber optic transmissions are extremely resistant to noise. Because light has a greater signal-carrying capacity than does electricity, the transmission capacity of fiber optic cable is substantially higher than either twisted-pair wire or coaxial cable.

The primary disadvantages of fiber optic cable are its high cost and difficulty in installation. High-quality thin fiberglass cables are expensive to produce. Plastic optical cable can be used for its lower cost but at some sacrifice in capacity. Connections between cables and devices require specialized and expensive hardware. Accidental breaks in fiber optic cable are also more difficult to locate and repair.

Various types of devices can augment the cables described thus far:

- Amplifiers
- Repeaters
- Line conditioners

An *amplifier* increases the strength of a signal. It is commonly used with signals propagated electrically over long distances. The strength of all electrical signals decreases as the signal passes through a wire. This decrease is primarily due to electrical resistance. Over a sufficiently long distance, the strength of a signal can decrease to a point where it cannot be distinguished from noise.

An amplifier increases the strength of an incoming signal. It is primarily used to extend the range of an electrical cable. However, the effective length over which the signal travels cannot be increased indefinitely. It is limited by two factors, one of which is noise and interference introduced during transmission. All electrical signals are subject to such interference. It becomes a part of the signal, indistinguishable from the message.

An amplifier amplifies whatever signal is fed to it. If that signal includes noise along with the intended message, the noise is amplified as well as the message. In addition, amplifiers are never perfect. Some distortion or noise is generally introduced in the amplification process. Thus, a signal that is amplified many times will contain noise from transmission interference as well as noise introduced by each stage of amplification.

A *repeater* fills much the same role as an amplifier, but operates on a different principle. Rather than amplifying whatever is sent to it, a repeater interprets the message it receives and retransmits, or repeats, it. Because of this mode of operation, noise is much less a problem with repeaters than with amplifiers. As long as the noise introduced since the last transmission does not cause a misinterpretation of the message, the retransmitted message will be noise-free. Thus, the noise introduced in one transmission stage is not generally carried into the next.

The term *line conditioner* is generically used to describe a number of devices. In general, a line conditioner suppresses noise or prevents damage to network components. An example of the latter function is to prevent strong power surges (e.g., as induced by a nearby lightning strike) from entering a network interface device.

Network interface hardware. Hardware devices are required to connect individual computers to a network transmission cable. The generic term for such a device is a *network interface unit*. However, the term is also used to describe other hardware. The following discussion will differentiate the various meanings of the term.

The network interface unit for a single computer is usually a printed circuit board (card) attached directly to the system bus. It is responsible for the following functions:

- Detection of incoming transmissions
- Generation of outgoing transmissions
- Low-level media access

Both incoming and outgoing network traffic require translation or data conversion functions between the bus protocol and the physical network protocol. These translation functions can include character framing, serial-to-parallel conversion, analog-to-digital conversion, and other translation functions.

The detection of incoming traffic in a bus network is accomplished by scanning a field within each packet header for a physical address. Each network interface unit is assigned a unique address during manufacture or installation. Packets that match the physical address are placed on the computer system bus. Other packets are simply ignored.

In ring networks, all incoming traffic must be recognized. If the packet is destined for the receiving device, the packet will be removed from the ring (not repeated) and placed on the computer system bus. If the packet contains a different physical address, it is simply echoed (repeated) onto the outgoing network connection.

The physical connection between a computer system's network interface unit and a bus network is normally implemented with a *transceiver*. Its function is to detect any data passing on the network bus and to generate an equivalent signal on the line to the network interface unit. In some types of bus networks, such as those using twisted-pair wire, the function of the network interface unit and the transceiver are combined in a single device. In this case, physical connection to the network bus is implemented with a *T connector*. As discussed earlier, the installation of a tap (a transceiver or T connector) causes a decrease in signal strength on the network bus. For this and other reasons, the number and placement of transceivers and T connectors on network bus cables are restricted to specified maximums and intervals of distance.

In some networks, individual computer systems are not connected directly to the network cable. Separate devices, also called *network interface units*, serve as intermediate communication devices. These devices frequently use a single network connection for several attached computers. Communication between the computers and the network interface unit is normally accomplished with conventional serial or parallel communication. This organization of components is often used when attaching many microcomputers to a single network connection.

A *bridge* is a device that connects one network cable to another. The exact function and operation of a bridge is dependent on the types of networks being connected. They are typically used to route traffic between a local-area network and a wide-area network. The bridge constantly scans passing traffic on each network for physical addresses that correspond to nodes on the other network. Any such messages are echoed onto the other network.

Standard Network Architectures

The IEEE has drafted a number of standards concerning the hardware, transmission method, and protocols of network communications and control. These are collectively referred to as the *IEEE 802 Standards*. Each of the standards is listed in Table 14.1.

TABLE 14.1 **The various IEEE network standards.**

IEEE Standard	Functional Description
802.2	Logical Link Control
802.3	CSMA/CD
802.4	Token Bus
802.5	Token Ring
802.6	DQDB
802.7	Broadband
802.8	Fiber Optic
802.9	Integrated

The IEEE 802.2 standard covers logical links between network nodes. It addresses issues such as routing, error control, and flow control. The functions of the logical link layer correspond roughly to the functions of the data link layer of the OSI model described earlier. All of the remaining IEEE 802 standards address alternative methods of implementing media access protocols (also part of the OSI data link layer) and the OSI physical layer.

The most common standards for local-area networks are 802.3 and 802.5. The 802.3 standard describes the physical implementation of bus networks using the CSMA/CD media-access protocol. The 802.5 standard describes the physical implementation of ring networks using the token-passing media-access strategy. Each of these standards has a number of subdefinitions, which correspond to the use of cabling, bit-encoding schemes, and other physical implementation parameters.

A number of commercial products are based on the IEEE standards. For example, Ethernet is based on one of the subdefinitions of 802.3. IBM's token ring network is based on a subdefinition of 802.5. Many other local-area network products also conform to one of the definitions within 802.3 or 802.5.

The standards numbered 802.6 through 802.9 cover physical implementations for wide-area networking. Some of these are well established and have existing commercial implementations. Others are simply plans for the future.

SUMMARY

The goal of distributed computing systems is to provide low-cost and/or high-quality services to users. Such systems allow the distribution of various computer resources among multiple computer systems. Resources that can be distributed include access to mass storage (data, programs, and temporary storage), I/O devices, and CPU services. Access to distributed resources is provided by a computer network. The network is a collection of hardware and software components that allows communication between computer systems.

Sharing hardware resources among computer systems often allows a reduction in total hardware cost. This can result from sharing access to expensive I/O devices, consolidating duplicate storage of data and files, or from distributing peak demands for CPU capacity among multiple computers. Disadvantages of distributed systems include the cost and complexity of network hardware and systems software. Networks require substantial investments in communication capacity and that capacity must be maintained and updated in a dynamic environment. Users also face a loss of control over resources that are shared with (and required by) others.

Network (distributed) services can be loosely classified as I/O device access, extended communication, file sharing, and distributed processing services. I/O device access and extended communication services are direct extensions of services typically provided in multi-user operating systems. Both of these services require modifications to normal operating service routines to allow users and resources to be located on various computer systems. Service requests by users or application programs (clients) must be redirected to the remote machines that provide those services (servers). The results of processing service requests must be routed back to client processes. This approach to resource sharing is called a client-server architecture.

Distributed file systems allow the sharing of files among computer systems. Their implementation requires modifications to operating system services for file manipulation. Accesses to remote files must be redirected to appropriate server processes, and I/O to and from those files must occur over the network. Distributed file systems can be implemented via explicit file transfer, transparent file transfer, and transparent file access. Explicit file transfer requires that a user know the location of a file and that the user explicitly request that the file be copied to the local machine for manipulation. Transparent file transfer also copies a file to the local machine, but does so without the user's direct knowledge. Transparent file access performs all read and write operations over the network.

Only those portions of the file actually needed are transferred between client and server. This method is the most common form of file sharing, but requires the most complex systems software support.

Distributed processing allows CPU resources to be shared among computers. It can be implemented in several ways, including remote login, remote process execution, and load sharing. Remote login allows the user to execute an interactive dialog with a remote computer. The local computer redirects all I/O to and from the remote computer. Remote process execution allows a user or application to request the execution of a single process on a remote machine. Input, output, and interprocess communication for that process are routed across the network. Load sharing allows any executing process to be moved from one CPU to another. It generally occurs without the knowledge of a user or application program. Load sharing requires that each machine have compatible hardware architecture and operating system services.

As with other systems software, network services are generally implemented in layers. A standard layering scheme has been defined by the ISO and is referred to as the Open Systems Integration (OSI) model. The layers of the OSI model are (from most logical to most physical) the application, presentation, session, transport, network, data link, and physical layers. Each layer utilizes the services of the layer below. Each layer is unaware of the implementation of the next layer and the existence of any lower layers. Each layer on a client machine has an equivalent layer on the server machine. Coordination between equivalent client and server layers is achieved via message passing over the network.

The application layer is the layer at which requests for network services are generated. Applications may include specific network utilities or embedded operating system services. The presentation layer is responsible for interactive I/O between client and server. It resolves incompatibilities between high-level I/O facilities such as graphic display and full-screen terminal functions. The session layer is responsible for initiating connections between client and server. It also establishes initial communication parameters.

The transport layer is responsible for the delivery of messages between client and server. It is responsible for the division of messages into packets and for the reassembly of packets into messages. The network layer is responsible for packet routing within the network. It chooses a communication path and, in intermediate network nodes, is responsible for packet forwarding. The data link layer is responsible for the transmission of individual bits of a packet. It also checks for errors and requests retransmissions, if necessary. The physical layer refers to the hardware implementation of the communication medium and the devices that directly interact with that medium. Issues such as timing, character framing, and digital/analog conversions are addressed in this layer.

Various options exist for implementing the physical layer of the network. Important implementation choices include network topology, access protocol, channel hardware, and communication capacity. There are many possible combinations of these parameters. To address this complexity, a number of standards have been developed by the IEEE. Each of these specifies an integrated set of choices in the above categories. These standards are collectively referred to as the IEEE 802 Standards.

Network topology refers to the physical connections between network nodes. Nodes are generally connected in either a ring or bus. A bus network utilizes a common communication channel with a single connection (tap) for each node. A node sends messages by placing them on the bus and they propagate in both directions simultaneously. Messages are received by detecting a packet with an appropriate address. Bus networks require relatively simple wiring and interface devices. However, they are limited in length and number of nodes due to noise and signal degradation.

A ring network connects each node to two others. The nodes form a ring, and packets travel in one direction around the ring from sender to receiver. Each node receives every packet on the network. It retransmits (repeats) the packet if its own address does not appear in the packet. Ring networks are not as susceptible to noise and signal degradation due to the use of active repeaters. However, they are more difficult to wire and maintain, and the ring can be broken by the failure of any network node.

When two packets are placed on the network simultaneously, the result—noise—is referred to as a collision. A media-access protocol ensures that collisions do not occur or that they are appropriately remedied. The two most common access protocols are Carrier Sense Multiple Access/Collision Detection (CSMA/CD) and token passing. CSMA/CD does not prevent collisions from occurring, but it does detect them. Rules for detection and retransmission ensure accurate communication. Token passing regulates network access by passing a special control packet (the token) among nodes. Only the node that possesses the token is allowed to place packets on the network. CSMA is less efficient than token passing due to network capacity wasted during collisions. Token passing is more efficient but is also substantially more complex.

Network channel hardware includes the medium for signal propagation and the devices that interact directly with it. Signals are generally propagated using twisted-pair, coaxial, or fiber optic cable. Fiber optic cable has the highest communication capacity; twisted-pair wire has the lowest. Twisted-pair wire is the cheapest to obtain and install, and fiber optic cable is the most expensive. Additional devices can be used to extend network length and to protect communication lines and equipment. An amplifier is used to increase the strength of analog signals. A repeater retransmits analog or digital signals at full strength. A bridge moves signals between networks. A line conditioner protects network devices from power surges.

Key Terms

amplifier
application layer
bridge
bus topology
Carrier Sense Multiple Access/ Collision Detection (CSMA/CD)
client
client-server architecture
coaxial cable
collision
data link layer
distributed file system
distributed processing
distributed system
explicit file transfer
fiber optic cable
IEEE 802 Standards
Institute of Electrical and Electronics Engineers (IEEE)
line conditioner
load sharing
local-area network
media-access protocol
media-access unit (MAU)
name server
network
network interface unit
network layer
network message
Open Systems Integration (OSI) model
packet
physical layer
presentation layer
remote login
remote process execution
repeater
ring topology
scheduler
server
session layer
spooler
star topology
T connector
tap
token
token passing
transceiver
transparent file access
transparent file transfer
transport layer
twisted-pair wire
wide-area network

Vocabulary Exercises

1. Standards for many types and aspects of network communication are defined in the _____.
2. A _____ is a special-purpose packet, representing the right to originate network messages.
3. _____ cabling is generally the least expensive alternative for network wiring.
4. The _____ layer determines the routing and addressing of packets.
5. A _____ increases the strength of a signal, including both the encoded message and any noise that might be present.
6. The _____ layer refers to programs that generate requests for network services.
7. _____ is used in some networks for physical node connections, as it is in cable television networks.

8. A physical connection between two networks is implemented using a _____.
9. The generic term for a node's connection to a bus network is a ____, which can be implemented physically with a _____ or _____.
10. _____ carries messages encoded as pulses of light.
11. In a network that supports _____, a node with an overburdened CPU could ask other nodes to perform processing tasks for it.
12. A _____ provides token initialization and recovery from node failures in a _____ network.
13. Using a _____ facility, a user connects to a remote system over a network as though he or she were using a directly attached terminal.
14. In a ring network, each node contains a _____ which retransmits incoming messages that are addressed to other nodes.
15. A _____ is a network node that supplies the physical network address associated with a node's symbolic name.
16. Messages to be transmitted on a network are normally broken up into _____ prior to physical transmission.
17. Under the _____ media-access strategy, collisions are allowed to occur, but they are detected and corrected.
18. The _____ layer is responsible for negotiating parameters for interactive I/O between a server and client.

Review Questions

1. What is a distributed system? What specific resources can be distributed?
2. Why do distributed computer systems frequently have lower total hardware costs than centralized systems of equal capacity?
3. Why is communication between devices in the same computer system usually less efficient than communication between devices across a network?
4. What types of services are normally provided by a network?
5. Define the term *client-server architecture.*
6. What is network transparency? Why is it desirable? What are its implications for the implementation of operating system services?

7. By what methods can a distributed file system be implemented? What are the comparative advantages and disadvantages of each method?
8. What options exist for implementing distributed processing?
9. What is load sharing? What does it require of the processors and operating systems among which load is shared?
10. Describe the function of each layer of the OSI network model.
11. Describe the movement of messages on a bus network. Describe the movement of messages on a ring network.
12. What are the comparative advantages and disadvantages of ring and bus networks?
13. What is a media-access unit?
14. Describe the operation of the CSMA/CD media-access protocol.
15. Describe the operation of the token passing media-access protocol.
16. What is the difference between an amplifier and a repeater?
17. What functions are normally implemented within the network interface hardware of a single computer system?

Research Problems

1. The Internet is a worldwide network that connects other networks and allows messages to be passed among them. Most universities and many government, commercial, and research organizations are connected to the Internet. Investigate the architecture of the Internet to answer the following questions: What methods are used to implement each of the OSI layers? What standard services are provided? By what mechanisms does the Internet allow highly diverse networks and computer systems to interact?
2. A U.S. government initiative has been proposed to connect every household to a nationwide network. This network would combine many of the functions of existing computer network, telephone communication, and cable television systems. Investigate some of the methods that have been proposed to implement such a network. What are the data transfer requirements of such a network? What physical transmission technologies will be required to connect households and other nodes? How can a single network combine so many kinds of communication?

15 Advanced Computer Architectures

Chapter Goals

- *Describe the limitations of von Neumann processor architecture.*
- *Describe various approaches to parallel processor computer architecture.*
- *Describe the architecture of database machines.*
- *Describe architectural requirements to support nonprocedural and symbolic processing.*

PARALLEL- AND MULTIPLE-PROCESSOR COMPUTERS

The demand for computer processing power has almost always accelerated as fast (or faster) than computer designers' and manufacturers' ability to provide that power at reasonable cost. Certain types of applications have led this rise in demand. These include many scientific applications (such as three-dimensional numeric modelling), data intensive applications (such as large online transaction processing systems), and graphically-intensive applications (such as real-time animation). These and other applications place extensive demands on the CPU, bus, memory, and secondary storage subsystems of a computer system based on the traditional von Neumann architecture.

Limitations of von Neumann Architecture

A von Neumann processor operates in *single-instruction, single data stream (SISD)* mode. This type of processor, also called a uniprocessor, handles one set of instructions and executes them sequentially. Each of these instructions operates on one or two operands, which must be loaded from memory (and results stored back to memory). This mode of operation creates bottlenecks for the applications previously mentioned. By far the most critical of these is the execution of one instruction at a time.

Each of the aforementioned applications exhibits a characteristic commonly called *parallelism*. For example, three-dimensional modelling and real-time animation all require the execution of many similar instructions on various portions of the model or image. Identical instructions are typically executed on a matrix of numbers representing different parts of the model or image.

For example, consider the matrix representation of an image fragment shown in Figure 15.1. Each number in the matrix corresponds to a pixel in the image. Assuming that the image is displayed in black and white, the numbers represent the intensity (brightness) of each pixel. Now imagine that a user viewing the image on a video display decides to brighten it. This can be accomplished by multiplying each number by a constant. For instance, if the user wanted the image to be twice as bright, each number in the matrix would be multiplied by two.

$$\begin{bmatrix} 0 & 0 & 30 & 50 & 112 & 90 & 45 & 2 & 0 & 0 \\ 0 & 15 & 25 & 67 & 122 & 110 & 53 & 4 & 1 & 0 \\ 4 & 23 & 44 & 83 & 110 & 121 & 67 & 23 & 8 & 4 \\ 1 & 6 & 39 & 67 & 104 & 119 & 48 & 39 & 14 & 3 \end{bmatrix}$$

FIGURE 15.1 A matrix representing a portion of an image. Each number corresponds to the brightness of a single pixel.

Note that each element in the matrix is multiplied by the same constant. In a von Neumann machine, this operation would be accomplished by loading a register with the constant and then successively loading a matrix element, executing a multiply instruction, and storing the result. At a minimum, the number of execution cycles required to perform the operation would be equal to the number of matrix elements plus one (to load the constant into a register). This method of implementing the operation is referred to as *sequential execution*.

Now imagine that the computer used to implement this operation has multiple processors, each accessing a common pool of memory locations and having at least one common register. The same operation could be performed with this machine via *parallel execution*. The constant could be stored in a common register, and each processor could execute a multiply instruction on a different matrix element from the common pool of memory locations.

The total number of execution cycles would be the same as with the sequential implementation, but the *elapsed time* necessary to execute those instructions would be substantially shorter due to the use of multiple processors. If the number of processors was at least equal to the number of matrix elements, the minimum elapsed time to perform the operation would be equivalent to two execution cycles (one to load the constant, one to execute the multiply instructions on all matrix elements simultaneously).

Types of Multiprocessor Architectures

In a multiprocessor computer, processor hardware units are hardwired together, usually within the same equipment cabinet or within the same computer facility. Processors linked in this way are said to be *closely coupled.* Types of closely coupled processor architecture include the following:

- Single-instruction, multiple data stream (SIMD)
- Multiple-instruction, single data stream (MISD)
- Multiple-instruction, multiple data stream (MIMD)

Array processors (or *vector processors*) operate in *single-instruction, multiple data stream (SIMD)* mode.[1] In this mode, several inputs can be assigned to different processors, as shown in Figure 15.2. The same instruction or type of transformation is applied to all inputs concurrently. Thus, one program sequence is applied concurrently to multiple inputs (or data streams), much as described in the image-brightening example.

[1] In this context, the terms *array* and *vector* refer to a matrix of numbers consisting of either a single column or a single row (i.e., a one-dimensional matrix).

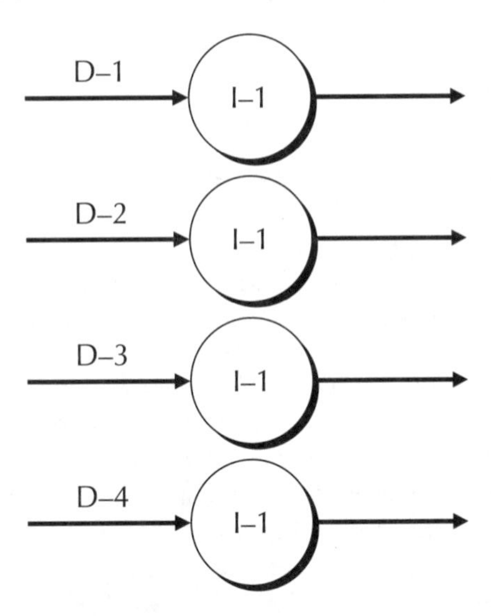

FIGURE 15.2 **The organization of an SIMD computer. Circles represent processors executing instruction I – *n* on data item D – *n*.**

Another multiprocessor configuration is *pipeline processing*. Computers that implement pipeline processing operate in *multiple-instruction, single data stream (MISD)* mode. Each processor within such a configuration is dedicated to a specialized function or set of functions. Such computers do not have a CPU in the traditional sense. Instead, they have a complex processing unit consisting of one general-purpose processor and one or more specialized processors.

In certain configurations, the processors are linked together in a fixed sequence. Database machines (discussed in a later section) are an example of this configuration. In general-purpose MISD computers, general- and special-purpose processors can be linked in a sequence that corresponds to a complex processing task. These links can be altered to address various types of complex processing.

The first processor in the sequence reads data from registers or memory locations. Its output is sent to the next processor in the sequence. As each processor completes its function, the result is passed to the next processor in the pipeline. The final processor in the sequence stores its output to a register or memory location. This method of operation is illustrated in Figure 15.3.

Pure pipeline-processing computers are unusual. But other types of computers (in particular, those that use vector processors) often incorporate specialized processors and pipeline processing. A single processor can be implemented in stages, with each stage linked in a fixed sequence. Pipelines can also connect general-purpose CPUs to specialized processors for I/O or secondary storage access.

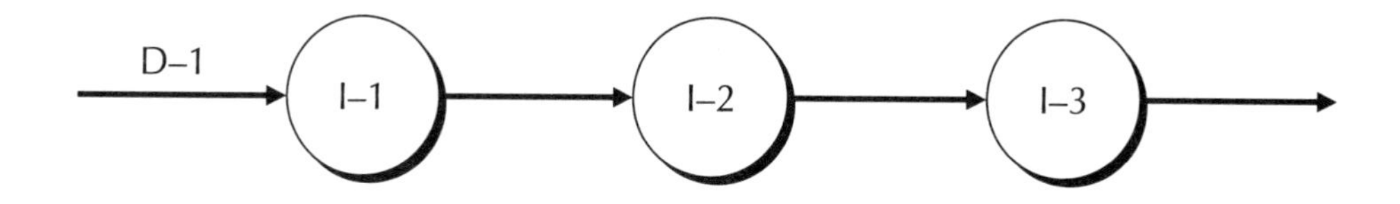

FIGURE 15.3 **The organization of an MISD computer. Circles represent processors executing instruction I – *n* on data item D – *n*.**

In *multiple-instruction, multiple data stream (MIMD)* mode processing, multiple inputs can be accepted, and different instructions can be executed on each of these inputs. This architecture usually consists of multiple general-purpose CPUs that can execute independent of one another, as shown in Figure 15.4. The complexity of MIMD mode places exceptional demands on systems software. Because MIMD computers are designed to handle a high degree of parallelism among processes, they are often called *parallel processors*. However, in a general sense, computers that implement array and pipeline processing can also be regarded as parallel architectures.

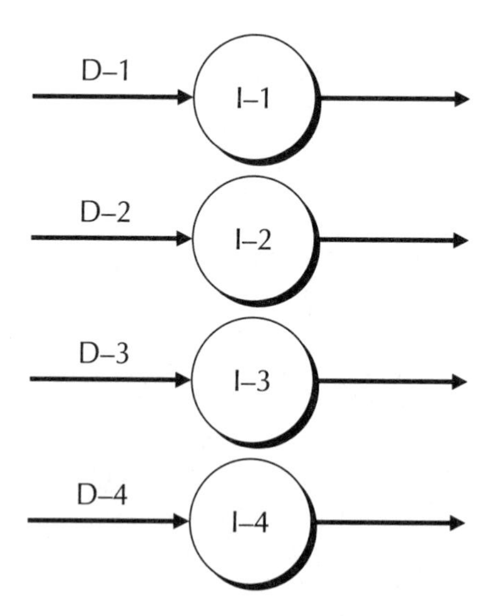

FIGURE 15.4 **The organization of an MIMD computer. Circles represent processors executing instruction I – *n* on data item D – *n*.**

VECTOR PROCESSORS

Computers with vector processors were among the first machines to be called *supercomputers*. They first appeared in the late 1960s and early 1970s in machines manufactured by Cray and Control Data Corporation, among others. Since that time, vector processing has found its way into lower classes of machines. Vector processing is typically an optional component in mainframes, some minicomputers, and many workstations. A machine with vector processing capabilities (but no other form of parallelism) is not considered a supercomputer by current standards.

The primary use for vector processing machines is to execute large numbers of simultaneous floating-point operations on matrices. As such, the processing power of machines that incorporate this architecture is normally measured in MFLOPS instead of MIPS. Typical applications that run on such machines today include simulation, high-resolution graphics, computer animation, and computer-aided design.

Hardware Features

The processor architecture of a vector processor differs substantially from that of a von Neumann CPU. Several key hardware features comprise this difference. These include

- Primitive processor units
- Pipelined execution
- Vector registers
- Overlapped memory access

Processor architecture. All numeric and logic operations are performed within the ALU of a von Neumann processor. The ALU is essentially "programmed" by the control unit each time an instruction is fetched and decoded. This "programming" serves to select a path through the ALU circuitry that corresponds to a particular processing function (add, multiply, etc.). Because all circuits are integrated into a single unit, only one pathway (or processing function) can be used in a single execution cycle.

In a machine that implements *primitive processor units*, individual circuits corresponding to basic instructions are placed in separate (but closely coupled) processors. Thus, the ALU is decomposed into a number of separate processors, with one or more dedicated to a single primitive processing function, as shown in Figure 15.5. Each of these processors can operate concurrently during a single execution cycle. Thus, a degree of parallelism is introduced into the machine as a whole.

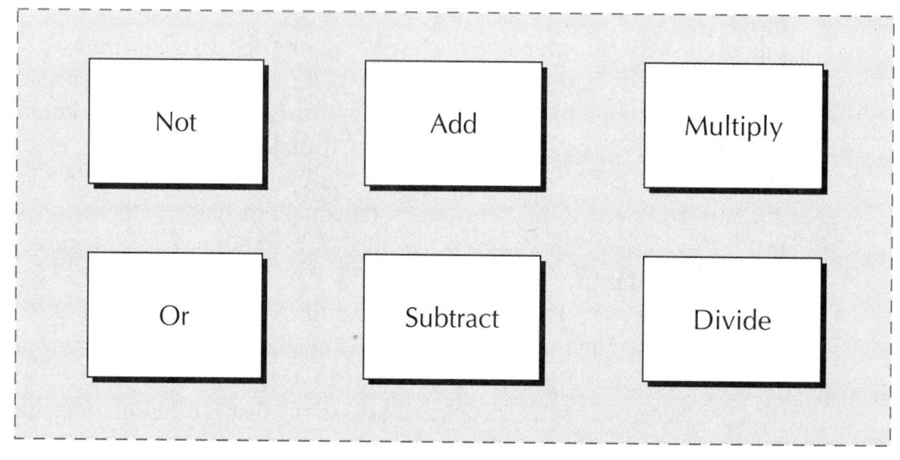

FIGURE 15.5 The ALU decomposed into a set of primitive processors.

Each processor must be capable of accessing operands (and storing results) to a common set of registers. However, a large number of registers would be required if each primitive processor was required to utilize registers as the source and target of all operations. To address this problem, individual processors can be connected together in a pipeline, as shown in Figure 15.6. That is, the output of one processor can be connected directly to the input of another.

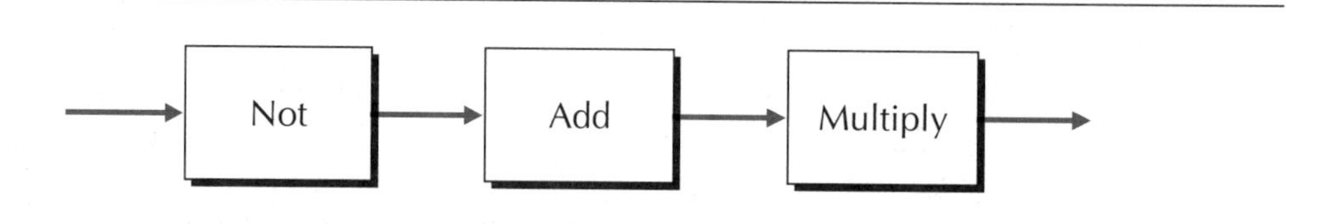

FIGURE 15.6 Three primitive processors connected in a pipeline.

An application must be programmed as a sequence of primitive operations to take advantage of this parallel arrangement of processors. Many matrix manipulation problems can be represented in such a manner. For example, consider the following sequence of primitive transformations:

1. Negate an operand.
2. Add the result to another operand.
3. Multiply the result by a constant.

This sequence can be programmed in pipeline fashion, as shown in Figure 15.6. When executing at steady state, the machine performs three processing functions at once.

Registers and memory access. One of the primary bottlenecks in most computers is the access path (bus) between the processor and memory. Input operands must be loaded into registers and results must be stored back to memory via this pathway. In a typical von Neumann machine, the bus width is equal to the processor word length, and access time is slower than processor cycle time.

Several methods can be used to address this bottleneck. Those commonly used with vector processors include

- Data and instruction caching
- Block transfers to vector registers
- Overlapped memory access

Caching was previously discussed in Chapter 6. Most modern computers make extensive use of memory caching; vector computers are no exception. In general, one cache is provided for instructions and another (often larger) cache is provided for data. Cache flushing and filling is normally implemented in hardware and (other than possible wait states) is completely transparent to the processor. The cache must be capable of delivering operands to the processor within a single processor cycle.

To gain efficiency and to simplify programming, vector processors normally implement a set of *vector registers*. These registers store a series of operands to be used by the processor. The access pathways between these registers and the data cache are designed to provide fast register loading. This can be accomplished by providing wide data paths between them (e.g., two- or four-word simultaneous transfer). It can also be addressed by performing multiple transfers in a single processor cycle, perhaps by implementing the cache and processor on the same chip.

Vector registers allow a single instruction or instruction sequence to operate on multiple data inputs in series. Such an instruction, referred to as a *vector instruction*, normally specifies the operation to be performed and the number of operands stored in vector registers. Upon receiving a vector instruction, a primitive processing unit will proceed sequentially through the vector registers. It will use the first register as input on the first cycle, the second register on the second cycle, and so forth.

A matrix is normally represented within memory as a sequential set of storage locations, each containing an integer or floating-point number. A vector processor takes advantage of this organization by performing block (or vector) transfers to vector registers. A vector load instruction causes the specified number of elements to be sequentially loaded into vector registers from sequential memory locations. Thus, multiple loads are performed as a result of a single instruction.

To maintain processor throughput, all activities (processing functions, load operations, and store operations) must occur simultaneously. Figure 15.7 shows the previous processing example expanded to include both load and store operations. In a vector processor, all five activities would take place on each processor cycle. Thus, processing units are kept continually busy.

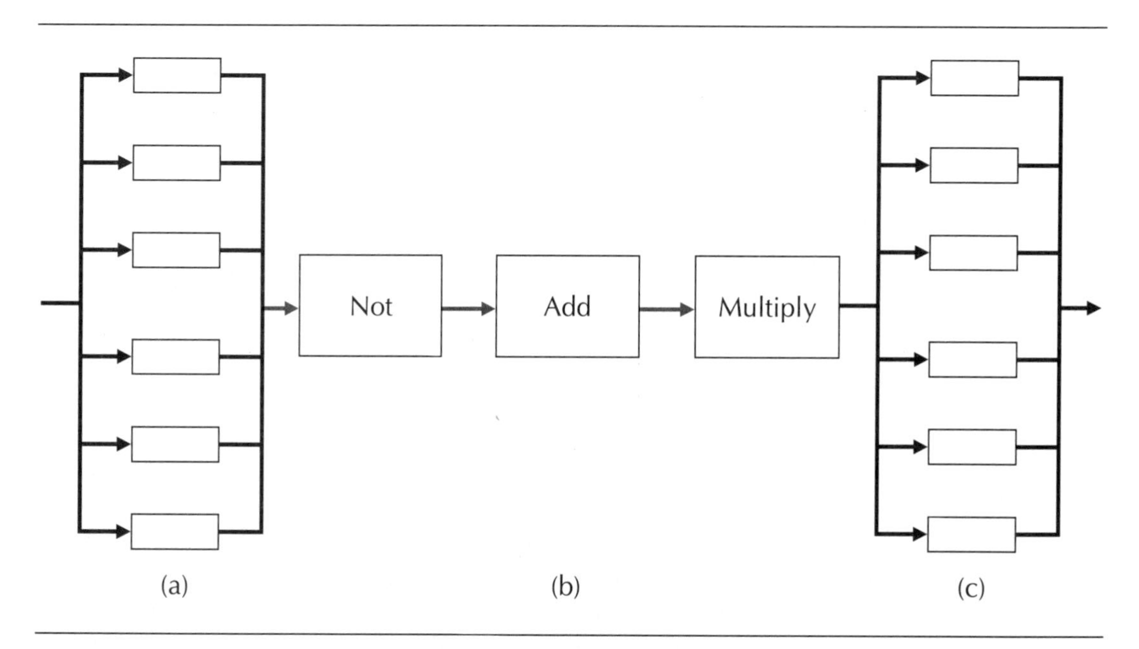

FIGURE 15.7 **(a) Operands are loaded into vector registers. The first processor (b) in the pipeline reads sequentially from the registers. Output of the final processor is sequentially routed to (c) vector registers and stored in memory.**

Implications for Software

Performance improvements for matrix operations on a vector processor are not automatic. They must be programmed directly into application software. A vector processor has a set of instructions dedicated to vector manipulation functions. These instructions must be contained within an application program in order to realize the benefits of vector architecture. Other aspects of the application must also be organized for maximal execution speed. For example, matrices should be stored in memory to facilitate sequential register loading and instruction execution. Instruction loops should be positioned within memory to eliminate swaps of the instruction cache.

The language used to write the application program must be capable of expressing matrix operations. In addition, the compiler (or interpreter) for that language must generate vector instructions based on those matrix operations and must properly align those instructions (and data) in memory. To address this, vector extensions have been incorporated into many programming languages that are commonly used for scientific programming. FORTRAN is by far the most predominant of these, although other languages are also used.

Another approach has emerged to address the problem of programming matrix operations on vector machines—that is, to provide a software translator to convert sequential (but potentially parallel) operations into vector operations. This can take the form of a language preprocessor or of an optimizing compiler. In the case of a language preprocessor, source code is scanned to identify sequential operations (loops containing identical instructions on sequentially stored data). These code sections are then translated into an equivalent matrix representation and compiled normally.

In the case of an optimizing compiler, object code is scanned to identify repetitive sequential instructions on sequential memory locations. These instructions are replaced by vector instructions. In either case, the result is an executable program that uses vector instructions in place of loops or sequences of single instructions.

TECHNOLOGY FOCUS: Intel I860. The Intel i860 is a single-chip microprocessor introduced in 1989. Intel markets the processor as a "Cray on a chip." The claim is not simply a marketing gimmick. The internal architecture of the processor is very similar to the architecture of the Cray 1 (one of the first vector processing computers).

A portion of the i860 processor components is shown in Figure 15.8. Separate processors are provided for addition and multiplication. All other (core) functions are implemented within a single general-purpose (core) processor. These three processors can execute concurrently, and pipelines between them can be constructed in virtually any combination through register I/O. The add and multiply processors can accept one or two 64-bit (double-precision) operands.

The Intel i860 carries processor decomposition one step further. Both the add and the multiply processors are implemented in three stages. These stages can operate concurrently, and each completes its function in a single cycle. All core instructions execute in a single cycle. Thus, up to seven functions (three add stages, three multiply stages, and one core instruction) can be executed simultaneously. Typically, the core processor will be used to load registers for use by the add and multiply processors.

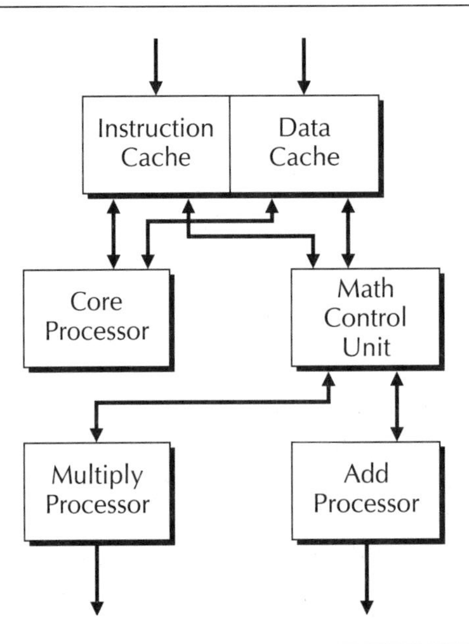

FIGURE 15.8 A functional overview of a portion of the Intel i860 microprocessor.

The processor contains two on-chip memory caches: one for data and one for instructions. The data cache is connected to the processing elements by a 128-bit data path. This allows two double-precision operands to be transferred to the processor simultaneously. Input from both caches can be stored to registers in a single processor cycle. Processing results can also be stored in one cycle concurrently with other processing. Thus, in any single cycle, the i860 can concurrently perform the following operations:

- Load two operands.
- Execute three stages of an add instruction.
- Execute three stages of a multiply instruction.
- Store two results to memory or registers.

The i860 processor is thus capable of performing 10 functions in a single processor cycle.

MULTIPROCESSOR CONFIGURATIONS

Although a vector processor can be described as a multiprocessor, each of the primitive processing units is not a CPU in the general sense. Each is capable of only one type of transformation, and all share a common control unit. An alternative approach to parallel processing is to provide multiple CPUs within a single computer system. A computer system with multiple CPUs is referred to as a *multiprocessor computer*.

Multiprocessor computers can be used to implement vector processing in a manner similar to that previously described. The processors can each be assigned different functions and can be pipelined together. Thus, a multiprocessor computer can simulate the operation of a vector computer. However, the existence of multiple (full-function) CPUs provides much greater flexibility (and capability) than is possible with a vector processor.

The existence of multiple CPUs within a single computer system presents a number of architectural problems and configuration possibilities. Architectural problems arise from the difficulty in distributing processing activity among several processors. Also, access to shared resources must be coordinated to prevent one processor from interfering with the others.

Master-Slave Configurations

The simplest multiprocessor configuration is a *master-slave architecture*. Within such an architecture, one CPU is designated the master, and all others are slaves. The master CPU is responsible for delegating processing tasks to the slave processors. It is also responsible for coordinating access to common resources, including memory, secondary storage, and I/O devices.

Recall that the functions of resource allocation and processor scheduling are the responsibility of the operating system. In a typical master-slave architecture, the master processor executes these portions of the operating system. Slave processors are relegated to executing user processes and/or interrupt handlers, under the direction of the master processor.

The role of master can also be defined through hardware capabilities provided exclusively to one CPU. Some of the capabilities that might be elusively provided to the master include

- Access to certain control registers
- Detection of interrupts and/or other bus control signals
- Detection of memory protection violations
- Detection of slave processor error conditions

The master-slave architecture is popular because it allows relatively simple processor coordination. It is typical of early (and many current) multiprocessor computer systems. Because all control and coordination is centralized, a minimum of communication overhead is required for control activities. Conflicts between processors are also limited. Memory conflicts are avoided by assigning entire user processes to individual CPUs. These processes normally operate in separate regions of memory, thus eliminating memory conflicts between CPUs.

Conflicts between processors for file and I/O services are minimized by tight centralized control of interrupt processing and system services. Interrupt handling is under the control of the master CPU, although it can delegate the processing of a particular interrupt to a slave. As long as the master ensures that only one slave processes a given interrupt, conflicts will be prevented.

The master-slave architecture is sometimes implemented with multiple computer systems. That is, multiple von Neumann (or vector) computers can be combined to act as a single computer system. Network connections are used in place of the bus connections to allow interprocessor communication. Such a system typically exhibits a great degree of specialization. A number of slave processors with limited (or no) I/O and secondary storage hardware can be used primarily for "number-crunching" activities. One or more computers can be dedicated as file or I/O servers.

Cooperating Peer Configurations

Many modern multiprocessing computers use a *cooperating peer architecture*. In such an architecture, no single CPU permanently assumes the role of the master. Under control of software (i.e., the operating system), it is possible to centralize control activities within a single processor. However, no hardware features such as routing of interrupts give any one CPU more capability than the others.

A cooperating peer organization is generally used to execute portions of a single-user process in parallel. In comparison, the master-slave architecture normally executes each user process sequentially (although several users' processes can be executed concurrently). Thus, parallelism is provided at the process level as well as at the system level.

The possibility for resource allocation or access conflicts is much greater with a cooperating peer architecture than with a master-slave architecture. In general, every processor has full bus access and control. Thus, it is possible for two processors to access the same memory location or process the same interrupt. As in the master-slave architecture, coordination is primarily the responsibility of systems software. However, the lack of a designated master means that processors must negotiate with one another for access to shared resources. Thus, systems software operates in a mode very similar to that of a client-server network architecture.

Implications for Software

As with vector processors, applications must be specifically programmed to take advantage of multiprocessing capabilities. Because multiprocessor computers are relatively new, the exact mechanisms by which this is accomplished vary considerably. As with vector processing, multiprocessing extensions are often provided to standardized languages. In particular, compilers for multiprocessing versions of FORTRAN and C are frequently provided by computer or operating system vendors. FORTRAN is commonly used to program parallel numeric and scientific applications. C is commonly used to program parallel systems software, including operating systems for multiprocessor computers.

To program an application for multiple processors, the programmer must identify sections of the code (portions of the application) that can execute in parallel. Synchronization of these code sections must also be specified, either by creating data pipelines or through the use of signals. The object code generated by the compiler contains this information (parallel code segments, pipeline requirements, and signals). The operating system will attempt to assign a processor to each parallel code segment in the application. It will also configure communication pathways to correspond to pipeline directives.

MASSIVELY PARALLEL ARCHITECTURES

The number of CPUs in a typical multiprocessor computer is eight or less. A *massively parallel computer system* extends this number to hundreds, thousands, or tens of thousands. Computer systems are currently available with as many as 65,535 (64K) independent processors, and larger machines are currently being designed. The complexity of interprocessor communication and control requires radical architectural changes, compared to more traditional multiprocessor computers.

Hardware Features

It would be virtually impossible to implement a memory subsystem (storage and bus) that could keep 64K processors busy simultaneously. The requirements for data transfer and access speed are simply beyond the capabilities of conventional electrical devices. As such, massively parallel machines normally provide a local memory region for each individual processor. This memory region is relatively small (typically 64 Kbytes or less) but sufficient to hold a large amount of data for a single processor.

As in vector processor and traditional multiprocessor computers, a mechanism is needed to construct pipelines between individual processing elements. Once again, the number of processors requires radical architectural approaches. Typically, a very wide data transfer pathway is implemented that is capable of a transfer rate of hundreds of thousands or millions of bytes per second. System-wide data pathways are often supplemented by local pathways that connect a subset of the processors (e.g., tens or hundreds of processors). Communication between processors in "close" proximity can occur over the local pathways; communication to "distant" processors uses the system pathway.

Data transfer over the system bus often uses an addressing scheme similar to that found in networks. That is, data is transmitted in packets; each packet carries the address of its destination. This increases the amount of data to be transferred—the address as well as the data is transmitted—but simplifies the communication network.

Because of the large number of processors and pathways, the probability of component failure within the computer system is relatively high. Provisions must be made within the hardware to detect failed components and to route processing and/or data flow to alternative locations. Ideally, the hardware and systems software should deal with component failures without discernible effect on currently executing application programs. This requires a great deal of sophistication on the part of both hardware and systems software.

Implications for Software

As with all parallel processing architectures, applications must be specifically programmed to take advantage of parallel processing capabilities. Parallel code sections must be identified and systems software must be informed of their existence, pipeline links, and signalling requirements. The programming is similar to that of more traditional multiprocessor computers, but the greater degree of parallelism increases execution speed by orders of magnitude. Also, data must be partitioned among the local processor memory regions.

Systems software for massively parallel architectures is substantially more complex than software for other parallel architectures. Not only must parallelism in applications be recognized and implemented, but it must be done for many applications executing concurrently. This is because many applications will not use all of the processors, because all of the processors that could be used to support an application might not be available. In essence, the operating system must create a virtual machine (i.e., allocate a subset of the processors and communication capacity) for each application process when it is loaded.

TECHNOLOGY FOCUS: The Connection Machine. The Connection Machine is a massively parallel computer system. An overview of the architecture is shown in Figure 15.9. It consists of one to four processing units, each of which contains a control unit and 16K individual processors. Each processor is a fully capable CPU with a small local memory region. The system is designed to use external computers for front-end processing such as I/O, secondary storage, and job preparation and submission. Later models rely less on these front-end computers for I/O and storage access than do earlier models.

Each of the processing elements is capable of a full range of logical and mathematical functions. All processors can communicate with one another over a system-wide data pathway. Additional pathways are also provided to allow faster communication between closely positioned processors. Depending on the exact

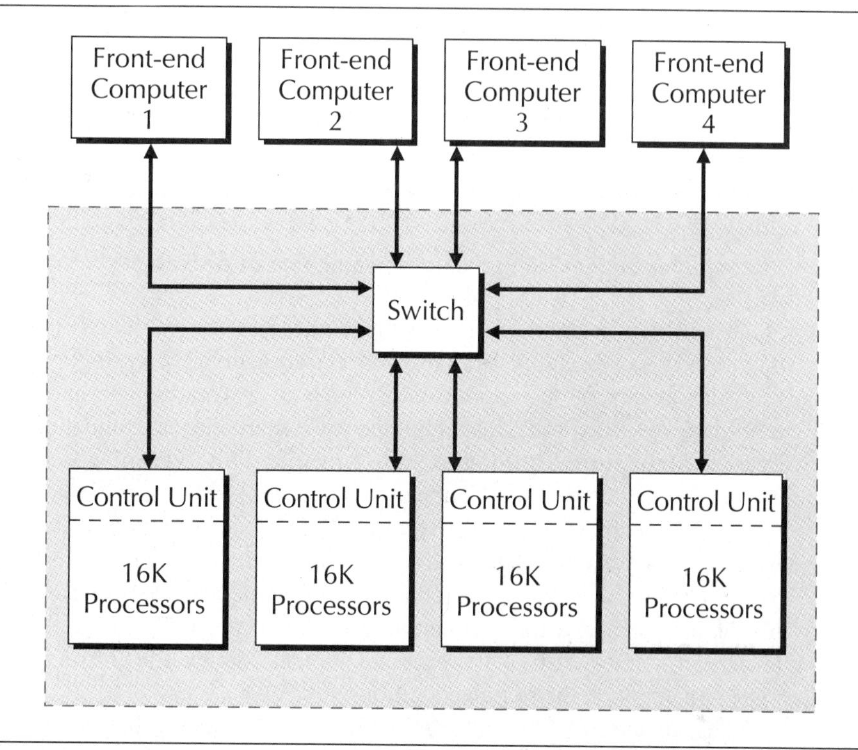

FIGURE 15.9 **A functional overview of the architecture of the Connection Machine.**

model, processors can be implemented in groups on a single chip or on entirely separate chips. The exact amount of memory per processor also varies between models.

Software development and execution on the Connection Machine uses many common tools. Unix is used for the operating system on later models of the machine. Programming languages include FORTRAN, C, and LISP. FORTRAN applications usually execute in SIMD mode using vector extensions to the ANSI standard language. C and LISP provide true MIMD programming environments through parallel programming extensions.

DATABASE MACHINES AND PROCESSORS

Database machines are back-end computers that are normally connected to other types of computers in a MISD pipeline, as shown in Figure 15.10. They can be implemented as stand-alone computers, network servers, intelligent storage controllers, or coprocessors within another computer architecture. Regardless of the exact architecture in which they are used, their function is the same: to take over many file-oriented processing functions from the primary CPU.

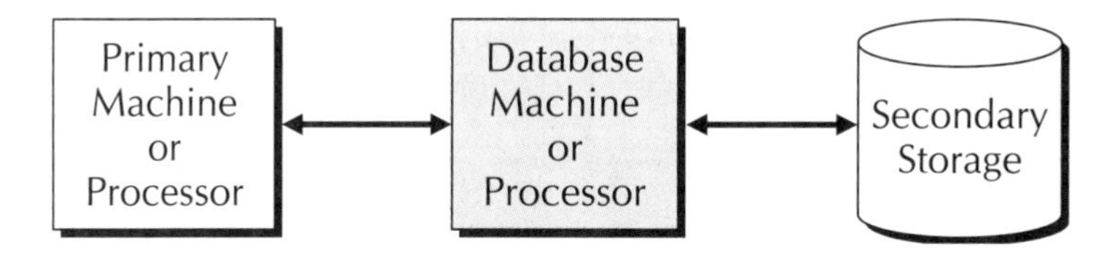

FIGURE 15.10 **A computer system using a database machine or processor.**

Much of information processing requires extensive search procedures for data stored on secondary storage. Online transaction processing systems are the most demanding of such applications. Such an application can have hundreds or thousands of user processes simultaneously searching and updating millions of bytes of data stored on disk or other media. This type of processing places extreme demands on a von Neumann computer system.

A von Neumann processor is normally the focus of all computer activity. All processing actions are performed by the CPU and, thus, all data must be available to the CPU. The normal procedure for implementing a file search on such a machine is to move the contents of the file to the CPU through intermediate buffers in memory. The CPU executes logical comparison instructions to determine a match with the search conditions.

This method of implementing a file search requires the movement of large amounts of data between secondary and primary storage. A great deal of research and development effort has been expended on efficient algorithms by which this movement can be minimized. Techniques such as indexing, hashing, and others limit the amount of data that must be transferred, but with very large files, even these techniques still require a substantial amount of data transfer.

The idea behind a database machine or processor is simple: offload file search processing and data transfer from the CPU and system bus to an auxiliary processor and bus. This machine or processor is then responsible for implementing a subset of file manipulation processing functions. The results of the file manipulation operation are then transferred from the database processor to the primary CPU.

Because the processor is specialized to database functions, a number of special architectural features well suited to this type of processing can be employed. Some of these include

- High-bandwidth communication channels
- Parallel processing
- Associative memory

Regardless of the type of processor, high bandwidth in the communication channel between the processor and secondary storage will always improve the performance of database processing functions. However, it might not be cost-effective to provide such bandwidth on the entire system bus. By isolating database processing to an auxiliary processor, the communication channel used for database processing can also be isolated. This normally takes the form of an auxiliary bus used strictly for transfer between secondary storage and the database processor.

Search procedures normally require that comparison operations be performed on all records in a file or index. If such procedures can be performed in parallel, the elapsed time necessary to complete the search is substantially reduced. Vector architectures are well suited to file-searching operations. Consider, for example, a search of an index where each search value is a single word in size. Index elements could be moved to vector registers in a block transfer. Comparison operations can then be performed in parallel on the vector registers.

Note that the architecture of a vector database processor need not be as complex as that of a normal vector processor. The majority of database processing requires only logical functions. Many of the instructions (and associated processor circuitry) found in a general-purpose vector processor can be omitted (e.g., complex math instructions).

Associative memory is a special type of RAM that is addressed by value rather than by location. It is sometimes called *content-addressable memory*. A processor issues a load request by requesting a certain data value and size of transfer. For example, the request might ask for any sequence of 40 bytes, where the first 4 bytes are equal to some constant. In essence, the memory searches itself for the constant and sequentially places any matching data blocks onto the bus. Content-addressable memory is a relatively new technology and is still very expensive, but it eliminates the need to use the processor for searching functions.

One trend in database machines and processors has been toward the inclusion of more complex processing functions. This has been driven by the widespread adoption of relational database management systems and of some standardization (e.g., SQL) in those systems. Some database processors and machines implement relational database manipulation commands as basic machine instructions. In the case of a database processor, database instructions are accepted from a host processor and only the final results are returned. All intermediate processing and data transfer to or from secondary storage is performed by the database processor, as shown in Figure 15.11. A similar arrangement can be implemented between specialized computers (i.e., a database machine and a general-purpose computer) via a network.

Although the trend toward using separate processors and/or computers for database and file processing is accelerating, the use of special-purpose architectures is

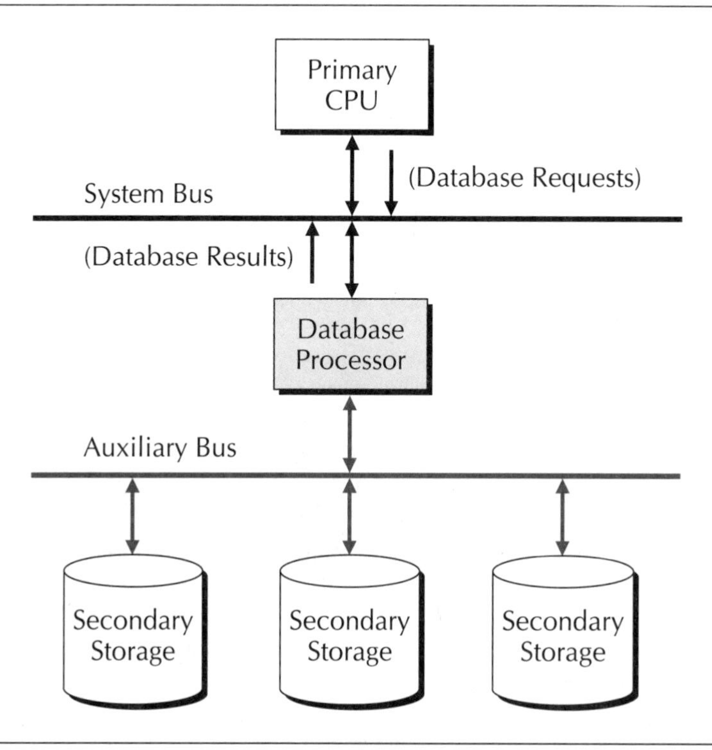

FIGURE 15.11 **The architecture of a machine using a database processor with an auxiliary bus for low-level database access.**

stagnant. Very few vendors provide machines with architectures that are highly specialized to database processing. Those that are provided are very expensive. This phenomenon can be attributed to two primary factors:

- The widespread use and relatively low cost of "standard" architecture machines
- The increased use of networks and distributed processing

The amount of capital required to design and manufacture a new specialized computer is formidable. It can only be justified by a high sale price and/or a large number of units sold. It can also only be justified if the machine provides substantially greater performance for its intended processing than more traditional architectures.

Although database machines using advanced architectures do provide high performance, their cost relative to traditional architecture machines is extremely high. In addition, the database processing power of traditional architectures can be substantially improved by relatively inexpensive hardware enhancements (e.g., intelligent disk controllers and large data caches). The rapidly decreasing cost and increasing power of traditional architecture machines makes them more cost efficient in all but the most demanding database environments.

A related issue is the increased use and sophistication of computer networks. When all processing was performed in a single large machine, large investments to justify database processors were easier to justify. When data is distributed around a network of smaller machines, the costs are less easy to justify. In a distributed system, it is generally more cost effective to use several cheap traditional architecture computers instead of one expensive database machine.

ADVANCED PROGRAMMING PARADIGMS

The capabilities of most programming languages correspond directly to the capabilities (and limitations) of a von Neumann processor. Thus, traditional programming languages are well suited to expressing procedures that can be performed on a von Neumann processor. Conversely, they tend to be poorly suited to processing problems for which a von Neumann processor is poorly suited.

A number of new programming approaches (paradigms) have been developed that extend programming possibilities beyond those imposed by the von Neumann architecture. All of these can produce code that will execute on von Neumann machines. However, the true power of these paradigms is generally realized when non-von Neumann architectures are available. This section discusses some of the new programming paradigms.

Nonprocedural Programming

A von Neumann processor utilizes a stored program to direct all processing actions. The instruction set does provide some flexibility in program sequence (e.g., looping, conditional branching, etc.), but the flow of control through the program is basically static. The control structure of such a program can be thought of as a gravity, or waterfall, model. Data enters at the top and "falls through" the instructions. Although different data elements take different paths to the bottom, they all wind up there eventually.

This basic approach to programming is also known as *procedural programming*. That is, the path of data through the instructions (processing elements) is predetermined and encoded into a standard procedure. On a more abstract level, procedural programming embeds the intelligent aspects of processing in a fixed sequence of instructions—a procedural knowledge representation. This intelligence cannot be used in any way other than that specified in the instruction sequence.

The advantages of procedural programming include predictability and repeatability. If the programmer has accounted for all possible variations in data inputs, the results of processing are completely determined before the data is processed. The procedure for processing the data is fixed. Therefore, the same data can be input to the procedure many times, and the processing results will always be the same.

Although predictability and repeatability are highly desirable for many types of processing, the traditional programming paradigms that ensure them are poorly suited to many other types of processing. A relevant analogy is the operation of a typical bureaucracy. To ensure predictability and repeatability, standard operating procedures are created. All processors (personnel) within the bureaucracy follow these standard procedures. To the extent the procedures account for all possibilities and the implementation is reliable, the system works well.

However, problems arise when unanticipated conditions must be dealt with. The standard operating procedures provide no guidance for processing the unanticipated conditions. The system grinds to a halt because no action has been predetermined. Personnel trained only in standard procedures are unable to alter their behavior to account for the unanticipated conditions. Chaos reigns until updated procedures are created to account for the new possibilities.

An alternative method for representing processing knowledge is *nonprocedural programming*, which attempts to provide more flexibility in the use of processing knowledge embedded within a program. This is accomplished by separating the statement of processing knowledge from the control flow of a program. The processing knowledge is said to be represented *declaratively*. That is, processing rules are stated directly, without embedding them in any fixed sequence. Rules are applied on an "as needed" basis as data is presented to the program.

The general format of a statement in a nonprocedural programming language is a set of statements of the form:[2]

If {Preconditions} Then {Actions} {Postconditions}

Although they are similar to IF statements in a traditional programming language, these statements are referred to as *rules*. A rule consists of a set of (pre)conditions that determines whether it is applicable to the data being processed, a set of actions to be performed, and a set of (post)conditions that are true if the rule is applicable. The set of rules comprising a program is called a *rule base*.

A condition can be a comparison of data values, the existence of certain stored data, or any number of other possibilities. One or more preconditions are checked for each rule. If they are found to be true, the associated actions are

[2] The curly brackets indicate the occurrence of zero or more elements.

performed and postconditions are asserted to be true. The primary differences between rules and IF statements are

- The rules are not evaluated in any particular order.
- Postconditions can be directly asserted.

These two characteristics form the basis of most nonprocedural programming languages.

The execution of a nonprocedural program begins when a data input is presented to the program. The rule base is searched to determine if precondition(s) for any rules are satisfied. For each applicable rule, the associated actions (if any) are performed, and postconditions (if any) are asserted. As a result of asserting postconditions, rules that were not previously applicable could now be applicable. Thus, each time postconditions are asserted, the preconditions of all rules must be reevaluated to determine if they are now applicable.

This style of programming and program execution is well suited to many types of intelligent decision-making problems. It mimics the way humans approach many problems; that is, they search for knowledge (rules) that are applicable to the problem. Application of these rules could require the execution of procedures and can produce new (or refined) knowledge that leads to the application of additional rules. Because of this correspondence to human problem-solving techniques, applications constructed in this manner are sometimes called *expert systems* or *intelligent systems*. The study of such systems is one branch of the field of study called *artificial intelligence*.

Hardware Architecture for Nonprocedural Programs

Several aspects of nonprocedural program execution are poorly suited to von Neumann processors. The most important of these are inefficiencies in searching the rule base and the concurrent evaluation of multiple rules. The rules in a complex rule base can number from hundreds to tens of thousands. At any given time, the preconditions of many rules can be satisfied. Searching for rules with satisfied preconditions is similar in many ways to file and database searching as previously described. It is best done in parallel for maximum efficiency.

Because multiple rules can be applicable concurrently (via satisfied preconditions), parallelism may also be applied to the execution of rule actions and the assertion of rule postconditions. Multiprocessor architectures are well suited to exploit this type of program parallelism. For example, if the preconditions of 10 rules are all satisfied, 10 processors could be used to execute all 10 rules simultaneously.

Current trends in hardware design for nonprocedural program execution emphasize both database processing (for rule base searching) and parallel rule processing. As the complexity and scope of intelligent systems increases, the need for such hardware also increases. This is especially true for complex intelligent systems that must solve problems or make decisions in real time (e.g., friend or foe identification in battlefield systems).

Symbolic Reasoning

Consider the following simple arithmetic problem:

$$\frac{1}{3} \times \frac{3}{1} = ?$$

Humans are taught a number of ways to solve this problem. The easiest (and most accurate) is

$$\frac{1}{3} \times \frac{3}{1} = \frac{1 \times 3}{3 \times 1} = \frac{3}{3} = 1$$

Another possibility is

$$\frac{1}{3} \times \frac{3}{1} = 0.\overline{333} \times \frac{3}{1} = 0.\overline{999}$$

The native capabilities of a digital computer are designed to solve the problem by the second method. That is because the computer must represent all data as numbers. This is the essence of *digital processing*. Humans are also capable of digital processing, although with higher error rates and slower processing times. However, they are also capable of *symbolic processing*—that is, the representation and manipulation of non-numeric data.

A human is able to correctly solve this problem because he or she can represent the data elements in non-numeric ways (i.e., as fractions). Processing rules can then be applied to this non-numeric representation and results generated. Much of the human processing that constitutes intelligent behavior is non-numeric in nature.

Consider, for example, the problem of recognizing the face of somebody you haven't seen in many years. You look for similarities such as skin tone, color of eyes, size of the nose, and height of the forehead. You attempt to compensate for the effects of aging or changes that the person made in hair style or makeup. Finally, you make a decision about which you are not entirely certain.

It is extremely difficult to program a digital computer to solve this problem. How do you represent facial features numerically? What is the numerical representation for thin; or for happy? How is eye color measured? What are the quantitative effects of aging or makeup on facial features? And is the final decision always either an unqualified yes (i.e., true) or no (i.e., false)?

Symbolic programming languages allow a programmer to define symbols and rules that apply to those symbols. Those symbols can be arbitrary strings, characters, or numbers. Rules can be specified that transform individual symbols into others (i.e., a replacement rule). Rules can also be specified to operate on groups (or patterns) of symbols. These rules are generally stated and executed as described earlier for nonprocedural programming.

Basic algebra problems are excellent examples of problems that are well suited to symbolic programming. Consider the following set of equations:

$$2X + 1 = 7$$
$$X + Y = 7$$

One possible method of solving for X is

$$2X + 1 = 7$$
$$2X + 1 - 1 = 7 - 1$$
$$2X = 6$$
$$\frac{2X}{2} = \frac{6}{2}$$
$$X = 3$$

The solution to this problem required performing some computational operations—for instance, subtracting 1 from 7. However, these computations were supplemented by the application of several symbol manipulation rules. For example, the division of $2X$ by 2 cannot be solved numerically, but can be solved symbolically.

Symbolic processing uses certain primitive processing operations over and over again. Examples of these include the replacement of one symbol with another, the concatenation of one set of symbols with another, and the division of one set of symbols into two or more sets. The general term for such operations is *list processing* (i.e., the transformation of lists of symbols). A number of programming languages have been developed based on list and nonprocedural programming.

The programming language LISP was developed in the 1960s to express list manipulation functions. It is the first (and most primitive) of the languages devoted to this programming paradigm. Later languages that embody these concepts include SNOBOL and Prolog. These languages, and others that have evolved from them, are still in use today in the development of intelligent systems.

Symbol Manipulation Hardware

The increasing use of intelligent applications written in LISP has created a demand for computers that are optimized to execute list processing instructions. This demand has been filled by the development of *LISP processors*. These are commonly incorporated as auxiliary processors in computer systems designed to execute LISP applications.

A LISP processor executes machine instructions corresponding to low-level LISP functions such as

- Removal of the first element of a list
- Removal of the last element of a list
- Concatenation of two lists
- Replacement of one list element with another

In essence, these low-level list processing functions serve as an assembly language for application programs.

Programs written in LISP can be executed directly on these processors. Other languages (e.g., Prolog) can be executed using a compiler that generates LISP primitives. The execution speed of applications on these processors is substantially faster than on von Neumann processors. As such, they have found widespread application in many applications that utilize artificial intelligence. Examples of such applications include model analysis, natural language translation, and visual recognition.

SUMMARY

The traditional von Neumann architecture is relatively inefficient when performing some types of processing functions, especially on large amounts of data. This inefficiency arises from the use of a single processor, centralization of all processing activity within that processor, and the digital representation of data. Types of applications most susceptible to inefficiency include those that manipulate large matrices (e.g., numerical modelling and graphics processing), those that search large files (e.g., online transaction processing systems), and those that manipulate symbols and lists (e.g., expert and artificially intelligent systems).

One approach to modifying the von Neumann architecture has been to introduce the ability to execute multiple simultaneous instructions. This approach is especially advantageous to applications that manipulate large matrices. Parallel instruction execution requires the use of multiple processors or processor elements. Multiprocessor architectures can be classified as single instruction, multiple data stream (SIMD); multiple instruction, single data stream (MISD); and multiple instruction, multiple data stream (MIMD).

Vector (or array) processors allow the execution of a single instruction on multiple data items simultaneously. Most vector architectures achieve parallelism by a combination of SIMD and MISD techniques. MISD techniques are utilized by dividing the arithmetic logic unit (ALU) into primitive processing units, one for each type of math or logic function. Each of these units executes its function in a single instruction cycle. Individual units can be chained together to implement complex processing functions. This chain of individual processing elements is called a pipeline.

SIMD architecture is employed through the use of vector registers and instructions. Multiple operands are loaded into a series of registers. A single instruction causes the processing pipeline to access these registers one at a time. Thus, a single instruction is executed on multiple data items in series. SIMD and MISD features are typically supplemented by extensive caching of instructions and data and the use of high-capacity channels between caches and main memory. The combination of these architectural features results in short execution cycle times and extremely efficient execution of vector operations.

MIMD architectures allow concurrent execution of instructions on various data inputs. They achieve this by utilizing multiple general-purpose processing units. Multiple processors can be organized in a master-slave or cooperating peer configuration. In a master-slave configuration, one processor is responsible for all control activities. Processing tasks are delegated to slave processors. This configuration minimizes control complexity arising from concurrent access to shared hardware resources. In a cooperating peer configuration, no one processor exercises full control. Individual processors must negotiate access to shared hardware resources.

Massively parallel architectures typically use hundreds or thousands of general-purpose processors. These architectures allow a high degree of concurrency both among and within application programs. Memory is usually segmented with a small main memory area dedicated to each processor. High-capacity communication channels are required to allow individual processors to exchange data and to coordinate their activities. Mechanisms must be provided to recover from processor failure due to the large number of processors in use.

In any parallel architecture, application software must be specially programmed to achieve the maximum efficiency implied by the architecture. For vector processors, data must be segmented into vectors or arrays and vector instructions and registers must be explicitly used. In multiprocessor configurations, individual sections of an application that can execute in parallel must be identified. The operating system must allocate these individual segments to separate processors and memory regions and must coordinate their activities. Application programs are generally developed using parallel extensions to common programming languages such as FORTRAN and C.

A database machine is a computer or processor dedicated to file and database processing functions. Many database operations require searching large amounts of data for particular key fields. In a von Neumann architecture, this requires that every data element be brought into the CPU and a comparison instruction executed. Database processors offload these processing steps to an auxiliary processor. Such processors typically use a high-capacity communication channel to secondary storage. Other architectural features of these processors may include a restricted instruction set, large data caches, and the use of associative memory.

Several modern programming paradigms also tax the capabilities of the von Neumann architecture. These include nonprocedural and symbolic programming. Nonprocedural programs consist of large sets of rules (i.e., a rule base) that may be applied to data inputs. Each rule can consist of preconditions, actions, and postconditions. Preconditions determine whether the rule may be applied. Actions are executed if reconditions are satisfied. Postconditions are asserted to be true after actions have been executed.

The execution of a nonprocedural program begins by searching the rule base for rules with satisfied preconditions. This process is similar to that of database or file searching. Thus, architectures optimized toward this type of processing may use architectural features commonly found in database processors or machines. Multiple rules can be applicable at any given time. Parallel architectures allow the actions of each of these rules to be executed simultaneously.

Symbolic programming allows the manipulation of data other than numbers. Data can include characters, strings, lists, and graphic images or patterns. Although such processing can be simulated on a digital processor, it is most efficiently addressed by a processor designed for symbol manipulation. A common programming language for symbol and list manipulation is LISP. LISP processors incorporate low-level LISP processing functions as primitive machine instructions. Computers based on these processors are well suited to efficient symbolic processing.

Key Terms

array
artificial intelligence
associative (content-addressable) memory
closely coupled processors
cooperating peer architecture
database machine
database processor
declarative knowledge representation
digital processing
expert (intelligent) system
LISP processor
list processing
massively parallel computer system
master-slave architecture
multiple instruction, multiple data stream (MIMD)
multiple instruction, single data stream (MISD)
multiprocessor computer
nonprocedural programming
parallel execution
parallel processor
parallelism
pipeline processing
primitive processor unit
procedural knowledge representation
procedural programming
rule
rule base
single instruction, multiple data stream (SIMD)
single instruction, single data stream (SISD)
sequential execution
supercomputer
symbolic processing
symbolic programming language
vector instruction
vector processor
vector register

Vocabulary Exercises

1. Most computers are only capable of _____, unlike humans, who are also capable of _____.
2. _____ memory is searched by content rather than by physical location.
3. In a _____ multiprocessor architecture, multiple general-purpose processors must negotiate access to shared resources.
4. In a _____ multiprocessor architecture, one processor coordinates the activities of all other processors.
5. A _____ processor repetitively executes the same instruction on operands stored in successive storage locations.
6. _____, or _____ processing, uses multiple processing elements to sequentially perform various operations on a single stream of data inputs.
7. A _____ processor performs its function on multiple operands stored in _____.
8. To effectively use the power of multiple processors, _____ within a program must be identified and specially coded.
9. A _____ computer system contains hundreds or thousands of general-purpose processors.
10. _____ states processing knowledge as rules rather than procedures or algorithms.

Review Questions

1. For what types of applications is parallel processing best suited?
2. How and why are matrix manipulations performed more efficiently on a vector processor than a von Neumann processor?
3. Explain the difference between SIMD, MISD, and MIMD processing.
4. What is pipelined execution? How does it lead to faster processing?
5. What are vector registers?
6. What is required of programming languages and compilers to effectively use a vector processor?

7. What are the possible architectural organizations of processors within a multiprocessor computer?
8. What are the comparative advantages and disadvantages of master-slave and cooperating peer multiprocessor architectures?
9. What is the typical implementation and organization of main memory in a massively parallel computer?
10. Why is the detection and correction of hardware failures more important in a massively parallel computer than in other types of computers?
11. What is a database machine? What special architectural features and technologies are likely to be used in a database machine?
12. What is nonprocedural programming? For what types of applications is it well suited?
13. What modifications to von Neumann architecture support the execution of nonprocedural programs?
14. What is symbolic processing? How does it differ from digital processing?

Research Problems

1. Due to the expense of special database processing architectures, some vendors have attempted to adapt multiprocessor and massively parallel architectures for database processing. These machines are typically designed to operate as database servers in a network or as back-end database processors. Examples of companies currently pursuing this strategy are NCube, Teradata, and NCR. Some database software vendors such as Oracle have adapted their products to take advantage of these architectures. Investigate one or more of these hardware vendors and their current products. To what extent do they incorporate features of older database processors such as associative memory? How do these architectures (hardware and software) apply parallel processing techniques to database processing?

2. Thinking Machines, Inc., maker of the Connection Machine, has publicly claimed that they can deliver a teraflop (one trillion floating-point operations per second) machine to anyone who can write a big enough check. They claim that such a machine would be a straightforward extension to their existing CM-5. Investigate the architecture of the CM-5. How can that architecture be "scaled up" to teraflop capabilities? What applications might use such capabilities?
3. Several U.S. government agencies such as the National Science Foundation and the Office of Technology Assessment have identified "grand challenge" research problems. These include mapping the human genome and accurate weather prediction. Investigate the grand challenge problems. To what extent will their solution rely on advances in computing power? What degree of advance in computing power will be required to meet them? When is such power likely to be commercially available?

16 Evaluation and Acquisition

Chapter Goals

- *Describe strategic issues in the acquisition of computer hardware and systems software.*
- *Describe the process of acquiring computer hardware and systems software.*
- *Describe tools and processes for evaluating application resource requirements and computer system performance.*
- *Describe issues to be considered when planning the physical installation and protection of computer hardware.*

STRATEGIC ISSUES

The acquisition of computer hardware and systems software should occur only in the context of a well-defined *strategic plan* for the organization as a whole. For the purposes of this chapter, a strategic plan is defined as a set of long-range goals and a plan to attain those goals. The planning horizon is typically three years and beyond. Minimally, the goals must include services to be provided and resources needed to provide the services. The strategic plan addresses the following issues with respect to achieving stated goals:

- Strategies for developing services and a market for them
- Strategies for acquiring sufficient resources for operations and growth
- Organizational structure and control

In all cases, strategic plans must address the basic question of "How do we get there from here?" with respect to the strategic goals.

The information system component of an organization is only one part of the entire organization. Its strategic plan is, therefore, but one part of the strategic plan for the organization as a whole. All portions of its plan must be evaluated in concert with those of other organizational units. This need for coordination is driven largely by the service nature of information systems in most organizations. That is, information systems are normally a means for supporting other organization units and functions such as customer service, accounting, and manufacturing. Thus, the strategic plan for information systems tends to follow rather than lead the strategic plans of other units in the organization.

Hardware/Software as Infrastructure

The resources devoted to most organizational activities can be roughly classified into two categories: *capital expenditures* and *operating expenditures*. Capital expenditures are used to purchase *capital resources (assets)*. These resources are expected to provide benefits for more than one operating period (i.e., beyond the current fiscal year). Examples of typical capital expenditures in an organization include buildings, land, equipment, and research and development costs. The expected useful lifetime of computer hardware and software has decreased in recent years due to rapid technological change. However, it is still long enough to be considered a capital resource.

Many capital resources provide benefits to a wide range of organizational units and functions. An office building, for example, provides benefit to all of the units and functions that are housed within it. Such resources, referred to as *infrastructure*, have the following characteristics:

- Service to a large and diverse set of users
- Difficulty in allocating costs to individual users

- Recurring need for new capital expenditures
- Significant operating costs for maintenance

Computer hardware and systems software that provide information system services are infrastructure. This is obvious in organizations that rely extensively on large computer systems used by many units within the organization. It is less obvious (but no less true) in organizations that have highly decentralized hardware and systems software. The strategic planning issues in information systems are thus similar to those in many infrastructure-based service organizations.

Examples of infrastructure-based service organizations include those that provide services such as communications, electrical power, and water. In general, the strategic issues that must be addressed in such an environment are

- What services will be provided?
- How will service users be charged?
- What infrastructure is required to provide the services?
- How can the infrastructure be operated, maintained, and improved at minimal cost?

As an example, consider the provision of communication services by local telephone companies. The primary strategic question to be addressed is, What types of services should be provided—basic telephone only, expanded telephone services, computer communication services, mobile services, information storage and retrieval? The answer to that question leads to decisions regarding the nature of the required infrastructure and its associated capital and operating costs, such as cellular transmission facilities, fiber optic transmission, total communication capacity, or devices required/available for user interaction with the network.

Standards

Providing infrastructure-based services to a wide variety of users requires a great deal of standardization. However, standardization tends to stifle innovation and to produce solutions that are suboptimal for some users. Once again, consider the local telephone system as an example. All users agree upon and abide by a number of standards in telephone service. These include allowable user devices, basic service availability, and standards for interacting with the infrastructure including line voltage, signal encoding, and others. The nature of infrastructure requires standardization in order to provide service at reasonable cost.

Standardization often causes problems for some users, especially those who demand services at or near the leading edge of technology. For example, there are still some telephone lines in this country that are not capable of touch-tone dialing. This limits the range of services available to users of those lines. As another example, consider the use of fiber optic communication lines. The lack of fiber optic connections in local telephone grids limits the use of high-speed computer communication. This in turn slows the trend toward employees working at home. The provider of an infrastructure-based service must constantly balance the benefits of standardization (reduced costs and simplified service) against its costs (stifled innovation and failure to meet the needs of some users).

It is particularly hard to deal with the standardization issue when considering computer hardware and systems software. There are an extremely large number of choices in both categories. In addition, although there has been some progress in hardware and software compatibility, it is still the exception rather than the rule. The issue is further complicated by the diverse set of components required for information processing within even a modest-sized organization.

Competitive Advantage

Discussion of computer hardware and systems software strictly as infrastructure ignores certain types of opportunities. Infrastructure management tends to concentrate on the provision of short-term services at minimal cost. Such an outlook tends to preclude major technical innovations as well as radical redefinition of services to be provided.

The term *competitive advantage* describes a state of affairs in which one organization employs resources to give it a significant advantage over its competitors. This can take a number of forms:

- The provision of services that others are unable to provide
- The provision of services of unusually high quality
- The provision of services at unusually low price
- The generation of services at unusually low cost

Computer hardware and systems software can be applied to achieve competitive advantage in any or all of these areas. Examples of applications include automated bank tellers, scanning grocery checkouts, computer integrated manufacturing, and many others.

Note that each of these examples was considered a competitive advantage at one time but is now commonplace. Unfortunately, this is the nature of applying technology for competitive advantage. Rapid technology changes and adoption by competitors severely restrict the useful life of most technology-based competitive advantages. Furthermore, there are substantial risks in pursuing competitive advantage through new technology. The costs of technology tend to decrease rapidly after introduction, especially with computer hardware and software. Early adopters also face the inefficiency of starting at the beginning of a learning curve. These factors often combine to create high costs with limited benefit. Late adopters might enjoy substantially reduced costs while enjoying most of the benefits.

THE ACQUISITION PROCESS

The process of acquiring computer hardware and systems software is an ongoing one in most organizations. New hardware and software can be acquired to

- Support entirely new applications
- Increase the capability to support existing applications
- Reduce the cost of supporting existing applications

The exact nature of the acquisition process depends on which of these cases (or a combination thereof) motivates the new acquisition. It also depends on a number of other factors, including

- The mix of applications that the hardware/software will support
- Existing plans for upgrade or change in those applications
- Requirements for compatibility with existing hardware and software
- Existing technical capabilities

Note that applications to be supported figure prominently in both the motivation for new acquisitions as well as factors for choosing among them. It must be remembered that hardware and software exist merely to support present and future applications. Therefore, planning for acquisition is little more than guesswork if a thorough understanding of present and anticipated application needs is lacking.

The acquisition process consists of the following steps:

1. Determine the applications that the hardware/software will support.
2. Specify detailed requirements in terms of hardware and software capability and capacity.
3. Draft a request for proposals and circulate it to potential vendors.
4. Evaluate the responses to the RFP.
5. Contract with a vendor (or vendors) for purchase, installation, and/or maintenance.

These steps are discussed in detail in the following sections.

Determining and Stating Requirements

As first discussed in Chapter 2, computer system performance is measured in terms of application tasks that can be performed within a given time frame. This can be measured in terms of throughput, response time, or some combination of the two. Thus, the first step in stating hardware/software requirements is a statement of application tasks to be performed. This is supplemented by stated performance requirements for application tasks.

Depending on the motivation for the proposed acquisition (new applications, growth of old applications, various techniques can be applicable to the determination of exact requirements. For existing applications, hardware and systems software requirements can be based on measurements of existing performance and resource consumption. Requirements for new applications are more difficult to derive. A number of techniques that can be used to determine requirements are discussed in later sections of this chapter.

Although application requirements form the primary basis for stated requirements, other factors must also be considered:

- Integration with existing hardware/software
- Availability of maintenance services
- Availability of training
- Physical parameters such as size, cooling requirements, or disk space required for systems software
- Availability of upgrades

These form an integral part of the overall requirements statement. Although some of these might be essential requirements (e.g., physical parameters), others are less absolute bases for differentiating among potential vendors.

Request for Proposal

A *request for proposal (RFP)* is a formal document sent to vendors. Its basic purpose is to state requirements and solicit proposals to meet those requirements. It is often, particularly in governmental purchasing, a legal document as well. Vendors rely on information and procedures specified in the RFP. Problems such as erroneous or incomplete information, failure to enforce deadlines, and failure to state all relevant requirements and procedures can lead to litigation.

The general outline of an RFP is as follows:

1. Identification of requestor
2. Format, content, and timing requirements for responses
3. Requirements
4. Evaluation criteria

The identification section identifies the organization that is requesting proposals. Identification should include the name of a person to whom questions can be addressed. Addresses, phone numbers, fax numbers, and the like should also be included.

The RFP should clearly state the procedural requirements for submitting a valid proposal. Where possible, an outline of a valid proposal should be given, with a statement of the contents of each section. In addition, deadlines for questions, proposal delivery, and other important events should be clearly stated.

The requirements statement comprises the majority of the RFP. Requirements should be categorized by type and listed completely. Relevant categories include

- Hardware/software capability
- Related services
- Warranties and guarantees
- Financial considerations

Requirements should be separated into those that are absolute (essential) and those that are optional or subject to negotiation. For example, minimum hardware capacity is generally stated as an absolute requirement, whereas some related services may merely be desirable.

Evaluation criteria are stated with as much specificity as possible. A point system or weighing scheme will often be used for the optional (or desirable) requirements. Weight can also be given to factors that are not stated as part of the requirements. Such factors might include the financial stability of the vendor and good (or bad) previous experiences with the vendor.

Evaluating Proposals

Proposal evaluation is a multistep process. The usual steps are these:

1. Determine minimal acceptability of each proposal.
2. Rank acceptable proposals.
3. Validate high-ranking proposals.

Each proposal must be evaluated to determine if it meets minimal criteria of acceptability. These criteria include meeting essential (or absolute) requirements, financial requirements, and deadlines. Proposals that fail to satisfy minimal criteria in any of these categories are eliminated from further consideration.

The remaining proposals must be ranked by evaluating the extent to which they exceed minimal requirements. This includes the provision of excess capability/capacity, satisfaction of optional requirements, and other factors. Measurements of subjective criteria such as compatibility, technical competence, and vendor stability are also considered at this stage.

A subset of the highly ranked proposals is chosen for validation. The subset should be relatively small due to the length and expense of validation procedures. To validate a proposal, the evaluator must determine the correctness of vendor claims and the ability of the vendor to meet commitments in the proposal. Note that the ranking process relies primarily on assertions by the vendor in the proposal. The validation stage is where the accuracy of those assertions is determined.

Various methods and sources of information are applicable to proposal validation. The most reliable of these is a *benchmark* of the proposed system with actual applications. A benchmark is a performance evaluation of application software (or test programs) using actual hardware and systems software under realistic processing conditions. In years past, benchmarking was often difficult to perform due to the expense of the hardware/software configurations and the length and cost of installation. Currently, this is less of a problem due to cheaper hardware, streamlined installation procedures, and fierce competition among vendors. For all but the largest systems, it is now common practice to deliver and install hardware and systems software for customer evaluation.

Alternatives to on-site benchmarking with actual application software include these:

- Benchmarking at alternative sites with actual applications
- Benchmarking of test applications
- Validation through published evaluations

With systems that are difficult to install, it might be possible to test applications at an alternative site on the premises of the vendor or of another customer.

It might not be possible to test configurations with actual applications if those applications are large, difficult to install, or not yet developed. In such a case, it might be possible to use standard benchmarking software or to construct benchmarking software that simulates the execution of the actual application. Many standard benchmarks are available, primarily for smaller computer systems and for scientific applications.

Care must be taken when using such benchmarks, especially if they are derived in isolation (i.e., if only one benchmark, testing one type of performance, is executed at a time). The performance of individual application programs can differ substantially from individually-derived benchmarks when several application programs are executed simultaneously. For standard configurations, benchmark results can sometimes be obtained from trade publications, periodicals, or independent testing agencies.

REQUIREMENTS DETERMINATION AND PERFORMANCE EVALUATION

This section discusses tools and methods for evaluating performance and determining performance requirements for a set of applications. In general, performance evaluation and system modelling must be concerned with the entire system as a whole. This includes all hardware components, operating software, and application software.

Of particular concern is the impact of operating software on overall performance. As discussed in earlier chapters, many aspects of systems software can be tuned for optimal performance. For example, I/O performance can usually be enhanced by allocating larger portions of main memory for buffers. However, such measures always represent a tradeoff of one type of performance against another. For example, allocating additional memory to I/O buffers can increase virtual memory swapping due to a reduction in memory available to hold active processes. A tradeoff that might be optimal for one set of applications and performance requirements may be unacceptable for another.

Additional problems arise from the interaction of a particular operating system software with the hardware. It is possible to use completely different operating software on some hardware platforms. For example, most Digital Equipment Corporation hardware uses Unix, ULTRIX, or VMS as the operating system. Many microcomputers use MS-DOS, OS/2, or Unix. The choice of operating software has a substantial impact on application performance and on available capabilities.

The discussions below ignore differences in operating software capability and assume that operating software parameters are optimally tuned for the intended application environment. The issue of selecting appropriate operating software for a particular hardware configuration is ignored. Thus, the techniques discussed assume that the hardware/software configuration is appropriate to the applications it will support.

Modelling Concepts

A *model* is an abstract representation of physical reality. Models can be of many types, including graphic representations, mathematical representations and combinations thereof. An artist's rendition (drawing) of the facade of a building is a purely graphical model. A blueprint of a building is a combination of a graphic and mathematical model. A set of simultaneous equations is a purely mathematical model.

All models limit the features of reality that are represented within the model. The process of choosing features to include in the model and of choosing an appropriate representation for those features is called *abstraction.* Various types of models abstract different features. For example, a drawing of a building facade abstracts visual features while ignoring many others such as construction materials and hidden support structure. A blueprint models physical placement of materials in great detail while ignoring many issues of external appearance. The validity of a model is dependent on the relative importance of features abstracted and ignored and the accuracy of representation. That is, a valid model accurately models all important features and ignores unimportant ones.

Models are used to make a wide variety of decisions, primarily because of the expense of building and testing actual systems. To be useful, a model must be easier to build, understand, and manipulate than the corresponding reality. For example, a blueprint is a useful model of a building because it is far easier to create and modify than is an actual building. A model must also be valid for its intended purpose. That is, the model must accurately represent all features that are important to the decisions the model will support.

A primary use of a model is as a prediction tool, as shown in Figure 16.1. When used for prediction, a model can be manipulated by modifying one or more model parameters. In a mathematical model, modification can be accomplished by changing the numerical values (parameters) assigned to one or more model features (variables). For example, an architect might change assumptions about the materials used to construct a certain part of a building. The new materials have a different mass (weight) than the old. The mass of the new materials can be used as a new parameter value for various predictions (e.g., weight of a floor or required support strength). A model can also be used for prediction by changing model components (as opposed to parameter values). For example, an architect might add or delete a wall and assess (predict) the effect of the change on the usability or structural integrity of the building.

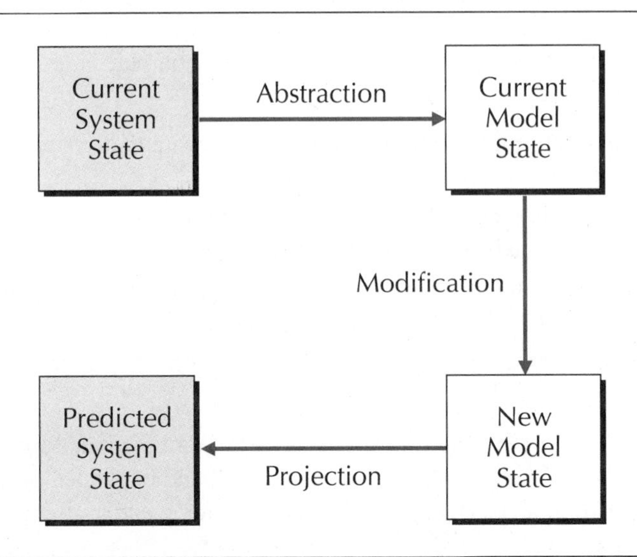

FIGURE 16.1 The process of model-based prediction. A model is created by abstracting relevant features of a system. Model parameters or components are modified, and the results are projected back to the system as a prediction.

Predictions based upon a model are made by projecting the results of model manipulation back onto the system originally abstracted. In other words, it is assumed that modification of model parameters and components will produce results that mirror the effects of those same modifications to an actual system. Every projection from a model to reality carries a risk of incorrect prediction. The projection could be invalid due to any number of factors, including incomplete abstraction, improper representation of features, and incorrect model manipulation.

Each type of model modification (parameter value or model component) introduces its own type of prediction risk. For parameter changes, it is possible to apply the model to parameter values outside of the predictive abilities of the model. A model is often valid only for a limited range (or set of values) of each parameter. Values outside of those ranges could result in model behavior that does not reflect reality. An example of this phenomenon was encountered in many economic models during the 1970s. Inflation levels far beyond normal historical ranges exposed weaknesses in models that used inflation as a parameter.

Computer System Models

With respect to computer hardware and software, both performance evaluation and requirements determination can be addressed through the use of predictive models. In fact, models are virtually required for an accurate assessment of any medium- or large-scale system. Models of computer hardware and software can be used for a number of specific purposes:

- Determining requirements for new hardware and systems software to support a new application
- Evaluating the ability of an existing configuration to handle a new application or additional workload from an existing application
- Determining the change in application performance resulting from changes in hardware and/or systems software
- Formulating long-range plans for expanding or acquiring new computer capacity

Variations of a generic modelling strategy can be applied to each of these scenarios. This strategy is outlined in Figure 16.2 and explained in detail below. It combines two models: an *application demand model* and a *resource availability model.*

Application demand models. A model of application demand for computer resources is often called a *workload model.* It describes the relationship between user-level processing activities and demands for hardware and systems software resources. Examples of user-level processing tasks include transaction updates, database queries, generation of reports, and batch file updates. Each of these processing tasks places different demands on hardware resources. Thus, each requires a separate model of resource utilization. For example, a database query will typically require many accesses to secondary storage with little use of the CPU. A batch update can require many storage accesses (e.g., for a large file) and/or many CPU cycles (e.g., due to complex calculations).

Model Development and Evaluation

1. List each distinct type of application processing task and its stated performance requirements.
2. Determine the hardware and systems software resource requirements for each task.
3. Determine the frequency and distribution of each task.
4. Construct a composite model of application demand.
5. Construct a model of hardware resource availability for each configuration of hardware and systems software under consideration.
6. Test each configuration model against the application demand model.

FIGURE 16.2 **Generic steps for analytical modelling of application demand and resource delivery.**

Resource demand by application tasks can be measured in many different ways. These can range from very high-level measurements, such as average number of CPU instructions executed, to detailed low-level measurements of individual machine actions and operating system service requests. A frequently used set of demand variables is shown below:

- Average number of CPU instructions executed
- Bytes read from and/or written to secondary storage
- Bytes read from and/or written to specific I/O devices
- Bytes read from and/or written to servers or clients over a network

CPU instructions include only those instructions executed to perform computational and logical processing functions. In practice, this is a difficult number to estimate. It is affected by the type of processing performed within the application, the efficiency of executable or interpreted code, and other factors. These factors are difficult to estimate for applications that have yet to be developed or tested. Estimates can be derived for existing applications through the use of program profiling, as described in a later section.

Bytes of secondary storage I/O represent demand for access to mass-storage devices and channels as well as the CPU (operating system service layer) overhead necessary to implement those accesses. Similarly, communication to and from I/O devices and network nodes includes the use of channel capacity and related CPU overhead. For these reasons, I/O to devices or files is frequently measured in terms of specific operating system service calls. For existing systems, these calls can be profiled in much the same manner as application code, in order to derive accurate estimates of CPU overhead.

A composite model of application resource demand is a combination of workload models for individual processing tasks and estimates of the frequency (or volume) and distribution of those tasks over time. Processing volume can be measured in transactions per minute or hour, reports per day or week, queries per minute, and many other units of measurement. Distribution measures show the expected norms and variations in frequency. For example, the volume of sales transactions can average 200 per hour in a normal week. However, typical volume could vary by time of day, day of week, time of year, or other factors. Because of these sources of variation, estimates of processing volume can rarely be stated accurately as a single number. Instead, they are normally stated as a combination of average, minimum, and maximum volumes, with explicitly stated assumptions regarding the timing and duration of each volume level.

Resource availability models. Resource availability can be modelled for either a hardware configuration or a configuration of both hardware and systems software. The latter approach is more accurate, although it introduces additional complexities and possibilities for error. The raw capabilities of hardware represent only a potential. For most applications, access to that capability is obtained exclusively through the systems software (most importantly, the operating system). Thus, the efficiency with which systems software manages and delivers hardware resources to applications is a primary factor determining resource availability.

Resource availability for a configuration of hardware and systems software must be expressed in the same terms as for application tasks. Thus, as discussed above, resource availability is generally measured in terms of CPU instructions and the ability to move data to and from secondary storage, I/O, and network devices. Units of measure for these various capacities must also use the same units of time.

CPU capability is generally measured in millions of instructions per second (MIPS). Several problems can arise from the use of this measure:

- The execution of long instructions by an application
- Wait states due to memory-access bottlenecks
- Incompatibility between the instruction sets of competing hardware

To simplify the model, application demand for CPU instructions is generally standardized to a least common denominator (e.g., 32-bit integer instructions). Configuration capabilities are also standardized to these "generic" instructions and wait states are assumed to be nonexistent.

For access to secondary storage and I/O devices, resource availability is generally measured in terms of *sustainable data transfer rates*. These rates depend on the characteristics of the devices, device controllers, and communication channels in use. Difficulties can be encountered when substantial differences exist between burst mode data transfer and sustainable data transfer rates. For example, the use of a caching disk controller can allow short bursts of extremely fast data transfer limited only by the speed of cache access. However, for large transfers that exceed cache capacity, the sustainable data transfer rate would be limited by the physical characteristics of the storage device. An average rate is sometimes used if it can be derived from detailed and exhaustive hardware tests.

Model Manipulation and Analysis

The use of mathematical models for application demand and resource availability allows well-established computational techniques to be applied to model evaluation. Computational approaches applicable to computer performance modelling include static and dynamic analysis. Each of these can be implemented by a number of specific computational techniques.

Static analysis refers to the use of computational techniques that assume that demand for resources (and the ability to supply them) are constant. Thus, in static analysis, variation over time in application demand or in the ability of a configuration to supply resources is ignored. In contrast, *dynamic analysis* explicitly accounts for variation over time in the values of model parameters. However, computational techniques for dynamic analysis are substantially more complex than those for static analysis.

Static analysis. When static analysis is used to evaluate models, all parameter values are assumed to be constant. Thus, application tasks are assumed to have a constant frequency and resource demand. Configurations of hardware and systems software are assumed to have a constant ability to supply computing resources. These assumptions are not terribly unrealistic with respect to resource delivery. However, they are generally very unrealistic with respect to application demand.

A static model may take many forms. Common techniques used for computer performance evaluation include simultaneous equations, linear programming, and nonlinear programming. For a simple static analysis, average resource demands for each application processing task are summed over an interval of time. For example, assume an application that processes two types of user tasks

(A and B). Assume further that resource demand for each transaction can be expressed by the following formulae:

$$A\text{: } 15000X_C + 300X_S + 200X_I$$

$$B\text{: } 10000X_C + 750X_S + 500X_I$$

where

X_C = thousands of CPU cycles
X_S = kilobytes of mass storage I/O
X_I = kilobytes of display device I/O

The composite resource demand model is simply the sum of these two formulae, weighted by their expected volumes over a stated time interval. Thus, if peak volumes for tasks A and B are 100 and 250 per hour, respectively, the peak composite resource demand is

$$A\text{: } 100(15000X_C + 300X_S + 200X_I) = 1500000X_C + 30000X_S + 20000X_I$$

$$B\text{: } 250(10000X_C + 650X_S + 500X_I) = 2500000X_C + 162500X_S + 125000X_I$$

$$\text{Composite: } = 4000000X_C + 192500X_S + 145000X_I$$

For each configuration under consideration, a resource availability model using the same variables, units of measure, and timing conventions must be constructed. Thus, the capacity of each configuration would be expressed as a set of maximum values for each resource class during a one-hour period. Note that these values are interdependent due to tradeoffs between various types of processing. System overhead associated with mass-storage and I/O accesses consumes CPU cycles and bus capacity. A model of resource availability must account for these complexities. For example, resource availability for a configuration might be stated as

$$X_C \leq 5000000000$$
$$X_b \leq 25000000$$
$$X_s \leq 8000000$$
$$X_i \leq 1000000$$
$$X_S = X_b + X_s + 200X_C$$
$$X_I = X_b + X_i + 150X_C$$

where

X_b = kilobytes of bus I/O
X_s = kilobytes of disk device I/O
X_i = kilobytes of display device I/O

The intermediate variables X_b, X_s, and X_i express maximum capacities for bus I/O, low-level disk I/O, and low-level display device I/O, respectively. Resource utilization for each unit of application I/O to a disk or display device is expressed as a formula describing its use of CPU, bus, and device resources. The resource availability model is thus a set of simultaneous equations that accounts for trade-offs among various types of application task processing.

The two models can be analyzed via a number of techniques. For example, a linear program can be constructed to test the feasibility of meeting application demand with a given configuration. More complex formulations can be used to determine detailed configuration requirements (e.g., number of disk drives or I/O channels) or to determine a least-cost configuration among several competing configurations.

Dynamic analysis. Note that parameter values in the examples of static analysis were all single-valued. In dynamic analysis, the distribution of values is explicitly modelled. This implies the use of more model information than with static analysis. In theory, this additional information will yield more accurate results. However, this additional information must be accurate to guarantee valid analyses of the model. Thus, although dynamic analysis promises more accurate predictions of system behavior, that accuracy is obtained at the expense of substantially more complex model inputs.

Dynamic analysis of a computer system is generally implemented via *simulation*. Simulation is, in turn, based on *queuing theory* and *statistics*. A *simulation model* is composed of processing tasks, processing elements, and a set of interconnections between them. Computer hardware and systems software are modelled as a linked set of server processes, as shown in Figure 16.3. Each server is dedicated to a specific processing function (e.g., CPU, disk I/O, and display device I/O). Each server accepts input from a queue of pending requests, as shown in Figure 16.4. Output from the server can leave the system entirely or can be directed to the input queue of another server.

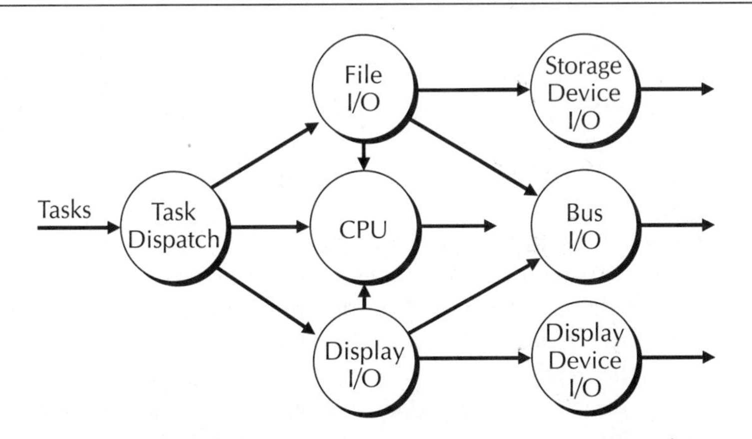

FIGURE 16.3 **A network diagram of a simulation model. The system is modelled as a set of server processes and input queues to those processes. Circles represent processes; arrows represent input and output queues.**

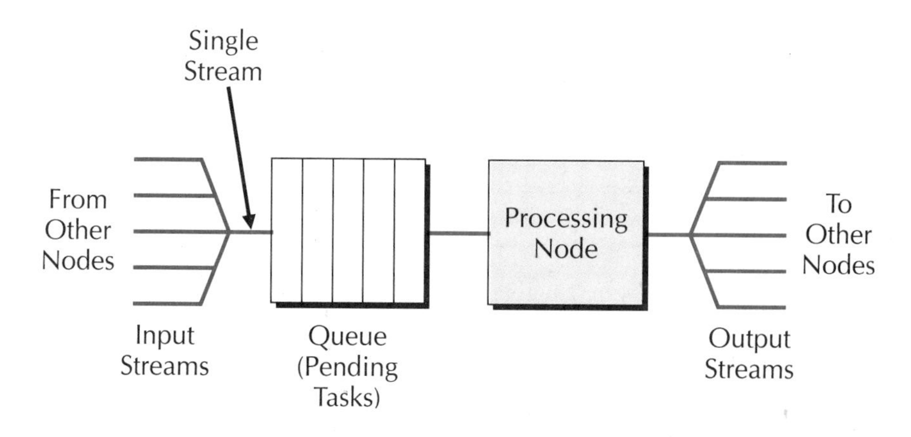

FIGURE 16.4 An individual processing node within a simulation model and its input and output queues.

The arrival rate of processing tasks to initial system nodes (the dispatcher, in this example) is stated as a statistical distribution. This distribution is typically expressed as a distribution type (e.g., normal or Poisson), mean, and standard deviation. The distribution is used to determine the arrival rate for each processing task. Queuing theory can be used to determine the composition of and arrival times for combined queues. Each processing node accepts requests from the input queue and satisfies those requests. The amount of time needed to satisfy a request depends on the request. The time needed by a processor to satisfy each type of request is stated as part of the model inputs. It, too, is often stated as a statistical distribution.

A number of software products are available to implement simulation models and provide analyses thereof. Commercially available packages include GPSS, SLAM II, and SIMSCRIPT. These packages accept a description of the modelled system as input. The operation of the system over time is simulated by advancing a computer-generated clock. Each time the clock is advanced, the program generates a set of task arrivals and hardware actions based on the statistical distributions of task arrival and hardware operation. The current state of the system (hardware states and queue sizes) is updated based on these calculations. Iterations continue until the simulation model reaches *steady-state* operation. This state simulates the normal of operation of the modelled system.

Reports showing many kinds of system information can be produced:

- Average queue size for each processor
- Average waiting time for each processing request

- Average processing time for each processing request
- Proportion of idle time for each processor

This information can be used to evaluate the performance of the system and/or to identify performance bottlenecks.

Measuring Resource Demand and Utilization

Whether dynamic or static models are used for performance evaluation, accurate model inputs are always problematic. Accurate model analysis depends on the quality of estimates for resource demand by application tasks. Yet for new application and systems, the derivation of these demand estimates is often little more than guesswork.

Some specific types of automated tools have been developed for monitoring resource demand and utilization. These techniques provide accurate model input data when considering upgrades or modifications to existing applications or hardware/software configurations. The information generated by these monitors describes the behavior of specific devices, resources, or subsystems over some period of time. These tools include

- Hardware monitors
- Software monitors
- Program profilers

A *monitor* is a program or hardware device that detects and reports processing or I/O activity. A *hardware monitor* is a device that is attached directly to the communication link between two hardware devices. It monitors the communication activity between the two devices and stores communication statistics or summaries. This data can be retrieved and printed in a report. Hardware monitors are often used to monitor the use of communication channels, disk drives, and network traffic in mainframe computers.

A *software monitor* is a program that detects and reports processing activity or requests. Software monitors are typically included within operating system service routines and can be activated or deactivated by a system administrator. They are used to monitor high-level processing requests (file open, read, write, and close) or low-level kernel routines (flushing of file or I/O buffers and virtual memory paging). When activated, a software monitor accumulates statistics of service utilization or processing activity in a file. Records are added to the file as requests are processed or events occur. Separate programs read these files to produce printed reports for the system administrator.

Monitors operate either continuously or intermittently. A continuously operating monitor records all activity as it occurs. A sampling monitor checks for activity on a periodic basis (e.g., 20 times per second). The advantage of a sampling monitor is that less computer resources are expended executing the monitor process itself. Another advantage is that less data is accumulated in files, which can become very large. Continuous utilization statistics can be estimated based on the sampled activity. Continuous monitors provide complete information on activity. However, their operation can consume excessive amounts of system resources, because every monitored action generates a recording of activity.

Monitors are typically used to identify performance bottlenecks and to determine maximal performance levels for hardware/software configurations. For example, monitoring I/O activity on all secondary storage channels can indicate that some disks are used continuously and others much less so. Based on this information, a system administrator might decide to move highly active files from heavily utilized disks to less utilized disks, or to reallocate disks among controllers or I/O channels. An entire system can be tested at full load to determine its maximum sustainable resource delivery.

A *program profiler* determines the resource or service utilization of a program during execution. Typically, a set of monitor subroutines are added to the program's executable image during link editing. As the program executes, these subroutines record service requests in a file. They can also record other statistics such as elapsed (wall clock) time to complete each service request, CPU time consumed by service calls, and CPU time consumed by program subroutines. This information can be used to derive a resource demand model for the application. It may also be used to identify segments of the program that may be inefficiently implemented.

Monitors and profilers are commonly used tools for performance evaluation and requirements determination. By providing accurate data on resource demands and utilization, they allow accurate modelling of computer system configurations. This, in turn, leads to accurate model analyses. Although these techniques are of little help when assessing entirely new systems, they are extremely valuable when considering upgrades or modifications to application programs, systems software, or computer hardware.

PHYSICAL ENVIRONMENT

The installation of computer hardware requires special attention to many aspects of its physical environment. Some of these are a matter of convenience and others are a matter of protection for the equipment itself. Particular issues to be addressed include

- Electrical power
- Heat dissipation
- Moisture

- Cable routing
- Fire protection

Each of these is discussed in detail below.

Electrical Power

Computer hardware is very sensitive to fluctuations in power levels. Processing circuitry such as the CPU and device controllers is designed to operate at a constant low power level. Fluctuations can cause momentary loss of operation (if power levels drop) or damage to electrical circuits (if power levels rise). Fluctuations can be of several types:

- Momentary power surges
- Momentary power sags
- Long-term voltage sags
- Total loss of power

Power surges can be caused by a number of events. Lightning strikes in power generation or transmission facilities tend to cause the most dangerous types of power surges. Similarly dangerous spikes can be caused by the failure of power transformers or other transmission equipment. These events lead to brief surges of very high intensity. Because the surges are brief, they might not engage standard protection devices such as fuses or circuit breakers before significant damage has occurred. Thus, standard fuses or breakers (e.g., those controlling the distribution of power to a floor or entire building) do not provide adequate protection for computer equipment.

Power sags normally occur when a device that requires a large amount of power is started. This can be seen in the home when devices such as air conditioners, refrigerators, and electric dryers are started, causing a momentary dimming of lights. Small power sags are almost always present when multiple devices share a single electrical circuit. Large power sags are a symptom of overloaded circuits. Particularly dangerous is the mixing of equipment with large variations in power requirements on a single circuit.

Longer term power sags are often caused by the power provider itself. The common term for this event is a *brownout*. Brownouts occur when the demand for electricity exceeds the generation and transmission capabilities of the provider. This commonly occurs during peak demand periods such as hot summer days. The power provider will temporarily reduce the voltage level on a system-wide basis to spread the available power evenly.

Most computer equipment is designed to operate reliably over a range of voltage levels. Transformers within computer equipment are designed to operate with

110 volts power input. However, they are typically constructed to tolerate variations of up to 10 percent. This characteristic provides some protection against surges and sags due to the startup and shutdown of equipment as well as brownouts.

Other provisions must be made to protect equipment against more severe variations. These include mechanisms to deal with powerful surges and total power loss. Equipment can be protected against high power surges by the use of a *surge protector*. These devices detect incoming power surges and quickly (within millionths of a second) divert them to ground. Surge protectors differ in the speed at which they react, the intensity of the surge that can be grounded, and whether the device can be reused after a surge. Surge protectors can be purchased as separate devices or as a component of an integrated device such as a line conditioner.

By itself, total power loss rarely causes damage to computer hardware. However, a tripped circuit breaker or blackout is often accompanied by one or more power surges. The primary problem with power loss lies in the loss of data. Data held in RAM, including process data areas, secondary storage buffers, and communication buffers, is lost when power is interrupted.

Protection against power loss requires the use of an auxiliary power source. This can take several forms, including the use of secondary circuits, auxiliary generators, battery backup, or some combination of these. An *uninterruptible power supply (UPS)* is a device (usually battery-based) that provides power to attached devices in the event of external power failure. These devices vary in their power-delivery capacity, switching time, and duration of operation between power loss and power restoration. Surge protection is normally incorporated into a UPS. This is somewhat of a necessity, because switching between external and internal power supplies can introduce surges.

Some UPSs, particularly those with short delivery times, are designed to work in concert with a computer's operating system. On detecting a power failure, these devices will generate an error interrupt. This informs the operating system of the power failure and provides a mechanism for initiating protective actions prior to a total loss of power. Typically the operating system will initiate a normal or emergency shutdown procedure when such an interrupt is detected.

Heat Dissipation

All electrical devices generate heat as a by-product of normal operation. Excessive heat can cause intermittent or total failure of electrical circuits. Thus, all computer equipment requires some means of *heat dissipation*. In equipment that generates little heat, vents in the equipment cabinet are normally sufficient to allow heat to dissipate. Care must be taken to ensure that vents do not become blocked, so as to allow a free movement of air through the cabinet.

Many hardware devices supplement venting with the forced movement of air through the unit. This is accomplished with one or more fans. Fans either force cool exterior air into the cabinet or draw hot interior air out. Either method requires at least two vents and a clear pathway for air movement. Vents are normally positioned at opposite corners of the cabinet to ensure that all components receive adequate cooling. Forced air cooling also requires some filtering to remove dust and other contaminants.

When heat is dissipated from an equipment cabinet, it collects in the room in which the cabinet is located. Some mechanism must be provided to dissipate heat from the room as well as the cabinets. This is especially true when many hardware devices are situated in a relatively small space. Normal room or building air conditioning may be enough, but it is generally supplemented by additional cooling capacity. This more effectively counteracts heat buildup and provides a measure of protection should the primary (building) cooling system fail.

In extreme cases, auxiliary cooling can be provided within an individual equipment cabinet. This can take the form of a refrigerant-based heat exchanger, a liquid cooling system, or even a liquid nitrogen system. Such measures are often used in equipment where semiconductor devices are operated at extremely high clock rates or access speeds, such as CPUs employing 100-MHz clocks or 20-nanosecond static RAM. Liquid cooling is used in some mainframes.

Moisture

Excessive moisture is an enemy of electrical circuitry due to the danger of short circuits (water conducting electricity between two otherwise unconnected conductors). Short circuits can lead to circuit damage and can be a fire hazard as well. Even when power is turned off, water can still damage computer equipment. Impurities in water are left as it evaporates. These impurities can corrode exposed electrical contacts. They can also cause corrosion of other hardware components such as disk platters, printed circuit boards, and metal cabinets.

Well-designed cabinets are one defense against the dangers of moisture. However, cabinets protect only against overt spills and leaks. Another protective measure is to mount cabinets (or devices within cabinets) above floor level. This minimizes the danger from roof leaks, broken pipes, and similar problems that can lead to standing water.

Protection must also be provided against condensation due to excessive humidity. Thus, excessive humidity levels must be avoided through direct control (e.g., a dehumidifier). Low humidity is also a problem. Excessively low humidity levels increase the buildup of static electricity. This increases the likelihood of circuit damage due to inadvertent static discharges. In general, the humidity level of a room containing computer equipment should be near 50 percent.

Cabling

Computer facilities must be designed to provide protection for data communication lines. Because the configuration of these lines changes frequently, ease of access is also very important. Computer facilities tend to deal with this problem in two ways: raised floor and dedicated cable conduits.

A raised floor will generally be used in a room that contains multiple hardware cabinets. The raised floor serves several purposes. The primary purpose is to provide an accessible location for cables connecting devices. The floor consists of a set of load-bearing supports, on which a grid of panel supports is laid. The surface of the floor consists of solid panels that can be easily installed or removed from the grid. Cables are typically routed in a straight line under walkway areas (i.e., between hardware cabinets). Thus, cabling is easily accessible by removing floor panels.

Secondary reasons for raised floors include protection from standing water and to allow the movement of chilled air. Several inches of water can accumulate without reaching the level of equipment cabinets. Moisture sensors are typically placed below the floor panels to detect the buildup of standing water. The space between the actual floor and the floor panels can also be used as a conduit for chilled air. When used in this manner, equipment cabinets are vented at the bottom and top. Chilled air is thus forced through the floor into the bottom of the cabinets, and heat is dissipated through the top.

Dedicated cabling conduits are normally used to provide cable access between rooms or floors of a building. To prevent electromagnetic interference, these conduits should not be used to route both electrical power and electrical data communication lines (fiber optic lines are usually immune to interference from electrical power lines). In addition, access panels should be provided at regular intervals to allow the addition, removal, or rerouting of cables. Conduits should also be shielded to limit external electromagnetic interference.

Fire Protection

Fire protection is an important consideration, both for safety of personnel and the protection of expensive computer hardware. As with cooling, the normal fire-protection mechanisms incorporated within buildings are inadequate for rooms that house computer hardware. In fact, such measures actually increase the danger to both personnel and equipment. This is primarily due to the extensive use of water (automatic sprinklers).

Carbon dioxide, fire retardant foams and powders, and halon gas are alternative methods of fire protection. Carbon dioxide is generally unacceptable because it is a hazard to humans and because it promotes condensation within computer equipment. Fire-retardant foams are unacceptable due to their moisture content, and powders generally have corrosive properties.

Most large computer facilities use halon gas. It does not promote condensation and also doesn't displace oxygen to the extent of carbon dioxide. This allows personnel adequate time to evacuate a room or floor. Unfortunately, halon gas is a chlorofluorocarbon-based product. As such, its production is scheduled to be phased out by the year 2000. No clear replacement has emerged as yet.

Fire detection is also a special problem within computer facilities. Electrical fires often do not generate heat or smoke as quickly as do conventional fires. Thus, normal detection equipment may be slow to react. Fast detection is an economic necessity. Fires within one item of computer equipment can quickly spread and/or cause damage to attached equipment through power surges.

Normal building fire detection equipment is typically supplemented within a computer room. Additional smoke detectors are generally placed near large concentrations of equipment. Smoke detectors should also be placed below raised floors to quickly detect fires in cabling.

Disaster Planning and Recovery

Because disasters such as fires, floods, and earthquakes cannot be totally avoided, plans must be made to recover from them. Disaster planning is especially critical in online systems and those in which extended downtime will cause extreme economic impact. A number of measures are normally taken:

- Periodic data backup and storage of backups at alternative sites
- Backup and storage of critical software at alternative sites
- Provision of duplicate or supplementary equipment at alternative sites
- Arrangements for leasing existing equipment at alternative sites (e.g., with another company or a service bureau)

The exact measures that are appropriate for any given installation are heavily dependent on local characteristics.

SUMMARY

The evaluation and acquisition of computer resources must occur in the context of a strategic plan. A strategic plan for an organization states services to be provided and a plan for acquiring and maintaining resources needed to provide those services. As one unit of the organization, an information system department's strategic plan addresses its contribution to service provision and resource acquisition.

Computer hardware and systems software provide resources to application programs throughout the organization. As such, these resources can be considered organizational infrastructure. Characteristics of infrastructure include service to a wide range of users, difficulty in allocating service costs to users, and a recurring need for large capital and operating expenditures. These characteristics determine many of the constraints under which computer resource acquisition must operate. These include a need to plan for growth and change in future service provision and the need to reliably provide services at low cost.

Management of an infrastructure requires attention to standardization of services (or lack thereof). In general, service provision is reliable and cost-efficient when services are standardized across many users. These advantages result from economies of scale and a broad base of learning and experience. Unfortunately, standardization tends to stifle innovation and creativity in both the use and generation of services. It is particularly damaging to attempt to use computer resources for competitive advantage. Thus, a manager of computer resources must constantly balance low cost and reliability against innovation and opportunities for competitive advantage.

The process of acquiring computer resources is standardized in most large organizations. The basis for acquiring computer resources is always the provision of those resources to application programs. Thus, a clear understanding of resource requirements for present and future applications is always the starting point for the acquisition process.

A request for proposal (RFP) is a formal document used in the acquisition process. An RFP specifies detailed requirements for computer resources, format and content of proposals, and criteria for evaluating those responses. Proposals are evaluated by a multistep process. The first step is to determine the minimal acceptability of the proposal. The proposal must meet all absolute requirements including performance, financial, and time requirements. Proposals that are deemed minimally acceptable are then ranked according to the evaluation criteria stated in the RFP. A small set of the highest ranked proposals are then validated. Based on the validation process, a vendor is selected that best meets the stated evaluation criteria.

Determination of application resource requirements and the performance of computer configurations may be accomplished by benchmarking or modelling. A benchmark is a system performance test under actual working conditions. Benchmarking is the most accurate form of performance testing, but often the most costly and difficult. The complexity of system installation and testing as well as large costs often make benchmarking an impractical alternative.

A model, an abstract representation of a system, is constructed by abstracting features from reality (i.e., an existing or proposed computer system). Modifications to model components are then made to simulate changes in a system or its environment. The state of the model after those changes is a prediction of the state of an actual system under similar circumstances.

Computer systems are generally modelled numerically. An integrated system model consists of a model of application resource demand and a model of resource availability. An application demand model is the sum of expected resource demands by all application-processing tasks under a stated set of conditions. A resource availability model describes the ability of a configuration of hardware and systems software to deliver computing resources to applications. It is expressed in the same terms (i.e., parameters, time periods, and units of measurement) as the corresponding application demand model.

Computer system models, and the analysis methods used on them, can be either static or dynamic. A static model or analysis method assumes that resource demand and supply are described by known constant values. Model evaluation can be performed as a solution of simultaneous equations through algebraic methods and/or various linear and nonlinear optimization algorithms. A dynamic model or analysis represents variations in resource demand and supply. Dynamic models are more complex models due to the additional information used. They also require more complex solution techniques. Dynamic computer models are normally constructed as statistically based simulation models and are solved by a computer-implemented simulation program.

The performance of existing applications and configurations may be tested by a number of monitoring methods. A hardware monitor measures communication between two hardware devices. Communication data is stored and may be analyzed at a later time to produce various utilization statistics. A software monitor measures the utilization of system services in the service layer or kernel. It records utilization data in a file for later analysis and report generation. A program profiler records the use of system resources for a specific program. It also stores utilization data in a file for later analysis.

The acquisition of computer hardware requires attention to details of its physical environment. Important issues to be considered include electrical power, heat dissipation, moisture, cable routing, and fire protection. Computer equipment is sensitive to fluctuations in power including power sags, power surges, brownouts, and power failure. Surge protectors protect computer equipment against brief amounts of excessive power transmission. Uninterruptible power supplies protect against problems due to power failure. Protection against power sags is built into most computer equipment. Power sags can be minimized by careful attention to power circuit design and to the allocation of electrical devices to individual power circuits.

All electrical equipment produces heat as a result of internal resistance. Within computing equipment, heat can also be generated by motors, display devices, and other equipment. Heat must be dissipated to protect electrical components. Most computer equipment contains some heat dissipation capability in the form of vented enclosures and fans. For concentrations of computing equipment, additional cooling measures must be used.

Moisture is dangerous to computer equipment due to the possibility of short circuits and component damage. In larger computer installations, equipment is normally placed on a raised floor to protect against water leaks. Moisture sensors are used below the floor to detect standing water. The raised floor also provides a convenient place to route communication cables. Humidity levels must also be controlled. Low humidity increases the buildup of static electrical charges; high humidity increases condensation.

Fire protection must be provided to protect both personnel and equipment. Standard methods of file protection are generally inadequate or dangerous in computer installations. Sprinklers cannot be used due to potential water damage and the negative interactions between electricity and water. Carbon dioxide cannot be used due to the danger to personnel and the danger of water condensation. Fire-retardant foams and powders generally contain chemicals that are harmful to electrical equipment. Halon gas is a common protection method. However, its production is scheduled to be eliminated in the United States, and no clear successor has yet been identified.

Key Terms

abstraction
application demand model
benchmark
brownout
capital resources
capital expenditures
competitive advantage
dynamic analysis
hardware monitor
heat dissipation
infrastructure
model
monitor
operating expenditures
power sag
power surge
program profiler
queuing theory
request for proposal (RFP)
resource availability model
simulation
simulation model
software monitor
static analysis
statistics
steady state
strategic plan
surge protector
sustainable data transfer rate
uninterruptible power supply (UPS)
workload model

Vocabulary Exercises

1. An estimate of resource delivery by a hardware/software configuration can be obtained by a tested or published _____.
2. _____ are expected to provide service over a period of years.
3. Integrated analysis of the operation of application software, systems software, and computer hardware requires a _____ model and a _____ model.
4. _____ uses _____ theory to determine the arrival of service requests.
5. Features and capabilities of an application or computer system are _____ to create a model.
6. _____ analysis explicitly considers variations in model parameters over time; _____ analysis ignores those variations.
7. A _____ detects and reports hardware or software actions.

8. Provision must be made to protect computer hardware against _____ and _____ in electrical power.
9. A _____ is an abstract representation of reality.
10. Long-range acquisition of computer hardware and software should be made in the context of a _____ for the entire organization.
11. The resource demands of an existing application can be measured with a _____.

Review Questions

1. What is infrastructure? In what ways do computer hardware and systems software qualify as infrastructure?
2. What are the basic strategic planning questions that are addressed with respect to infrastructure?
3. What are the advantages and disadvantages of standardization in computer hardware and systems software?
4. What is a request for proposal? How are responses to a request for proposal evaluated?
5. How is computer performance evaluation related to applications?
6. What are some purposes of performance evaluation?
7. What are the essential elements of a simplified performance model? How can these elements be organized?
8. What is the difference between static analysis and dynamic analysis? What are the comparative advantages and disadvantages of each?
9. What steps are used in model-based requirement determination and performance evaluation?
10. What parameters might be included in a measurement of application resource demand?
11. What is a monitor? List the various types of monitors and the information they provide.
12. Why are conventional methods of fire protection inadequate or dangerous for computer equipment?
13. What problems associated with electrical power must be considered in planning the physical environment of computer hardware?

Appendix A

American Standard Code for Information Interchange (ASCII)

Decimal Value	Character	Description
0	NUL	Null
1	SOH	Start of Heading
2	STX	Start of Text
3	ETX	End of Text
4	EOT	End of Transmission
5	ENQ	Enquiry
6	ACK	Acknowledge
7	BEL	Bell (or Beep)
8	BS	Backspace
9	HT	Horizontal Tab
10	LF	Line Feed
11	VT	Vertical Tab
12	FF	Form Feed
13	CR	Carriage Return
14	SO	Shift Out
15	SI	Shift In
16	DLE	Data Link Escape
17	DC1	Device Control 1
18	DC2	Device Control 2
19	DC3	Device Control 3
20	DC4	Device Control 4
21	NAK	Negative Acknowledge
22	SYN	Synchronous Idle
23	ETB	End of Transmission Block
24	CAN	Cancel
25	EM	End Medium

Decimal Value	Character	Description
26	SUB	Substitute
27	ESC	Escape
28	FS	File Separator
29	GS	Group Separator
30	RS	Record Separator
31	US	Unit Separator
32		Blank Space
33	!	Exclamation Point
34	"	Double Quote
35	#	Pound (or Number) Sign
36	$	Dollar Sign
37	%	Percent
38	&	Ampersand
39	'	Right Single Quote (or Apostrophe)
40	(	Left Parenthesis
41	)	Right Parenthesis
42	*	Asterisk
43	+	Plus
44	,	Comma
45	-	Dash (or Minus)
46	.	Period (or Decimal Point)
47	/	Forward Slash
48	0	
49	1	
50	2	
51	3	

Decimal Value	Character	Description
52	4	
53	5	
54	6	
55	7	
56	8	
57	9	
58	:	Colon
59	;	Semicolon
60	<	Less Than
61	=	Equal To
62	>	Greater Than
63	?	Question Mark
64	@	"At" Sign
65	A	
66	B	
67	C	
68	D	
69	E	
70	F	
71	G	
72	H	
73	I	
74	J	
75	K	
76	L	
77	M	
78	N	
79	O	
80	P	
81	Q	

Decimal Value	Character	Description
82	R	
83	S	
84	T	
85	U	
86	V	
87	W	
88	X	
89	Y	
90	Z	
91	[	Left Square Bracket
92	\	Backward Slash
93	]	Right Square Bracket
94	^	Caret
95	_	Underscore
96	‘	Left Single Quote
97	a	
98	b	
99	c	
100	d	
101	e	
102	f	
103	g	
104	h	
105	i	
106	j	
107	k	
108	l	
109	m	
110	n	
111	o	

Decimal Value	Character	Description
112	p	
113	q	
114	r	
115	s	
116	t	
117	u	
118	v	
119	w	

Decimal Value	Character	Description
120	x	
121	y	
122	z	
123	{	Left Curly Bracket
124	|	Vertical Bar
125	}	Right Curly Bracket
126	~	Tilde (or Negation)
127	DEL	Delete (or Rubout)

Appendix B

Special Notes on Units of Measure

Units of measure for capacity and performance are generally stated in bits and bytes, abbreviated *b* and *B*, respectively. For large-capacity measurements, the symbols can be modified by a prefix indicating the order of magnitude, as described below.

Performance and density measures generally include an abbreviation of a rate. Typically, rates are spoken or written as *per* some unit. These are abbreviated as the letter *p* followed by one or more letters indicating the unit. The most commonly encountered units are

s	second
m	minute
i	inches

Thus, for example, *bps* is an abbreviation of bits per second, *ppm* is an abbreviation of pages per minute, and *tpi* is an abbreviation of tracks per inch.

Prefixes indicating magnitude can be added to virtually any capacity, density, or performance measure. The most common prefixes and their interpretations are

n	nano (one billionth)
m	milli (one millionth)
K	kilo (1024)
M	mega (1024^2)
G	giga (1024^3)

Thus, for example, *MB* is an abbreviation of megabyte, *KHz* is an abbreviation of kiloHertz, and *ns* is an abbreviation of nanosecond.

Glossary

A

Absolute loader A program or subroutine that controls the transfer of instructions from input files into designated locations in memory and passes control of the computer to the program.

Absolute address An address reference within a program that refers to a specific physical memory location.

Absolute addressing The assumption that all address references within a program correspond to actual physical memory locations.

Absolute value The value of a positive number or the product of (result of multiplying) a negative number and –1.

Abstraction 1. With respect to software, the conceptual difference between the physical reality of computer hardware and the logical view of that same hardware provided by systems software and/or a programming language. 2. With respect to computer system modelling, the process of selecting and representing computing resources and application demands within a model.

Access arm A movable device that positions the read/write head over the recording surface of a disk platter.

Access controls Restrictions and limitations on access to resources by users and/or processes, as implemented in systems software.

Access time The elapsed time between the initiation of a read or write command and completion of the function.

Acyclic graph directory A directory structure in which any node (files or directories) can be related to any other node unless that relationship creates a self-referencing relationship (i.e., a cycle).

Acknowledge (ACK) An ASCII control character used by receiver to indicate the successful receipt of transmitted data.

ADD operation A CPU instruction that causes the arithmetic sum of two numbers to be computed and stored in a register or memory location.

Address The physical location of data in main memory, secondary storage, or within a memory-mapped I/O device.

Address resolution As performed by the control unit, the conversion of address references within program instructions into their corresponding physical memory locations.

Address mapping A synonym for *address resolution*.

Address bus The lines (subset) of a bus used to transmit the address of a storage location to be read from or written to.

Addressable memory The maximum amount of memory that can be physically addressed by the CPU. This value is normally limited by the number of bits used to represent an address.

Algorithm A series of processing steps that describe the solution to a problem. An algorithm establishes the sequence of instructions to be followed in a computer program.

American National Standards Institute (ANSI) A governmental body that promulgates numerous standards, including many for computer-related issues (e.g., programming languages).

American Standard Code for Information Interchange (ASCII) A standard coding scheme used to represent character data and a limited set of I/O device control functions.

Amplifier A hardware device used to increase the strength of an analog signal in a communication channel.

Amplitude The magnitude of wave peaks of an analog waveform.

Amplitude modulation (AM) A method for encoding data values as variations in the amplitude (volume) of a carrier signal.

Analog signal 1. A signal that varies continuously in one or more signal characteristics (e.g., frequency or amplitude). 2. A signal in which a continuous range of data values are encoded as continuous variations in one or more signal characteristics.

Analog-to-digital (A/D) conversion The translation of an analog signal into a digital signal of corresponding or similar value.

Analysis model A model of user requirements for information processing. It is often presented using a data flow diagram and supporting written descriptions.

AND operation A processing operation in which bit value pairs of 1/1 are combined to yield 1. All other pairs are combined to yield 0.

Append operation The addition of a data item or record to the end of a sequence of stored data (e.g., adding a record to the end of a file).

Append mode A mode of file output in which records are appended to the end of a file as they are written.

Application layer A generic term referring to application programs (software), as related to systems software and computer hardware.

Application demand model A model of the demand for computer resources by one or more application programs.

Application generator Software that submits prepared program formats to a programmer via a series of menus and prompts. The programmer selects appropriate formats and adds parameter specifications to fit application requirements. Program coding is generated automatically.

Application program A program that addresses a user's single specific need or a narrowly defined class of information processing tasks.

Application programmer A person who creates and/or maintains application software.

Application software A generic term referring to the class of software that consists of all application programs.

Application tool A program designed for use by end users that addresses one or more general types of information processing (e.g., a spreadsheet program).

Archive A backup copy of one or more files from secondary storage.

Arithmetic logic unit (ALU) The circuitry within a central processing unit that performs computation and comparison operations.

Arithmetic shift The use of a shift operation to perform arithmetic calculations (e.g., division and multiplication).

Array A group of consecutive storage areas in memory that are identified by the same name and whose elements are accessed through subscripts.

Array processor A synonym for *vector processor*.

Artificial intelligence (AI) The field of study concerned with computer-based implementation of one or more aspects of intelligent human behavior (e.g., image recognition or learning).

Assembler A translator program that generates machine instructions or object code based on assembly language input.

Assembly language A programming language in which alphanumeric codes (*mnemonics*) represent computer instructions and addresses.

Association for Computing Machinery (ACM) A professional organization for computer scientists, programmers, and engineers.

Associative memory A type of high-performance memory that accesses its contents by value rather than address (location). See also **Content-addressable memory**.

Asynchronous transmission A low-cost, slow-speed method of data transmission in which sending and receiving devices function autonomously.

Attribute A field (or column) within a database relation.

Audio response unit A device that generates spoken messages in response to computer-generated control signals.

Average access time The average (mean) elapsed time between the initiation of a read or write command and completion of the function. The average is computed across all storage locations within the storage device.

B

Back-end CASE tool 1. A synonym for *code generator.* 2. A software tool that creates program instructions from a symbolic or textual description of data processing requirements.

Band A portion of a signal's message-carrying capacity, typically expressed as a range of frequencies (e.g., the band between 200 and 1000 Hz).

Bandwidth The range of signal frequencies that can be transmitted over a transmission medium or carrier signal.

Bar code scanner A device that optically scans surfaces for numbers encoded in sequences of dark bands on a light background.

Base address The first physical address in which a group of data items or instructions are stored.

Base register The register in which a base address is stored, as typically used by the control unit when performing offset address resolution.

Baseband A local-area network transmission method in which multiple digital signals are carried over a single channel through time-division multiplexing.

Batch 1. A collection of records (or other input data) assembled into a group of manageable size for processing purposes. 2. A synonym of batch processing. 3. A mode of operation in which all data inputs are processed without interruption.

Batch processing A mode of operation in which a process executes without the user's simultaneous input or control.

Baud rate A unit of measurement of data transmission that indicates the binary units of information transmitted per second.

Benchmark Representative program segment and data file to be used as a system performance test and selection criteria.

Binary A numbering system for computers that is based on powers (exponents) of two.

Binary-coded decimal (BCD) An early coding method that uses strings of six bits.

Binary editor An editor that allows the addition, deletion, or modification of bit patterns within a file.

Binary mode A data communication mode in which a digital signal is interpreted as a series of individual bit values (as opposed to characters encoded as bytes).

Binary signal A signal that carries data items that have only two possible values (e.g., zero or one).

Binary synchronous communication (BSC) A protocol that provides for character-oriented transmission of data in blocks.

Binder A synonym for *link editor.*

Biphase A term describing a signal coding scheme (e.g., Manchester Coding) in which a digit is encoded as a combination of both positive and negative signal states.

Bisync A synonym for *binary synchronous communication.*

Bit 1. The value represented in one position of a binary number. 2. A number that can have a value of zero or one. 3. An abbreviation of Binary digIT.

Bit map In graphic processing applications, the array of pixel values residing in computer memory for a given image.

Bit position The position or location of a single bit within a byte, word, or bit string.

Bit string A meaningful group or sequence of binary digits.

Bit time The time interval during which the value of a single bit is present within a signal.

Bits per second (bps) A measure of data transmission speed and/or capacity.

Block 1. A series of logical records that are grouped on a storage device for processing efficiency. 2. The unit of data transfer between a storage device and other computer hardware. 3. A synonym for *module.*

Block chaining The storage of a large collection of data in randomly dispersed blocks, correlated by a set of links (pointers).

Block check character (BCC) An extra eight-bit character that is sent after each block. Typically used with various forms of redundancy checking for detecting transmission errors.

Blocked state The state of a process that is waiting for the completion of a processing event.

Blocking factor The number of logical records grouped within a single physical record on a storage device.

Boolean (data type) A data item that can have only the values "true" or "false."

Boolean logic A logical combinatorial system, developed by English mathematician George Boole, which represents logical relationships between entities.

Boot program A synonym for *bootstrap loader.*

Bootstrap loader A simple program that loads and executes the first program after powering on a computer system.

Branch instruction A command that causes the normal sequential execution of processing to be changed so that an alternative sequence, or program branch, is executed instead.

Bridge A hardware device that transfers data communication signals between two signal carriers. It is typically used to connect independent data communication networks.

Broadband 1. A communication carrier with a wide frequency range and high capacity. 2. A local-area network transmission that is analog, one-way, and is based on frequency-division multiplexing.

Broadcast mode A type of transmission in which the same message is sent to all devices on a network simultaneously.

Brownout A reduction in overall electrical voltage by a power provider, usually due to temporary excess of power demand over power supply.

Bubble memory A type of high-speed random access memory capable of retaining data in the absence of electrical power.

Buffer A memory area or device that temporarily stores data to support processing or communication operations. It is generally used to efficiently resolve speed or timing differences between sending and receiving devices.

Buffering The process of interposing a *buffer* between a program and data read from or written to secondary storage and I/O devices.

Burn in The physical process of permanently storing instructions and/or data in a ROM device.

Bus A transmission medium (or communication channel) physically shared by multiple devices within a computer or network.

Bus cycle The period of time required to perform one data transfer operation on a bus.

Bus master A device attached to a bus that is capable of regulating access to the bus by other devices.

Bus protocol A set of signal and processing conventions that defines the means by which a device accesses a bus.

Bus slave A device that can access the bus only in response to an explicit command by a bus master.

Bus topology With respect to data communication networks, a network implementation that utilizes a bus as a signal medium.

Byte 1. A string of eight bits. 2. A unit of measure for storage device capacity. 3. The unit of storage used to hold a single character.

Byte count A count of the number of bytes in a transmitted data block.

Byte count-oriented (transmission) protocol Any synchronous data transmission protocol that uses byte counts as a component of an error-detection mechanism.

Bytes per inch (Bpi) A measure of storage density typically used for magnetic tapes.

Bytes per second (Bps) A measure of data transmission speed and/or capacity.

C

Cable television (CATV) A specific method of encoding and transmitting electrical signals.

Cache An area of high-speed memory that holds portions of data also held within another storage device. It is used to improve the average speed of read operations from its associated device.

Cache controller A hardware device that controls input and output to a cache.

Cache hit An access to a storage location in a storage device that is also present in the device's cache.

Cache miss An access to a storage location in a storage device that is not also present in the device's cache. A cache miss normally requires that a cache swap be performed.

Cache swap An exchange of a portion of the data held within a cache to or from its associated storage device.

Capital expenditures Funds expended to obtain *capital resources.*

Capital resources Resources that are expected to provide benefits beyond the current operating period (e.g., the fiscal year).

Carrier The wave or signal that carries encoded messages within a communication channel.

Carrier-sense multiple-access/collision detection (CSMA/CD) In data communication, a method in which nodes sense the availability of a communication channel to avoid collisions among competing signals.

CASE statement A control statement in a higher level programming language that chooses among several alternate processing paths, based on a multivalued condition.

Cathode ray tube (CRT) 1. A specific technology for implementing video display using beams of electrons that excite phosphor(s) in a vacuum tube. 2. A generic term for a video display device.

Central processing unit (CPU) The hardware component of a computer that executes processing instructions.

Character array 1. A data structure (array) in which each element is a single character. 2. One method of implementing (storing) a string.

Channel A synonym for *input/output channel.*

Character framing A serial data communication technique that groups bit patterns into bytes (coded characters) for purposes of transmission and error detection.

Character-oriented transmission Any data transmission method in which characters (or character-length data items) are the basic units of data transmission.

Characters per second (cps) A measure of output speed typically used for impact printers.

Check bit The bit position in a data-recording pattern used for automatic checking of the validity of a character (e.g., parity checking).

Checkpoint A verification step in an application program or application system. Examples include the validation of data inputs through redundant data entry or batch total comparisons.

Child process A process created by another executing process (the parent), and which executes under direct control of the parent process.

Chip A synonym for *integrated circuit (IC).*

Chromatic resolution A term referring to the number of bits used to describe the value of each pixel in a graphic display and, thus, the maximum number of colors that can be represented.

Classical system development An older method of software development that relies primarily on system and program flowcharts to model both user processing requirements and the physical characteristics of a system that meets those requirements.

Client A program or computer that requests services from another program or computer via network communication.

Client-server architecture The organization of software and hardware into clients (service requesters) and servers (service providers) that interact via a communication network.

Clock rate The time required for a processor to fetch and execute its shortest single processing instruction. It is generally measured and stated in megahertz (MHz).

Close (operation) An operation in which the relationship between a file and a process is severed—typically by flushing any buffers in use and deleting the file control block.

Closely coupled processors Refers to a high degree of interaction among processors that usually share a common memory area.

Coaxial cable A transmission medium composed of a single strand of wire (the carrier) surrounded by a braided return wire and a tough plastic outer coating.

Code 1. A synonym for instructions in a program. 2. The representation of a value by a set of symbols or signals. 3. The act of creating program instructions.

Coder A synonym for *programmer*.

Code checker A software tool that searches program source code for errors and reports those that are found.

Code generator A term usually applied to software tools that produce source, object, or executable code, based on a general description of processing requirements.

Code segment As stored in main memory, the portion of a program's executable image that contains only CPU instructions (as opposed to data values).

Coding scheme A set of rules (or equivalence relationships) for assigning values or signal characteristics to symbols (e.g., ASCII is a numeric coding scheme for character values).

Collating sequence The order of symbols produced by sorting them according to their numeric code values.

Collision The result of attempted simultaneous transmission by two nodes on a shared communication channel.

Command argument A parameter (input data or instruction) to an operating system command.

Command interpreter A component of systems software that interactively accepts user commands and executes corresponding processes.

Command language A specific set of syntactic and semantic rules describing a language that directs the actions of an operating system.

Command layer A component of systems software that accepts user commands and executes corresponding processes. It encompasses both batch and interactive methods of program and system control.

Command line interface A type of interactive command language interface in which a user directs operating system actions via typed characters, words, and phrases.

Command option A command input that modifies the actions of another command input.

Comment A text string, text line, or sequence of text lines in a program that contains explanatory information. Comments are not instructions and are usually ignored by assemblers, compilers, and interpreters.

Common business-oriented language (COBOL) A third-generation programming language used primarily for writing business application software.

Common wire A single return wire shared by one or more signal wires in a multichannel configuration.

Communication buffer An area of primary storage within a sending or receiving device (or device controller) used to store data temporarily before or after transmission.

Communication channel A general term describing the combination of a signal medium, carrier, or communication protocol. It may also refer to these elements in combination with sending and receiving devices.

Communication protocol A general term describing the signalling and processing conventions by which messages are transmitted, media access is regulated, and transmission errors are detected.

Compact disk (CD) A circular platter that optically stores data bits.

Compact disk read-only memory (CD-ROM) A compact disk containing data that cannot be altered once written.

Compaction The process of reallocating storage locations assigned to a file to eliminate empty storage locations.

Compensating errors Errors within the same message that counterbalance one another, thus "fooling" a method of error detection.

Competitive advantage The use of resources to provide better or cheaper services than others provide with those same resources.

Compiler A program that translates a set of source code instructions into object code in batch mode.

Compiler library A set of predefined object code modules that a compiler substitutes for source language statements.

Complement A number substituted for another number to which it has an inverse relationship. For example, in the binary number system, zero is the complement of one.

Complementary arithmetic The implementation of arithmetic operations using complementary values of one or more operands.

Complementary metal oxide semiconductors (CMOS) A class of semiconductors typically used to implement random access memory (RAM).

Complete path The name of an access path, identified within the directory hierarchy, that begins at the level of the root node and proceeds through all nodes along a path to the desired data item.

Complex (machine) instruction A processor instruction that results in the execution of multiple-processor primitive actions. Complex instructions can also be of variable length.

Complex instruction set computer (CISC) A computer that uses a CPU that executes complex instructions. Contrast to *reduced instruction set computer (RISC).*

Compression algorithm An algorithm by which data inputs can be translated into equivalent outputs that consume less storage space or communication channel capacity.

Compression ratio The ratio of the size of data inputs to and outputs from a compression algorithm.

Compute bound Refers to a condition in which computational requirements of a process saturate or exceed the capacity of a processor.

Computer-assisted software engineering (CASE) The use of software tools to support the analysis, design, and/or development of programs.

Computer-assisted software engineering (CASE) tool An integrated set of software tools that support the analysis, design, and/or development of programs.

Computer engineering The field of study encompassing the implementation and organization of computer hardware devices.

Computer operator A person responsible for controlling computer operations and responding to abnormal processing circumstances.

Computer science The study of the implementation, organization, and application of computer software and hardware resources.

Computer system An integrated set of computer hardware devices and associated systems software.

Concurrent processing A general term describing the execution of multiple processes, either simultaneously (with multiple processors) or within the same immediate time frame (through interleaved instruction execution on a single processor).

Conditional branch A branching operation triggered by data content or by sensing of a triggering mechanism such as a flag.

Conditional macro A macro that is executed only if a stated condition is true.

Console A unit that represents a standard input and output device for an entire computer system, and through which the user or operator exercises global control over system functions.

Consolidator A synonym for *link editor.*

Content-addressable memory A type of physical memory that locates storage contents by data values rather than physical addresses.

Contiguous allocation The storage of a set of data elements in sequential (adjacent) physical storage locations.

Continuous high idle signal A series of flags transmitted by a sending device to maintain a continuous bit stream between frames in synchronous communication transmissions.

Control bus The portion of a bus used to transmit command and status signals between devices.

Control characters Signals, coded in a manner similar to printable characters, that control specialized functions within storage and I/O devices (e.g., ASCII character codes 0 through 31).

Control code A command to a storage or I/O device encoded as a control character, possibly followed by one or more additional characters.

Control structure A programming language statement that describes the selection or repetition of other program instructions (e.g., an IF or WHILE statement).

Control unit A subcomponent of a central processing unit that is responsible for data movement, instruction decoding, and control of the arithmetic logic unit and the bus.

Controller A processor that exercises exclusive control over a device or set of devices.

Cooperating peer architecture A multiprocessor or multicomputer architecture in which nodes have equal responsibility, control, and access to resources.

Coprocessor A general term describing a processor that executes specialized or time-consuming operations, freeing a central processing unit to perform other tasks.

Coupler A device that transforms TTL signals within a computer to zero crossing signals for digital transmission.

Coupling The degree to which modules within a program are interdependent and interact among each other.

CPU bound A synonym for *compute bound*, with particular respect to a central processing unit.

Cross assembler An assembler that translates object or machine code for one computer system into object or machine code for another computer system.

Cross compiler A compiler that can produce object code for multiple computer systems.

Current directory The directory that serves as the origin for file and directory search operations.

Cursor A visual symbol on a video display (e.g., a colored box or underline) that indicates the current input or output position.

Cycle time 1. The time required to execute a single processor instruction. 2. The minimum time interval between two accesses to the same storage location. 3. The time interval between two maximum or minimum values of a signal parameter.

Cycles per second A unit of measurement for frequency or processor speed.

Cyclical redundancy check (CRC) A technique for detecting errors in data transmission; it generates check characters that double the number of bit characters transmitted for each block. For example, 16-bit check characters are generated to check transmissions containing 8-bit characters.

D

Daisy wheel printer An impact printer that uses characters stored on individual spokes of a print wheel.

Data bus The portion of a bus used to transmit data between devices.

Data compression The reduction of storage or communication capacity required for data by the application of a compression algorithm.

Data declaration A programming language statement that declares the existence, name, type, and other characteristics of a data element or structure.

Data definition A synonym for *data declaration.*

Data dictionary A stored set of data declarations or definitions.

Data driven A general term describing any device that displays output based on display list or vector input.

Data element 1. A unit of data that has meaning within an application program. 2. A unit of data within a database management system or data flow diagram.

Data end flag A bit string used to indicate the end of a data frame in synchronous transmission.

Data flow diagram (DFD) A graphic model of processing requirements composed of data sources, data sinks, processes, files (data stores), and data flows. It is generally created during the analysis phase of the structured system development life cycle.

Data link escape (DLE) An ASCII control character used to indicate a change between character-oriented and transparent (binary) data transmission modes.

Data link layer The layer of network systems software responsible for establishing and maintaining data transmission links to other network nodes.

Data manipulation tool An approximate synonym for *database management system (DBMS)*.

Data model A set of descriptions and definitions that identify the logical structure of a set of files or a database.

Data start flag A bit string used for clock synchronization and to indicate the start of a data frame in synchronous transmission.

Data structure A data item composed of multiple primitive data elements. Examples of data structures include linked lists and records.

Data transfer rate The rate at which data can be transmitted between one device and another.

Data type A category of data corresponding to one of the primitive data types that can be defined within a CPU instruction set (e.g., integer, real, character, Boolean, and address).

Database A stored set of data elements, organized to facilitate access and manipulation by multiple users and/or application programs, generally through the manipulation of a network, or relational data model.

Database administration The actions by which database integrity, reliability, security, and accessibility are ensured.

Database machine A stand-alone computer that provides database access services (e.g., read, write, update, and search) to other computers via a communication network.

Database management system (DBMS) A set of systems software programs that provide access to and administration of a database.

Database processor 1. A synonym for *database machine*. 2. One processor in a multiprocessor computer system dedicated to database processing operations.

Debugger A software tool or set of tools that assist a programmer in locating and correcting program errors through actual or simulated execution.

Declarative knowledge representation 1. The statement of processing rules without the specification of a particular order for the application of those rules. 2. The statement of processing rules independent of an algorithm or procedure.

Decoding (instruction) The process of separating an instruction into its component parts (op code and operands) and routing data and control signals to their appropriate destinations (e.g., registers and the ALU). Instruction decoding is performed by the control unit of a CPU.

Decryption The opposite of *encryption*.

Dedicated mode A term describing a mode of operation in which a device controller controls a single high-speed storage or I/O device.

Delete operation 1. The removal of a data element from within a data structure. 2. The removal of a record in a file. 3. The removal of a file by deleting its directory entry and deallocating its storage.

Device controller A special-purpose processor that controls the physical actions of a storage or I/O device.

Device directory A stored list of filenames and allocated storage locations for a single storage device.

Device driver A component of the operating system kernel that is dedicated to control of and communication with a single hardware device.

Device status A general term describing the ability of a device to respond to service requests. Typical status values are *ready, busy,* and *error.*

Dialogue A pattern of interactive communication between a user and computer software.

Direct file organization A method of physically organizing file contents in which the value of a field determines the physical location of the corresponding record.

Digital computer A computer that represents and manipulates data as digital (usually binary) signals.

Digital data communication message protocol (DDCMP) A byte-oriented synchronous protocol for high-speed data transmission.

Digital processing A general term describing the manipulation of digital signals.

Digital signal A signal in which a discrete (countable) set of values can be encoded. With respect to computer hardware, the terms *digital signal* and *binary signal* are usually synonymous.

Digital-to-analog (D/A) conversion The translation of a digital signal into an analog signal or corresponding or similar value.

Digitizer 1. A device that captures the position (row and column) of a pointing device as input data. 2. A device that converts printed images into a corresponding stream of digitally encoded pixels.

Direct access A general term describing storage devices that allow data records to be accessed directly, or in any desired order, because of physical relationships between read/write mechanisms and recording patterns on the storage media. Examples include random access memory and disk storage devices.

Direct memory access (DMA) A method of data transfer that allows direct data movement between main memory and a storage or I/O device, bypassing the central processing unit.

Direct memory access (DMA) controller A dedicated processor that handles the processing overhead of data transfers during direct memory access operations.

Directive A command to an assembler or compiler embedded within source code, but not considered a part of the program instructions.

Discrete signal A synonym for *digital signal*.

Diskette A small, removable magnetic disk storage medium encased in a protective cover.

Dispatching The act of giving a process control of the central processing unit.

Displacement A synonym for *offset*.

Display attribute A term describing a special feature of printed or displayed output (e.g., color, bold intensity, or inverse video).

Display driven A general term describing the generation of output via signals specifically oriented toward the physical implementation and operation of the display device.

Display list A group of high-level commands that describe graphic and/or textual images.

Display object A primitive graphic object, as described and/or manipulated within a display list.

Distortion The undesirable transformation of a signal during the communication process, possibly resulting in a misinterpretation of the message encoded in the signal.

Distributed file system The storage and provision of access to files on multiple computer systems via a communication network.

Distributed processing 1. A general term describing the distribution of application processing over multiple computers that are connected by a computer network. 2. The ability to execute processes on any of a set of machines via a communication network.

Distributed system A computer system consisting of multiple hardware and software elements connected via a communication network.

Dot-matrix printer An impact printer that generates printed characters using a matrix of print pins.

Dots per inch (dpi) A measure of print or display resolution (pixel density) for a printing or video display device.

Double precision Representation of a numeric value with twice the usual number of bit positions for accuracy in calculations.

Doubly linked list A linked list in which element order (ascending and descending) is represented with two sets of links or pointers.

Dyadic instruction An instruction that uses two operands as input.

Dynamic analysis A general term describing the use of models and/or model analysis techniques that incorporate variability and random events in the representation of system behavior.

Dynamic processing A job-queuing technique that allows entire processes to be moved in and out of main memory to achieve interleaved execution.

Dynamic resource allocation A general term describing the allocation of resources to processes as they are requested or needed.

E

Edit operation Any operation in which the content of a stored data item or structure is permanently altered.

Editor A software utility program used to create or modify the contents of files.

Electroluminescent display A display device that generates light by the application of electrical power to solid-state elements.

Electromagnetic interference (EMI) Disruption to computer operation or digital communication, often caused by electric motors or magnetic devices.

Electronically erasable programmable read-only memory (EEPROM) A (normally) read-only memory chip that can be programmed (written), erased, and reprogrammed (rewritten) under processor control.

Encryption Alteration or encoding of signals to prevent unauthorized access to data. Encryption is normally implemented as an algorithmic transformation of coded data based on an encryption key.

Encryption key A code that controls the encryption of data signals.

End of file (EOF) A control character that indicates the end of a file.

End of line (EOL) A control character that indicates the end of a line of text.

End of text (ETX) An ASCII control character that indicates the end of a data block (see *Synchronous data transmission*).

Erasable programmable read-only memory (EPROM) A (normally) read-only memory chip that can be erased and reprogrammed (rewritten).

Escape sequence A command to a storage or I/O device consisting of multiple characters, in which the first character is an escape (ASCII 27) character.

Even parity An error detection method that uses a parity bit. The parity bit is given a value of one if the number of one-valued bits in a character is even.

Excess notation A system of coding integers as bit strings, in which the leftmost bit is 0 if the integer is positive, 1 if the integer is negative or zero.

Exclusive-OR (XOR) operation A processing operation in which bit value pairs of 0/0 are combined to yield 1. All other pairs are combined to yield 0. This is the opposite of an OR operation.

Executable code Program instructions that are ready for loading and execution.

Execution 1. The process of carrying out an instruction in the central processing unit and completing the operation it specifies. 2. The same process as applied to all instructions in a program.

Execution cycle The portion of a central processing unit cycle in which an instruction is executed and results stored to registers or memory locations.

Expert system A program or software system that mimics human decision-making behavior in a narrow area of expertise.

Explicit file transfer The transfer of files between machines via a communication network as a result of an explicit user request.

Explicit prioritization Process scheduling through explicitly stated priority levels and a corresponding scheduling algorithm.

Exponent 1. A portion of the internal representation of a real number. 2. A numeric value representing the power (or exponentiation) of another value.

Expression Within a program, a formula consisting of constants and/or symbolic references (variable names).

Extended Binary-Coded Decimal Interchange Code (EBCDIC) A standard coding system for representing data in an eight-bit byte format, most commonly used within IBM mainframe computers.

F

Father The original version of a file after updates have been applied to generate a new version (the son).

Fetch cycle The portion of a central processing unit cycle in which an instruction is loaded into a register and decoded.

Fiber optic cable A transmission medium for optical signal propagation, generally consisting of one or more strands of plastic or glass sheathed in a protective plastic coating.

Field 1. A basic, processing-oriented unit of data consisting of a primitive data element or a data structure (e.g, a string). 2. The data items of which records (in a file) are composed.

File 1. A collection of related records. 2. A fundamental unit of data storage on secondary storage devices.

File allocation table An index that records the correlation between files and storage device locations.

File backup 1. A copy of a file made to allow recovery in case of loss of or damage to the original. 2. The act of creating such a copy.

File close The operation of disassociating a file with an active process.

File control block A data structure created by the operating system when a file is opened. It is used to store information about the file, including current position, the number and location of I/O buffers, and the current state of each buffer.

File control layer A systems software layer that accepts file manipulation requests from application programs and translates them into corresponding low-level processing commands to the CPU and secondary storage devices.

File control system A synonym for *file control layer.*

File create The operation of creating a directory entry for a file and allocating storage space to it.

File delete The act of deleting a directory entry for a file and deallocating its storage space.

File destruct An operation in which a file delete is immediately followed by overwriting the deallocated storage locations with null values.

File header Descriptive data about a file that the operating system retains and writes at the beginning of the designated storage area.

File lock An operation that assigns exclusive access to a file to a single process.

File management system A synonym for *file control layer.*

File migration A management technique for secondary storage, in which older versions of a file are automatically moved to slower and/or less costly storage media (e.g., magnetic tape).

File open The process of associating a file with an active process. Typically, data buffers are allocated and information is added to process and file control data structures maintained by the operating system.

File packing The process of compacting the contents of a file.

File specification The written description of the structure and content of a file on a data flow diagram.

File type An indication of the content of a file (e.g., text, binary, indexed, and executable).

File undelete The operation of recovering a previously deleted file by reinitializing its directory entry and reclaiming its previously allocated storage locations (and their contents).

File-by-file dump A specific type of backup (copy) operation for files. For each existing file, a new file is first created on a new storage media. Each record in the existing file is then individually copied to the new file.

Firmware Programs that have been encoded permanently into the circuits of microchips (e.g., read-only memory chips).

First come, first served (FCFS) A processing order in which processes are served in order of their arrival.

First in, first out (FIFO) A processing order equivalent to *first come, first served.*

Fixed-length instruction A term referring to one member of an instruction set in which all instructions contain the same number of bits.

Flip-flop circuit An electrical circuit that can be switched between two states, thus representing the binary values zero and one.

Floating-point notation A notation system for representing real numbers within computers.

Floating-point operation The execution of any processor instruction utilizing floating-point (real) numbers as data inputs.

Floppy disk A synonym for *diskette*.

Font A named set of display formats (similar in appearance) for printable characters and/or symbols.

Forms-based interface A style of user interface in which the user is interactively prompted for input by the display of forms on a video display device.

FORmula TRANslator (FORTRAN) A third-generation programming language primarily used to write scientific and mathematical application software.

Formulaic problems Processing problems that can be represented as formulas and solved by stepwise computational operations.

Fourth-generation development environment An integrated set of tools for developing and testing programs written in a fourth-generation (programming) language.

Fourth-generation language (4GL) A programming language that supports nonprocedural programming, database manipulation, and advanced I/O capabilities.

Fragmentation A general term describing a situation in which storage locations allocated to a single file, program, or other storage entity are scattered throughout a storage device in noncontiguous locations.

Frame With respect to data communications, a synonym for *packet*.

Frequency The number of complete waveform transitions (e.g., change from positive to negative to positive energy peak) that occur in one second.

Frequency division multiplexing (FDM) A data communication technique in which a single broadband channel is partitioned into multiple subchannels (frequency bands), each capable of carrying a separate data stream.

Frequency modulation (FM) A method for encoding data values as variations in the frequency of a carrier signal.

Frequency range 1. The range of frequencies over which a carrier signal can be varied. 2. The range of signal frequencies that can be transmitted over a transmission medium or communication channel.

Front-end CASE tool A tool that supports the creation and modification of models describing data processing requirements.

Full backup A backup operation in which all files in a directory or storage device are copied to a backup storage device or medium.

Full duplex A type of transmission that permits simultaneous two-way communication through the use of two separate signals.

Fully qualified file reference The combination of a filename and a complete path name.

G

Gallium arsenide A substance used to implement both electrical and optical microchips.

Gas plasma display A display technology that generates light by applying electricity to trapped bubbles of neon gas.

Gate The general term for electronic switching devices that perform processing operations.

General graph directory A directory structure in which any node (files or directories) can be related to any other node.

General-purpose processor A processor with capabilities that can be applied to perform a wide variety of processing tasks.

Generative translator A general term describing a program that generates machine code from program statements containing symbolic references.

Grandfather The version of a file prior to the father version.

Graph structure directory A directory structure in which allowable relationships between directories can be represented as a graph.

Gray-scale value A numeric value representing the level of brightness for each pixel.

Grosch's Law An outdated statement of the mathematical relationship between computer size and cost per unit of instruction execution. The law states that cost per executed instruction decreases as computer system size increases.

H

Half duplex A transmission method in which sending and receiving terminals exchange information in one direction at a time over a single communication channel.

Halt instruction An instruction that terminates the current sequence of execution.

Handshaking A general term describing the procedure by which two devices negotiate and coordinate access to a communication channel.

Hard-copy terminal A hardware device combining keyboard input with printed output, as used to support interactive processing.

Hard disk 1. A storage medium consisting of a rigid (usually metal) platter coated with a metallic oxide substance, on which data is recorded as patterns of magnetic spots. 2. A synonym for *hard disk drive.*

Hard disk drive A storage device incorporating one or more hard disk platters, read/write heads, and low-level control circuitry.

Hardware independence A general term describing the independence of a program or processing method from the physical details of computer system hardware.

Hardware monitor A program or device that records and reports processing and/or communication activity within or between hardware devices.

Hash table A table of hash key values and associated storage locations.

Hashing algorithm An algorithm that converts key field values of a record into hash key values.

Header 1. A general term for the description of stored information contained in the first storage location of a storage volume, file, or other set of data. 2. With respect to assembly language programs, a title record that lists all modules in a file referenced within the code segments, the symbolic names of any references to other modules, and any data that are shared among modules.

Heat dissipation The act of conducting heat away from a device, thus reducing its temperature.

Hertz (Hz) A unit of measure for current or signal frequency. One hertz is one cycle per second.

Hewlett-Packard Graphics Language (HPGL) A specific display list language used to describe printed output for Hewlett-Packard laser printers and plotters.

Hexadecimal A term referring to numbers in the base 16 system of notation, under which digit values range from 0 to 9 and from A to F (corresponding to decimal values of 0 to 15).

Hierarchic data model A database model in which records can be related to one another in a hierarchy or inverted tree.

Hierarchic directory structure A multilevel system of directories in which directories and files can be related to one another in a hierarchy or inverted tree.

High-level data link control (HDLC) A bit-oriented synchronous communication protocol that is often used for high-speed communication within mainframe networks.

High-level programming language A programming language that uses symbolic statements to represent computer operations and memory addresses and in which a single instruction generates multiple machine instructions.

High order bit A synonym for *most significant digit.*

Home directory The primary directory associated with (and owned by) a single user.

I

Icon A visual (graphic) representation of a processing object or command as displayed on a video display device.

IEEE 802 Standards A group of standards for data communication networks promulgated by the Institute of Electrical and Electronics Engineers.

IF statement A high-level programming language statement that executes a set of instructions only if a stated condition is true.

Implementation model A model of both user processing requirements and a specific method of satisfying those requirements.

Inches per second (ips) A measure of the speed at which a tape is moved past the read/write head of a tape drive.

Incremental backup A backup operation in which only those files that have been altered since the most recent backup are copied to a backup storage device or medium.

Index A stored set of paired data items. The first data item is a key value and the second is a pointer or address field that indicates the location of the data item possessing (or corresponding to) that key value.

Indexed file access A method of accessing records in a file, in which a requested record is specified by the value of a key field, rather than by position.

Indexed file organization A method of file organization in which both records and an index to those records are maintained as a single unit.

Indirect addressing A general term describing any addressing method where memory references made within a program do not necessarily correspond to physical memory storage locations. In general, the control unit must calculate physical storage locations, based on the program's memory reference and an indirect addressing formula or algorithm.

Infrastructure A general term describing facilities that provide services that are pervasively available and commonly used by a large number of users.

Initiator Within a SCSI bus, the device that is currently designated the bus master.

Input/output (I/O) A general term referring to actions and mechanisms by which data and commands are communicated between a user and software or by which data and commands are communicated between computer systems and/or their components.

Input/output bound Refers to a condition in which input/output requirements of a process saturate or exceed the capacity of I/O devices and/or channels.

Input/output cache An area of primary storage within an I/O device, secondary storage device, or device controller.

Input/output channel A special-purpose processor dedicated to an I/O port that allows many devices to share access to (and the capacity of) the port. The term also refers to a specific hardware feature of IBM mainframe computer systems.

Input/output device Any hardware device that implements or facilitates communication between computer systems, computer system components, or users and software.

Input/output port 1. The set of signal lines, control logic, and hardware units that facilitate communication among the central processing unit and device controllers. 2. A tap or connection point on a computer system bus.

Input/output redirection The act of rerouting process input or output to alternate devices or storage locations.

Input/output wait state A term describing idle processor cycles due to a wait for data transmission from a secondary storage or input/output device.

Insert operation The addition of data to a stored set of data in a storage location or ordinal position other than the first or last.

Institute of Electrical and Electronics Engineers (IEEE) A professional organization that promulgates many standards for computer hardware and systems software.

Instruction A command or input code that triggers a processing operation.

Instruction code A synonym for *operation (op) code*.

Instruction cycle A synonym for *fetch cycle.*

Instruction explosion A general term describing the correspondence of a single high-level programming language statement to multiple central processing unit instructions.

Instruction pointer A register that stores the address of the next instruction to be fetched from main memory.

Instruction register A register that holds an instruction prior to decoding by the control unit.

Instruction set The set of all instructions that can be executed by a central processing unit.

Integer A whole number or a value that does not have a fractional part.

Integer arithmetic Arithmetic operations on integers.

Integrated circuit (IC) A device incorporating circuitry and semiconductor components within a single unit, usually a miniature chip. Circuit elements and components are created as part of the same manufacturing procedures.

Intelligent system A program or set of programs that mimic one or more aspects of intelligent human behavior.

Interactive A mode of program operation under which a computer prompts a user for input and immediately reacts to that input.

Interleaved execution A general term describing scheduling techniques under which a processor alternates execution among multiple active processes.

International Alphabet Number 5 (IA5) The international equivalent of ASCII, used to transmit text in languages other than English.

International Standards Organization (ISO) An international body with functions similar to those of the American National Standards Institute.

Interpreter A generative translator that translates and executes high-level program statements, one at a time.

Interprocess communication The communication of data values between two or more modules within a single program.

Interprocess synchronization The coordination of two or more processes through the exchange of data and/or signals.

Interrecord gap A blank section of a tape between two physical records.

Interrupt A signal to the CPU that some event requires its attention.

Interrupt code The numerically coded value of an interrupt, indicating the type of event that has occurred.

Interrupt handler A program that is executed in response to an interrupt.

Interrupt register A special register in the control unit that stores interrupt codes to initiate interrupt processing at the machine level.

Interrupt table An index of interrupt codes and interrupt handler starting addresses in memory.

Interrupts enabled A mode of central processing unit operation in which interrupts are received and processed as they occur.

Interrupts masked A mode of central processing unit operation in which one or more interrupts are ignored.

J

Job A grouping, or packaging, of processes into a single processing stream.

Job control language (JCL) The language of an operating system used to identify programs, execution sequences, and resource requirements for batch processing.

Job queue An ordered set of programs awaiting execution.

Job step Within a batch processing environment, an independent segment of a job that has been subdivided into a separate processing unit or step.

Job stream A sequence of programs or steps that make up a single processing job.

Join operation An operation on the contents of a relational database, in which the tuples of two relations are joined based on equivalent values of a common attribute.

Journaling A method of file update in which all changes to file contents are also written immediately to a separate journal. The journal is used to recover lost data in the event of a system crash.

Jump A synonym for *branch instruction.*

K

Kernel 1. The portion of the operating system that directly interacts with computer hardware. 2. The lowest (innermost) layer of an operating system.

Key 1. An access control field that uniquely identifies a record or classifies it as a member of a category of records within a file. 2. In cryptography, a sequence of symbols that control the operations of *encryption* and *decryption.*

L

Label An alphanumeric constant used to identify a line, instruction, or data item (variable) within a program.

Large-scale integration (LSI) A term describing semiconductor devices (chips) containing thousands of transistors or other primitive electrical components.

Laser 1. An acronym of Light Amplification by Stimulated Emission of Radiation. 2. A beam of strong coherent light.

Laser printer A printer that operates using a laser to charge areas of a photoconductive drum and paper.

Last in, first out (LIFO) A processing order in which the most recent arrival is served or processed first. (The opposite of *first come, first served*).

Least significant digit The rightmost digit in a bit string, or the bit occupying the position of smallest magnitude.

Least significant byte The byte within a multibyte data item that contains the digits of least (lowest) magnitude.

Level of abstraction One of a series of levels within a top-down process that breaks a problem or processing task into increasingly more detailed subproblems.

Library call 1. A service request by an application program that causes the execution of an operating system program (a library program). 2. A programming language statement that causes the execution of a compiler or interpreter library program.

Line conditioner 1. A hardware device that ensures constant voltage and amperage in an electrical power circuit. 2. A device that ensures a stable and error-free signal in a communication network.

Line printer An impact printer that prints an entire line of text at one time.

Linear address space A general term referring to the logical view of a secondary storage or I/O device as a sequential set of storage locations.

Linear density A measure of the proximity or spacing of recorded bits on a storage medium.

Linear list A set of data elements or structures in sequential storage locations.

Linear search A method of searching a series of data items in which items are sequentially read until the desired item is located.

Lines per minute (lpm) A measure of output speed typically used for impact and/or line printers.

Link A synonym for *pointer.*

Link editor A program that accounts for and reconciles all address references within and among modules and replaces those references with a single, consistent scheme of relative addresses.

Link map A listing of module and data memory addresses produced by a link editor.

Linkage editor A synonym for *link editor.*

Linked allocation A method of noncontiguous allocation within which physically separated blocks are correlated by a set of links or pointers to form a single set of storage locations.

Linked list An unordered linear list in which each data item contains a pointer to the next or prior logical element in the list.

Linker A synonym for *link editor.*

Linking loader A utility program that combines the functions of a link editor and a loader.

Liquid crystal display (LCD) A display device that uses liquid crystals that can be changed from transparent to opaque by the application of electricity.

LISP processor A processor that implements low-level list manipulation functions of the LISP programming language as processor instructions.

List processing A general term describing any operation that manipulates the content of a list.

List Processing (LISP) A programming language that describes processing operations as list manipulation operations.

Literal A numeric or an alphanumeric constant, the value of which is defined by the characters of which it is composed.

Load operation The act of copying a word from main memory to a register.

Load sharing The distribution of processor tasks among multiple processors in a single computer or among multiple computers via a communication network.

Load-and-go assembler A machine code generator that acts on a single program that will be loaded to absolute addresses.

Loader A systems software program that copies application programs from secondary storage into main memory for execution.

Loading The act of copying an application program from secondary storage into main memory for execution.

Local-area network (LAN) Linking of computers, peripherals, and office machines by hardwiring within a limited area, such as a single building or an office floor.

Logical A general term describing characteristics of objects and methods that are apparent to users and/or application programs, as opposed to the physical objects and methods upon which they are based.

Logical access An access to a storage location expressed in terms of a linear address space.

Logical delete The deletion of a data item by deleting address references (e.g., pointers) to it without actually deallocating or overwriting the corresponding storage location.

Logical file structure The internal organization of a file, as seen and manipulated by a user and/or application program.

Logical file view A synonym for *logical file structure*.

Logical model A model of user processing requirements that is independent of physical methods of satisfying those requirements, as used in the structured system development life cycle.

Logical record A record formatted and described in terms of an application or user view.

Long integer A double precision representation of an integer.

Longitudinal redundancy check (LRC) A scheme of parity checking in which parity bits are determined based on equivalent bit positions in a group of characters.

Lost update A modification to a file that is never applied to physical storage due to interference caused by another (simultaneous) update and the use of buffered storage access.

Low-order bit A synonym for *least significant digit*.

M

Machine code A program consisting of instructions that can be executed by computer hardware.

Machine independence A synonym for *hardware independence*.

Machine language A language consisting solely of instructions from the instruction set of a specific hardware configuration.

Machine primitive One action from the set of most basic actions of which a processor or hardware component is capable (e.g., the OR, AND, and NOT instructions of a central processing unit).

Machine state The current processing state of the central processing unit, as represented by the values currently held in its registers.

Macro A general term describing a symbolic constant (e.g., an alphanumeric name) used to identify a sequence of instructions in a program.

Magneto-optical disk A secondary storage device that reads and writes data bits by a combination of magnetic and optical technology.

Main memory A general term applied to devices that implement primary storage within a computer system.

Mainframe A large computer system designed to support hundreds of users and processes simultaneously.

Mainframe channel The combination of a mainframe computer bus port and a dedicated I/O processor attached to that port.

Manchester coding A method of signalling that represents a bit value of one as a positive half bit followed by a negative half bit.

Mantissa 1. A portion of the internal representation of a real number. 2. A stored number that is multiplied by a power of two to derive the intended value.

Mark sensor An input device that recognizes printed marks (e.g., bars or shapes) at predetermined locations on an input document.

Massively parallel computer system A computer system with hundreds, thousands, or tens of thousands of central processing units.

Master-slave architecture A multiprocessor or multicomputer architecture in which one processor or computer controls the actions of all others.

Media-access protocol A set of signalling and processing conventions by which access to a communication channel is regulated.

Media-access unit (MAU) A control device for token ring networks that serves as a central wiring point, provides network initialization, and reconfigures network connections in the event of a failed network node.

Medium-scale integration (MSI) A term describing semiconductor devices (chips) containing hundreds of transistors or other primitive electrical components.

MegaHertz (MHz) A measurement of wave or clock frequency; one million cycles per second.

Memory A general term describing a collection of storage locations, usually referring to primary storage within a computer system.

Memory allocation The allocation of primary storage resources to competing processes.

Memory map A tabular representation of the storage locations assigned to data items and program segments.

Memory-mapped input/output A method of data transfers to I/O ports in which communication to I/O devices is implemented through primary storage locations.

Memory/storage hierarchy A term describing the set of possible data storage devices in a computer system and the cost/performance relationships among those devices.

Message With respect to communication networks, a command, request, or response sent from one network node to another.

Microchip Miniaturized circuitry and electronic components integrated on a small silicon base.

Microcomputer A computer designed to meet low-intensity processing needs for a single user.

Microprocessor A microchip on which all of the components of a central processing unit are implemented.

Millions of floating-point operations per second (MFLOPS) A measure of processor or computer system speed in terms of the number of floating-point operations executed per second.

Millions of instructions per second (MIPS) A measure of processor or computer system speed in terms of the number of CPU instructions executed per second.

Minicomputer A computer system designed to meet the processing needs of a small to medium-sized group of users.

Mirror-image backup A file copy that is identical to the original.

Mnemonics Symbolic representations of instructions or data items composed of alphabetic characters and other special symbols.

Model An abstract representation of a system or other physical reality.

Modem 1. A contracted form of the term MOdulator-DEModulator. 2. A device that translates analog signals into digital signals (and vice versa), allowing computer hardware to utilize telephone lines for data communication.

Modulate The act of varying the amplitude, frequency, or phase of a signal.

MOdulator-DEModulator See *Modem*.

Module In a program, a set of instructions that are typically executed as a unit (e.g., a subroutine).

Module assembler An assembly language implementation that is designed for use with linker/loader software.

Monadic instruction An instruction that causes transformations of bits within a single operand.

Monitor 1. A general term describing any video display device. 2. A hardware or software element that monitors and reports processing or communication activity.

Most significant byte The byte within a multibyte data item that contains the digits of greatest (highest) magnitude.

Most significant digit The leftmost digit in a bit string, or the bit occupying the position of greatest magnitude.

Move operation The act of copying the contents of one register to another register.

Multichannel Any transmission mode that utilizes two or more communication paths.

Multileaving A technique by which a receiver can reverse the direction of transmission in a half-duplex line at any time.

Multimode PCM A digital transmission method that combines pulse code modulation with frequency-division multiplexing.

Multiple instruction/multiple data stream (MIMD) A term describing a multiprocessor or computer that can execute more than one instruction on different data inputs simultaneously.

Multiple instruction/single data stream (MISD) A term describing the use of multiple processing elements operating in series on a single stream of data.

Multiple master bus A bus in which more than one attached device can control access to the bus (i.e., be a *bus master*).

Multiplexing Merging two or more signals for transmission over a single communication channel.

Multiprocessor (computer) A computer system containing more than one central processing unit.

Multiprogramming A method of memory allocation and process execution that allows concurrent execution of multiple programs held simultaneously in memory.

Multitasking A method of memory allocation and process execution that is similar to multiprogramming, but without the requirement that an entire program be held in memory.

N

Name server A network node that responds to requests to supply the physical or logical network address corresponding to the symbolic name of another network node.

NAND gate A gate combining the primitive functions of NOT and AND.

Needs analysis 1. An estimate of the computing resource requirements for an application program or set of application programs. 2. A detailed description of user processing requirements, generally created during the survey or analysis phase of the system development life cycle.

Negative acknowledge (NAK) An ASCII control character used by a receiver to indicate the unsuccessful receipt of transmitted data.

Network 1. In general, a system in which communication is possible from one location to one or more remote locations. 2. In data communications, the linking of computers and peripheral devices across distances through the use of communication carriers. 3. A data structure in which child data structures can have multiple children and multiple parents.

Network data model A database model in which any record can be directly related to any number of other records.

Network interface unit A general term describing the hardware device by which one or more computing devices interact with a network transmission medium.

Network layer The layer of network systems software responsible for routing packets to their destination.

Network message A command, signal, or data item, communicated between two processes via the facilities of a network.

Noise Electronic interference on a transmission channel that degrades or distorts the signal, possibly resulting in loss of misinterpretation of data.

Noncontiguous allocation Any storage allocation scheme under which multiple storage locations allocated to a filer or a program need not be physically adjacent to one another.

Nonmaskable interrupt A phrase referring to interrupt signals that demand immediate access to the processor and arise from emergency conditions such as a loss of power or failure of a memory device.

Nonpreemptive scheduling A scheduling method in which a process retains control of the CPU until it terminates execution.

Nonprocedural programming 1. A programming language or paradigm that utilizes declarative knowledge representation. 2. A programming language or paradigm in which all physical details of processing need not be explicitly stated by the programmer prior to program translation or execution. 3. Creation of program instructions using a code generator.

Nonvolatile storage A term describing storage devices that retain their contents indefinitely (e.g., magnetic or optical disk).

O

Object code Machine code of a module or program that contains unresolved external references and/or requires relocation or link editing.

Octal Base-8 notation that uses a three-bit binary number.

Odd parity An error-detection method that uses a parity bit. The parity bit is given a value of one if the number of one-valued bits in a character is odd.

Offset A constant that must be added to a program memory reference to derive the corresponding physical memory address.

Offset register A register in which an offset is stored, as typically used by the control unit when resolving program memory references.

Online A term describing a mode of software execution in which user input is repetitively accepted and processed for an extended or indefinite period of time. See also *Interactive processing*).

Open operation See *File open*.

Open System Integration (OSI) model A conceptual architecture for computer networks that consists of seven predefined layers.

Operand As used within an instruction, an input value or the memory address of an input value.

Operating expenditures Funds expended during the current operating period (e.g., fiscal year) to support normal operations.

Operating system A set of software programs that manage and control access to computer hardware resources.

Operation A synonym for *instruction*.

Operation (op) code The numeric code by which a central processing unit recognizes and executes an instruction.

Optical character recognition (OCR) A general term applied to optical methods and devices by which printed characters are recognized as computer system input.

Optical disk A secondary storage device that uses rotating disks of reflective material that can be read by a reflected laser beam.

Optical scanner A device that can scan printed graphic inputs and convert them to computer system input.

Optical sensor A device that senses optical input, usually by a reflected laser.

OR operation A processing operation in which bit value pairs of Ø/Ø are combined to yield Ø. All other pairs are combined to yield 1.

Overflow A condition that occurs when the output bit string of a processing operation exceeds the size of a register, resulting in truncation.

Overlapped processing An early form of processing in which a computer simultaneously inputs one record, processes another, and outputs still another, thus overlapping the functions of input, process, and output.

P

Packet A fundamental unit of data communication in a computer network.

Packet header A message that provides information for message routing and for the reassembly of multiple-packet messages by the receiver.

Packing A synonym for *compaction.*

Page A small fixed size portion of a program, swapped between primary and secondary storage under virtual memory management.

Page fault The condition that occurs when a program references a memory location in a page not currently held in primary storage.

Page table A table of pages and information about them maintained by an operating system that implements virtual memory management.

Pages per minute (ppm) A measure of output speed typically used for laser printers.

Parallel execution The processing of more than one instruction at a time through multiple processors or processing elements.

Parallel processor A general term referring to a processor or computer with multiple processing elements that can execute simultaneously.

Parallel transmission The transmission of multiple signals simultaneously over multiple transmission media or channels.

Parallelism A general term describing the occurrence of multiple simultaneous processing or input/output operations.

Parent process A process that initiates and controls the execution of another (child) process.

Parity bit The channel or bit position in a data recording pattern used for automatic checking of the validity of the remainder of the bit string. See *even parity* and *odd parity.*

Parity check The act of validating the value of a parity bit.

Partitioned memory The division of primary storage into multiple segments, usually for the purpose of efficient memory allocation to multiple active processes.

Password A protected word or string of characters that identifies or authenticates a user, a specific resource, or an access type.

Path name A symbolic name describing a directory, either with respect to the current working directory, the user's home directory, or the root directory.

Personal computer (PC) A synonym for *microcomputer.*

Phase difference A timing difference between two identical waveforms.

Phase-shift modulation A modulating scheme under which binary values are encoded as varying intervals by which one waveform leads or lags the other.

Phoneme An individual vocal sound, comprising a primitive component of human speech.

Photoelectric cell A device that generates electrical current in response to light energy.

Physical A general term describing characteristics of objects and methods that are implemented within hardware, as opposed to the corresponding logical objects and methods that are apparent to users and/or application programs.

Physical access A general term describing the physical actions required to implement an access to or from a storage or I/O device.

Physical delete The act of physically deleting data from a storage device, usually by overwriting it with null values.

Physical file structure The physical organization of data in a file on a storage device.

Physical layer The layer of network systems software responsible for physical transmission of signals that represent data.

Physical memory A term referring to the physical capacity of actual primary storage in a computer system (contrast to *virtual memory*).

Physical model A model of information processing requirements containing information about specific hardware and software components to meet those processing requirements.

Physical record A unit of physical data transfer to or from a storage device.

Pipe A software mechanism by which the data output of one process is routed to another process as its input.

Pipeline processing A multiprocessor configuration in which complex functions are implemented with a set of specialized processors, the output data of one processor being routed as input to the next processor.

Pixel 1. An abbreviation of the term *picture element*. 2. A single unit of data in a graphic image.

Plotter A device that generates printed output through the movement of paper and one or more pens.

Pointer 1. A data element that contains the address (location in a storage device) of another data element. 2. A device used to input positional data or control the location of a cursor.

Polarization 1. A term referring to the alignment of ferrous particles along the lines of force of a nearby electromagnetic field. 2. A term referring to the parallel alignment of photon paths within a beam of light.

Pop operation The process of removing an item from the top of a stack.

Positional number system A system in which numeric values are represented as multiples of one another according to the placement of symbols within a string.

Power sag A momentary reduction in the voltage or amperage of electrical power.

Power surge A momentary increase in the voltage or amperage of electrical power.

Preemptive scheduling A scheduling method that allows a process of higher priority to interrupt and suspend a process of lower priority.

Presentation layer The layer of network systems software responsible for input and output to I/O devices (e.g., video display terminals).

Primary storage Within a computer system, high-speed storage that is accessed directly by the central processing unit. It is generally used to hold currently active programs and data immediately needed by those programs.

Primitive processor unit A component of a pipeline or MISD processor that can perform only one or a small number of possible data processing operations.

Printed circuit board (PCB) An electrical component containing many electrical devices on a board in which intercircuit wiring is embedded.

Procedural knowledge representation The representation of processing knowledge within a specific algorithm or procedure by which that knowledge is applied to data inputs.

Procedural programming A program paradigm based on *procedural knowledge representation.*

Process 1. To transform input data by the application of processing operations. 2. A program or program fragment. 3. A representation of data or information processing on a data flow diagram.

Process control block (PCB) A data structure maintained by an operating system containing information about a currently active process.

Process specification 1. The written or graphical description of a process on a data flow diagram. 2. A synonym for *program specification.*

Processor Any device capable of performing transformation operations on one or more data inputs.

Program A sequence of processing instructions.

Program counter A synonym for *instruction pointer.*

Program cycle A set of program instructions that can be executed repetitively.

Program development tool A class of software tool used to assist in the development of programs. Examples include text editors, compilers, interpreters, and debuggers.

Program flowchart A graphic representation of the flow of control between program components.

Program profiler A software utility that monitors and reports the activities and resource utilization of another program during execution.

Program specification 1. The written description of a program on a system flowchart. 2. A written description of processing requirements to be implemented in a program.

Program status word (PSW) A multiple-byte bit string held in a separate register within the control unit to monitor execution.

Program translator A program that translates instructions in one programming language or instruction set into equivalent instructions in another programming language or instruction set. Examples include compilers, interpreters, and assemblers.

Program verifier A synonym for *debugger* or *code checker.*

Programmable read-only memory (PROM) A (normally) read-only memory chip that is manufactured blank and can be programmed (written) once.

Programmer A person who creates and/or maintains programs.

Programming language Any language in which computer processing functions or instructions can be expressed.

PROgramming with LOGic (PROLOG) A programming language in which first-order logic is used to state rules of processing.

Project operation A relational database operation in which a new relation is created by copying a subset of the attributes of another relation.

Protocol A formal set of rules that govern the exchange of data over a communication channel.

Protocol transfer A handshaking approach that implements two-way communication under program control in both devices.

Pseudo-instruction A mnemonic that has meaning in an assembly language but is not included in the target machine's instruction set.

Pulse-code modulation (PCM) A signal-coding method under which bit values correspond to bursts of light.

Push operation The process of adding an item to the top of a stack.

Q

Queue 1. A list of data items to which, through controlled procedures, additions are made only at the tail and from which deletions are made only at the head. 2. A waiting line. 3. In computer processing, a set of programs and/or data in secondary storage awaiting processing.

Queuing theory A technique that provides a mathematical basis for calculating the mean response time of a processing system and for describing system behavior.

R

Radio frequency (RF) Electromagnetic radiation propagated through space.

Radix The base of a number system, as 10 is for the decimal number system.

Radix point The symbol usually referred to as the decimal point (.) in the decimal number system.

Random access A synonym for *direct access.*

Random access memory (RAM) The common term for semiconductor devices used to implement primary storage.

Rapid prototyping A system development methodology in which programs are rapidly developed and iteratively refined.

Raster display A device that produces video or printed output from sequences of horizontal scan lines.

Read-only memory (ROM) A read-only memory chip with preprogrammed contents.

Read operation An access operation for retrieval of data from an input or storage device.

Read/write head The mechanism within tape and disk units that reads data from and writes data to the storage medium.

Ready state The state of a process that is waiting only for access to the central processing unit.

Real resource A computer resource that exists physically.

Real-time scheduling Any scheduling method that guarantees the execution of a program segment within a stated time interval.

Record 1. A data structure composed of data items relating to a single entity such as a person or a transaction. 2. A unit of data transfer. 3. The primary component data structure of a file.

Record locking A method to prevent lost updates by preventing multiple users or processes from simultaneously updating the same record.

Recording density The closeness or spacing of bit positions on a storage medium, typically as measured in bytes or tracks per inch.

Recursive macro Within an assembly language program, a macro that calls one or more other macros.

Red/green/blue (RGB) The primary colors used to produce displayed color output.

Reduced instruction set computer (RISC) A computer with an architecture that provides no complex instructions.

Refresh rate The rate at which the content of a video display is updated.

Register Within a central processing unit, a high-speed storage location that can hold a single word.

Relation A tabular data structure that forms the basis of the relational database model.

Relational data model A database model in which individual files are related to one another by redundant (key) fields.

Relative addressing A specific form of indirect addressing in which program memory references are offsets to a base register.

Relative path The name of an access path that begins at the level of the current directory.

Relocatable code 1. Program instructions with address values that are changeable or not absolute. 2. A synonym for *object code*.

Relocating loader A utility that can reallocate programs within memory and recalculate addresses in those programs.

Relocation The act of moving instructions and data from one memory region to another.

Relocator A systems software program that performs relocation.

Remote job entry (RJE) A specialized form of batch processing under which batches of data are captured and stored at remote points, then transmitted to a central processing site at scheduled times.

Remote login A user's ability to start an interactive session on a remote computer using another computer and a communication network.

Remote process execution The execution of a process on a remote computer via a request from another computer transmitted over a communication network.

Repeater In data communication, a device used to relay signals for transmission over long distances.

Repetition 1. The repetitive execution of a set of program instructions. 2. A programming language statement or construct that causes repetitive execution of a set of program instructions.

Request for proposal (RFP) A formal document stating hardware and/or software requirements and requesting proposals from vendors for supplying those requirements.

Requirements analysis A synonym for *system analysis*.

Resistance A phenomenon in electrical conductors whereby electrical power is converted to heat, thus reducing signal strength.

Resource availability model A model of the computing resources that can be provided by a given combination of computer hardware and systems software.

Resource declaration Within a job control language, a statement of the resources that will be required to execute a job or a job step.

Response time The average delay between a request from a user or application program and the receipt of a processing result.

Return wire The path in a communication channel that completes an electrical circuit between the sending and the receiving device.

Revolutions per minute (RPM) A measure of the rotational speed of a rotating device (e.g., a disk drive).

Ring topology A network configuration in which each network node is connected to two other network nodes, with the entire set of connections and nodes forming a ring.

Root directory The topmost directory (or entry point) in a hierarchical directory structure.

Rotate operation A modified shift operation in which bit values shifted beyond one end of a bit string are placed at the opposite end of a bit string.

Rotational delay The waiting time for the desired sector of a disk to rotate beneath a read/write head.

Rule An independent statement of processing knowledge consisting of preconditions, actions, and postconditions.

Rule base A set of rules comprising a program.

Run queue The set of process control blocks for all currently active processes.

Run-time library A set of predefined executable programs used by an interpreter during program translation and execution.

Running state The state of a process that is currently executing within the central processing unit.

S

Sampling 1. The process of measuring and digitally encoding one or more parameters of an analog signal at regular time intervals. 2. The process of digitally encoding sound waves.

Scalar A term referring to data types that possess a set of values with an ordinal relationship or a predefined sequence.

Scan code A coded output generated by a keyboard for interpretation by a keyboard controller or a processor.

Scanning laser A laser that is automatically swept back and forth over a predefined viewing area (e.g., as used in a bar-code reader).

Scheduler A program within the operating system that controls process states and access to the CPU.

Scheduling The process of determining and implementing process priorities for access to hardware resources.

Search path The set of directories that are automatically searched for program and/or data files.

Secondary storage A general term describing large amounts of relatively slow-speed storage within a computer system.

Select operation A relational database operation in which a subset of tuples from a relation are copied to a new relation. The subset is determined by state criteria (e.g., the value(s) of one or more attributes).

Selection A program control construct that chooses processing statements for execution on the basis of condition tests.

Semiconductor A material with resistance properties that can be tailored between those of a conductor and an insulator by adding chemical impurities.

Sequence An implicit control structure in which processes are executed in sequential order, with each process awaiting the completion of the previous process prior to commencing execution.

Sequential execution The execution of program instructions in the order in which they are stored.

Sequential file access An access technique whereby data items are read or written in an order corresponding to their position within allocated storage locations.

Sequential file organization A method of file organization in which contents are stored sequentially and contiguously in allocated storage locations.

Serial transmission A method of transmission whereby individual bits are sequentially transmitted over a single communication channel.

Server A general term describing a network node that responds to service requests from other network nodes.

Service call An application program's request for an operating system service.

Service layer A portion of the operating system that accepts service calls and translates them into low-level requests to the kernel.

Session layer The layer of network systems software responsible for establishing connections with server processes on other network nodes.

Set 1. An unordered group of related objects. 2. In a network database management system, a logical data structure linking two or more record types.

Shell A term describing the user interface (or command layer) of an operating system.

Shift operation A monadic operation whereby individual bits of a bit string are moved left or right by a stated number of positions. Empty positions are filled with zeros, and bit values that shift beyond the bounds of the bit string are truncated.

Shortest time remaining A scheduling strategy that assigns highest priority to processes with shortest time remaining until completion.

Sign bit A bit (typically the high-order bit) that indicates the sign of a numeric value.

Signed magnitude notation A notation that indicates the sign of the binary value.

Signal Any means by which a message can be transmitted from one location to another.

Signal event The period of time during which a signal parameter is modulated to encode one unit (e.g., a bit) of data transmission.

Signal-to-noise (S/N) ratio The relationship between the power of a carrier signal and the power of the noise in the communication channel.

Signal wire A component of an electrical communication channel (circuit) used to transmit message-carrying signals.

Simple (machine) instruction A processor instruction that results in the execution of only one primitive processor action (e.g., load, store, add).

Simplex A term describing unidirectional data communication.

Simulation The process of examining the behavior of a system by creating and manipulating a model of the system.

Simulation model 1. A model of a system that simulates the behavior of a system. 2. A specific type of model that represents variability in system inputs and mathematically models behavior based on those inputs.

Sine wave A type of analog waveform that varies continually rather than in abrupt distinct increments.

Single instruction/multiple data stream (SIMD) A mode of processor operation in which a single instruction is simultaneously executed on multiple data inputs.

Single instruction/single data stream (SISD) A mode of processor operation in which a single instruction is executed on a single data input.

Single-job processing A mode of program execution in which one data record passes through the system completely before the next record is input.

Single-level directory A directory structure in which all directories are at the device level, and one directory exists for each device.

Single-mode pulse code modulation Refers to the modulation scheme used if an optical fiber carries only one signal.

Single tasking An operating system in which only one process can be active at a time.

Singly linked list A linked list in which element pairs are connected with a single pointer.

Skew A term describing timing differences in the arrival of signals transmitted simultaneously on parallel communication channels.

Software monitor A program that monitors and reports the utilization of a software resource.

Son A file containing the results of updating a master file (the father) based on the contents of a transaction file.

Sorted list A list in which the component elements are ordered based on their value or on the value(s) of one or more component data items.

Source code Instructions in a high-level programming language.

Spatial resolution The sharpness of a display, corresponding to the number of horizontal and vertical pixels within a fixed area of display or output surface.

Spawning A term describing the creation of a child process by a parent process.

Special-purpose processor 1. A processor with a limited set of processing functions. 2. A processor capable of executing only a single program.

Speech recognition The process of recognizing human speech as computer system input.

Speech synthesis The process of generating human speech based on character or textual input.

Spooler 1. An acronym of Shared Peripheral Operation On Line. 2. A software utility that allows access to a single output device by multiple processes through the use of storage buffers.

Square wave A general term describing digital waveforms.

Stack 1. A list of data items maintained in last-in, first-out order. 2. A set of registers or memory locations used to store the register values of temporarily suspended processes.

Stack overflow A condition that occurs when an attempt is made to add data to a stack that is already at its maximum capacity.

Stack pointer A register containing the address of the current top of the stack (in memory).

Stacked job processing A processing technique used for stacking multiple jobs for sequential execution.

Start bit A bit that is always zero (it has no setting) used at the beginning of each character in asynchronous serial data transmission.

Start of header (SOH) An ASCII control character that indicates the beginning of a header field. See *synchronous data transmission.*

Start of text (STX) An ASCII control character that indicates the beginning of a data block. See *synchronous data transmission.*

Static analysis The analysis of a model in which variability in input parameters is not explicitly considered.

Statistics A branch of mathematics concerned with the description of phenomena that exhibit variable and/or random behavior.

Status code A signal sent from a device to a central processing unit to indicate its ability or readiness to respond to a command.

Stop bit A bit appended to the end of a byte in asynchronous serial transmission.

Storage input/output control layer A layer of systems software responsible for controlling data input to and output from storage devices.

Store operation An operation in which the contents of a register are copied to a memory location.

Stored addressing information Information describing the physical organization of data, written to a storage volume when it is initialized or formatted under control of the operating system.

Strategic plan A long-range plan stating the services to be provided by an organization and the means for obtaining and using needed resources.

String A set of related character data elements, usually stored as a list or an array.

Structure chart A graphic representation of the hierarchical organization of processing functions (modules) in a program or a system.

Structured query language (SQL) A standardized language for data manipulation operations in a relational database management system.

Structured system development life cycle A particular form of system development life cycle in which analysis is clearly separated from design, and in which specific system modelling tools are used.

Subcarrier An analog signal of a specific frequency that is transmitted to represent a specific data value.

Supercomputer A computer designed for very fast numeric processing, typically implemented with some form of parallel processing.

Superconductor A conductive material that displays little or no resistance to the flow of electrons.

Supervisor An operating system program that controls the content of the instruction pointer and, therefore, processing priorities and access to the CPU.

Surge protector A hardware device that eliminates electrical power surges.

Sustainable data transfer rate The maximum data transfer rate that can be sustained by a device or communication channel during large or lengthy data transmissions.

Swap space An area of secondary storage that holds virtual memory pages that cannot fit into primary storage.

Switch A device that exists in one of two states and can be instructed to alternate between them.

Symbol table A list of program and data names and the storage locations allocated to them.

Symbolic debugger A debugger that can report program execution in terms of symbolic module and data names used within program source code.

Symbolic processing A general term describing the manipulation of symbolic (as opposed to numeric) data.

Symbolic programming language A programming language designed to express symbolic processing instructions.

Synchronous A general term describing the precise (timed) coordination of two independent devices or programs. The term is typically used to describe methods or protocols for data transmission and interprocess communication.

Synchronous data link control (SDLC) A bit-oriented, synchronous communication protocol that represents actual standards for high-speed communication within mainframe networks.

Synchronous idle A handshaking method some protocols use to set the internal clocks within sending and receiving devices.

Synchronous transmission A reliable, high-speed method of data transmission in which data is sent in packets and in timed sequences.

Syntax A set of rules of construction for a programming language.

System administration A general term describing the managerial actions required to ensure efficient, reliable, and secure computer system operation.

Systems analysis The process of ascertaining and documenting information processing requirements.

Systems analyst A person who performs systems analysis tasks.

System call A synonym for *service call.*

System clock A digital circuit that generates timing pulses and transmits them to other devices within a computer.

Systems design The process of selecting and organizing hardware and software elements to satisfy stated processing requirements.

Systems designer A person who performs systems design tasks.

System development life cycle (SDLC) A formal process for systems analysis, design, implementation, maintenance, and evaluation.

Systems development tool 1. A software tool designed to aid in the creation of groups of programs comprising an entire systems. 2. An approximate synonym for *computer-assisted software engineering (CASE) tool.*

System flowchart A graphic model of manual processes, software elements, files, and the flow of data and control among them.

Systems implementation The process of software construction, installation, and testing on a given hardware configuration.

System loader A synonym for *bootstrap loader.*

System model A formal symbolic representation of processing requirements, software elements, and/or hardware elements.

Systems programmer A person who creates and/or maintains systems software.

Systems software A general term describing software programs that perform hardware interface, resource management, or application support functions.

Systems survey The process of initially ascertaining user processing requirements and determining the feasibility of meeting those requirements with computer-based solutions.

T

T connector A specific type of tap used to connect network nodes to a communication bus or network.

Tap A general term describing a connection to a bus communication channel.

Tape header A record, typically written at the beginning of a tape, that contains information describing the physical storage parameters of the tape (e.g., recording density or record length).

Target language The output language of a compiler or other generative translator.

Teletype An input/output device consisting of a keyboard and a character printer.

Terminal A synonym for *video display terminal.*

Terminal capability (TERMCAP) A data structure that stores information to use specialized display and control functions for a particular video display terminal.

Text editor A program that allows a user to create and/or modify files containing character data.

Third-generation language (3GL) A general term applied to high-level programming languages that do not possess advanced capabilities for interactive input/output, database processing, or nonprocedural programming.

Threshold The value of a signal parameter at which the interpretation of the signal value changes.

Throughput The volume of work that a computer system can perform within a given time.

Time-division multiplexing (TDM) In digital communication, a technique that subdivides data transmissions into separate, discrete packets and transmits them in interleaved fashion within a single communication channel.

Time sharing An approach to processing in which users at remote sites gain access to a computer through online terminals to concurrently share computing resources.

Time slice One or more processor cycles allocated to one of a competing set of processes.

Token A control packet that regulates access to a data communication network.

Token passing A media-access protocol that employs a token passed between network nodes.

Track-to-track seek time The amount of time required to move a disk read/write head between two adjacent tracks.

Tracks per inch (tpi) A partial measure of storage density, primarily used with magnetic disk drives.

Transaction 1. An event or system input that initiates the execution of one or more program segments. 2. A real-world event from which data input is derived (e.g., a sale). 3. A general term describing any change to stored data (e.g., the addition of a record to a file).

Transaction logging A synonym for *journaling*.

Transceiver A device that detects signals passing through a transmission medium and generates an equivalent signal on another transmission medium.

Transistor A solid-state electrical switch that forms the basic component of most computer processing circuitry.

Transistor-to-transistor logic (TTL) A digital signalling method.

Transmission medium The physical communication path through which a signal is propagated.

Transparent file access Access to files stored on a remote machine over a communication network, without the explicit knowledge or control of a user or application program.

Transparent file transfer Transfer of a file between computers over a communication network, without the explicit knowledge or control of a user or application program.

Transparent mode A synonym for *binary mode*.

Transport layer The layer of network systems software responsible for converting network messages into packets or other data structures suitable for transmission.

Traverse To exhaustively read each component of a data structure. The term is typically used in conjunction with hierarchical data structures (e.g., trees).

Tree structure directory A synonym for *hierarchic directory structure*.

Tuple One row (set of attribute values) within a relation.

Twisted-pair wire A signal propagation medium consisting of two electrical conductors continuously twisted around one another.

Two-level directory A simple directory structure in which a device or medium (root) directory contains multiple user directories, each of which contains one or more files.

Two-pass assembler An assembler that correlates external references by stepping through the source code twice.

Two's complement A notation system in which the complement of a bit string is formed by substituting 0 for all values of 1 and 1 for all values of 0.

U

Unblocked The storage of a single logical record within a physical record (block).

Unconditional branch A synonym for *branch instruction.*

Underflow A condition in which an error arises if the radix point of a processing result cannot be moved far enough to the left.

Uninterruptible power supply (UPS) A device that provides electrical power when normal power inputs are interrupted.

Unresolved reference A reference within an object code module to a symbolic name not defined within that module.

Update in place A method of altering the contents of a record or field in which a new value overwrites the old value.

Update operation The process of modifying a set of data items to reflect the addition, change, and/or deletion of data items.

User A person who directly or indirectly interacts with a software program.

User directory 1. A directory belonging to a single user. 2. The topmost directory in a hierarchical directory structure belonging to a single user.

Utility program A program that performs a common function (e.g., file access or storage initialization). Utility programs are typically provided by the operating system either as separate software tools or as part of the service layer.

V

Vacuum tube An electrical device consisting of electrical components contained within a glass tube, from which all gases have been removed.

Validity checking A general term describing any method of error detection in data transmission.

Variable-length instruction An instruction that can have a variable number of operands and/or operands of variable length.

Vector 1. A line segment that has direction and length. 2. A one-dimensional array.

Vector instruction An instruction that is executed sequentially on all elements of a vector or array.

Vector list A list of line descriptions for the input or output of graphic images.

Vector processor A processor capable of executing vector instructions.

Vector register A register that is automatically loaded with subsequent elements of a vector or array on each processor cycle.

Vertical redundancy check (VRC) A type of bit-count parity checking.

Very large-scale integration (VLSI) A term describing semiconductor devices (chips) containing tens of thousands of transistors or other primitive electrical components.

Victim The memory page chosen to be swapped to secondary storage under virtual memory management.

Video display terminal (VDT) An input/output device comprised of a keyboard and a video display device.

Virtual machine A set of hardware resources available to a user of a virtual memory timesharing system.

Virtual memory management A mode of operating system memory management in which secondary storage is used to extend the capacity of primary storage.

Virtual resource A term defining a set of capabilities or resources that are apparent to a user but are not necessarily available to the user at a given time.

Volatile storage A term describing storage devices that cannot retain their contents indefinitely (e.g., random access memory).

Volatility The rate of change in a set of data items.

Volume directory A synonym for *device directory*.

von Neumann machine Named after its designer, John von Neumann, a computer system that contains a single processor that sequentially executes single instructions.

W

Wait state A processor cycle that is unused due to a delay in accessing an input/output or storage device.

Wide-area network (WAN) A general term describing a data communication network that spans large physical devices (e.g., greater than one mile).

Wideband A description of a communication link with high capacity.

Word The unit of data processed by a central processing unit instruction.

Working directory A synonym for *current directory*.

Working storage A general term describing the use of primary storage to hold data needed quickly by the central processing unit.

Workload model A synonym for *application demand model*.

Workstation A powerful microcomputer designed to support demanding numerical and/or graphical processing tasks.

Write once, read many (WORM) A type of optical storage medium that is manufactured blank and can be written once.

Write operation The process of encoding data within a storage device.

Write protection The protection of a set of data elements or a storage medium against accidental or malicious alteration.

X

XON/XOFF A pair of control characters or signals (typically ASCII 17 and 19) used to start and stop the flow of data in asynchronous data transmission.

Z

Zero-crossing signal A digital signal that has both positive and negative states.

Index

A

B

C

D

E

F

G

N

O

P

Q

R

S

T

U

V

W

X

Z